Applied Linear Regression Models

$$Y'_{1\times1} Y = \Sigma Y^2$$

$$X'_{2\times2} X = \begin{matrix} n & \Sigma X \\ \Sigma X & \Sigma X^2 \end{matrix}$$

$$(X'X)^{-1} = \begin{bmatrix} \Sigma X^2 & -\Sigma X \\ -\Sigma X & n \end{bmatrix} \frac{1}{n\Sigma(X-\bar{X})^2}$$

$$X'Y_{2\times1} = \begin{matrix} \Sigma Y \\ \Sigma XY \end{matrix}$$

$$e = Y - \hat{Y} = Y - Xb$$

$$\left(\frac{1}{n}\right) Y' 1 1' Y = \frac{(\Sigma Y)^2}{n} = n\bar{Y}^2$$

$$b = (X'X)^{-1} X'Y \qquad (X'X)b = X'Y$$

$$SSTO = Y'Y - \left(\frac{1}{n}\right) Y' 1 1' Y \; (n-1) \qquad Y'1 = 1'Y = \Sigma Y$$
$$ {}_{1\times1} \qquad {}_{1\times1}$$

$$SSE = Y'Y - b'X'Y \quad or \quad \underline{e}'\underline{e} = \Sigma e^2 \; (n-p)$$

$$SSR = b'X'Y - \left(\frac{1}{n}\right) Y' 1 1' Y \; (p-1)$$
$$ SST - SSE$$

$$\hat{\underline{Y}} = \underline{Xb}$$

Applied Linear Regression Models

John Neter
University of Georgia

William Wasserman
Syracuse University

Michael H. Kutner
Emory University

1983

RICHARD D. IRWIN, INC.
Homewood, Illinois 60430

SQUARE MATRIX
|
SYMMETRIC
|
DIAGONAL MATRIX $\quad$ off $= 0$
$\qquad$ others = on diag
|
IDENTITY (on all 1's) $\qquad$ SCALAR on = same
λI

ISBN 0-256-02547-9

Library of Congress Catalog Card No. 82–82121

Printed in the United States of America

4 5 6 7 8 9 0 MP 0 9 8 7 6

To
Dorothy, Ron, David
Cathy, Christopher, Timothy,
   Randall, Erin, Fiona
Nancy, Michelle, Allison

Preface

Applied Linear Regression Models is a revision of the regression portion of *Applied Linear Statistical Models*. The publication of a separate book which offers a revised and updated treatment of regression models fills an important need in view of the many significant developments in regression analysis in recent years. The remainder of *Applied Linear Statistical Models* is now being revised and will be published upon completion of the revision.

Linear regression models are widely used today in business administration, economics, and the social, health, and biological sciences. Successful application of these models requires a sound understanding of both the underlying theory and the practical problems which are encountered in using the models in real-life situations. While *Applied Linear Regression Models* is basically an applied book, it seeks to blend theory and applications effectively, avoiding the extremes of presenting theory in isolation and of giving elements of applications without the needed understanding of the theoretical foundations.

Applied Linear Regression Models differs from *Applied Linear Statistical Models* in a number of important respects.

1. We have added some important new topics, including detection of multicollinearity, ridge regression, detection of influential observations, and nonlinear regression. In recent years, noteworthy new developments in the detection of multicollinearity and influential observations have taken place, and a current text in regression analysis needs to cover these topics adequately. In addition, we have added some specialized topics, such as P-values of statistical tests and the method of least absolute deviations.

2. We have also reorganized and expanded a number of topics, including

weighted regression, selection of independent variables, and normal probability plots. At the same time, we have made extensive revisions of other materials on the basis of classroom experience to improve the clarity of the presentation.

3. The scope of the examples has been expanded to include applications from the health and biological sciences, in addition to applications from management, economics, and the social sciences. In all cases, an application can be readily understood by the general reader, regardless of background.

4. We have introduced a number of computer-generated plots to demonstrate the usefulness of computer graphics in regression analysis.

5. We have added two extensive real-world data sets that can be employed in a variety of ways.

6. Finally, we have substantially expanded the problem materials at the ends of the chapters and have grouped them into three categories, namely *Problems, Exercises,* and *Projects.* The *Problems* category includes basic problems and questions, the *Exercises* category includes conceptual and theoretical questions, and the *Projects* category includes problems utilizing large data sets and/or involving extensive calculations and analysis.

We have included in this book not only the more conventional topics in regression but also have taken up topics that are frequently slighted though important in practice. Thus, we devote a full chapter to indicator variables, covering both dependent and independent indicator variables. Another chapter takes up computer-assisted selection procedures for obtaining a "best" set of independent variables to be employed in the regression model. The use of residual analysis for examining the aptness of the model is a recurring theme throughout this book. So is the use of remedial measures that may be helpful when the model is not appropriate. In the analysis of the results of a study, we emphasize the use of estimation procedures, rather than tests, because estimation is often more meaningful in practice. Also, since practical problems seldom are concerned with a single estimate, we stress the use of simultaneous estimation procedures.

Theoretical ideas are presented to the degree needed for good understanding in making sound applications. Proofs are given in those instances where we feel they serve to demonstrate an important method of approach. Emphasis is placed on a thorough understanding of the models, particularly the meaning of the model parameters, since such understanding is basic to proper applications. A wide variety of case examples is presented to illustrate the use of the theoretical principles, to show the great diversity of applications of regression models, and to demonstrate how analyses are carried out for different problems.

We use "Notes" and "Comments" sections in each chapter to present additional discussion and matters related to the mainstream of development. In this way, the basic ideas in a chapter are presented concisely and without distraction. Similarly, optional "Topics" chapters supplement chapters containing the main development and present a variety of additional topics that in most cases can be omitted without loss of continuity.

Applications of regression models frequently require extensive computations. We take the position that a computer is available in most applied work. Further, almost every computer user has access to program packages for regression analysis. Hence, we explain the basic mathematical steps in fitting a regression model but do not dwell on computational details. This approach permits us to avoid many complex formulas and enables us to focus on basic principles. We make extensive use in this text of computer capabilities for performing computations and illustrate a variety of computer printouts and explain how these are used for analysis.

A selection of problems is provided at the end of each chapter (excepting Chapter 1). Here the reader can reinforce his or her understanding of the methodology and use the concepts learned to analyze data. We have been careful to supply data-analysis problems that typify genuine applications. In most problems the calculations are best handled on a calculator or computer, and we urge that this avenue be used when possible.

We assume that the reader of *Applied Linear Regression Models* has had an introductory course in statistical inference, covering the material outlined in Chapter 1. Should some gaps in the reader's background exist, he or she can read the relevant portions of an introductory text, or the instructor of the class may use supplemental materials for covering the missing segments. Chapter 1 is primarily intended as a reference chapter of basic statistical results for continuing use as the reader progresses through the book.

Calculus is not required for reading *Applied Linear Regression Models*. In a number of instances we use calculus to demonstrate how some important results are obtained, but these demonstrations are confined to supplementary comments or notes and can be omitted without any loss of continuity. Readers who do know calculus will find these comments and notes in natural sequence so that the benefits of the mathematical developments are obtained in their immediate context. Some basic elements of matrix algebra are needed for multiple regression. Chapter 6 introduces these elements of matrix algebra in the context of simple regression for easy learning.

Applied Linear Regression Models is intended for use in undergraduate or graduate courses in regression analysis and in second courses in applied statistics. The extent to which material presented in this text is used in a particular course depends upon the amount of time available and the objectives of the course. The basic elements of regression are covered in Chapters 2, 3, 4, 5 (Sections 5.1–5.4 only), 6, 7, 8, and 12. Chapters 9, 10, 11, 13, 14, and 15 can be covered as time permits and interests dictate.

This book can also be used for self-study by persons engaged in the fields of business administration, economics, and the social, health, and biological sciences who desire to obtain competence in the application of regression models.

A book such as this cannot be written without substantial assistance from others. We are indebted to the many contributors who have developed the theory and practice discussed in this book. We also would like to acknowledge appreciation to our students who helped us in a variety of ways to fashion the method of presentation contained herein. We are grateful to the many users of *Applied Linear Statistical Models* who provided us with comments and suggestions based on their teaching with this text. We are also indebted to Professors James E. Holstein, University of Missouri, and David L. Sherry, University of West Florida, who carefully reviewed *Applied Linear Statistical Models* to provide suggestions for this volume. Robert L. Vogel assisted us diligently in the checking of the manuscript, for which we are most appreciative. Michael J. Lynn prepared the computer-generated plots using a Zeta model 3600 plotter, and George Cotsonis and Shizuki Yamamoto assisted us in checking of calculations and in other ways. Almost all of the typing was done by Rebecca Baggett, who ably handled the preparation of a difficult manuscript. We are most grateful to all of these persons for their help and assistance.

Finally, our families bore patiently the pressures caused by our commitment to complete this revision. We are appreciative of their understanding.

John Neter
William Wasserman
Michael H. Kutner

Contents

1. Some basic results in probability and statistics, 1

1.1 Summation and product operators, 1
1.2 Probability, 2
1.3 Random variables, 3
1.4 Normal probability distribution and related distributions, 6
1.5 Statistical estimation, 9
1.6 Inferences about population mean—Normal population, 10
1.7 Comparisons of two population means—Normal populations, 13
1.8 Inferences about population variance—Normal population, 16
1.9 Comparisons of two population variances—Normal populations, 17

Part I Basic regression analysis, 21

2. Linear regression with one independent variable, 23

2.1 Relations between variables, 23
2.2 Regression models and their uses, 26
2.3 Regression model with distribution of error terms unspecified, 31
2.4 Estimation of regression function, 35
2.5 Estimation of error terms variance σ^2, 46
2.6 Normal error regression model, 48
2.7 Computer inputs and outputs, 51

3. Inferences in regression analysis, 60

3.1 Inferences concerning β_1, 60
3.2 Inferences concerning β_0, 68
3.3 Some considerations on making inferences concerning β_0 and β_1, 70
3.4 Interval estimation of $E(Y_h)$, 72
3.5 Prediction of new observation, 76
3.6 Considerations in applying regression analysis, 82
3.7 Case when X is random, 83
3.8 Analysis of variance approach to regression analysis, 84
3.9 Descriptive measures of association between X and Y in regression model, 96
3.10 Computer output, 99

4. Aptness of model and remedial measures, 109

4.1 Residuals, 109
4.2 Graphic analysis of residuals, 111
4.3 Tests involving residuals, 122
4.4 F test for lack of fit, 123
4.5 Remedial measures, 132
4.6 Transformations, 134

5. Simultaneous inferences and other topics in regression analysis—I, 147

5.1 Joint estimation of β_0 and β_1, 147
5.2 Confidence band for regression line, 154
5.3 Simultaneous estimation of mean responses, 157
5.4 Simultaneous prediction intervals for new observations, 159
5.5 Regression through the origin, 160
5.6 Effect of measurement errors, 164
5.7 Weighted least squares, 167
5.8 Inverse predictions, 172
5.9 Choice of X levels, 175

Part II General regression and correlation analysis, 183

6. Matrix approach to simple regression analysis, 185

6.1 Matrices, 185
6.2 Matrix addition and subtraction, 190
6.3 Matrix multiplication, 192
6.4 Special types of matrices, 196

6.5 Linear dependence and rank of matrix, 199

6.6 Inverse of a matrix, 200

6.7 Some basic theorems for matrices, 204

6.8 Random vectors and matrices, 205

6.9 Simple linear regression model in matrix terms, 208

6.10 Least squares estimation of regression parameters, 210

6.11 Analysis of variance results, 212

6.12 Inferences in regression analysis, 216

6.13 Weighted least squares, 219

6.14 Residuals, 220

7. Multiple regression—I, 226

7.1 Multiple regression models, 226

7.2 General linear regression model in matrix terms, 237

7.3 Least squares estimators, 238

7.4 Analysis of variance results, 239

7.5 Inferences about regression parameters, 242

7.6 Inferences about mean response, 244

7.7 Predictions of new observations, 246

7.8 An example—Multiple regression with two independent variables, 247

7.9 Standardized regression coefficients, 261

7.10 Weighted least squares, 263

8. Multiple regression—II, 271

8.1 Multicollinearity and its effects, 271

8.2 Decomposition of *SSR* into extra sums of squares, 282

8.3 Coefficients of partial determination, 286

8.4 Testing hypotheses concerning regression coefficients in multiple regression, 289

8.5 Matrix formulation of general linear test, 293

9. Polynomial regression, 300

9.1 Polynomial regression models, 300

9.2 Example 1—One independent variable, 305

9.3 Example 2—Two independent variables, 313

9.4 Estimating the maximum or minimum of a quadratic regression function, 317

9.5 Some further comments on polynomial regression, 319

10. Indicator variables, 328

10.1 One independent qualitative variable, 328

10.2 Model containing interaction effects, 335

10.3 More complex models, 339

10.4 Other uses of independent indicator variables, 343

10.5 Some considerations in using independent indicator variables, 351

10.6 Dependent indicator variable, 354

10.7 Linear regression with dependent indicator variable, 357

10.8 Logistic response function, 361

11. Multicollinearity, influential observations, and other topics in regression analysis—II, 377

11.1 Reparameterization to improve computational accuracy, 377

11.2 Problems of multicollinearity, 382

11.3 Variance inflation factors and other methods of detecting multicollinearity, 390

11.4 Ridge regression and other remedial measures for multicollinearity, 393

11.5 Identification of outlying observations, 400

11.6 Identification of influential observations and remedial measures, 407

12. Selection of independent variables, 417

12.1 Nature of problem, 417

12.2 Example, 419

12.3 All possible regression models, 421

12.4 Stepwise regression, 430

12.5 Selection of variables with ridge regression, 436

12.6 Implementation of selection procedures, 436

13. Autocorrelation in time series data, 444

13.1 Problems of autocorrelation, 445

13.2 First-order autoregressive error model, 448

13.3 Durbin-Watson test for autocorrelation, 450

13.4 Remedial measures for autocorrelation, 454

14. Nonlinear regression, 466

14.1 Linear, intrinsically linear, and nonlinear regression models, 466

14.2 Example, 469

14.3 Least squares estimation in nonlinear regression, 470

14.4 Inferences about nonlinear regression parameters, 480

14.5 Learning curve example, 483

15. Normal correlation models, 491

15.1 Distinction between regression and correlation models, 491

15.2 Bivariate normal distribution, 492

15.3 Conditional inferences, 496
15.4 Inferences on ρ_{12}, 501
15.5 Multivariate normal distribution, 505

Appendix tables, 515

SENIC data set, 533

SMSA data set, 537

Index, 543

1

Some basic results in probability and statistics

This chapter contains some basic results in probability and statistics. It is intended as a reference chapter to which you may refer as you read this book. Sometimes, specific references to results in this chapter are made in the text. At other times, you may wish to refer on your own to particular results in this chapter as you feel the need.

You may prefer to scan the results on probability and statistical inference in this chapter before reading Chapter 2, or you may proceed directly to the next chapter.

1.1 SUMMATION AND PRODUCT OPERATORS

Summation operator

The summation operator Σ is defined as follows:

$$(1.1) \qquad \sum_{i=1}^{n} Y_i = Y_1 + Y_2 + \cdots + Y_n$$

Some important properties of this operator are:

(1.2a) $$\sum_{i=1}^{n} k = nk \qquad \text{where } k \text{ is a constant}$$

(1.2b) $$\sum_{i=1}^{n} (Y_i + Z_i) = \sum_{i=1}^{n} Y_i + \sum_{i=1}^{n} Z_i$$

(1.2c) $$\sum_{i=1}^{n} (a + cY_i) = na + c \sum_{i=1}^{n} Y_i \qquad \text{where } a \text{ and } c \text{ are constants}$$

The double summation operator $\Sigma\Sigma$ is defined as follows:

(1.3) $$\sum_{i=1}^{n} \sum_{j=1}^{m} Y_{ij} = \sum_{i=1}^{n} (Y_{i1} + \cdots + Y_{im})$$
$$= Y_{11} + \cdots + Y_{1m} + Y_{21} + \cdots + Y_{2m} + \cdots + Y_{nm}$$

An important property of the double summation operator is:

(1.4) $$\sum_{i=1}^{n} \sum_{j=1}^{m} Y_{ij} = \sum_{j=1}^{m} \sum_{i=1}^{n} Y_{ij}$$

Product operator

The product operator Π is defined as follows:

(1.5) $$\prod_{i=1}^{n} Y_i = Y_1 \cdot Y_2 \cdot Y_3 \cdots Y_n$$

1.2 PROBABILITY

Addition theorem

Let A_i and A_j be two events defined on a sample space. Then:

(1.6) $$P(A_i \cup A_j) = P(A_i) + P(A_j) - P(A_i \cap A_j)$$

where $P(A_i \cup A_j)$ denotes the probability of either A_i or A_j or both occurring; $P(A_i)$ and $P(A_j)$ denote, respectively, the probability of A_i and the probability of A_j; and $P(A_i \cap A_j)$ denotes the probability of both A_i and A_j occurring.

Multiplication theorem

Let $P(A_i|A_j)$ denote the conditional probability of A_i occurring, given that A_j has occurred. This conditional probability is defined as follows:

$$(1.7) \qquad P(A_i|A_j) = \frac{P(A_i \cap A_j)}{P(A_j)} \qquad P(A_j) \neq 0$$

The multiplication theorem states:

$$(1.8) \qquad P(A_i \cap A_j) = P(A_i)P(A_j|A_i)$$
$$= P(A_j)P(A_i|A_j)$$

Complementary events

The complementary event of A_i is denoted by $\bar{A_i}$. The following results for complementary events are useful:

$$(1.9) \qquad P(\bar{A_i}) = 1 - P(A_i)$$

$$(1.10) \qquad P(\overline{A_i \cup A_j}) = P(\bar{A_i} \cap \bar{A_j})$$

1.3 RANDOM VARIABLES

Throughout this section, we assume that the random variable Y assumes a finite number of outcomes. (If Y is a continuous random variable, the summation process is replaced by integration.)

Expected value

Let the random variable Y assume the outcomes $Y_1, \ldots, Y_k$ with probabilities given by the probability function:

$$(1.11) \qquad f(Y_s) = P(Y = Y_s) \qquad s = 1, \ldots, k$$

The expected value of Y is defined:

$$(1.12) \qquad E(Y) = \sum_{s=1}^{k} Y_s f(Y_s)$$

An important property of the expectation operator E is:

$$(1.13) \qquad E(a + cY) = a + cE(Y) \qquad \text{where } a \text{ and } c \text{ are constants}$$

Special cases of this are:

$$(1.13a) \qquad E(a) = a$$

$$(1.13b) \qquad E(cY) = cE(Y)$$

$$(1.13c) \qquad E(a + Y) = a + E(Y)$$

Variance

The variance of the random variable Y is denoted by $\sigma^2(Y)$ and is defined as follows:

(1.14)
$$\sigma^2(Y) = E\{[Y - E(Y)]^2\}$$

An equivalent expression is:

(1.14a)
$$\sigma^2(Y) = E(Y^2) - [E(Y)]^2$$

The variance of a linear function of Y is frequently encountered. We denote the variance of $a + cY$ by $\sigma^2(a + cY)$ and have:

(1.15) $\qquad \sigma^2(a + cY) = c^2\sigma^2(Y) \qquad$ where a and c are constants

Special cases of this result are:

(1.15a)
$$\sigma^2(a + Y) = \sigma^2(Y)$$

(1.15b)
$$\sigma^2(cY) = c^2\sigma^2(Y)$$

Joint, marginal, and conditional probability distributions

Let the joint probability function for the two random variables Y and Z be denoted by $g(Y, Z)$:

(1.16) $\quad g(Y_s, Z_t) = P(Y = Y_s \cap Z = Z_t) \qquad s = 1, \ldots, k; t = 1, \ldots, m$

The marginal probability function of Y, denoted by $f(Y)$, is:

(1.17a)
$$f(Y_s) = \sum_{t=1}^{m} g(Y_s, Z_t) \qquad s = 1, \ldots, k$$

and the marginal probability function of Z, denoted by $h(Z)$, is:

(1.17b)
$$h(Z_t) = \sum_{s=1}^{k} g(Y_s, Z_t) \qquad t = 1, \ldots, m$$

The conditional probability function of Y, given $Z = Z_t$, is:

(1.18a) $\qquad f(Y_s|Z_t) = \dfrac{g(Y_s, Z_t)}{h(Z_t)} \qquad h(Z_t) \neq 0; s = 1, \ldots, k$

and the conditional probability function of Z, given $Y = Y_s$, is:

(1.18b) $\qquad h(Z_t|Y_s) = \dfrac{g(Y_s, Z_t)}{f(Y_s)} \qquad f(Y_s) \neq 0; t = 1, \ldots, m$

Covariance

The covariance of Y and Z is denoted by $\sigma(Y, Z)$ and is defined:

$$(1.19) \qquad \sigma(Y, Z) = E\{[Y - E(Y)][Z - E(Z)]\}$$

An equivalent expression is:

$$(1.19a) \qquad \sigma(Y, Z) = E(YZ) - [E(Y)][E(Z)]$$

The covariance of $a_1 + c_1Y$ and $a_2 + c_2Z$ is denoted by $\sigma(a_1 + c_1Y, a_2 + c_2Z)$, and we have:

$$(1.20) \qquad \sigma(a_1 + c_1Y, a_2 + c_2Z) = c_1c_2\sigma(Y, Z) \qquad \text{where } a_1, a_2, c_1, c_2 \text{ are constants}$$

Special cases of this are:

$$(1.20a) \qquad \sigma(c_1Y, c_2Z) = c_1c_2\sigma(Y, Z)$$

$$(1.20b) \qquad \sigma(a_1 + Y, a_2 + Z) = \sigma(Y, Z)$$

By definition, we have:

$$(1.21) \qquad \sigma(Y, Y) = \sigma^2(Y)$$

where $\sigma^2(Y)$ is the variance of Y.

Independent random variables

(1.22) Random variables Y and Z are independent if and only if:

$$g(Y_s, Z_t) = f(Y_s)h(Z_t) \qquad s = 1, \ldots, k; \ t = 1, \ldots, m$$

If Y and Z are independent random variables:

$$(1.23) \qquad \sigma(Y, Z) = 0 \qquad \text{when } Y, Z \text{ are independent}$$

(In the special case where Y and Z are jointly normally distributed, $\sigma(Y, Z) = 0$ implies that Y and Z are independent.)

Functions of random variables

Let $Y_1, \ldots, Y_n$ be n random variables. Consider the function Σa_iY_i where the a_i are constants. We then have:

$$(1.24a) \qquad E\left(\sum_{i=1}^{n} a_iY_i\right) = \sum_{i=1}^{n} a_iE(Y_i) \qquad \text{where the } a_i \text{ are constants}$$

$$(1.24b) \qquad \sigma^2\left(\sum_{i=1}^{n} a_iY_i\right) = \sum_{i=1}^{n}\sum_{j=1}^{n} a_ia_j\sigma(Y_i, Y_j) \qquad \text{where the } a_i \text{ are constants}$$

Specifically, we have for $n = 2$:

(1.25a) $E(a_1Y_1 + a_2Y_2) = a_1E(Y_1) + a_2E(Y_2)$

(1.25b) $\sigma^2(a_1Y_1 + a_2Y_2) = a_1^2\sigma^2(Y_1) + a_2^2\sigma^2(Y_2) + 2a_1a_2\sigma(Y_1, Y_2)$

If the random variables Y_i are independent, we have:

(1.26) $$\sigma^2\left(\sum_{i=1}^{n} a_iY_i\right) = \sum_{i=1}^{n} a_i^2\sigma^2(Y_i)$$ when the Y_i are independent

Special cases of this are:

(1.26a) $\sigma^2(Y_1 + Y_2) = \sigma^2(Y_1) + \sigma^2(Y_2)$ when Y_1, Y_2 are independent

(1.26b) $\sigma^2(Y_1 - Y_2) = \sigma^2(Y_1) + \sigma^2(Y_2)$ when Y_1, Y_2 are independent

When the Y_i are independent random variables, the covariance of two linear functions Σa_iY_i and Σc_iY_i is:

(1.27) $$\sigma\left(\sum_{i=1}^{n} a_iY_i, \sum_{i=1}^{n} c_iY_i\right) = \sum_{i=1}^{n} a_ic_i\sigma^2(Y_i)$$ when the Y_i are independent

Central limit theorem

(1.28)   If $Y_1, \ldots, Y_n$ are independent random observations from a population with probability function $f(Y)$ for which $\sigma^2(Y)$ is finite, the sample mean $\bar{Y}$:

$$\bar{Y} = \frac{\sum_{i=1}^{n} Y_i}{n}$$

is approximately normally distributed when the sample size n is reasonably large, with mean $E(Y)$ and variance $\sigma^2(Y)/n$.

1.4 NORMAL PROBABILITY DISTRIBUTION AND RELATED DISTRIBUTIONS

Normal probability distribution

The density function for a normal random variable Y is:

(1.29) $$f(Y) = \frac{1}{\sqrt{2\pi}\sigma}\exp\left[-\frac{1}{2}\left(\frac{Y - \mu}{\sigma}\right)^2\right]$$ $-\infty < Y < +\infty$

where μ and σ are the two parameters of the normal distribution and $\exp(a)$ denotes e^a.

The mean and variance of a normal random variable Y are:

(1.30a) $$E(Y) = \mu$$

(1.30b) $$\sigma^2(Y) = \sigma^2$$

Function of normal random variable. A linear function of a normal random variable Y has the following property:

(1.31)   If Y is a normal random variable, the transformed variable $Y' = a + cY$ (a and c are constants) is normally distributed, with mean $a + cE(Y)$ and variance $c^2\sigma^2(Y)$.

Standard normal variable. The standard normal variable z:

(1.32) $z = \dfrac{Y - \mu}{\sigma}$ where Y is a normal random variable

is normally distributed, with mean 0 and variance 1. We denote this as follows:

(1.33) z is $N(0, 1)$

Mean   Variance

Table A–1 in the Appendix contains the cumulative probabilities A for percentiles $z(A)$ where:

(1.34) $$P\{z \le z(A)\} = A$$

For instance, when $z(A) = 2.00$, $A = .9772$. Because the normal distribution is symmetrical about 0, when $z(A) = -2.00$, $A = 1 - .9772 = .0228$.

Function of independent normal random variables. Let $Y_1, \ldots, Y_n$ be independent normal random variables. We then have:

(1.35)   When $Y_1, \ldots, Y_n$ are independent normal random variables, the linear combination $a_1Y_1 + a_2Y_2 + \cdots + a_nY_n$ is normally distributed, with mean $\Sigma a_i E(Y_i)$ and variance $\Sigma a_i^2 \sigma^2(Y_i)$.

χ^2 distribution

Let $z_1, \ldots, z_\nu$ be ν independent standard normal variables. We then define:

(1.36) $\chi^2(\nu) = z_1^2 + z_2^2 + \cdots + z_\nu^2$ where the z_i are independent

The χ^2 distribution has one parameter, ν, which is called the *degrees of freedom* (*df*). The mean of the χ^2 distribution with ν degrees of freedom is:

(1.37) $$E[\chi^2(\nu)] = \nu$$

Table A–3 in the Appendix contains percentiles of various χ^2 distributions. We define $\chi^2(A; \nu)$ as follows:

$$(1.38) \qquad P\{\chi^2(\nu) \leq \chi^2(A; \nu)\} = A$$

Suppose $\nu = 5$. The 90th percentile of the χ^2 distribution with 5 degrees of freedom is $\chi^2(.90; 5) = 9.24$.

t distribution

Let z and $\chi^2(\nu)$ be independent random variables (standard normal and χ^2, respectively). We then define:

$$(1.39) \qquad t(\nu) = \frac{z}{\left[\dfrac{\chi^2(\nu)}{\nu}\right]^{1/2}} \qquad \text{where } z \text{ and } \chi^2(\nu) \text{ are independent}$$

The t distribution has one parameter, the *degrees of freedom* ν. The mean of the t distribution with ν degrees of freedom is:

$$(1.40) \qquad E[t(\nu)] = 0$$

Table A–2 in the Appendix contains percentiles of various t distributions. We define $t(A; \nu)$ as follows:

$$(1.41) \qquad P\{t(\nu) \leq t(A; \nu)\} = A$$

Suppose $\nu = 10$. The 90th percentile of the t distribution with 10 degrees of freedom is $t(.90; 10) = 1.372$. Because the t distribution is symmetrical about 0, we have $t(.10; 10) = -1.372$.

F distribution

Let $\chi^2(\nu_1)$ and $\chi^2(\nu_2)$ be two independent χ^2 random variables. We then define:

$$(1.42) \qquad F(\nu_1, \nu_2) = \frac{\chi^2(\nu_1)}{\nu_1} \div \frac{\chi^2(\nu_2)}{\nu_2} \qquad \text{where } \chi^2(\nu_1) \text{ and } \chi^2(\nu_2) \text{ are independent}$$

Numerator Denominator
df   df

The F distribution has two parameters, the *numerator degrees of freedom* and the *denominator degrees of freedom*, here ν_1 and ν_2, respectively.

Table A–4 in the Appendix contains percentiles of various F distributions. We define $F(A; \nu_1, \nu_2)$ as follows:

$$(1.43) \qquad P\{F(\nu_1, \nu_2) \leq F(A; \nu_1, \nu_2)\} = A$$

Suppose $\nu_1 = 2$, $\nu_2 = 3$. The 90th percentile of the F distribution with 2 and 3 degrees of freedom, respectively, in the numerator and denominator is $F(.90; 2, 3) = 5.46$.

Percentiles below 50 percent can be obtained by utilizing the relation:

(1.44)
$$F(A; \nu_1, \nu_2) = \frac{1}{F(1 - A; \nu_2, \nu_1)}$$

Thus, $F(.10; 3, 2) = 1/F(.90; 2, 3) = 1/5.46 = .183$.

The following relation exists between the t and F random variables:

(1.45a)
$$[t(\nu)]^2 = F(1, \nu)$$

and the percentiles of the t and F distributions are related as follows:

(1.45b)
$$[t(.5 + A/2; \nu)]^2 = F(A; 1, \nu)$$

1.5   STATISTICAL ESTIMATION

Properties of estimators

(1.46)   An estimator $\hat{\theta}$ of the parameter θ is *unbiased* if:
$$E(\hat{\theta}) = \theta$$

(1.47)   An estimator $\hat{\theta}$ is a *consistent estimator* of θ if:
$$\lim_{n \to \infty} P(|\hat{\theta} - \theta| \geq \varepsilon) = 0 \qquad \text{for any } \varepsilon > 0$$

(1.48)   An estimator $\hat{\theta}$ is a *sufficient estimator* of θ if the conditional joint probability function of the sample observations, given $\hat{\theta}$, does not depend on the parameter θ.

(1.49)   An estimator $\hat{\theta}$ is a *minimum variance estimator* of θ if for any other estimator θ^*:
$$\sigma^2(\hat{\theta}) \leq \sigma^2(\theta^*) \qquad \text{for all } \theta^*$$

Maximum likelihood estimators

The method of maximum likelihood is a general method of finding estimators. Suppose we are sampling a population whose probability function $f(Y; \theta)$ involves one parameter, θ. Given independent observations $Y_1, \ldots, Y_n$, the joint probability function of the sample observations is:

(1.50a)
$$g(Y_1, \ldots, Y_n) = \prod_{i=1}^{n} f(Y_i; \theta)$$

When this joint probability function is viewed as a function of θ, with the observations given, it is called the *likelihood function* $L(\theta)$.

(1.50b)
$$L(\theta) = \prod_{i=1}^{n} f(Y_i; \theta)$$

Maximizing $L(\theta)$ with respect to θ yields the maximum likelihood estimator of θ. Under quite general conditions, maximum likelihood estimators are consistent and sufficient.

Least squares estimators

The method of least squares is another general method of finding estimators. The sample observations are assumed to be of the form (for the case of a single parameter θ):

$$(1.51) \qquad Y_i = f_i(\theta) + \varepsilon_i \qquad i = 1, \ldots, n$$

where $f_i(\theta)$ is a known function of the parameter θ and the ε_i are random variables, usually assumed to have expectation $E(\varepsilon_i) = 0$.

With the method of least squares, for the given sample observations, the sum of squares:

$$(1.52) \qquad Q = \sum_{i=1}^{n} [Y_i - f_i(\theta)]^2$$

is considered as a function of θ. The least squares estimator of θ is obtained by minimizing Q with respect to θ. In many instances, least squares estimators are unbiased and consistent.

1.6 INFERENCES ABOUT POPULATION MEAN—NORMAL POPULATION

We have a random sample of n observations $Y_1, \ldots, Y_n$ from a normal population with mean μ and standard deviation σ. The sample mean and sample standard deviation are:

$$(1.53a) \qquad \bar{Y} = \frac{\sum_{i} Y_i}{n}$$

$$(1.53b) \qquad s = \left[\frac{\sum_{i} (Y_i - \bar{Y})^2}{n-1} \right]^{1/2}$$

and the estimated standard deviation of the sampling distribution of $\bar{Y}$ is:

$$(1.53c) \qquad s(\bar{Y}) = \frac{s}{\sqrt{n}}$$

We then have:

$(1.54) \qquad \dfrac{\bar{Y} - \mu}{s(\bar{Y})}$ is distributed as t with $n - 1$ degrees of freedom when the random sample is from a normal population.

Interval estimation

The confidence limits for μ with a confidence coefficient of $1 - \alpha$ are obtained by means of (1.54):

(1.55) $$\bar{Y} \pm t(1 - \alpha/2; n - 1)s(\bar{Y})$$

Example 1. Obtain a 95 percent confidence interval for μ when:

$$n = 10 \qquad \bar{Y} = 20 \qquad s = 4$$

We require:

$$s(\bar{Y}) = \frac{4}{\sqrt{10}} = 1.265 \qquad t(.975; 9) = 2.262$$

so that the confidence limits are $20 \pm 2.262(1.265)$. Hence, the 95 percent confidence interval for μ is:

$$17.1 \le \mu \le 22.9$$

Tests

One-sided and two-sided tests concerning the population mean μ are constructed by means of (1.54), based on the test statistic:

(1.56) $$t^* = \frac{\bar{Y} - \mu_0}{s(\bar{Y})}$$

Table 1.1 contains the decision rules for each of three possible cases, with the risk of making a Type I error controlled at α.

TABLE 1.1 Decision rules for tests concerning mean μ of normal population

Alternatives	Decision Rule
	(a)
$H_0: \mu = \mu_0$	If $\lvert t^* \rvert \le t(1 - \alpha/2; n - 1)$, conclude H_0
$H_a: \mu \ne \mu_0$	If $\lvert t^* \rvert > t(1 - \alpha/2; n - 1)$, conclude H_a
	where:
	$t^* = \dfrac{\bar{Y} - \mu_0}{s(\bar{Y})}$
	(b)
$H_0: \mu \ge \mu_0$	If $t^* \ge t(\alpha; n - 1)$, conclude H_0
$H_a: \mu < \mu_0$	If $t^* < t(\alpha; n - 1)$, conclude H_a
	(c)
$H_0: \mu \le \mu_0$	If $t^* \le t(1 - \alpha; n - 1)$, conclude H_0
$H_a: \mu > \mu_0$	If $t^* > t(1 - \alpha; n - 1)$, conclude H_a

Example 2. Choose between the alternatives:

$$H_0: \mu \leq 20$$
$$H_a: \mu > 20$$

when α is to be controlled at .05 and:

$$n = 15 \qquad \bar{Y} = 24 \qquad s = 6$$

We require:

$$s(\bar{Y}) = \frac{6}{\sqrt{15}} = 1.549$$
$$t(.95; 14) = 1.761$$

so that the decision rule is:

If $t^* \leq 1.761$, conclude H_0
If $t^* > 1.761$, conclude H_a

Since $t^* = (24 - 20)/1.549 = 2.58 > 1.761$, we conclude H_a.

Example 3. Choose between the alternatives:

$$H_0: \mu = 10$$
$$H_a: \mu \neq 10$$

when α is to be controlled at .02 and:

$$n = 25 \qquad \bar{Y} = 5.7 \qquad s = 8$$

We require:

$$s(\bar{Y}) = \frac{8}{\sqrt{25}} = 1.6$$
$$t(.99; 24) = 2.492$$

so that the decision rule is:

If $|t^*| \leq 2.492$, conclude H_0
If $|t^*| > 2.492$, conclude H_a

where the symbol $|\ |$ stands for the absolute value. Since $|t^*| = |(5.7 - 10)/1.6|$ $= |-2.69| = 2.69 > 2.492$, we conclude H_a.

P-value for sample outcome. The P-value for a sample outcome is the probability that the sample outcome could have been more extreme than the observed one when $\mu = \mu_0$. Large P-values support H_0 while small P-values support H_a. A test can be carried out by comparing the P-value with the specified α risk. If the P-value equals or is greater than the specified α, H_0 is concluded. If the P-value is less than α, H_a is concluded.

Example 4. In Example 2, $t^* = 2.58$. The P-value for this sample outcome is the probability $P[t(14) > 2.58]$. From Table A–2, we find $t(.985; 14) = 2.415$ and $t(.990; 14) = 2.624$. Hence, the P-value is between .010 and .015. In fact, it can be shown to be .011. Thus, for $\alpha = .05$, H_a is concluded.

Example 5. In Example 3, $t^* = -2.69$. We find from Table A–2 that $P[t(24) < -2.69]$ is between .005 and .0075. In fact, it can be shown to be .0064. Because the test is two-sided and the t distribution is symmetrical, the two-sided P-value is twice the one-sided value, or $2(.0064) = .013$. Hence, for $\alpha = .02$, we conclude H_a.

Relation between tests and confidence intervals. There is a direct relation between tests and confidence intervals. For example, the two-sided confidence limits (1.55) can be used for testing:

$$H_0: \mu = \mu_0$$
$$H_a: \mu \neq \mu_0$$

If μ_0 is contained within the $1 - \alpha$ confidence interval, then the two-sided decision rule in Table 1.1a, with level of significance α, will lead to conclusion H_0, and vice versa. If μ_0 is not contained within the confidence interval, the decision rule will lead to H_a, and vice versa.

There are similar correspondences between one-sided confidence intervals and one-sided decision rules.

1.7 COMPARISONS OF TWO POPULATION MEANS—NORMAL POPULATIONS

Independent samples

There are two normal populations, with means μ_1 and μ_2, respectively, and with the same standard deviation σ. The means μ_1 and μ_2 are to be compared on the basis of independent samples for each of the two populations:

$$\text{Sample 1: } Y_1, \ldots, Y_{n_1}$$
$$\text{Sample 2: } Z_1, \ldots, Z_{n_2}$$

Estimators of the two population means are the sample means:

(1.57a)
$$\bar{Y} = \frac{\sum_i Y_i}{n_1}$$

(1.57b)
$$\bar{Z} = \frac{\sum_i Z_i}{n_2}$$

and an estimator of $\mu_1 - \mu_2$ is $\bar{Y} - \bar{Z}$.

An estimator of the common variance σ^2 is:

(1.58)
$$s^2 = \frac{\sum_i (Y_i - \bar{Y})^2 + \sum_i (Z_i - \bar{Z})^2}{n_1 + n_2 - 2}$$

and an estimator of $\sigma^2(\bar{Y} - \bar{Z})$, the variance of the sampling distribution of $\bar{Y} - \bar{Z}$, is:

(1.59)
$$s^2(\bar{Y} - \bar{Z}) = s^2 \left[\frac{1}{n_1} + \frac{1}{n_2} \right]$$

We have:

(1.60) $\dfrac{(\bar{Y} - \bar{Z}) - (\mu_1 - \mu_2)}{s(\bar{Y} - \bar{Z})}$ is distributed as t with $n_1 + n_2 - 2$ degrees

of freedom when the two independent samples come from normal populations with the same standard deviation.

Interval estimation. The confidence limits for $\mu_1 - \mu_2$ with confidence coefficient $1 - \alpha$ are obtained by means of (1.60):

(1.61) $(\bar{Y} - \bar{Z}) \pm t(1 - \alpha/2; n_1 + n_2 - 2)s(\bar{Y} - \bar{Z})$

Example 6. Obtain a 95 percent confidence interval for $\mu_1 - \mu_2$ when:

$$n_1 = 10 \quad \bar{Y} = 14 \quad \Sigma(Y_i - \bar{Y})^2 = 105$$
$$n_2 = 20 \quad \bar{Z} = 8 \quad \Sigma(Z_i - \bar{Z})^2 = 224$$

We require:

$$s^2 = \frac{105 + 224}{10 + 20 - 2} = 11.75$$

$$s^2(\bar{Y} - \bar{Z}) = 11.75 \left(\frac{1}{10} + \frac{1}{20} \right) = 1.7625$$

$$s(\bar{Y} - \bar{Z}) = 1.328$$

$$t(.975; 28) = 2.048$$

$$3.3 = (14 - 8) - 2.048(1.328) \le \mu_1 - \mu_2 \le (14 - 8) + 2.048(1.328) = 8.7$$

Tests. One-sided and two-sided tests concerning $\mu_1 - \mu_2$ are constructed by means of (1.60). Table 1.2 contains the decision rules for each of three possible cases, based on the test statistic:

(1.62)
$$t^* = \frac{\bar{Y} - \bar{Z}}{s(\bar{Y} - \bar{Z})}$$

with the risk of making a Type I error controlled at α.

TABLE 1.2 Decision rules for tests concerning means μ_1 and μ_2 of two normal populations ($\sigma_1 = \sigma_2 = \sigma$)

Alternatives	Decision Rule
(a)	
$H_0: \mu_1 = \mu_2$	If $\lvert t^* \rvert \leq t(1 - \alpha/2; n_1 + n_2 - 2)$, conclude H_0
$H_a: \mu_1 \neq \mu_2$	If $\lvert t^* \rvert > t(1 - \alpha/2; n_1 + n_2 - 2)$, conclude H_a
	where: $$t^* = \frac{\bar{Y} - \bar{Z}}{s(\bar{Y} - \bar{Z})}$$
(b)	
$H_0: \mu_1 \geq \mu_2$	If $t^* \geq t(\alpha; n_1 + n_2 - 2)$, conclude H_0
$H_a: \mu_1 < \mu_2$	If $t^* < t(\alpha; n_1 + n_2 - 2)$, conclude H_a
(c)	
$H_0: \mu_1 \leq \mu_2$	If $t^* \leq t(1 - \alpha; n_1 + n_2 - 2)$, conclude H_0
$H_a: \mu_1 > \mu_2$	If $t^* > t(1 - \alpha; n_1 + n_2 - 2)$, conclude H_a

Example 7. Choose between the alternatives:

$$H_0: \mu_1 = \mu_2$$
$$H_a: \mu_1 \neq \mu_2$$

when α is to be controlled at .10 and the data are those of Example 6. We require $t(.95; 28) = 1.701$, so that the decision rule is:

$$\text{If } \lvert t^* \rvert \leq 1.701, \text{ conclude } H_0$$
$$\text{If } \lvert t^* \rvert > 1.701, \text{ conclude } H_a$$

Since $\lvert t^* \rvert = \lvert (14 - 8)/1.328 \rvert = \lvert 4.52 \rvert = 4.52 > 1.701$, we conclude H_a.

The one-sided P-value here is the probability $P[t(28) > 4.52]$. We see from Table A–2 that this P-value is less than .0005. In fact, it can be shown to be .00005. Hence, the two-sided P-value is .0001. For $\alpha = .10$, the appropriate conclusion therefore is H_a.

Paired observations

When the observations in the two samples are paired (e.g., attitude scores Y_i and Z_i for the ith sample employee before and after a year's experience on the job), we use the differences:

$$(1.63) \qquad W_i = Y_i - Z_i \qquad i = 1, \ldots, n$$

in the fashion of a sample from a single population. Thus, when the W_i can be treated as observations from a normal population, we have:

(1.64) $\dfrac{\bar{W} - (\mu_1 - \mu_2)}{s(\bar{W})}$ is distributed as t with $n - 1$ degrees of freedom

when the differences W_i can be considered to be observations from a normal population and:

$$\bar{W} = \frac{\sum\limits_i W_i}{n}$$

$$s^2(\bar{W}) = \frac{\sum\limits_i (W_i - \bar{W})^2}{n - 1} \div n$$

1.8 INFERENCES ABOUT POPULATION VARIANCE—NORMAL POPULATION

When sampling from a normal population, the following holds for the sample variance s^2 where s is defined in (1.53b):

(1.65) $\dfrac{(n - 1)s^2}{\sigma^2}$ is distributed as χ^2 with $n - 1$ degrees of freedom when

the random sample is from a normal population.

Interval estimation

The lower confidence limit L and the upper confidence limit U in a confidence interval for the population variance σ^2 with confidence coefficient $1 - \alpha$ are obtained by means of (1.65):

(1.66) $L = \dfrac{(n - 1)s^2}{\chi^2(1 - \alpha/2; n - 1)} \qquad U = \dfrac{(n - 1)s^2}{\chi^2(\alpha/2; n - 1)}$

Example 8. Obtain a 98 percent confidence interval for σ^2, using the data of Example 1 ($n = 10$, $s = 4$).

We require:

$$s^2 = 16 \qquad \chi^2(.01; 9) = 2.09 \qquad \chi^2(.99; 9) = 21.67$$

$$6.6 = \frac{9(16)}{21.67} \leq \sigma^2 \leq \frac{9(16)}{2.09} = 68.9$$

Tests

One-sided and two-sided tests concerning the population variance σ^2 are constructed by means of (1.65). Table 1.3 contains the decision rule for each of three possible cases, with the risk of making a Type I error controlled at α.

TABLE 1.3   Decision rules for tests concerning variance σ^2 of normal population

Alternatives	*Decision Rule*
(a)	
$H_0\colon \sigma^2 = \sigma_0^2$	If $\chi^2(\alpha/2; n-1) \leq \dfrac{(n-1)s^2}{\sigma_0^2} \leq \chi^2(1-\alpha/2; n-1)$,
	conclude H_0
$H_a\colon \sigma^2 \neq \sigma_0^2$	Otherwise conclude H_a
(b)	
$H_0\colon \sigma^2 \geq \sigma_0^2$	If $\dfrac{(n-1)s^2}{\sigma_0^2} \geq \chi^2(\alpha; n-1)$, conclude H_0
$H_a\colon \sigma^2 < \sigma_0^2$	If $\dfrac{(n-1)s^2}{\sigma_0^2} < \chi^2(\alpha; n-1)$, conclude H_a
(c)	
$H_0\colon \sigma^2 \leq \sigma_0^2$	If $\dfrac{(n-1)s^2}{\sigma_0^2} \leq \chi^2(1-\alpha; n-1)$, conclude H_0
$H_a\colon \sigma^2 > \sigma_0^2$	If $\dfrac{(n-1)s^2}{\sigma_0^2} > \chi^2(1-\alpha; n-1)$, conclude H_a

1.9   COMPARISONS OF TWO POPULATION VARIANCES—NORMAL POPULATIONS

Independent samples are selected from two normal populations, with means and variances of μ_1 and σ_1^2 and μ_2 and σ_2^2, respectively. Using the notation of Section 1.7, the two sample variances are:

$$(1.67a) \qquad s_1^2 = \frac{\sum_i (Y_i - \bar{Y})^2}{n_1 - 1}$$

$$(1.67b) \qquad s_2^2 = \frac{\sum_i (Z_i - \bar{Z})^2}{n_2 - 1}$$

We have:

$$(1.68) \qquad \frac{s_1^2}{\sigma_1^2} \div \frac{s_2^2}{\sigma_2^2} \text{ is distributed as } F(n_1 - 1, n_2 - 1) \text{ when the two inde-}$$

pendent samples come from normal populations.

Interval estimation

The lower and upper confidence limits L and U for σ_1^2/σ_2^2 with confidence coefficient $1 - \alpha$ are obtained by means of (1.68):

$$(1.69) \quad L = \frac{s_1^2}{s_2^2} \frac{1}{F(1 - \alpha/2; n_1 - 1, n_2 - 1)}$$

$$U = \frac{s_1^2}{s_2^2} \frac{1}{F(\alpha/2; n_1 - 1, n_2 - 1)}$$

Example 9. Obtain a 90 percent confidence interval for σ_1^2/σ_2^2 when the data are:

$$n_1 = 16 \qquad n_2 = 21$$
$$s_1^2 = 54.2 \qquad s_2^2 = 17.8$$

We require:

$$F(.05; 15, 20) = 1/F(.95; 20, 15) = 1/2.33 = .429$$
$$F(.95; 15, 20) = 2.20$$

$$1.4 = \frac{54.2}{17.8} \frac{1}{2.20} \leq \frac{\sigma_1^2}{\sigma_2^2} \leq \frac{54.2}{17.8} \frac{1}{.429} = 7.1$$

Tests

One-sided and two-sided tests concerning σ_1^2/σ_2^2 are constructed by means of (1.68). Table 1.4 contains the decision rules for each of three possible cases, with the risk of making a Type I error controlled at α.

TABLE 1.4 Decision rules for tests concerning variances σ_1^2 and σ_2^2 of two normal populations

Alternatives	Decision Rule
	(a)
H_0: $\sigma_1^2 = \sigma_2^2$	If $F(\alpha/2; n_1 - 1, n_2 - 1) \leq \dfrac{s_1^2}{s_2^2}$
	$\leq F(1 - \alpha/2; n_1 - 1, n_2 - 1)$, conclude H_0
H_a: $\sigma_1^2 \neq \sigma_2^2$	Otherwise conclude H_a
	(b)
H_0: $\sigma_1^2 \geq \sigma_2^2$	If $\dfrac{s_1^2}{s_2^2} \geq F(\alpha; n_1 - 1, n_2 - 1)$, conclude H_0
H_a: $\sigma_1^2 < \sigma_2^2$	If $\dfrac{s_1^2}{s_2^2} < F(\alpha; n_1 - 1, n_2 - 1)$, conclude H_a
	(c)
H_0: $\sigma_1^2 \leq \sigma_2^2$	If $\dfrac{s_1^2}{s_2^2} \leq F(1 - \alpha; n_1 - 1, n_2 - 1)$, conclude H_0
H_a: $\sigma_1^2 > \sigma_2^2$	If $\dfrac{s_1^2}{s_2^2} > F(1 - \alpha; n_1 - 1, n_2 - 1)$, conclude H_a

Example 10. Choose between the alternatives:

$$H_0: \sigma_1^2 = \sigma_2^2$$
$$H_a: \sigma_1^2 \neq \sigma_2^2$$

when α is to be controlled at .02 and the data are those of Example 9. We require:

$$F(.01; 15, 20) = 1/F(.99; 20, 15) = 1/3.37 = .297$$
$$F(.99; 15, 20) = 3.09$$

so that the decision rule is:

$$\text{If } .297 \leq \frac{s_1^2}{s_2^2} \leq 3.09, \text{ conclude } H_0$$

Otherwise conclude H_a

Since $s_1^2/s_2^2 = 54.2/17.8 = 3.04$, we conclude H_0.

PART I

Basic regression analysis

2

Linear regression with one independent variable

Regression analysis is a statistical tool that utilizes the relation between two or more quantitative variables so that one variable can be predicted from the other, or others. For example, if one knows the relation between advertising expenditures and sales, one can predict sales by regression analysis once the level of advertising expenditures has been set.

In Part I of this book, we take up regression analysis when a single predictor variable is used for predicting the variable of interest. In this chapter specifically, we consider the basic ideas of regression analysis and discuss the estimation of the parameters of the regression model.

2.1 RELATIONS BETWEEN VARIABLES

The concept of a relation between two variables, such as between family income and family expenditures for housing, is a familiar one. We distinguish between a *functional* relation and a *statistical* relation, and consider each of these in turn.

Functional relation between two variables

A functional relation between two variables is expressed by a mathematical formula. If X is the *independent* variable and Y the *dependent* variable, a functional relation is of the form:

$$Y = f(X)$$

Given a particular value of X, the function f indicates the corresponding value of Y.

Example. Consider the relation between dollar sales (Y) of a product sold at a fixed price and number of units sold (X). If the selling price is $2 per unit, the relation is expressed by the equation:

$$Y = 2X$$

This functional relation is shown in Figure 2.1. Number of units sold and dollar sales during three recent periods (while the unit price remained constant at $2) were as follows:

Period	Number of Units Sold	Dollar Sales
1	75	$150
2	25	50
3	130	260

These observations are plotted also in Figure 2.1. Note that all fall directly on the line of functional relationship. This is characteristic of all functional relations.

FIGURE 2.1 Example of functional relation

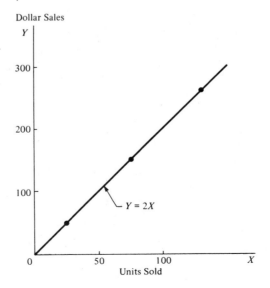

Statistical relation between two variables

A statistical relation, unlike a functional relation, is not a perfect one. In general, the observations for a statistical relation do not fall directly on the curve of relationship.

Example 1. A certain spare part is manufactured by the Westwood Company once a month in lots which vary in size as demand fluctuates. Table 2.1, page 36, contains data on lot size and number of man-hours of labor for 10 recent production runs performed under similar production conditions. These data are plotted in Figure 2.2a. Man-hours are taken as the *dependent* or *response* variable Y, and lot size as the *independent* or *predictor* variable X. The plotting is done as before. For instance, the first production run results are plotted as $X = 30$, $Y = 73$.

FIGURE 2.2   Statistical relation between lot size and number of man-hours—Westwood Company example

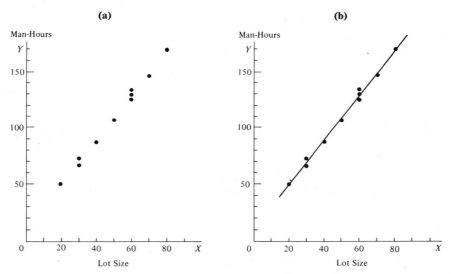

Figure 2.2a clearly suggests that there is a relation between lot size and number of man-hours, in the sense that the larger the lot size, the greater tends to be the number of man-hours. However, the relation is not a perfect one. There is a scattering of points, suggesting that some of the variation in man-hours is not accounted for by lot size. For instance, two production runs (1 and 8) consisted of 30 parts, yet they required somewhat different numbers of man-hours. Because of the scattering of points in a statistical relation, Figure 2.2a is called a *scatter diagram* or *scatter plot*. In statistical terminology, each point in the scatter diagram represents an *observation* or *trial*.

In Figure 2.2b, we have plotted a line of relationship which describes the statistical relation between man-hours and lot size. It indicates the general tendency by which man-hours vary with changes in lot size. Note that most of the points do not fall directly on the line of statistical relationship. This scattering of points around the line represents variation in man-hours which is not associated with the lot size, and which is usually considered to be of a random nature. Statistical relations can be highly useful, even though they do not have the exactitude of a functional relation.

Example 2. Figure 2.3 presents data on age and level of a steroid in plasma for 17 healthy females between 8 and 25 years old. The data strongly suggest that the statistical relationship is *curvilinear* (not linear). The curve of relationship has also been drawn in Figure 2.3. It implies that as age becomes increasingly higher, steroid level increases up to a point and then begins to decline. Note again the scattering of points around the curve of statistical relationship, typical of all statistical relations.

FIGURE 2.3 Curvilinear statistical relation between age and steroid level in healthy females aged 8 to 25

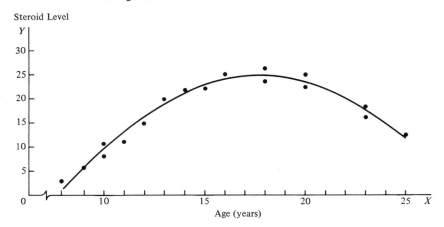

2.2 REGRESSION MODELS AND THEIR USES

Basic concepts

A regression model is a formal means of expressing the two essential ingredients of a statistical relation:

1. A tendency of the dependent variable Y to vary with the independent variable or variables in a systematic fashion.
2. A scattering of observations around the curve of statistical relationship.

These two characteristics are embodied in a regression model by postulating that:

1. In the population of observations associated with the sampled process, there is a probability distribution of Y for each level of X.
2. The means of these probability distributions vary in some systematic fashion with X.

Example. Consider again the Westwood Company lot size example. The number of man-hours Y is treated in a regression model as a random variable. For each lot size, there is postulated a probability distribution of Y. Figure 2.4 shows such a probability distribution for $X = 30$, which is the lot size for the first production run in Table 2.1. The actual number of man-hours Y (73 in our example in Table 2.1) is then viewed as a random selection from this probability distribution.

Figure 2.4 also shows probability distributions of Y for lot sizes $X = 50$ and $X = 70$. Note that the means of the probability distributions have a systematic relation to the level of X. This systematic relationship is called the *regression function of Y on X*. The graph of the regression function is called the *regression curve*. Note that in Figure 2.4 the regression function is linear. This would imply for our example that the expected (mean) number of man-hours varies linearly with lot size.

There is of course no a priori reason why man-hours need be linearly related to lot size. Figure 2.5 shows another possible regression model for our example.

FIGURE 2.4 Pictorial representation of linear regression model

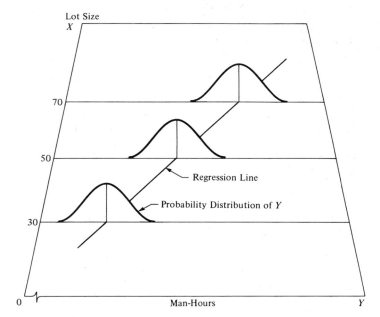

Here the regression function is curvilinear, with a shape reflecting economies of scale with larger lot sizes. Figure 2.5 differs in orientation from Figure 2.4 in that the X and Y axes are plotted conventionally in Figure 2.5. While this makes it not quite as easy to view the probability distributions, the orientation of Figure 2.5 shows the regression curve in the perspective to be utilized from here on.

FIGURE 2.5 Pictorial representation of curvilinear regression model

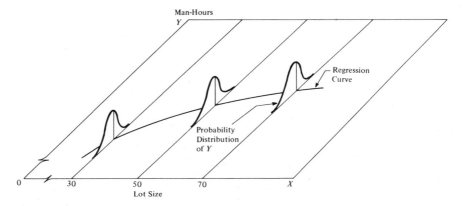

Regression models may differ in the form of the regression function as in Figures 2.4 and 2.5, in the shape of the probability distributions of the Y's, and in still other ways. Whatever the variation, the concept of a probability distribution of Y for given X is the formal counterpart to the empirical scatter in a statistical relation. Similarly, the regression curve, which describes the relation between the means of the probability distributions and X, is the counterpart to the general tendency of Y to vary with X systematically in a statistical relation.

Note

The expressions ''independent variable'' or ''predictor variable'' for X and ''dependent variable'' or ''response variable'' for Y in a regression model simply are conventional labels. There is no implication that Y causally depends on X in a given case. No matter how strong the statistical relation, no cause-and-effect pattern is necessarily implied by the regression model. In some applications, an independent variable actually is dependent causally on the response variable, as when we estimate temperature (the response) from the height of mercury (the independent variable) in a thermometer.

Regression models with more than one independent variable. Regression models may contain more than one independent variable.

1. In an application of regression analysis pertaining to 67 branch offices of a consumer finance chain, the regression model contained direct operating cost for the year just ended as the response variable and four independent variables—average size of loan outstanding during the year, average num-

ber of loans outstanding, total number of new loan applications processed, and office salary scale index.

2. In a tractor purchase study, the response variable was volume (in horse-power) of tractor purchases in each sales territory of a farm equipment firm. There were nine independent variables, including average age of tractors on farms in the territory, number of farms in the territory, and a quantity index of crop production in the territory.

3. In a medical study of short children, the response variable was the peak plasma growth hormone level. There were 14 independent variables, including age, sex, height, weight, and 10 skinfold measurements.

The features represented in Figures 2.4 and 2.5 must be extended into further dimensions when there is more than one independent variable. With two independent variables X_1 and X_2, for instance, a probability distribution of Y for each (X_1, X_2) combination is assumed by the regression model. The systematic relation between the means of these probability distributions and the independent variables X_1 and X_2 is then given by a regression surface.

Construction of regression models

Selection of independent variables. Since reality must be reduced to manageable proportions whenever we construct models, only a limited number of independent or predictor variables can—or should—be included in a regression model for any situation of interest. A central problem therefore is that of choosing, for a regression model, a set of independent variables which is ''good'' in some sense for the purposes of the analysis. A major consideration in making this choice is the extent to which a chosen variable contributes to reducing the remaining variation in Y after allowance is made for the contributions of other independent variables that have tentatively been included in the regression model. Other considerations include the importance of the variable as a causal agent in the process under analysis; the degree to which observations on the variable can be obtained more accurately, or quickly, or economically than on competing variables; and the degree to which the variable can be preset by management. In Chapter 12, we shall discuss procedures and problems in choosing the independent variables to be included in a regression model.

Functional form of regression equation. The choice of the functional form of the regression equation is related to the choice of the independent variables. Sometimes, relevant theory may indicate the appropriate functional form. Learning theory, for instance, may indicate that the regression function relating unit production costs to the number of previous times the item has been produced should have a specified shape with particular asymptotic properties.

More frequently, however, the functional form of the regression equation is not known in advance and must be decided upon once the data have been collected and analyzed. Thus, linear or quadratic regression functions are often used as satisfactory first approximations to regression functions of unknown nature. Indeed, these simple types of regression functions may be used even when theory provides the relevant functional form, notably when the known form is highly complex but can be reasonably approximated by a linear or quadratic regression function. Figure 2.6a illustrates a case where a complex regression function may be reasonably approximated by a linear regression function. Figure 2.6b provides an example where two linear regression functions may be used "piecewise" to approximate a complex regression function.

FIGURE 2.6 Uses of linear regression function to approximate complex regression functions

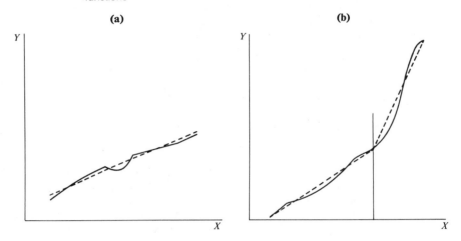

Scope of model. In formulating a regression model, we usually need to restrict the coverage of the model to some interval or region of values of the independent variable or variables. The scope is determined either by the design of the investigation or by the range of data at hand. For instance, a company studying the effect of price on sales volume investigated six price levels, ranging from $4.95 to $6.95. Here, the scope of the model would be limited to price levels ranging from near $5 to near $7. The shape of the regression function would be in serious doubt substantially outside this range because the investigation provided no evidence as to the nature of the statistical relation below $4.95 or above $6.95.

Uses of regression analysis

Regression analysis serves three major purposes: (1) description, (2) control, and (3) prediction, as illustrated by the three examples cited earlier. The tractor purchase study served a descriptive purpose. In the study of branch office operating costs, the purpose was administrative control; management was able by developing a usable statistical relation between costs and independent variables in the system, to set cost standards for each branch office in the company chain. In the medical study of short children, the purpose was prediction. Clinicians were able to use the statistical relation to predict growth hormone deficiencies in short children using simple measurements of the children.

The several purposes of regression analysis frequently overlap in practice. The Westwood Company lot size example provides a case in point. Knowledge of the relation between lot size and man-hours in past production runs enables management to predict the man-hour requirements for the next production run of given lot size, for purposes of cost estimation and production scheduling. After the run is completed, management can compare the actual man-hours against the predicted hours for purposes of administrative control.

2.3 REGRESSION MODEL WITH DISTRIBUTION OF ERROR TERMS UNSPECIFIED

Formal statement of model

In Part I of this book, we consider a basic regression model where there is only one independent variable and the regression function is linear. The model can be stated as follows:

(2.1) $$Y_i = \beta_0 + \beta_1 X_i + \varepsilon_i$$

where:

Y_i is the value of the response variable in the ith trial

β_0 and β_1 are parameters

X_i is a known constant, namely, the value of the independent variable in the ith trial

ε_i is a random error term with mean $E(\varepsilon_i) = 0$ and variance $\sigma^2(\varepsilon_i) = \sigma^2$; ε_i and ε_j are uncorrelated so that the covariance $\sigma(\varepsilon_i, \varepsilon_j) = 0$ for all $i, j; i \neq j$

$i = 1, \ldots, n$

Model (2.1) is said to be *simple, linear in the parameters,* and *linear in the independent variable.* It is "simple" in that there is only one independent variable, "linear in the parameters" because no parameter appears as an exponent or is multiplied or divided by another parameter, and "linear in the independent variable" because this variable appears only in the first power. A model which is linear in the parameters and the independent variable is also called a *first-order model.*

Important features of model

1. The observed value of Y in the ith trial is the sum of two components: (1) the constant term $\beta_0 + \beta_1 X_i$ and (2) the random term ε_i. Hence, Y_i is a random variable.

2. Since $E(\varepsilon_i) = 0$, it follows from (1.13c) that:

$$E(Y_i) = E(\beta_0 + \beta_1 X_i + \varepsilon_i) = \beta_0 + \beta_1 X_i + E(\varepsilon_i) = \beta_0 + \beta_1 X_i$$

Note that $\beta_0 + \beta_1 X_i$ plays the role of the constant a in theorem (1.13c).

Thus, the response Y_i, when the level of X existing in the ith trial is X_i, comes from a probability distribution whose mean is:

$$(2.2) \qquad\qquad E(Y_i) = \beta_0 + \beta_1 X_i$$

We therefore know that the regression function for model (2.1) is:

$$(2.3) \qquad\qquad E(Y) = \beta_0 + \beta_1 X$$

since the regression function relates the means of the probability distributions of Y for any given X to the level of X.

3. The observed value of Y in the ith trial exceeds or falls short of the value of the regression function by the error term amount ε_i.

4. The error terms ε_i are assumed to have constant variance σ^2. It therefore follows that the variance of the response Y_i is:

$$(2.4) \qquad\qquad \sigma^2(Y_i) = \sigma^2$$

since, using theorem (1.15a), we have:

$$\sigma^2(\beta_0 + \beta_1 X_i + \varepsilon_i) = \sigma^2(\varepsilon_i) = \sigma^2$$

Thus, model (2.1) assumes that the probability distributions of Y have the same variance σ^2, regardless of the level of the independent variable X.

5. The error terms are assumed to be uncorrelated. Hence, the outcome in any one trial has no effect on the error term for any other trial—as to whether it is positive or negative, or small or large. Since the error terms ε_i and ε_j are uncorrelated, so are the responses Y_i and Y_j.

6. In summary, model (2.1) implies that the response variable observations Y_i come from probability distributions whose means are $E(Y_i) = \beta_0 + \beta_1 X_i$ and whose variances are σ^2, the same for all levels of X. Further, any two observations Y_i and Y_j are uncorrelated.

Example

Suppose that regression model (2.1) is applicable for the Westwood Company lot size application and is as follows:

$$Y_i = 9.5 + 2.1 X_i + \varepsilon_i$$

Figure 2.7 contains a presentation of the regression function:

$$E(Y) = 9.5 + 2.1 X$$

Suppose that in the ith trial, a lot of $X_i = 45$ units is produced and the actual number of man-hours is $Y_i = 108$. In that case, the error term value is $\varepsilon_i = +4$, for we have:

$$E(Y_i) = 9.5 + 2.1(45) = 104$$

and:

$$Y_i = 108 = 104 + 4$$

Figure 2.7 displays the probability distribution of Y when $X = 45$, and indicates from where in this distribution the observation $Y_i = 108$ came. Note again that the error term ε_i is simply the deviation of Y_i from its mean value $E(Y_i)$.

FIGURE 2.7 Illustration of linear regression model (2.1)

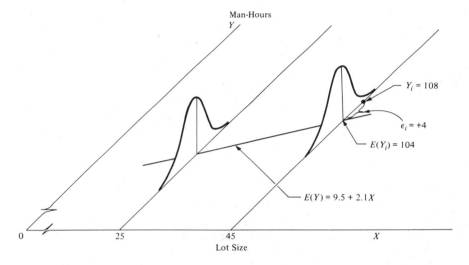

Figure 2.7 also shows the probability distribution of Y when $X = 25$. Note that this distribution exhibits the same variability as the probability distribution when $X = 45$, in conformance with the requirements of model (2.1).

Meaning of regression parameters

The parameters β_0 and β_1 in regression model (2.1) are called *regression coefficients*. β_1 is the slope of the regression line. It indicates the change in the mean of the probability distribution of Y per unit increase in X. The parameter β_0 is the Y intercept of the regression line. If the scope of the model includes $X = 0$, β_0 gives the mean of the probability distribution of Y at $X = 0$. When the scope of the model does not cover $X = 0$, β_0 does not have any particular meaning as a separate term in the regression model.

Example. Figure 2.8 shows the regression function:

$$E(Y) = 9.5 + 2.1X$$

for the previous Westwood Company lot size example. The slope $\beta_1 = 2.1$ indicates that an increase of one unit in lot size leads to an increase in the mean of the probability distribution of Y of 2.1 man-hours.

FIGURE 2.8 Meaning of linear regression parameters

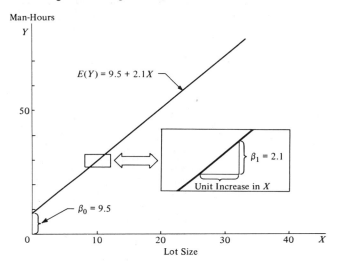

The intercept $\beta_0 = 9.5$ indicates the value of the regression function at $X = 0$. However, since the linear regression model was formulated to apply to lot sizes ranging from 20 to 80 units, β_0 does not have any intrinsic meaning of its own. In particular, it does not necessarily indicate the average setup time for the process (the average man-hours before actual output begins). A model with a curvilinear regression function and some different value of β_0 than that in the linear model might well be required if the scope of the model were to extend to lot sizes down to zero.

Alternative versions of model

Sometimes it is convenient to write model (2.1) in somewhat different, though equivalent, forms. Let X_0 be a *dummy variable* identically equal to one. Then, we can write (2.1) as follows:

$$(2.5) \qquad Y_i = \beta_0 X_0 + \beta_1 X_i + \varepsilon_i \qquad \text{where } X_0 \equiv 1$$

Another modification sometimes helpful is to use for the independent variable the deviation $X_i - \bar{X}$ rather than X_i. To leave model (2.1) unchanged, we need to write:

$$Y_i = \beta_0 + \beta_1(X_i - \bar{X}) + \beta_1\bar{X} + \varepsilon_i$$
$$= (\beta_0 + \beta_1\bar{X}) + \beta_1(X_i - \bar{X}) + \varepsilon_i$$
$$= \beta_0^* + \beta_1(X_i - \bar{X}) + \varepsilon_i$$

Thus, an alternative model version is:

(2.6) $$Y_i = \beta_0^* + \beta_1(X_i - \bar{X}) + \varepsilon_i$$

where:

(2.6a) $$\beta_0^* = \beta_0 + \beta_1\bar{X}$$

We shall use models (2.1), (2.5), and (2.6) interchangeably as convenience dictates.

2.4 ESTIMATION OF REGRESSION FUNCTION

Obtaining needed sample data

Ordinarily, we do not know the values of the regression parameters β_0 and β_1 in model (2.1) and need to estimate them from sample data. Such sample data may be obtained by experimental or nonexperimental means. We shall briefly consider each in turn.

Sometimes, it is possible to conduct a controlled experiment to provide data from which the regression parameters can be estimated. Consider, for instance, an insurance company that wishes to study the relation between productivity of its clerks in processing claims and amount of training. Five clerks selected at random are trained for two weeks, five for three weeks, five for four weeks, and five for five weeks, and the productivity of the clerks is then observed. These data on length of training (X) and productivity (Y) to serve as a basis for estimating the regression parameters are *experimental data*.

Often it is not practical or feasible to conduct controlled experiments, in which case *nonexperimental data*, also called *observational data*, will need to be utilized. Such data are obtained without controlling the independent variable of interest. For example, public health officials wishing to study the relation between age of person (X) and number of days of illness last year (Y) would probably use data obtained from health records or from a survey of the population since they cannot assign ages at random to persons. Such data are nonexperimental data since the independent variable is not controlled. Similarly, the Westwood Company in our earlier lot size example needed to rely on nonexperimental data since the lot size at any given time was dictated by the demand for the product, which was not under the control of the company.

Once the data have been obtained, either by experiment or nonexperimentally, they can be assembled in a table such as Table 2.1 for the Westwood

Company example. We shall denote the (X, Y) observations for the first trial as (X_1, Y_1), for the second trial (X_2, Y_2), and in general for the ith trial (X_i, Y_i) where $i = 1, \ldots, n$. For the data in Table 2.1, $X_1 = 30$, $Y_1 = 73$, and so on, and $n = 10$.

TABLE 2.1 Data on lot size and number of man-hours—Westwood Company example

Production Run i	Lot Size X_i	Man-Hours Y_i
1	30	73
2	20	50
3	60	128
4	80	170
5	40	87
6	50	108
7	60	135
8	30	69
9	70	148
10	60	132

Method of least squares

To find "good" estimators of the regression parameters β_0 and β_1, we shall employ the method of least squares. For each sample observation (X_i, Y_i), the method of least squares considers the deviation of Y_i from its expected value:

$$(2.7) \qquad Y_i - (\beta_0 + \beta_1 X_i)$$

In particular, the method of least squares requires that we consider the sum of the n squared deviations. This criterion is denoted by Q:

$$(2.8) \qquad Q = \sum_{i=1}^{n} (Y_i - \beta_0 - \beta_1 X_i)^2$$

According to the method of least squares, the estimators of β_0 and β_1 are those values b_0 and b_1, respectively, that minimize the criterion Q for the given sample observations (X_i, Y_i).

Example. Figure 2.9a contains a scatter plot of the sample data of Table 2.1 for the Westwood Company example. In Figure 2.9b is plotted a fitted regression line using the arbitrary estimates:

$$b_0 = 30$$
$$b_1 = 0$$

Also shown in Figure 2.9b are the deviations $Y_i - 30 - (0)X_i$. Note that each deviation corresponds to the vertical distance between Y_i and the fitted regression line. Clearly, the fit is poor. Hence, the deviations are large and so are the

FIGURE 2.9 Example of deviations from different fitted regression lines

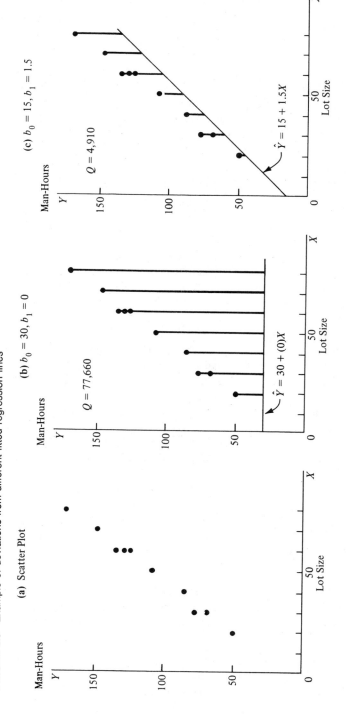

squared deviations. The sum Q of the squared deviations is (observations in ascending order):

$$Q = (50 - 30)^2 + (69 - 30)^2 + \cdots + (170 - 30)^2 = 77{,}660$$

Figure 2.9c contains the deviations $Y_i - b_0 - b_1 X_i$ for the estimates $b_0 = 15$, $b_1 = 1.5$. Here, the fit is better (though still not good), the deviations are much smaller, and hence the sum of the squared deviations is reduced to $Q = 4{,}910$. Thus, a better fit of the regression line to the data corresponds to a smaller sum Q.

The objective of the method of least squares is to find estimates b_0 and b_1 for β_0 and β_1, respectively, for which Q is a minimum. In a certain sense, to be discussed shortly, these estimates will provide a "good" fit of the linear regression function.

Least squares estimators. The estimators b_0 and b_1 which satisfy the least squares criterion could be found by a trial and error procedure. However, this is not necessary since it can be shown that the values b_0 and b_1 which minimize Q for any particular set of sample data are given by the following simultaneous equations:

(2.9a) $$\Sigma Y_i = nb_0 \quad + b_1 \Sigma X_i$$

(2.9b) $$\Sigma X_i Y_i = b_0 \Sigma X_i + b_1 \Sigma X_i^2$$

The equations (2.9a) and (2.9b) are called *normal equations;* b_0 and b_1 are called *point estimators* of β_0 and β_1, respectively.

The quantities ΣY_i, ΣX_i, and so on in (2.9) are calculated from the sample observations (X_i, Y_i). The equations then can be solved simultaneously for b_0 and b_1. Alternatively, b_0 and b_1 can be obtained directly as follows:

(2.10a) $$b_1 = \frac{\Sigma X_i Y_i - \dfrac{(\Sigma X_i)(\Sigma Y_i)}{n}}{\Sigma X_i^2 - \dfrac{(\Sigma X_i)^2}{n}} = \frac{\Sigma(X_i - \bar{X})(Y_i - \bar{Y})}{\Sigma(X_i - \bar{X})^2}$$

(2.10b) $$b_0 = \frac{1}{n}(\Sigma Y_i - b_1 \Sigma X_i) = \bar{Y} - b_1 \bar{X}$$

where $\bar{X}$ and $\bar{Y}$ have the usual meaning.

Note

The normal equations (2.9) can be derived by calculus. For given sample observations (X_i, Y_i), the quantity Q in (2.8) is a function of β_0 and β_1. The values of β_0 and β_1 which minimize Q can be derived by differentiating (2.8) with respect to β_0 and β_1. We obtain:

$$\frac{\partial Q}{\partial \beta_0} = -2\Sigma(Y_i - \beta_0 - \beta_1 X_i)$$

$$\frac{\partial Q}{\partial \beta_1} = -2\Sigma X_i(Y_i - \beta_0 - \beta_1 X_i)$$

We then set these partial derivatives equal to zero, using b_0 and b_1 to denote the particular values of β_0 and β_1, respectively, which minimize Q:

$$-2\Sigma(Y_i - b_0 - b_1 X_i) = 0$$
$$-2\Sigma X_i(Y_i - b_0 - b_1 X_i) = 0$$

Simplifying, we obtain:

$$\sum_{i=1}^{n} (Y_i - b_0 - b_1 X_i) = 0$$

$$\sum_{i=1}^{n} X_i(Y_i - b_0 - b_1 X_i) = 0$$

Expanding out, we have:

$$\Sigma Y_i - nb_0 - b_1 \Sigma X_i = 0$$
$$\Sigma X_i Y_i - b_0 \Sigma X_i - b_1 \Sigma X_i^2 = 0$$

from which the normal equations (2.9) are obtained by rearranging terms.

A test of the second partial derivatives will show that a minimum is obtained with the least squares estimators b_0 and b_1.

Properties of least squares estimators. An important theorem, called the *Gauss-Markov theorem,* states:

(2.11)   Under the conditions of model (2.1), the least squares estimators b_0 and b_1 in (2.10) are unbiased and have minimum variance among all unbiased linear estimators.

This theorem, which is proven in the next chapter, states first that both b_0 and b_1 are unbiased estimators. Hence:

$$E(b_0) = \beta_0$$
$$E(b_1) = \beta_1$$

so that neither estimator tends to overestimate or underestimate systematically.

Second, the theorem states that the sampling distributions of b_0 and b_1 have smaller variability than those of any other estimators belonging to a particular class of estimators. Thus, the least squares estimators are more precise than any of these other estimators. The class of estimators for which the least squares estimators are "best" consists of all unbiased estimators which are linear functions of the observations $Y_1, \ldots, Y_n$. The estimators b_0 and b_1 are such linear functions of the Y's. Consider, for instance, b_1. We have from (2.10a):

$$b_1 = \frac{\Sigma(X_i - \bar{X})(Y_i - \bar{Y})}{\Sigma(X_i - \bar{X})^2}$$

It will be shown in (3.5) that this expression is equal to:

$$b_1 = \frac{\Sigma(X_i - \bar{X})Y_i}{\Sigma(X_i - \bar{X})^2} = \Sigma k_i Y_i$$

where:

$$k_i = \frac{X_i - \bar{X}}{\Sigma(X_i - \bar{X})^2}$$

Since the k_i are known constants (because the X_i are known constants), b_1 is a linear combination of the Y_i and hence is a linear estimator.

In the same fashion, it can be shown that b_0 is a linear estimator.

Among all linear estimators that are unbiased then, b_0 and b_1 have the smallest variability in repeated samples in which the X levels remain unchanged.

Example. To illustrate the calculation of the least squares estimators b_0 and b_1, we will use the Westwood Company case discussed earlier. The sample data are given in Table 2.1 and plotted in Figure 2.9a. Table 2.2 gives the basic results required to calculate b_0 and b_1. We have: $\Sigma Y_i = 1,100$, $\Sigma X_i = 500$, $\Sigma X_i Y_i = 61,800$, $\Sigma X_i^2 = 28,400$, $n = 10$. Using (2.10) we obtain:

$$b_1 = \frac{\Sigma X_i Y_i - \dfrac{(\Sigma X_i)(\Sigma Y_i)}{n}}{\Sigma X_i^2 - \dfrac{(\Sigma X_i)^2}{n}} = \frac{61,800 - \dfrac{(500)(1,100)}{10}}{28,400 - \dfrac{(500)^2}{10}} = 2.0$$

$$b_0 = \frac{1}{n}(\Sigma Y_i - b_1 \Sigma X_i) = \frac{1}{10}[1,100 - 2.0(500)] = 10.0$$

Thus, we estimate that the mean number of man-hours increases by 2.0 hours for each unit increase in lot size.

TABLE 2.2 Basic calculations to obtain b_0 and b_1—Westwood Company example

Y_i	X_i	$X_i Y_i$	X_i^2	(for later use) Y_i^2
73	30	2,190	900	5,329
50	20	1,000	400	2,500
128	60	7,680	3,600	16,384
170	80	13,600	6,400	28,900
87	40	3,480	1,600	7,569
108	50	5,400	2,500	11,664
135	60	8,100	3,600	18,225
69	30	2,070	900	4,761
148	70	10,360	4,900	21,904
132	60	7,920	3,600	17,424
Total 1,100	500	61,800	28,400	134,660

Point estimation of mean response

Estimated regression function. Given sample estimators b_0 and b_1 of the parameters in the regression function (2.3):

$$E(Y) = \beta_0 + \beta_1 X$$

it is natural that we estimate the regression function as follows:

(2.12) $$\hat{Y} = b_0 + b_1 X$$

where $\hat{Y}$ (read Y hat) is the value of the estimated regression function at the level X of the independent variable.

We will call a *value* of the response variable a *response* and will call $E(Y)$ the *mean response*. Thus, the mean response is the mean of the probability distribution of Y corresponding to the level X of the independent variable. $\hat{Y}$ then is a point estimator of the mean response when the level of the independent variable is X. It can be shown as an extension of the Gauss-Markov theorem (2.11) that $\hat{Y}$ is an unbiased estimator of $E(Y)$, with minimum variance in the class of unbiased linear estimators.

For the observations in the sample, we will call $\hat{Y}_i$:

(2.13) $$\hat{Y}_i = b_0 + b_1 X_i \qquad i = 1, \ldots, n$$

the *fitted value* for the ith observation. Thus, the fitted value $\hat{Y}_i$ is to be viewed in distinction to the *observed value* Y_i.

Example. For the Westwood Company case, we found that the least squares estimates of the regression coefficients were:

$$b_0 = 10.0 \qquad b_1 = 2.0$$

Hence, the estimated regression function is:

$$\hat{Y} = 10.0 + 2.0X$$

If we are interested in the mean number of man-hours when the lot size is $X = 55$, our point estimate would be:

$$\hat{Y} = 10.0 + 2.0(55) = 120$$

Thus, we would estimate that the mean number of man-hours for production runs of $X = 55$ units is 120. We interpret this to mean that if many runs of size 55 are produced under the conditions of the 10 runs in the sample, the mean labor time for these many runs is about 120 hours. Of course, the labor time for any one run of size 55 is likely to fall above or below the mean response because of inherent variability in the system, as represented by the error term in the model.

Figure 2.10 contains a computer plot of the estimated regression function $\hat{Y} = 10.0 + 2.0X$, as well as the original data. Note the improved fit of the least squares regression line over the arbitrary lines in Figure 2.9. Indeed, the criterion

Q for the least squares regression line now is only $Q = 60$, as will be shown shortly, a much smaller value than the values of Q for the arbitrary fitted lines in Figure 2.9.

FIGURE 2.10 Observations and least squares regression line for Westwood Company example: $b_0 = 10.0$, $b_1 = 2.0$

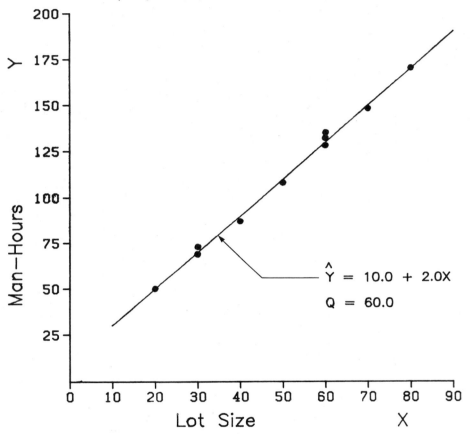

Fitted values for the sample data are obtained by substituting the X values in the sample into the estimated regression equation. For example, for our sample data, $X_1 = 30$. Hence, the fitted value is:

$$\hat{Y}_1 = 10.0 + 2.0(30) = 70$$

This compares with the observed man-hours of $Y_1 = 73$. Table 2.3 contains all the observed and fitted values for the Westwood Company data.

Alternative model (2.6). If the alternative regression model (2.6):

$$Y_i = \beta_0^* + \beta_1(X_i - \bar{X}) + \varepsilon_i$$

TABLE 2.3 Fitted values, residuals, and squared residuals—Westwood Company example

Observation Number i	Lot Size X_i	Man-Hours Y_i	Estimated Mean Response $\hat{Y}_i$	Residual $(Y_i - \hat{Y}_i) = e_i$	Squared Residual $(Y_i - \hat{Y}_i)^2 = e_i^2$
1	30	73	70	+3	9
2	20	50	50	0	0
3	60	128	130	−2	4
4	80	170	170	0	0
5	40	87	90	−3	9
6	50	108	110	−2	4
7	60	135	130	+5	25
8	30	69	70	−1	1
9	70	148	150	−2	4
10	60	132	130	+2	4
Total	500	1,100	1,100	0	60

is to be utilized, the least squares estimator b_1 of β_1 is the same as before. The least squares estimator of $\beta_0^* = \beta_0 + \beta_1\bar{X}$ is, using (2.10b):

$$(2.14) \qquad b_0^* = b_0 + b_1\bar{X} = (\bar{Y} - b_1\bar{X}) + b_1\bar{X} = \bar{Y}$$

Hence, the estimated regression equation for alternative model (2.6) is:

$$(2.15) \qquad \hat{Y} = \bar{Y} + b_1(X - \bar{X})$$

In our Westwood Company example, $\bar{Y} = 1,100/10 = 110$ and $\bar{X} = 500/10 = 50$ (Table 2.2). Hence, the estimated regression equation in alternative form is:

$$\hat{Y} = 110.0 + 2.0(X - 50)$$

For our sample data, $X_1 = 30$; hence, we estimate the mean response to be:

$$\hat{Y}_1 = 110.0 + 2.0(30 - 50) = 70$$

which, of course, is identical to our earlier result.

Residuals

The ith *residual* is the difference between the observed value Y_i and the corresponding fitted value $\hat{Y}_i$. Denoting this residual by e_i, we can write:

$$(2.16) \qquad e_i = Y_i - \hat{Y}_i = Y_i - b_0 - b_1X_i$$

Figure 2.11 shows the 10 residuals for the Westwood Company example. The magnitudes of the residuals are shown by the vertical lines between an observation and the fitted value on the estimated regression line. The residuals are calculated in Table 2.3 above.

FIGURE 2.11 Least squares regression line and residuals—Westwood
Company example (observed values and residuals not plotted
to scale)

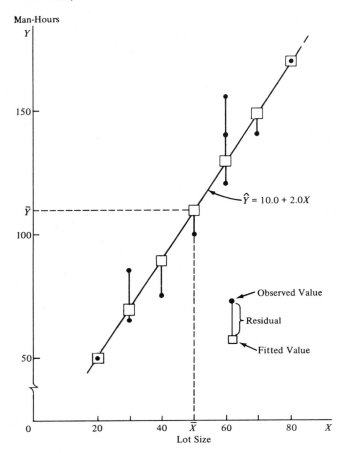

We need to distinguish between the model error term value $\varepsilon_i = Y_i - E(Y_i)$ and the residual $e_i = Y_i - \hat{Y}_i$. The former involves the vertical deviation of Y_i from the unknown population regression line, and hence is unknown. On the other hand, the residual is the observed vertical deviation of Y_i from the fitted regression line.

Residuals are highly useful for studying whether a given regression model is appropriate for the data at hand. We shall discuss this use in Chapter 4.

Properties of fitted regression line

The regression line fitted by the method of least squares has a number of properties worth noting:

1. The sum of the residuals is zero:

(2.17)
$$\sum_{i=1}^{n} e_i = 0$$

This property can be proven easily. We have:

$$\Sigma e_i = \Sigma(Y_i - b_0 - b_1 X_i)$$
$$= \Sigma Y_i - nb_0 - b_1 \Sigma X_i = 0$$

by the first normal equation (2.9a). Table 2.3 illustrates this property for our earlier example. Rounding errors may, of course, be present in any particular case.

2. The sum of the squared residuals, Σe_i^2, is a minimum. This was the requirement to be satisfied in deriving the least squares estimators of the regression parameters.

3. The sum of the observed values Y_i equals the sum of the fitted values $\hat{Y}_i$:

(2.18)
$$\sum_{i=1}^{n} Y_i = \sum_{i=1}^{n} \hat{Y}_i$$

This condition is implicit in the first normal equation (2.9a):

$$\Sigma Y_i = nb_0 + b_1 \Sigma X_i$$
$$= \Sigma b_0 + \Sigma b_1 X_i = \Sigma(b_0 + b_1 X_i) = \Sigma \hat{Y}_i$$

It follows from (2.18) that the mean of the $\hat{Y}_i$ is the same as the mean of the Y_i, namely, $\bar{Y}$.

4. The sum of the weighted residuals is zero when the residual in the ith trial is weighted by the level of the independent variable in the ith trial:

(2.19)
$$\sum_{i=1}^{n} X_i e_i = 0$$

This follows from the second normal equation (2.9b):

$$\Sigma X_i e_i = \Sigma X_i(Y_i - b_0 - b_1 X_i)$$
$$= \Sigma X_i Y_i - b_0 \Sigma X_i - b_1 \Sigma X_i^2 = 0$$

5. The sum of the weighted residuals is zero when the residual in the ith trial is weighted by the fitted value of the response variable for the ith trial:

(2.20)
$$\sum_{i=1}^{n} \hat{Y}_i e_i = 0$$

This property is a consequence of (2.17) and (2.19).

6. The regression line always goes through the point $(\bar{X}, \bar{Y})$. This can be readily seen from the alternative form of the estimated regression line in (2.15). If $X = \bar{X}$, we have:

$$\hat{Y} = \bar{Y} + b_1(X - \bar{X}) = \bar{Y} + b_1(\bar{X} - \bar{X}) = \bar{Y}$$

Figure 2.11 demonstrates this property for our lot size example.

2.5 ESTIMATION OF ERROR TERMS VARIANCE σ^2

The variance σ^2 of the error terms ε_i in the regression model (2.1) needs to be estimated for a variety of purposes. Frequently, we would like to obtain an indication of the variability of the probability distributions of Y. Further, as we shall see in the next chapter, a variety of inferences concerning the regression function and the prediction of Y require an estimate of σ^2.

Point estimator of σ^2

Single population. To lay the basis for developing an estimator of σ^2 for the regression model (2.1), let us consider for a moment the simpler problem of sampling from a single population. In obtaining the sample variance s^2, we begin by considering the deviation of an observation Y_i from the estimated mean $\bar{Y}$, squaring it, and then summing all such deviations:

$$\sum_{i=1}^{n} (Y_i - \bar{Y})^2$$

Such a sum is called a *sum of squares*. The sum of squares is then divided by the degrees of freedom associated with it. This number is $n - 1$ here, because one degree of freedom is lost by using the estimate $\bar{Y}$ instead of the population mean μ. The resulting estimator is the usual sample variance:

$$s^2 = \frac{\sum_{i=1}^{n} (Y_i - \bar{Y})^2}{n - 1}$$

which is an unbiased estimator of the variance σ^2 of an infinite population. The sample variance is often called a *mean square,* because a sum of squares has been divided by the appropriate number of degrees of freedom.

Regression model. The logic of developing an estimator of σ^2 for the regression model is the same as when sampling a single population. Recall in this connection from (2.4) that the variance of each observation Y_i is σ^2, the same as that of each error term ε_i. We again need to calculate a sum of squared deviations, but must recognize that the Y_i come from different probability distributions with different means, depending upon the level X_i. Thus, the deviation of an observation Y_i must be calculated around its own estimated mean $\hat{Y}_i$. Hence, the deviations are the residuals:

$$Y_i - \hat{Y}_i = e_i$$

and the appropriate sum of squares, denoted by SSE, is:

$$(2.21) \qquad SSE = \sum_{i=1}^{n} (Y_i - \hat{Y}_i)^2 = \sum_{i=1}^{n} (Y_i - b_0 - b_1 X_i)^2 = \sum_{i=1}^{n} e_i^2$$

where SSE stands for *error sum of squares* or *residual sum of squares*.

The sum of squares SSE has $n - 2$ degrees of freedom associated with it. Two degrees of freedom are lost because both β_0 and β_1 had to be estimated in obtaining $\hat{Y}_i$. Hence, the appropriate mean square, denoted by MSE, is:

$$(2.22) \qquad MSE = \frac{SSE}{n - 2} = \frac{\Sigma(Y_i - \hat{Y}_i)^2}{n - 2}$$

$$= \frac{\Sigma(Y_i - b_0 - b_1 X_i)^2}{n - 2} = \frac{\Sigma e_i^2}{n - 2}$$

where MSE stands for *error mean square* or *residual mean square*.

It can be shown that MSE is an unbiased estimator of σ^2 for the regression model (2.1):

$$(2.23) \qquad E(MSE) = \sigma^2$$

An estimator of the standard deviation σ is simply the positive square root of MSE.

Alternative computational formulas

There are a number of alternative computational formulas for SSE. Three of these are as follows:

$$(2.24a) \qquad SSE = \Sigma Y_i^2 - b_0 \Sigma Y_i - b_1 \Sigma X_i Y_i$$

$$(2.24b) \qquad SSE = \Sigma(Y_i - \bar{Y})^2 - \frac{[\Sigma(X_i - \bar{X})(Y_i - \bar{Y})]^2}{\Sigma(X_i - \bar{X})^2}$$

$$(2.24c) \qquad SSE = \left[\Sigma Y_i^2 - \frac{(\Sigma Y_i)^2}{n}\right] - \frac{\left[\Sigma X_i Y_i - \dfrac{\Sigma X_i \Sigma Y_i}{n}\right]^2}{\Sigma X_i^2 - \dfrac{(\Sigma X_i)^2}{n}}$$

Comments

1. Formula (2.24a) is useful if b_0 and b_1 have already been calculated. Otherwise, (2.24b) and (2.24c) are more direct.

2. In (2.24a), the estimates b_0 and b_1 should be carried to a large number of digits in order to yield reliable results for SSE.

3. To obtain (2.24a), recall that by (2.21) we have:

$$SSE = \Sigma(Y_i - b_0 - b_1 X_i)^2$$

Thus:

$$SSE = \Sigma Y_i^2 - 2b_0\Sigma Y_i - 2b_1\Sigma X_iY_i + nb_0^2 + 2b_0b_1\Sigma X_i + b_1^2\Sigma X_i^2$$
$$= \Sigma Y_i^2 - 2b_0\Sigma Y_i - 2b_1\Sigma X_iY_i + b_0(nb_0 + b_1\Sigma X_i)$$
$$+ b_1(b_0\Sigma X_i + b_1\Sigma X_i^2)$$

The expressions in parentheses are equal to ΣY_i and ΣX_iY_i, respectively, by the normal equations (2.9). Substituting these terms within the parentheses yields an expression which reduces directly to (2.24a).

4. None of the three alternative formulas explicitly provides the residuals e_i. As noted earlier, the residuals are useful for studying the appropriateness of the model.

Example

Returning to our Westwood Company lot size example, we will calculate SSE by (2.21). The residuals were obtained earlier in Table 2.3. This table also shows the squared residuals. From these results, we obtain:

$$SSE = 60$$

Since $10 - 2 = 8$ degrees of freedom are associated with SSE, we find:

$$MSE = \frac{60}{8} = 7.5$$

Finally, a point estimate of σ, the standard deviation of the probability distribution of Y for any X, is $\sqrt{7.5} = 2.74$ man-hours.

Consider again the case where the lot size is $X = 55$ units. We estimated earlier that the probability distribution of Y for this lot size has a mean of 120 man-hours. Now, we have the additional information that the standard deviation of this distribution is estimated to be 2.74 man-hours.

If we wished to use, say, (2.24a) for calculating SSE, we would need ΣY_i^2. This sum is calculated in Table 2.2. We then obtain, using the results in Table 2.2 and the estimates $b_0 = 10.0$, $b_1 = 2.0$:

$$SSE = \Sigma Y_i^2 - b_0\Sigma Y_i - b_1\Sigma X_iY_i$$
$$= 134,660 - 10.0(1,100) - 2.0(61,800) = 60$$

which is, of course, the same result (except sometimes for rounding errors) as obtained earlier.

2.6 NORMAL ERROR REGRESSION MODEL

No matter what may be the functional form of the distribution of ε_i (and hence of Y_i), the least squares method provides unbiased point estimators of β_0 and β_1 which have minimum variance among all unbiased linear estimators. To set up interval estimates and make tests, however, we do need to make an assumption about the functional form of the distribution of the ε_i. The standard assumption is that the error terms are normally distributed, and we will adopt it here. A normal error term greatly simplifies the theory of regression analysis and is justifiable in many real world situations where regression analysis is applied.

Normal error model

The normal error model is as follows:

(2.25)
$$Y_i = \beta_0 + \beta_1 X_i + \varepsilon_i$$

where:

> Y_i is the observed response in the ith trial
>
> X_i is a known constant, the level of the independent variable in the ith trial
>
> β_0 and β_1 are parameters
>
> ε_i are independent $N(0, \sigma^2)$
>
> $i = 1, \ldots, n$

Comments

1. The symbol $N(0, \sigma^2)$ stands for "normally distributed, with mean 0 and variance σ^2."

2. The normal error model (2.25) is the same as the regression model (2.1) with unspecified error distribution, except that model (2.25) assumes that the errors ε_i are normally distributed.

3. Because model (2.25) assumes that the errors are normally distributed, the assumption of uncorrelatedness of the ε_i in model (2.1) becomes one of independence in the normal error model.

4. Model (2.25) implies that the Y_i are independent normal random variables, with mean $E(Y_i) = \beta_0 + \beta_1 X_i$ and variance σ^2. Figure 2.4 (p. 27) pictures this normal error model. Each of the probability distributions of Y there is normally distributed, with constant variability, and the regression function is linear.

5. A major reason why the normality assumption for the error terms is justifiable in many situations is that the error terms frequently represent the effects of many factors omitted explicitly from the model, which do affect the response to some extent and which vary at random without reference to the independent variable X. For instance, in the lot size example, such factors as time lapse since the last production run, particular machines used, season of the year, and personnel employed, could vary more or less at random from run to run, independent of lot size. Also, there might be random measurement errors in recording Y. Insofar as these random effects have a degree of mutual independence, the composite error term ε_i representing all these factors would tend to comply with the central limit theorem and the error term distribution would approach normality as the number of factor effects becomes large.

A second reason why the normality assumption for the error terms is frequently justifiable is that the estimation and testing procedures to be discussed in the next chapter are based on the t distribution, which is not sensitive to moderate departures from normality. Thus, unless the departures from normality are serious, particularly with respect to skewness, the actual confidence coefficients and risks of errors will be close to the levels for exact normality.

Estimation of parameters by method of maximum likelihood

When the functional form of the probability distribution of the error terms is specified, estimators of the parameters β_0, β_1, and σ^2 can be obtained by the *method of maximum likelihood*. This method utilizes the joint probability distribution of the sample observations. When this joint probability distribution is viewed as a function of the parameters, given the particular sample observations, it is called the *likelihood function*. The likelihood function for the normal error model (2.25), given the sample observations $Y_1, \ldots, Y_n$, is:

$$(2.26) \quad L(\beta_0, \beta_1, \sigma^2) = \prod_{i=1}^{n} \frac{1}{(2\pi\sigma^2)^{1/2}} \exp\left[-\frac{1}{2\sigma^2}(Y_i - \beta_0 - \beta_1 X_i)^2\right]$$

$$= \frac{1}{(2\pi\sigma^2)^{n/2}} \exp\left[-\frac{1}{2\sigma^2}\sum_{i=1}^{n}(Y_i - \beta_0 - \beta_1 X_i)^2\right]$$

The values of β_0, β_1, and σ^2 which maximize this likelihood function are the maximum likelihood estimators. These are:

	Parameter	*Maximum Likelihood Estimator*	
	β_0	b_0	same as (2.10b)
(2.27)	β_1	b_1	same as (2.10a)
	σ^2	$\hat{\sigma}^2 = \dfrac{\Sigma(Y_i - \hat{Y}_i)^2}{n}$	

Thus, the maximum likelihood estimators of β_0 and β_1 are the same estimators as provided by the method of least squares. The maximum likelihood estimator $\hat{\sigma}^2$ is biased, and ordinarily the unbiased estimator *MSE* as given in (2.22) is used. Note that the unbiased estimator *MSE* differs but slightly from the maximum likelihood estimator $\hat{\sigma}^2$, especially if n is not small:

$$(2.28) \qquad\qquad MSE = \frac{n}{n-2}\hat{\sigma}^2$$

Comments

1. Since the maximum likelihood estimators b_0 and b_1 are the same as the least squares estimators, they have the properties of all least squares estimators:

a. They are unbiased.
b. They have minimum variance among all unbiased linear estimators.

In addition, the maximum likelihood estimators b_0 and b_1 for the normal error model (2.25) have other desirable properties:

c. They are consistent, as defined in (1.47).
d. They are sufficient, as defined in (1.48).
e. They are minimum variance unbiased; i.e., they have minimum variance in the class of all unbiased estimators (linear or otherwise).

Thus, for the normal error model, the estimators b_0 and b_1 have many desirable properties.

2. We find the values of β_0, β_1, and σ^2 which maximize the likelihood function L in (2.26) by taking partial derivatives of L with respect to β_0, β_1, and σ^2, equating each of the partials to zero, and solving the system of equations thus obtained. We can work with $\log_e L$, rather than L, because both L and $\log_e L$ are maximized for the same values of β_0, β_1, and σ^2:

$$(2.29) \qquad \log_e L = -\frac{n}{2}\log_e 2\pi - \frac{n}{2}\log_e \sigma^2 - \frac{1}{2\sigma^2}\Sigma(Y_i - \beta_0 - \beta_1 X_i)^2$$

Partial differentiation of this logarithmic likelihood is much easier; it yields:

$$\frac{\partial(\log_e L)}{\partial\beta_0} = \frac{1}{\sigma^2}\Sigma(Y_i - \beta_0 - \beta_1 X_i)$$

$$\frac{\partial(\log_e L)}{\partial\beta_1} = \frac{1}{\sigma^2}\Sigma X_i(Y_i - \beta_0 - \beta_1 X_i)$$

$$\frac{\partial(\log_e L)}{\partial\sigma^2} = -\frac{n}{2\sigma^2} + \frac{1}{2\sigma^4}\Sigma(Y_i - \beta_0 - \beta_1 X_i)^2$$

We now set these partial derivatives equal to zero, replacing β_0, β_1, and σ^2 by the estimators b_0, b_1, and $\hat{\sigma}^2$. We obtain after some simplification:

$$(2.30a) \qquad\qquad \Sigma(Y_i - b_0 - b_1 X_i) = 0$$

$$(2.30b) \qquad\qquad \Sigma X_i(Y_i - b_0 - b_1 X_i) = 0$$

$$(2.30c) \qquad\qquad \frac{\Sigma(Y_i - b_0 - b_1 X_i)^2}{n} = \hat{\sigma}^2$$

Formulas (2.30a) and (2.30b) are identical to the earlier least squares normal equations (2.9), and (2.30c) is the biased estimator of σ^2 given earlier in (2.27).

2.7 COMPUTER INPUTS AND OUTPUTS

Regression calculations used to be tedious chores, especially when the number of observations was large and when there were several independent variables. Today computers can be used quite easily to perform regression calculations, with one of many available packaged programs. Also, a number of calculators contain regression routines.

The inputting of data will vary from program to program. With some, the X and Y observations are entered as separate sets. For our Westwood Company data in Table 2.1, one form of data input would be:

X_i	Y_i
30	73
20	50
etc.	etc.

In some other cases, the inputting of the data would be in the form: X_1, Y_1, X_2, Y_2, etc. For our example, this input form would be: 30, 73, 20, 50, etc.

The computer output will also vary from one program package to another. Figure 2.12 illustrates a typical output format when the linear regression model is fitted to the Westwood Company data in Table 2.1 by the SPSS computer program (Ref. 2.1). The computed values of b_0, b_1, and $\sqrt{MSE}$ are annotated in Figure 2.12, and agree with our earlier results. In subsequent chapters, we shall explain the additional output in Figure 2.12.

FIGURE 2.12 Segment of computer output for regression run on Westwood Company data (SPSS, Ref. 2.1)

```
                        VARIABLES
1 SIZE                2 HOURS
    30.0000              73.0000
    20.0000              50.0000
    60.0000             128.0000
    80.0000             170.0000
    40.0000              87.0000
    50.0000             108.0000
    60.0000             135.0000
    30.0000              69.0000
    70.0000             148.0000
    60.0000             132.0000
```

DEPENDENT VARIABLE..   HOURS

VARIABLE(S) ENTERED ON STEP NUMBER   1..   SIZE

MULTIPLE R   0.99780
R SQUARE   0.99561

STANDARD ERROR   2.73861 ← $\sqrt{MSE}$
------------------ VARIABLES IN THE EQUATION ------------------
VARIABLE   B   STD ERROR B   F
SIZE   2.000000 ← b_1 0.04697   1813.333
(CONSTANT)   10.00000 ← b_0

VARIABLE   MEAN   STANDARD DEV   CASES
SIZE   50.0000   19.4365   10
HOURS   110.0000   38.9587   10

ANALYSIS OF VARIANCE   DF   SUM OF SQUARES   MEAN SQUARE
REGRESSION   1.   13600.00000   13600.00000
RESIDUAL   8.   60.00000   7.50000
```

---

# PROBLEMS

**2.1.** Refer to the sales volume example on page 24. Suppose that the number of units sold is measured accurately but clerical errors are frequently made in determining the dollar sales. Would the relation between the number of units sold and dollar sales still be a functional one? Discuss.

**2.2.** The members of a health spa pay annual membership dues of $300 plus a charge of $2 for each visit to the spa. Let $Y$ denote the total dollar cost for the year for a
```

member and X the number of visits by the member during the year. Express the relation between X and Y mathematically. Is it a functional or a statistical relation?

2.3. Experience with a certain type of plastic indicates that a relation exists between the hardness (measured in Brinell units) of items molded from the plastic (Y) and the elapsed time since termination of the molding process (X). It is proposed to study this relation by means of regression analysis. A participant in the discussion objects, pointing out that the hardening of the plastic "is the result of a natural chemical process that doesn't leave anything to chance, so the relation must be mathematical and regression analysis is not appropriate." Evaluate this objection.

2.4. In Table 2.1, the lot size X is the same in production runs 1 and 8 but the man-hours Y differ. What feature of regression model (2.1) is illustrated by this?

2.5. When asked to state the simple linear regression model, a student wrote it as follows: $E(Y_i) = \beta_0 + \beta_1 X_i + \varepsilon_i$. Do you agree?

2.6. Consider the normal error regression model (2.25). Suppose that the parameter values are $\beta_0 = 200$, $\beta_1 = 5.0$, and $\sigma = 4$.
   a. Plot this normal error regression model in the fashion of Figure 2.7. Show the distributions of Y for $X = 10$, 20, and 40.
   b. Explain the meaning of the parameters β_0 and β_1. Assume that the scope of the model includes $X = 0$.

2.7. In a simulation exercise, regression model (2.1) applies with $\beta_0 = 100$, $\beta_1 = 20$, and $\sigma^2 = 25$. An observation on Y will be made for $X = 5$.
   a. Can you state the exact probability that Y will fall between 195 and 205? Explain.
   b. If the normal error regression model (2.25) is applicable, can you now state the exact probability that Y will fall between 195 and 205? If so, state it.

2.8. In Figure 2.7, suppose another observation were obtained at $X = 45$. Would $E(Y)$ for this new observation still be 104? Would the Y value for this new observation again be 108?

2.9. A student in accounting enthusiastically declared: "Regression is a very powerful tool. We can isolate fixed and variable costs by fitting a linear regression model, even when we have no data for small lots." Discuss.

2.10. An analyst in a large corporation studied the relation between current annual salary (Y) and age (X) for the 46 computer programmers presently employed in the company. She concluded that the relation is curvilinear, reaching a maximum at 47 years. Does this imply that the salary for a programmer increases until age 47 and then decreases? Explain.

2.11. The regression function relating production output by an employee after taking a training program (Y) to the production output before the training program (X) is $E(Y) = 20 + .95X$, where X ranges from 40 to 100. An observer concludes that the training program does not raise production output on the average because β_1 is not greater than 1.0. Comment.

2.12. Evaluate the following statement: "For the least squares method to be fully valid, it is required that the distribution of Y is normal."

2.13. A person states that b_0 and b_1 in the fitted regression equation (2.12) can be estimated by the method of least squares. Comment.

2.14. According to (2.17), $\Sigma e_i = 0$ when model (2.1) is fitted to a set of n observations by the method of least squares. Is it also true that $\Sigma \varepsilon_i = 0$? Comment.

2.15. Grade point average. The director of admissions of a small college administered a newly designed entrance test to 20 students selected at random from the new freshman class in a study to determine whether a student's grade point average (GPA) at the end of the freshman year (Y) can be predicted from the entrance test score (X). The results of the study follow. Assume that the first-order regression model (2.1) is appropriate.

i:	1	2	3	4	5	6	7	8	9	10
TEST X_i:	5.5	4.8	4.7	3.9	4.5	6.2	6.0	5.2	4.7	4.3
GPA Y_i:	3.1	2.3	3.0	1.9	2.5	3.7	3.4	2.6	2.8	1.6

i:	11	12	13	14	15	16	17	18	19	20
X_i:	4.9	5.4	5.0	6.3	4.6	4.3	5.0	5.9	4.1	4.7
Y_i:	2.0	2.9	2.3	3.2	1.8	1.4	2.0	3.8	2.2	1.5

Summary calculational results are: $\Sigma X_i = 100.0$, $\Sigma Y_i = 50.0$, $\Sigma X_i^2 = 509.12$, $\Sigma Y_i^2 = 134.84$, $\Sigma X_i Y_i = 257.66$.

a. Obtain the least squares estimates of β_0 and β_1, and state the estimated regression function.

b. Plot the estimated regression function and the data. Does the estimated regression function appear to fit the data well?

c. Obtain a point estimate of the mean freshman GPA when the entrance test score is $X = 5.0$.

d. What is the point estimate of the change in the mean response when the entrance test score increases by one point?

2.16. Calculator maintenance. The Tri-City Office Equipment Corporation sells an imported desk calculator on a franchise basis and performs preventive maintenance and repair service on this calculator. The data below have been collected from 18 recent calls on users to perform routine preventive maintenance service; for each call, X is the number of machines serviced and Y is the total number of minutes spent by the service person. Assume that the first-order regression model (2.1) is appropriate.

i:	1	2	3	4	5	6	7	8	9
X_i:	7	6	5	1	5	4	7	3	4
Y_i:	97	86	78	10	75	62	101	39	53

i:	10	11	12	13	14	15	16	17	18
X_i:	2	8	5	2	5	7	1	4	5
Y_i:	33	118	65	25	71	105	17	49	68

Summary calculational results are: $\Sigma Y_i = 1{,}152$, $\Sigma X_i = 81$, $\Sigma(Y_i - \bar{Y})^2 = 16{,}504$, $\Sigma(X_i - \bar{X})^2 = 74.5$, $\Sigma(X_i - \bar{X})(Y_i - \bar{Y}) = 1{,}098$.

a. Obtain the estimated regression function.

b. Plot the estimated regression function and the data. How well does the estimated regression function fit the data?

c. Interpret b_0 in your estimated regression function. Does b_0 provide any relevant information here? Explain.

d. Obtain a point estimate of the mean service time when $X = 5$ machines are serviced.

2.17. **Airfreight breakage.** A substance used in biological and medical research is shipped by airfreight to users in cartons of 1,000 ampules. The data below, involving 10 shipments, were collected on the number of times the carton was transferred from one aircraft to another over the shipment route (X) and the number of ampules found to be broken upon arrival (Y). Assume that the first-order regression model (2.1) is appropriate.

i:	1	2	3	4	5	6	7	8	9	10
X_i:	1	0	2	0	3	1	0	1	2	0
Y_i:	16	9	17	12	22	13	8	15	19	11

a. Obtain the estimated regression function. Plot the estimated regression function and the data. Does a linear regression function appear to give a good fit here?

b. Obtain a point estimate of the expected number of broken ampules when $X = 1$ transfer is made.

c. Estimate the increase in the expected number of ampules broken when there are 2 transfers as compared to 1 transfer.

d. Verify that your fitted regression line goes through the point $(\bar{X}, \bar{Y})$.

2.18. **Plastic hardness.** Refer to Problem 2.3. Twelve batches of the plastic were made, and from each batch one test item was molded and the hardness measured at some specific point in time. The results are shown below; X is elapsed time in hours, and Y is hardness in Brinell units. Assume that the first-order regression model (2.1) is appropriate.

i:	1	2	3	4	5	6	7	8	9	10	11	12
X_i:	32	48	72	64	48	16	40	48	48	24	80	56
Y_i:	230	262	323	298	255	199	248	279	267	214	359	305

a. Obtain the estimated regression function. Plot the estimated regression function and the data. Does a linear regression function appear to give a good fit here?

b. Obtain a point estimate of the mean hardness when $X = 48$ hours.

c. Obtain a point estimate of the change in mean hardness when X increases by one hour.

2.19. Refer to **Grade point average** Problem 2.15.
a. Obtain the residuals e_i. Do they sum to zero in accord with (2.17)?
b. Estimate σ^2 and σ. In what units is σ expressed?

2.20. Refer to **Calculator maintenance** Problem 2.16.
a. Obtain the residuals e_i and the sum of the squared residuals Σe_i^2. What is the relation between the sum of the squared residuals here and the quantity Q in (2.8)?
b. Obtain point estimates of σ^2 and σ. In what units is σ expressed?

2.21. Refer to **Airfreight breakage** Problem 2.17.
a. Obtain the residual for the first observation. What is its relation to ε_1?
b. Compute Σe_i^2 and MSE. What is estimated by MSE?

2.22. Refer to **Plastic hardness** Problem 2.18.
    a. Obtain the residuals e_i. Do they sum to zero in accord with (2.17)?
    b. Estimate σ^2 and σ. In what units is σ expressed?

2.23. **Muscle mass.** A person's muscle mass is expected to decrease with age. To explore this relationship in women, a nutritionist randomly selected four women from each 10-year age group, beginning with age 40 and ending with age 79. The results follow; X is age, and Y is a measure of muscle mass. Assume that the first-order regression model (2.1) is appropriate.

i:	1	2	3	4	5	6	7	8
X_i:	71	64	43	67	56	73	68	56
Y_i:	82	91	100	68	87	73	78	80

i:	9	10	11	12	13	14	15	16
X_i:	76	65	45	58	45	53	49	78
Y_i:	65	84	116	76	97	100	105	77

    a. Obtain the estimated regression function. Plot the estimated regression function and the data. Does a linear regression function appear to give a good fit here? Does your plot support the anticipation that muscle mass decreases with age?
    b. Obtain the following: (1) a point estimate of the difference in the mean muscle mass for women differing in age by one year, (2) a point estimate of the mean muscle mass for women aged $X = 60$ years, (3) the value of the residual for the eighth observation, (4) a point estimate of σ^2.

2.24. **Robbery rate.** A criminologist studying the relationship between population density and robbery rate in medium-sized U.S. cities collected the following data for a random sample of 16 cities; X is the population density of the city (number of people per unit area), and Y is the robbery rate last year (number of robberies per 100,000 people). Assume that the first-order regression model (2.1) is appropriate.

i:	1	2	3	4	5	6	7	8
X_i:	59	49	75	54	78	56	60	82
Y_i:	209	180	195	192	215	197	208	189

i:	9	10	11	12	13	14	15	16
X_i:	69	83	88	94	47	65	89	70
Y_i:	213	201	214	212	205	186	200	204

    a. Obtain the estimated regression function. Plot the estimated regression function and the data. Does the linear regression function appear to give a good fit here? Discuss.
    b. Obtain point estimates of the following: (1) the difference in the mean robbery rate in cities that differ by one unit in population density, (2) the mean robbery rate last year in cities with population density $X = 60$, (3) ε_{10}, (4) σ^2.

EXERCISES

2.25. Refer to regression model (2.1). Assume that $X = 0$ is within the scope of the model. What is the implication for the regression function if $\beta_0 = 0$ so that the model is $Y_i = \beta_1 X_i + \varepsilon_i$? How would the regression function plot on a graph?

2.26. Refer to regression model (2.1). What is the implication for the regression function if $\beta_1 = 0$ so that the model is $Y_i = \beta_0 + \varepsilon_i$? How would the regression function plot on a graph?

2.27. Refer to **Plastic hardness** Problem 2.18. Suppose one test item was molded from a single batch of plastic and the hardness of this one item was measured at 12 different points in time. Would the error term in the model for this case still reflect the same effects as for the experiment initially described? Would you expect the error terms for the different points in time to be uncorrelated? Discuss.

2.28. Derive the expression for b_1 in (2.10a) from the normal equations in (2.9).

2.29. (Calculus needed.) Refer to the model $Y_i = \beta_0 + \varepsilon_i$ in Exercise 2.26. Using the method of least squares, derive the estimator of β_0 for this model.

2.30. Prove that the least squares estimator of β_0 obtained in Exercise 2.29 is unbiased.

2.31. Prove the result in (2.20)—that the sum of the residuals weighted by the fitted values is zero.

2.32. Refer to Table 2.1 for the Westwood Company example. When asked to present a point estimate of the expected man-hours for runs of 60 pieces, a person gave the estimate 131.7 because this is the mean number of man-hours in the three runs of size 60 in the study. A critic states that this person's approach "throws away" most of the data in the study because observations on lot sizes other than 60 are ignored. Comment.

2.33. In **Airfreight breakage** Problem 2.17, the least squares estimates are $b_0 = 10.20$ and $b_1 = 4.00$, and $\Sigma e_i^2 = 17.60$. Evaluate the least squares criterion Q in (2.8) for the estimates: (1) $b_0 = 9$, $b_1 = 3$; (2) $b_0 = 11$, $b_1 = 5$. Is the criterion Q larger for these estimates than for the least squares estimates?

2.34. Two observations on Y were obtained at each of three X levels—namely, $X = 5$, $X = 10$, and $X = 15$.
   a. Show that the least squares regression line fitted to the *three* points $(5, \bar{Y}_1)$, $(10, \bar{Y}_2)$, and $(15, \bar{Y}_3)$, where $\bar{Y}_1$, $\bar{Y}_2$, and $\bar{Y}_3$ denote the means of the Y observations at the three X levels, is identical to the least squares regression line fitted to the original six observations.
   b. In this study, could the error term variance σ^2 be estimated without fitting a regression line? Explain.

2.35. In the Westwood Company example, the observations at $X = 20$ and $X = 80$ fall directly on the fitted regression line (Table 2.3 and Figure 2.2b). If these two observations were deleted, would the least squares regression line fitted to the remaining eight observations be changed? [*Hint:* What is the contribution of the two observations to the least squares criterion Q in (2.8)?]

2.36. (Calculus needed.) Refer to the model $Y_i = \beta_1 X_i + \varepsilon_i$, $i = 1, \ldots, n$, in Exercise 2.25.

   a. Find the least squares estimator of β_1.

   b. Assume that the error terms ε_i are independent $N(0, \sigma^2)$ and that σ^2 is known. State the likelihood function for the n sample observations and obtain the maximum likelihood estimator of β_1. Is it the same as the least squares estimator?

   c. Show that the maximum likelihood estimator is unbiased.

2.37. Shown below are the number of galleys of type set (X) and the dollar cost of correcting typographical errors (Y) in a random sample of recent orders handled by a firm specializing in technical printing. Assume that the regression model $Y_i = \beta_1 X_i + \varepsilon_i$ is appropriate, with normally distributed independent error terms whose variance is $\sigma^2 = 16$.

i:	1	2	3	4	5	6
X_i:	7	12	4	14	25	30
Y_i:	128	213	75	250	446	540

   a. State the likelihood function for the six observations, for $\sigma^2 = 16$.

   b. Evaluate the likelihood function for $\beta_1 = 17, 18$, and 19. For which of these β_1 values is the likelihood function largest?

   c. The maximum likelihood estimator is $b_1 = \Sigma X_i Y_i / \Sigma X_i^2$. Find the maximum likelihood estimate. Are your results in part (b) consistent with this estimate?

PROJECTS

2.38. Refer to the **SMSA** data set (pp. 537–41). The number of active physicians in an SMSA (Y) is expected to be related to total population, land area, and total personal income. Assume that the first-order regression model (2.1) is appropriate in each case.

   a. Regress the number of active physicians in turn on each of the three independent variables. State the estimated regression functions.

   b. Plot the three estimated regression functions and data on separate graphs. Does a linear regression relation appear to provide a good fit in each of the three cases?

   c. Calculate MSE for each of the three cases. Which independent variable leads to the smallest variability around the fitted regression line?

2.39. Refer to the **SMSA** data set.

   a. For each geographic region, regress total serious crimes in an SMSA (Y) against total population (X). Assume that the first-order regression model (2.1) is appropriate for each region. State the estimated regression functions.

   b. Are the estimated regression functions similar for the four regions? Discuss.

   c. Calculate MSE for each region. Is the variability around the fitted regression line approximately the same for the four regions? Discuss.

2.40. Refer to the **SENIC** data set (pp. 533–36). The average length of stay in a hospital (Y) is anticipated to be related to infection risk, available facilities and services, and routine chest X-ray ratio. Assume that the first-order regression model (2.1) is appropriate in each case.

  a. Regress average length of stay on each of the three independent variables. State the estimated regression functions.
  b. Plot the three estimated regression functions and data on separate graphs. Does a linear relation appear to provide a good fit in each of the three cases?
  c. Calculate *MSE* for each of the three cases. Which independent variable leads to the smallest variability around the fitted regression line?

2.41. Refer to the **SENIC** data set.
  a. For each geographic region, regress average length of stay in hospital (Y) against infection risk (X). Assume that the first-order regression model (2.1) is appropriate for each region. State the estimated regression functions.
  b. Are the estimated regression functions similar for the four regions? Discuss.
  c. Calculate *MSE* for each region. Is the variability around the fitted regression line approximately the same for the four regions? Discuss.

CITED REFERENCE

  2.1 Nie, N. H.; C. H. Hull; J. G. Jenkins; K. Steinbrenner; and D. H. Bent. *SPSS: Statistical Package for the Social Sciences.* 2d ed. New York: McGraw-Hill, 1975.

3

Inferences in regression analysis

In this chapter, we first take up inferences concerning the regression parameters β_0 and β_1, considering both interval estimation of these parameters and tests about them. We then discuss interval estimation of the mean $E(Y)$ of the probability distribution of Y, for given X, and prediction intervals for a new observation Y, given X. Finally, we take up the analysis of variance approach to regression analysis.

Throughout this chapter, and in the remainder of Part I unless otherwise stated, we assume that the normal error model (2.25) is applicable. This model is:

$$(3.1) \qquad Y_i = \beta_0 + \beta_1 X_i + \varepsilon_i$$

where:

β_0 and β_1 are parameters
X_i are known constants
ε_i are independent $N(0, \sigma^2)$

3.1 INFERENCES CONCERNING β_1

Frequently, we are interested in drawing inferences about β_1, the slope of the regression line in model (3.1). For instance, a market research analyst studying the relation between sales (Y) and advertising expenditures (X) may wish to obtain an interval estimate of β_1 because it will provide information as to how many additional sales dollars, on the average, are generated by an additional dollar of advertising expenditure.

At times, tests concerning β_1 are of interest, particularly one of the form:

$$H_0: \beta_1 = 0$$
$$H_a: \beta_1 \neq 0$$

The reason for interest in testing whether or not $\beta_1 = 0$ is that $\beta_1 = 0$ indicates that there is no linear association between Y and X. Figure 3.1 illustrates the case when $\beta_1 = 0$ for the normal error model (3.1). Note that the regression line is horizontal and that the means of the probability distributions of Y are therefore all equal, namely:

$$E(Y) = \beta_0 + (0)X = \beta_0$$

Since model (3.1) assumes normality of the probability distributions of Y with constant variance, and since the means are equal when $\beta_1 = 0$, it follows that the probability distributions of Y are identical when $\beta_1 = 0$. This is shown in Figure 3.1. Thus, $\beta_1 = 0$ for model (3.1) implies not only that there is no linear association between Y and X but also that there is no relation of any type between Y and X, since the probability distributions of Y are then identical at all levels of X.

Before discussing inferences concerning β_1 further, we need to consider the sampling distribution of b_1, our point estimator of β_1.

FIGURE 3.1   Model (3.1) when $\beta_1 = 0$

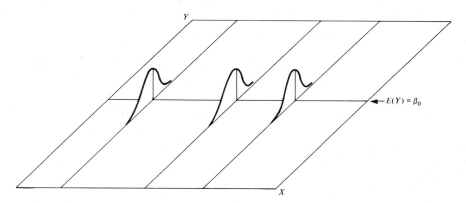

Sampling distribution of b_1

The point estimator b_1 was given in (2.10a) as follows:

(3.2)
$$b_1 = \frac{\Sigma(X_i - \bar{X})(Y_i - \bar{Y})}{\Sigma(X_i - \bar{X})^2}$$

The sampling distribution of b_1 refers to the different values of b_1 that would be obtained with repeated sampling when the levels of the independent variable X are held constant from sample to sample.

(3.3)   For model (3.1), the sampling distribution of b_1 is normal, with mean and variance:

(3.3a) $$E(b_1) = \beta_1$$

(3.3b) $$\sigma^2(b_1) = \frac{\sigma^2}{\Sigma(X_i - \bar{X})^2}$$

To show this, we need to recognize that b_1 is a linear combination of the observations Y_i.

b_1 as linear combination of the Y_i. We will show that b_1, as defined in (3.2), can be written:

$$b_1 = \Sigma k_i Y_i$$

Here the k_i are constants; hence, b_1 is a linear combination of the Y_i. We first prove:

(3.4) $$\Sigma(X_i - \bar{X})(Y_i - \bar{Y}) = \Sigma(X_i - \bar{X})Y_i$$

This follows since:

$$\Sigma(X_i - \bar{X})(Y_i - \bar{Y}) = \Sigma(X_i - \bar{X})Y_i - \Sigma(X_i - \bar{X})\bar{Y}$$

But $\Sigma(X_i - \bar{X})\bar{Y} = \bar{Y}\Sigma(X_i - \bar{X}) = 0$ since $\Sigma(X_i - \bar{X}) = 0$. Hence, (3.4) holds.
We now express b_1, using (3.4):

$$b_1 = \frac{\Sigma(X_i - \bar{X})(Y_i - \bar{Y})}{\Sigma(X_i - \bar{X})^2} = \frac{\Sigma(X_i - \bar{X})Y_i}{\Sigma(X_i - \bar{X})^2}$$

We can rewrite this as follows:

(3.5) $$b_1 = \Sigma k_i Y_i$$

where:

(3.5a) $$k_i = \frac{X_i - \bar{X}}{\Sigma(X_i - \bar{X})^2}$$

Observe that the k_i are fixed quantities since the X_i are fixed.

Note

The quantities k_i have a number of interesting properties that will be used later:

1.

(3.6) $$\Sigma k_i = 0$$

because $\Sigma k_i = \Sigma(X_i - \bar{X})/\Sigma(X_i - \bar{X})^2 = 0/\Sigma(X_i - \bar{X})^2 = 0$.

2.

(3.7)
$$\Sigma k_i X_i = 1$$

because $\Sigma k_i X_i = \Sigma(X_i - \bar{X})X_i/\Sigma(X_i - \bar{X})^2 = \Sigma(X_i - \bar{X})(X_i - \bar{X})/\Sigma(X_i - \bar{X})^2$
$= \Sigma(X_i - \bar{X})^2/\Sigma(X_i - \bar{X})^2 = 1$. The result $\Sigma(X_i - \bar{X})(X_i - \bar{X}) = \Sigma(X_i - \bar{X})X_i$
is obtained in the same way as (3.4).

3.

(3.8)
$$\Sigma k_i^2 = \frac{1}{\Sigma(X_i - \bar{X})^2}$$

because:

$$\Sigma k_i^2 = \Sigma \left[\frac{(X_i - \bar{X})}{\Sigma(X_i - \bar{X})^2} \right]^2 = \frac{1}{[\Sigma(X_i - \bar{X})^2]^2} \Sigma(X_i - \bar{X})^2$$

$$= \frac{1}{\Sigma(X_i - \bar{X})^2}$$

Normality. We return now to the sampling distribution of b_1 for the normal error model (3.1). The normality of the sampling distribution of b_1 follows at once from the fact that b_1 is a linear combination of the Y_i. The Y_i are independently, normally distributed according to model (3.1), and theorem (1.35) states that a linear combination of independent normal random variables is normally distributed.

Mean. The unbiasedness of the point estimator b_1, stated earlier in the Gauss-Markov theorem (2.11), is easy to show:

$$E(b_1) = E(\Sigma k_i Y_i) = \Sigma k_i E(Y_i) = \Sigma k_i(\beta_0 + \beta_1 X_i)$$
$$= \beta_0 \Sigma k_i + \beta_1 \Sigma k_i X_i$$

Hence, by (3.6) and (3.7) $E(b_1) = \beta_1$.

Variance. The variance of b_1 can be derived readily. We need only remember that the Y_i are independent random variables, each with variance σ^2, and that the k_i are constants. Hence, we obtain by (1.26):

$$\sigma^2(b_1) = \sigma^2(\Sigma k_i Y_i) = \Sigma k_i^2 \sigma^2(Y_i)$$
$$= \Sigma k_i^2 \sigma^2 = \sigma^2 \Sigma k_i^2$$

$$= \sigma^2 \frac{1}{\Sigma(X_i - \bar{X})^2}$$

The last step follows from (3.8).

Estimated variance. We can estimate the variance of the sampling distribution of b_1:

$$\sigma^2(b_1) = \frac{\sigma^2}{\Sigma(X_i - \bar{X})^2}$$

by replacing the parameter σ^2 with the unbiased estimator of σ^2, namely, MSE:

$$(3.9) \qquad s^2(b_1) = \frac{MSE}{\Sigma(X_i - \bar{X})^2} = \frac{MSE}{\Sigma X_i^2 - \dfrac{(\Sigma X_i)^2}{n}}$$

The point estimator $s^2(b_1)$ is an unbiased estimator of $\sigma^2(b_1)$. Taking the square root, we obtain $s(b_1)$, our point estimator of $\sigma(b_1)$.

Note

We stated in theorem (2.11) that b_1 has minimum variance among all unbiased linear estimators of the form:

$$\hat{\beta}_1 = \Sigma c_i Y_i$$

where the c_i are arbitrary constants. We shall now prove this. Since $\hat{\beta}_1$ must be unbiased, the following must hold:

$$E(\hat{\beta}_1) = E(\Sigma c_i Y_i) = \Sigma c_i E(Y_i) = \beta_1$$

Now $E(Y_i) = \beta_0 + \beta_1 X_i$ by (2.2) so that the above condition becomes:

$$E(\hat{\beta}_1) = \Sigma c_i(\beta_0 + \beta_1 X_i) = \beta_0 \Sigma c_i + \beta_1 \Sigma c_i X_i = \beta_1$$

For the unbiasedness condition to hold, the c_i must follow the restrictions:

$$\Sigma c_i = 0 \qquad \Sigma c_i X_i = 1$$

Now the variance of $\hat{\beta}_1$ is by (1.26):

$$\sigma^2(\hat{\beta}_1) = \Sigma c_i^2 \sigma^2(Y_i) = \sigma^2 \Sigma c_i^2$$

Let us define $c_i = k_i + d_i$, where the k_i are the least squares constants in (3.5) and the d_i are arbitrary constants. We can then write:

$$\sigma^2(\hat{\beta}_1) = \sigma^2 \Sigma c_i^2 = \sigma^2 \Sigma(k_i + d_i)^2 = \sigma^2 [\Sigma k_i^2 + \Sigma d_i^2 + 2\Sigma k_i d_i]$$

We know that $\sigma^2 \Sigma k_i^2 = \sigma^2(b_1)$ from our proof above. Further, $\Sigma k_i d_i = 0$; this follows from the restrictions on the k_i and the c_i. Hence, we have:

$$\sigma^2(\hat{\beta}_1) = \sigma^2(b_1) + \sigma^2 \Sigma d_i^2$$

Note that the smallest value of Σd_i^2 is zero. Hence, the variance of $\hat{\beta}_1$ is at a minimum when $\Sigma d_i^2 = 0$. But this can only occur if all $d_i = 0$, which implies $c_i \equiv k_i$. Thus, the least squares estimator b_1 has minimum variance among all unbiased linear estimators.

Sampling distribution of $(b_1 - \beta_1)/s(b_1)$

Since b_1 is normally distributed, we know that the standardized statistic $(b_1 - \beta_1)/\sigma(b_1)$ is a standard normal variable. Ordinarily, of course, we need to estimate $\sigma(b_1)$ by $s(b_1)$, and hence are interested in the distribution of the standardized statistic $(b_1 - \beta_1)/s(b_1)$. An important theorem in statistics states:

$$(3.10) \qquad \frac{(b_1 - \beta_1)}{s(b_1)} \quad \text{is distributed as } t(n - 2) \text{ for model (3.1)}$$

Intuitively, this result should not be unexpected. We know that if the observations Y_i come from the same normal population, $(\bar{Y} - \mu)/s(\bar{Y})$ follows the t distribution with $n - 1$ degrees of freedom. The estimator b_1, like $\bar{Y}$, is a linear combination of the observations Y_i. The reason for the difference in the degrees of freedom is that two parameters (β_0 and β_1) need to be estimated for the regression model; hence, two degrees of freedom are lost here.

Note

We can show that $(b_1 - \beta_1)/s(b_1)$ is distributed as t with $n - 2$ degrees of freedom by relying on the following theorem:

(3.11)   For model (3.1), SSE/σ^2 is distributed as χ^2 with $n - 2$ degrees of freedom, and is independent of b_0 and b_1.

First, let us rewrite $(b_1 - \beta_1)/s(b_1)$ as follows:

$$\frac{b_1 - \beta_1}{\sigma(b_1)} \div \frac{s(b_1)}{\sigma(b_1)}$$

The numerator is a standard normal variable z. The nature of the denominator can be seen by first considering:

$$\frac{s^2(b_1)}{\sigma^2(b_1)} = \frac{\dfrac{MSE}{\Sigma(X_i - \bar{X})^2}}{\dfrac{\sigma^2}{\Sigma(X_i - \bar{X})^2}} = \frac{MSE}{\sigma^2} = \frac{\dfrac{SSE}{n-2}}{\sigma^2}$$

$$= \frac{SSE}{\sigma^2(n-2)} \sim \frac{\chi^2(n-2)}{n-2}$$

where the symbol $\sim$ stands for "is distributed as." The last step follows from (3.11). Hence, we have:

$$\frac{b_1 - \beta_1}{s(b_1)} \sim \frac{z}{\sqrt{\dfrac{\chi^2(n-2)}{n-2}}}$$

But by theorem (3.11), z and χ^2 are independent, since z is a function of b_1 and b_1 is independent of $SSE/\sigma^2 \sim \chi^2$. Hence, by definition (1.39), it follows that:

$$\frac{b_1 - \beta_1}{s(b_1)} \sim t(n-2)$$

This result places us in a position to readily make inferences concerning β_1.

Confidence interval for β_1

Since $(b_1 - \beta_1)/s(b_1)$ follows a t distribution, we can make the following probability statement:

(3.12) $P\{t(\alpha/2; n-2) \leq (b_1 - \beta_1)/s(b_1) \leq t(1 - \alpha/2; n-2)\} = 1 - \alpha$

Here, $t(\alpha/2; n-2)$ denotes the $(\alpha/2)100$ percentile of the t distribution with

$n - 2$ degrees of freedom. Because of the symmetry of the t distribution, it follows that:

$$(3.13) \qquad t(\alpha/2; n - 2) = -t(1 - \alpha/2; n - 2)$$

Rearranging the inequalities in (3.12) and using (3.13), we obtain:

$$(3.14) \qquad P\{b_1 - t(1 - \alpha/2; n - 2)s(b_1) \leq \beta_1$$
$$\leq b_1 + t(1 - \alpha/2; n - 2)s(b_1)\} = 1 - \alpha$$

Since (3.14) holds for all possible values of β_1, the $1 - \alpha$ confidence limits for β_1 are:

$$(3.15) \qquad b_1 \pm t(1 - \alpha/2; n - 2)s(b_1)$$

Example. Let us return to the Westwood Company lot size example of Chapter 2. Management wishes an estimate of β_1 with a 95 percent confidence coefficient. We summarize in Table 3.1 the needed results obtained earlier. First, we need to obtain $s(b_1)$:

$$s^2(b_1) = \frac{MSE}{\Sigma(X_i - \bar{X})^2} = \frac{7.5}{3,400} = .002206$$

and:

$$s(b_1) = .04697$$

For a 95 percent confidence coefficient, we require $t(.975; 8)$. From Table A–2 in the Appendix, we find $t(.975; 8) = 2.306$. The 95 percent confidence interval, by (3.15), then is:

$$2.0 - 2.306(.04697) \leq \beta_1 \leq 2.0 + 2.306(.04697)$$
$$1.89 \leq \beta_1 \leq 2.11$$

Thus, with confidence coefficient .95, we estimate that the mean number of man-hours increases by somewhere between 1.89 and 2.11 for each increase of one part in the lot size.

TABLE 3.1 Results for Westwood Company example obtained in Chapter 2

$n = 10$	$\bar{X} = 50$
$b_0 = 10.0$	$b_1 = 2.0$
$\hat{Y} = 10.0 + 2.0 \, X$	$SSE = 60$
$\Sigma \, X_i^2 = 28,400$	$MSE = 7.5$

$$\Sigma \, X_i^2 - \frac{(\Sigma \, X_i)^2}{n} = \Sigma \, (X_i - \bar{X})^2 = 3,400$$

$$\Sigma \, X_i \, Y_i - \frac{\Sigma \, X_i \, \Sigma \, Y_i}{n} = \Sigma \, (X_i - \bar{X})(Y_i - \bar{Y}) = 6,800$$

$$\Sigma \, (Y_i)^2 - \frac{(\Sigma \, Y_i)^2}{n} = \Sigma \, (Y_i - \bar{Y})^2 = 13,660$$

Note

In Chapter 2, we noted that the scope of a regression model is restricted ordinarily to some interval of values of the independent variable. This is particularly important to keep in mind when using estimates of the slope β_1. In our lot size example, a linear regression model appeared appropriate for lot sizes between 20 and 80, the range of the independent variable in the recent past. It may not be reasonable to use the estimate of the slope to infer the effect of lot size on number of man-hours far outside this range since the regression relation may not be linear there.

Tests concerning β_1

Since $(b_1 - \beta_1)/s(b_1)$ is distributed as t with $n - 2$ degrees of freedom, tests concerning β_1 can be set up in ordinary fashion using the t distribution.

Example 1: Two-sided test. Suppose a cost analyst in the Westwood Company is interested in testing whether or not there is a linear association between man-hours and lot size, using regression model (3.1). The two alternatives then are:

$$(3.16) \qquad \begin{aligned} H_0: \beta_1 &= 0 \\ H_a: \beta_1 &\neq 0 \end{aligned}$$

If the analyst wishes to control the risk of a Type I error at .05, he could indeed conclude H_a at once by referring to the 95 percent confidence interval for β_1 constructed earlier, since the interval does not include 0.

An explicit test of the alternatives (3.16) is based on the test statistic:

$$(3.17) \qquad t^* = \frac{b_1}{s(b_1)}$$

The decision rule with this test statistic when controlling the level of significance at α is:

$$(3.17a) \qquad \begin{aligned} &\text{If } |t^*| \leq t(1 - \alpha/2; n - 2), \text{ conclude } H_0 \\ &\text{If } |t^*| > t(1 - \alpha/2; n - 2), \text{ conclude } H_a \end{aligned}$$

For the Westwood Company example, where $\alpha = .05$, $b_1 = 2.0$, and $s(b_1) = .04697$, we require $t(.975; 8) = 2.306$. Thus, the decision rule for testing alternatives (3.16) is:

$$\begin{aligned} &\text{If } |t^*| \leq 2.306, \text{ conclude } H_0 \\ &\text{If } |t^*| > 2.306, \text{ conclude } H_a \end{aligned}$$

Since $|t^*| = |2.0/.04697| = 42.58 > 2.306$, we conclude H_a, that $\beta_1 \neq 0$ or that there is a linear association between man-hours and lot size.

The P-value for the sample outcome is obtained by finding the probability $P[t(8) > t^* = 42.58]$. We see from Table A–2 that this probability is less than .0005. Indeed, it can be shown to be almost 0, to be denoted by 0+. Thus, the two-sided P-value is $2(0+) = 0+$. Since the two-sided P-value is less than the specified level of significance $\alpha = .05$, we could conclude H_a directly.

Example 2: One-sided test. If the analyst had wished to test whether or not β_1 is positive, controlling the level of significance at $\alpha = .05$, the alternatives would have been:

$$H_0: \beta_1 \leq 0$$
$$H_a: \beta_1 > 0$$

and the decision rule based on test statistic (3.17) would have been:

$$\text{If } t^* \leq t(1 - \alpha; n - 2), \text{ conclude } H_0$$
$$\text{If } t^* > t(1 - \alpha; n - 2), \text{ conclude } H_a$$

For $\alpha = .05$, we require $t(.95; 8) = 1.860$. Since $t^* = 42.58 > 1.860$, we would conclude H_a, that β_1 is positive.

This same conclusion could be reached directly from the one-sided P-value, which was noted in Example 1 to be $0+$. Since this P-value is less than .05, we would conclude H_a.

Comments

1. Many computer packages and scientific publications commonly report the P-value together with the value of the test statistic. In this way, one can conduct a test at any desired level of significance α by comparing the P-value with the specified level α. Users of computer packages need to be careful to ascertain whether one-sided or two-sided P-values are furnished.

2. Occasionally, it is desired to test whether or not β_1 equals some specified nonzero value β_{10}, which may be a historical norm, the value for a comparable process, or an engineering specification. For such a test, the appropriate test statistic is:

$$(3.18) \qquad t^* = \frac{b_1 - \beta_{10}}{s(b_1)}$$

The decision rule to be used for the alternatives:

$$H_0: \beta_1 = \beta_{10}$$
$$H_a: \beta_1 \neq \beta_{10}$$

is still (3.17a), but it now is based on t^* defined in (3.18).

Note that test statistic (3.18) simplifies to test statistic (3.17) when the test involves $H_0: \beta_1 = \beta_{10} = 0$.

3.2 INFERENCES CONCERNING β_0

As noted in Chapter 2, there are only infrequent occasions when we wish to make inferences concerning β_0, the intercept of the regression line. These occur when the scope of the model includes $X = 0$.

Sampling distribution of b_0

The point estimator b_0 was given in (2.10b) as follows:

$$(3.19) \qquad\qquad b_0 = \bar{Y} - b_1\bar{X}$$

The sampling distribution of b_0 refers to the different values of b_0 that would be obtained with repeated sampling when the levels of the independent variable X are held constant from sample to sample.

(3.20) For model (3.1), the sampling distribution of b_0 is normal, with mean and variance:

$$(3.20a) \qquad E(b_0) = \beta_0$$

$$(3.20b) \qquad \sigma^2(b_0) = \sigma^2 \frac{\Sigma X_i^2}{n\Sigma(X_i - \bar{X})^2} = \sigma^2\left[\frac{1}{n} + \frac{\bar{X}^2}{\Sigma(X_i - \bar{X})^2}\right]$$

The normality of the sampling distribution of b_0 follows because b_0, like b_1, is a linear combination of the observations Y_i. The results for the mean and variance of the sampling distribution of b_0 can be obtained in similar fashion as those for b_1.

An estimator of $\sigma^2(b_0)$ is obtained by replacing σ^2 by its point estimator MSE:

$$(3.21) \qquad s^2(b_0) = MSE\frac{\Sigma X_i^2}{n\Sigma(X_i - \bar{X})^2} = MSE\left[\frac{1}{n} + \frac{\bar{X}^2}{\Sigma(X_i - \bar{X})^2}\right]$$

The square root, $s(b_0)$, is an estimator of $\sigma(b_0)$.

Sampling distribution of $(b_0 - \beta_0)/s(b_0)$

Analogous to theorem (3.10) for b_1, there is a theorem for b_0 that states:

$$(3.22) \qquad \frac{b_0 - \beta_0}{s(b_0)} \text{ is distributed as } t(n - 2) \text{ for model (3.1)}$$

Hence, confidence intervals for β_0 and tests concerning β_0 can be set up in ordinary fashion, using the t distribution.

Confidence interval for β_0

The $1 - \alpha$ confidence limits for β_0 are obtained in the same manner as those for β_1 derived earlier. They are:

$$(3.23) \qquad\qquad b_0 \pm t(1 - \alpha/2; n - 2)s(b_0)$$

Example. As noted earlier, the scope of the model for the Westwood Company example does not extend to lot sizes of $X = 0$. Hence, the regression parameter β_0 may not have intrinsic meaning. If, nevertheless, a 90 percent confidence interval for β_0 were desired, we would proceed by finding $t(.95; 8)$ and $s(b_0)$. From Table A–2, we find $t(.95; 8) = 1.860$. Using the earlier results summarized in Table 3.1, we obtain by (3.21):

$$s^2(b_0) = MSE \frac{\Sigma X_i^2}{n\Sigma(X_i - \overline{X})^2} = (7.5)\frac{28,400}{10(3,400)} = 6.26471$$

or:

$$s(b_0) = 2.50294$$

Hence, the 90 percent confidence interval for β_0 is:

$$10.0 - 1.860(2.50294) \leq \beta_0 \leq 10.0 + 1.860(2.50294)$$
$$5.34 \leq \beta_0 \leq 14.66$$

We caution again that this confidence interval does not necessarily provide meaningful information. For instance, it does not necessarily provide information about the "setup" costs of producing a lot of parts (costs incurred in setting up the production process, no matter what is the lot size), since we are not certain whether a linear regression model is appropriate when the scope of the model is extended to $X = 0$.

3.3 SOME CONSIDERATIONS ON MAKING INFERENCES CONCERNING β_0 AND β_1

Effect of departures from normality

If the probability distributions of Y are not exactly normal but do not depart seriously, the sampling distributions of b_0 and b_1 will be approximately normal, and the use of the t distribution will provide approximately the specified confidence coefficient or level of significance. Even if the distributions of Y are far from normal, the estimators b_0 and b_1 generally have the property of *asymptotic normality*—their distributions approach normality under very general conditions as the sample size increases. Thus, with sufficiently large samples, the confidence intervals and decision rules given earlier still apply even if the probability distributions of Y depart far from normality. For large samples, the t value is, of course, replaced by the z value for the standard normal distribution.

Interpretation of confidence coefficient and risks of errors

Since model (3.1) assumes that the X_i are known constants, the confidence coefficient and risks of errors are interpreted with respect to taking repeated samples in which the X observations are kept at the same levels as in the observed sample. For instance, we constructed a confidence interval for β_1 with a confidence coefficient of .95 in the Westwood Company example. This coefficient is interpreted to mean that if many independent samples are taken where the levels of X (the lot sizes) in the first sample are repeated in these other samples and a 95 percent confidence interval is constructed for each sample, 95 percent of the intervals will contain the true value of β_1.

Spacing of the X levels

Inspection of formulas (3.3b) and (3.20b) for the variances of b_1 and b_0, respectively, indicates that for given n and σ^2, these variances are affected by the spacing of the X levels in the observed data. For example, the more the spread in the X levels, the larger is the quantity $\Sigma(X_i - \bar{X})^2$ and the smaller is the variance of b_1. We will discuss in Section 5.9 how the X observations should be spaced in experiments where spacing can be controlled.

Power of tests

The power of tests on β_0 and β_1 can be obtained from Table A–5 in the Appendix, which contains charts of the power function of the t test. Consider, for example, the general decision problem:

$$(3.24) \qquad \begin{aligned} H_0: \beta_1 &= \beta_{10} \\ H_a: \beta_1 &\neq \beta_{10} \end{aligned}$$

for which the general test statistic (3.18) is employed:

$$(3.24a) \qquad t^* = \frac{b_1 - \beta_{10}}{s(b_1)}$$

and the decision rule for level of significance α is:

$$(3.24b) \qquad \begin{aligned} &\text{If } |t^*| \leq t(1 - \alpha/2; n - 2), \text{ conclude } H_0 \\ &\text{If } |t^*| > t(1 - \alpha/2; n - 2), \text{ conclude } H_a \end{aligned}$$

The power of the test is the probability that the decision rule will lead to conclusion H_a when H_a in fact holds. Specifically, the power is given by:

$$(3.25) \qquad \text{Power} = P\{|t^*| > t(1 - \alpha/2; n - 2)|\delta\}$$

where δ is a measure of *noncentrality*—i.e., how far the true value of β_1 is from β_{10}:

$$(3.26) \qquad \delta = \frac{|\beta_1 - \beta_{10}|}{\sigma(b_1)}$$

Table A–5 presents the power of the two-sided t test (in percent) for $\alpha = .01$ and $\alpha = .05$, for various degrees of freedom df. To illustrate the use of this table, let us return to the Westwood Company example where we tested:

$$\begin{aligned} H_0: \beta_1 &= \beta_{10} = 0 \\ H_a: \beta_1 &\neq \beta_{10} = 0 \end{aligned}$$

Suppose we wish to know the power of the test when $\beta_1 = .25$. To ascertain this, we need to know σ^2, the variance of the error terms. Assume that $\sigma^2 = 10.0$ so that $\sigma^2(b_1)$ for our example would be:

$$\sigma^2(b_1) = \frac{\sigma^2}{\Sigma(X_i - \bar{X})^2} = \frac{10.0}{3,400} = .002941$$

or $\sigma(b_1) = .05423$. Then $\delta = |.25 - 0| \div .05423 = 4.6$. We enter the graph for $\alpha = .05$ (the level of significance used in the test) and approximate visually the curve for eight degrees of freedom. Reading the ordinate at $\delta = 4.6$, we obtain 97 percent approximately. Thus, if $\beta_1 = .25$, the probability would be about .97 that we would be led to conclude H_a ($\beta_1 \neq 0$). In other words, if $\beta_1 = .25$, we would be almost certain to conclude that there is a relation between man-hours and lot size.

The power of tests concerning β_0 can be obtained from Table A–5 in completely analogous fashion. For one-sided tests, Table A–5 should be entered so that one half the level of significance shown there is the level of significance of the one-sided test.

3.4 INTERVAL ESTIMATION OF $E(Y_h)$

In regression analysis, one of the major goals usually is to estimate the mean for one or more probability distributions of Y. Consider, for example, a study of the relation between level of piecework pay (X) and worker productivity (Y). The mean productivity at high and medium levels of piecework pay may be of particular interest for purposes of analyzing the benefits obtained from an increase in the pay. As another example, the Westwood Company may be interested in the mean response (mean number of man-hours) for lot sizes of $X = 40$ parts, $X = 55$ parts, and $X = 70$ parts for purposes of choosing appropriate lot sizes for production.

Let X_h denote the level of X for which we wish to estimate the mean response. X_h may be a value which occurred in the sample, or it may be some other value of the independent variable within the scope of the model. The mean response when $X = X_h$ is denoted by $E(Y_h)$. Formula (2.12) gives us the point estimator $\hat{Y}_h$ of $E(Y_h)$:

$$(3.27) \qquad \hat{Y}_h = b_0 + b_1 X_h$$

We consider now the sampling distribution of $\hat{Y}_h$.

Sampling distribution of $\hat{Y}_h$

The sampling distribution of $\hat{Y}_h$, like the earlier sampling distributions discussed, refers to the different values of $\hat{Y}_h$ which would be obtained if repeated samples were selected, each holding the levels of the independent variable X constant, and calculating $\hat{Y}_h$ for each sample.

(3.28)   For model (3.1), the sampling distribution of $\hat{Y}_h$ is normal, with mean and variance:

(3.28a) $$E(\hat{Y}_h) = E(Y_h)$$

(3.28b) $$\sigma^2(\hat{Y}_h) = \sigma^2 \left[\frac{1}{n} + \frac{(X_h - \bar{X})^2}{\Sigma(X_i - \bar{X})^2} \right]$$

Normality.   The normality of the sampling distribution of $\hat{Y}_h$ follows directly from the fact that $\hat{Y}_h$ is a linear combination of the observations Y_i.

Mean.   To prove that $\hat{Y}_h$ is an unbiased estimator of $E(Y_h)$, we proceed as follows:

$$E(\hat{Y}_h) = E(b_0 + b_1 X_h) = E(b_0) + X_h E(b_1)$$
$$= \beta_0 + \beta_1 X_h$$

by (3.3a) and (3.20a).

Variance.   First, we show that b_1 and $\bar{Y}$ are uncorrelated and hence, for model (3.1), independent:

(3.29) $$\sigma(\bar{Y}, b_1) = 0$$

where $\sigma(\bar{Y}, b_1)$ denotes the covariance between $\bar{Y}$ and b_1. We begin with the definitions:

$$\bar{Y} = \Sigma \left(\frac{1}{n} \right) Y_i$$
$$b_1 = \Sigma k_i Y_i$$

where k_i is as defined in (3.5a). We now use theorem (1.27), with $a_i = 1/n$ and $c_i = k_i$; remember that the Y_i are independent random variables:

$$\sigma(\bar{Y}, b_1) = \Sigma \left(\frac{1}{n} \right) k_i \sigma^2(Y_i) = \frac{\sigma^2}{n} \Sigma k_i$$

But we know from (3.6) that $\Sigma k_i = 0$. Hence, the covariance is 0.

Now we are ready to find the variance of $\hat{Y}_h$. We shall use the estimator in the alternative form (2.15):

$$\sigma^2(\hat{Y}_h) = \sigma^2(\bar{Y} + b_1[X_h - \bar{X}])$$

Since $\bar{Y}$ and b_1 are independent and X_h and $\bar{X}$ are constants, we obtain:

$$\sigma^2(\hat{Y}_h) = \sigma^2(\bar{Y}) + (X_h - \bar{X})^2 \sigma^2(b_1)$$

Now $\sigma^2(b_1)$ is given in (3.3b) and:

$$\sigma^2(\bar{Y}) = \frac{\sigma^2(Y_i)}{n} = \frac{\sigma^2}{n}$$

Hence:

$$\sigma^2(\hat{Y}_h) = \frac{\sigma^2}{n} + (X_h - \bar{X})^2 \frac{\sigma^2}{\Sigma(X_i - \bar{X})^2}$$

which, upon a slight rearrangement of terms, yields (3.28b).

Note the effect of the term $(X_h - \bar{X})^2$ on $\sigma^2(\hat{Y}_h)$. The further X_h is from $\bar{X}$, the greater is the quantity $(X_h - \bar{X})^2$ and the larger is the variance of $\hat{Y}_h$. An intuitive explanation of this effect is found in Figure 3.2. Shown there are two sample regression lines, based on two samples for the same set of X values. The two regression lines are assumed to go through the same $(\bar{X}, \bar{Y})$ point to isolate the effect of interest, namely, the effect of variation in the estimated slope b_1 from sample to sample. Note that at X_1, near $\bar{X}$, the fitted values $\hat{Y}_1$ for the two sample regression lines are close to each other. At X_2, which is far from $\bar{X}$, the situation is different. Here, the fitted values $\hat{Y}_2$ differ substantially. Thus, variation in the slope b_1 from sample to sample has a much more pronounced effect on $\hat{Y}_h$ for X levels far from the mean $\bar{X}$ than for X levels near $\bar{X}$. Hence, the variation in the $\hat{Y}_h$ values from sample to sample will be greater when X_h is far from the mean than when X_h is near the mean.

FIGURE 3.2 Effect on $\hat{Y}_h$ of variation in b_1 from sample to sample in two samples with same means $\bar{Y}$ and $\bar{X}$

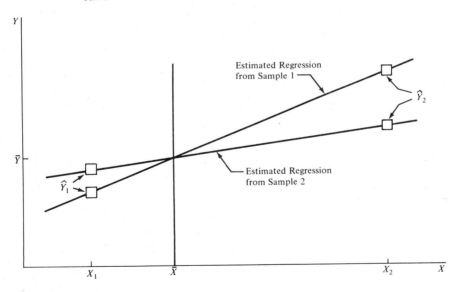

When *MSE* is substituted for σ^2 in (3.28b), we obtain $s^2(\hat{Y}_h)$, the estimated variance of $\hat{Y}_h$:

(3.30)
$$s^2(\hat{Y}_h) = MSE \left[\frac{1}{n} + \frac{(X_h - \bar{X})^2}{\Sigma(X_i - \bar{X})^2} \right]$$

The estimated standard deviation of $\hat{Y}_h$ then is $s(\hat{Y}_h)$, the square root of $s^2(\hat{Y}_h)$.

Sampling distribution of $[\hat{Y}_h - E(Y_h)]/s(\hat{Y}_h)$

Since we have encountered the t distribution in each type of inference for regression model (3.1) considered up to this point, it should not be surprising that:

(3.31) $\dfrac{\hat{Y}_h - E(Y_h)}{s(\hat{Y}_h)}$ is distributed as $t(n - 2)$ for model (3.1)

Hence, all inferences concerning $E(Y_h)$ are carried out in the usual fashion with the t distribution. We illustrate the construction of confidence intervals, since in practice these are more frequently used than tests.

Confidence interval for $E(Y_h)$

A confidence interval for $E(Y_h)$ is constructed in the standard fashion, making use of the t distribution as indicated by theorem (3.31). The $1 - \alpha$ confidence limits are:

(3.32) $$\hat{Y}_h \pm t(1 - \alpha/2; n - 2)s(\hat{Y}_h)$$

Example 1. Returning to the Westwood Company lot size example, let us find a 90 percent confidence interval for $E(Y_h)$ when the lot size is $X_h = 55$ parts. Using the earlier results in Table 3.1, we find the point estimate $\hat{Y}_h$:

$$\hat{Y}_{55} = 10.0 + 2.0(55) = 120$$

Next, we need to find the estimated standard deviation $s(\hat{Y}_h)$. We obtain, using (3.30):

$$s^2(\hat{Y}_{55}) = 7.5\left[\frac{1}{10} + \frac{(55 - 50)^2}{3,400}\right] = .80515$$

so that:

$$s(\hat{Y}_{55}) = .89730$$

For a 90 percent confidence coefficient, we require $t(.95; 8) = 1.860$. Hence, our confidence interval with confidence coefficient .90 is by (3.32):

$$120 - 1.860(.89730) \le E(Y_{55}) \le 120 + 1.860(.89730)$$
$$118.3 \le E(Y_{55}) \le 121.7$$

We conclude with confidence coefficient .90 that the mean number of man-hours required when lots of 55 parts are produced is somewhere between 118.3 and 121.7.

Example 2. Suppose the Westwood Company wishes to estimate $E(Y_h)$ when $X_h = 80$ parts with a 90 percent confidence interval. We require:

$$\hat{Y}_{80} = 10.0 + 2.0(80) = 170$$

$$s^2(\hat{Y}_{80}) = 7.5\left[\frac{1}{10} + \frac{(80-50)^2}{3,400}\right] = 2.73529$$

$$s(\hat{Y}_{80}) = 1.65387$$

$$t(.95; 8) = 1.860$$

Hence, the 90 percent confidence interval is:

$$170 - 1.860(1.65387) \le E(Y_{80}) \le 170 + 1.860(1.65387)$$

$$166.9 \le E(Y_{80}) \le 173.1$$

Note that this confidence interval is somewhat wider than that for example 1, since the X_h level here ($X_h = 80$) is substantially farther from the mean $\bar{X} = 50$ than the X_h level for example 1 ($X_h = 55$).

Comments

1. Since the X_i are known constants in model (3.1), the interpretation of confidence intervals and risks of errors in inferences on the mean response is in terms of taking repeated samples in which the X observations are at the same levels as in the sample actually taken. We noted this same point earlier in connection with inferences on β_0 and β_1.

2. We see from formula (3.28b) that for given sample results, the variance of $\hat{Y}_h$ is smallest when $X_h = \bar{X}$. Thus, in an experiment to estimate the mean response at a particular level X_h of the independent variable, the precision of the estimate will be greatest if (everything else remaining equal) the observations on X are spaced so that $\bar{X} = X_h$.

3. When the sample size is large, the t value in the confidence limits (3.32) may be replaced by the standard normal z value, since the t distribution approaches the standard normal distribution with increasing sample size.

4. The usual relationship between confidence intervals and tests applies in inferences concerning the mean response. Thus, the two-sided confidence limits (3.32) can be utilized for two-sided tests concerning the mean response at X_h. Alternatively, a regular decision rule can be set up.

5. Confidence limits (3.32) apply when a single mean response is to be estimated from the sample. We discuss in Chapter 5 how to proceed when a number of mean responses are to be estimated from the same sample.

3.5 PREDICTION OF NEW OBSERVATION

We consider now the prediction of a new observation Y corresponding to a given level X of the independent variable. In our Westwood Company illustration, for instance, the next lot to be produced consists of 55 parts and management wishes to predict the number of man-hours for this particular lot. As another example, an economist has estimated the regression relation between company sales and number of persons 16 or more years old, based on data for the past 10 years. Given a reliable demographic projection of the number of persons 16 or more years old for next year, the economist wishes to predict next year's company sales.

The new observation on Y is viewed as the result of a new trial, independent of the trials on which the regression analysis is based. We shall denote the level of X for the new trial as X_h and the new observation on Y as $Y_{h(new)}$. Of course, we assume that the underlying regression model applicable for the basic sample data continues to be appropriate for the new observation.

The distinction between estimation of the mean response $E(Y_h)$, discussed in the preceding section, and prediction of a new response $Y_{h(new)}$, discussed now, is basic. In the former case, we estimate the *mean* of the distribution of Y. In the present case, we predict an *individual outcome* drawn from the distribution of Y. Of course, the great majority of individual outcomes deviate from the mean response, and this must be allowed for in the procedure for predicting $Y_{h(new)}$.

Prediction interval when parameters known

To illustrate the nature of a *prediction interval* for a new observation $Y_{h(new)}$ in as simple a fashion as possible, we shall first assume that all regression parameters are known. Later we shall drop this assumption and make appropriate modifications.

Suppose that the Westwood Company plans to produce a lot of $X_h = 40$ parts in a few weeks and that the relevant parameters of the regression model are known to be:

$$\beta_0 = 9.5 \qquad \beta_1 = 2.1$$
$$E(Y) = 9.5 + 2.1X$$
$$\sigma^2 = 10.0$$

Thus, for $X_h = 40$ parts, we have:

$$E(Y_{40}) = 9.5 + 2.1(40) = 93.5$$

Figure 3.3 shows the probability distribution of Y for $X_h = 40$ parts. Its mean is $E(Y_{40}) = 93.5$, and its standard deviation is $\sigma = \sqrt{10.0} = 3.162$. Further, the distribution is normal in accord with model (3.1).

Suppose we were to predict that the number of man-hours for the next lot of $X_h = 40$ parts will be between:

$$E(Y_{40}) \pm 3\sigma$$
$$93.5 \pm 3(3.162)$$

so that the prediction interval would be:

$$84.0 \le Y_{40(new)} \le 103.0$$

Since 99.7 percent of the area in a normal probability distribution falls within three standard deviations from the mean, the probability is .997 that this prediction interval will give a correct prediction for the next production run of 40 parts.

The basic idea of a prediction interval is thus to choose a range in the distribution of Y wherein most of the observations will fall, and to declare that the next

FIGURE 3.3 Prediction of $Y_{h(new)}$ when parameters known

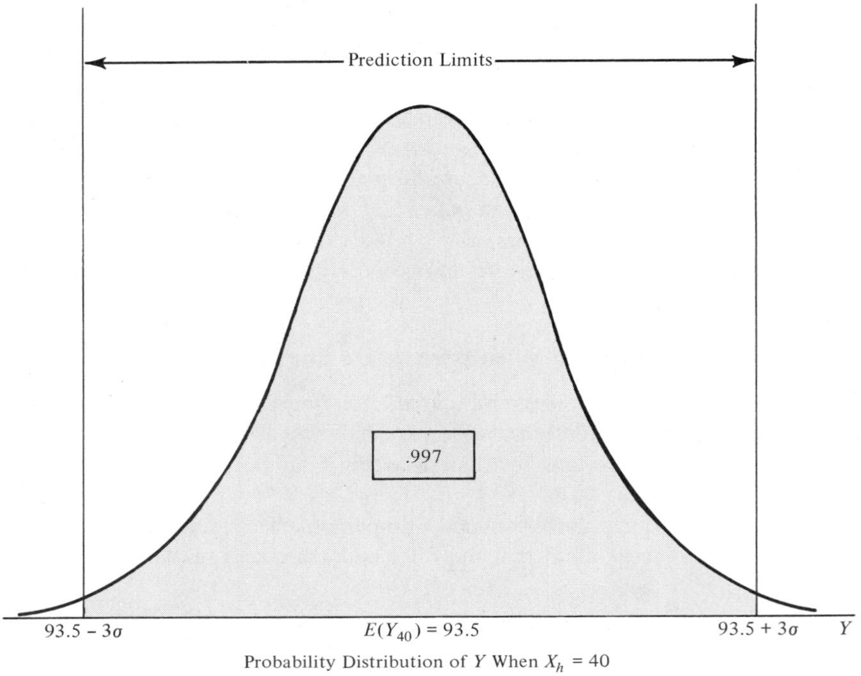

Probability Distribution of Y When $X_h = 40$

observation will fall in this range. The usefulness of the prediction interval depends, as always, on the width of the interval and the needs for precision by the user.

In general, when the regression parameters are known, the $1 - \alpha$ prediction limits for $Y_{h(new)}$ are:

$$(3.33) \qquad E(Y_h) \pm z(1 - \alpha/2)\sigma$$

In centering the limits around $E(Y_h)$, we obtain the narrowest interval consistent with the specified probability of a correct prediction.

Prediction interval for $Y_{h(new)}$ when parameters unknown

When the regression parameters are unknown, they must be estimated. The mean of the distribution of Y is estimated by $\hat{Y}_h$, as usual, and the variance of the distribution of Y is estimated by MSE. We cannot, however, simply use the prediction limits (3.33) with the parameters replaced by the corresponding point estimators. The reason is illustrated intuitively in Figure 3.4. Shown there are two probability distributions of Y, corresponding to the upper and lower limits of a confidence interval for $E(Y_h)$. In other words, the distribution of Y could be located as far left as the one shown, as far right as the other one shown, or

FIGURE 3.4   Prediction of $Y_{h(new)}$ when parameters unknown

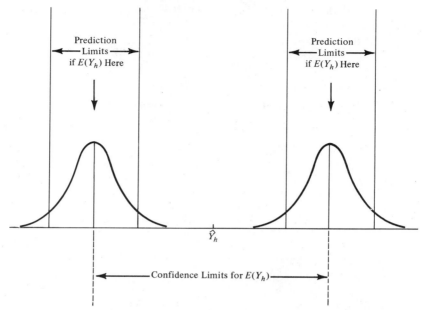

anywhere in between. Since we do not know the mean $E(Y_h)$ and only estimate it by a confidence interval, we cannot be certain of the location of the distribution of Y.

Figure 3.4 also shows the prediction limits for each of the two probability distributions of Y presented there. Since we cannot be certain of the location of the distribution of Y, prediction limits for $Y_{h(new)}$ clearly must take account of two elements, as shown in Figure 3.4:

1.   Variation in possible location of the distribution of Y.
2.   Variation within the probability distribution of Y.

Prediction limits for a new observation Y at a given level X_h are obtained by means of the following theorem:

$$(3.34) \qquad \frac{\hat{Y}_h - Y}{s(Y_{h(new)})} \text{ is distributed as } t(n-2) \text{ for model (3.1)}$$

Note that the standardized statistic (3.34) uses the point estimator $\hat{Y}_h$ in the numerator rather than the true mean $E(Y_h)$ because the true mean is unknown and cannot be used in making a prediction. The estimated standard deviation $s(Y_{h(new)})$ in the denominator of the standardized statistic will be defined shortly.

From theorem (3.34), it follows in the usual fashion that the $1 - \alpha$ prediction limits for a new observation are [for instance, compare (3.34) to (3.10) and relate $\hat{Y}_h$ to b_1 and Y to β_1]:

$$(3.35) \qquad \hat{Y}_h \pm t(1 - \alpha/2; n - 2)s(Y_{h(new)})$$

The variance of the numerator of the standardized statistic (3.34) is readily obtained, utilizing the independence of the new observation Y and the original sample observations on which $\hat{Y}_h$ is based. We shall denote the variance of the numerator by $\sigma^2(Y_{h(new)})$, and we obtain:

$$(3.36) \qquad \sigma^2(Y_{h(new)}) = \sigma^2(\hat{Y}_h - Y) = \sigma^2(\hat{Y}_h) + \sigma^2$$

Note that $\sigma^2(Y_{h(new)})$ has two components:

1. The variance of the sampling distribution of $\hat{Y}_h$.
2. The variance of the distribution of Y.

An unbiased estimator of $\sigma^2(Y_{h(new)})$ is:

$$(3.37) \qquad s^2(Y_{h(new)}) = s^2(\hat{Y}_h) + MSE$$

which can be expressed, using (3.30), as follows:

$$(3.37a) \qquad s^2(Y_{h(new)}) = MSE\left[1 + \frac{1}{n} + \frac{(X_h - \bar{X})^2}{\Sigma(X_i - \bar{X})^2}\right]$$

Example. Suppose that the Westwood Company wishes to predict the number of man-hours required in the forthcoming production run of size 55 with a 90 percent prediction interval, and that the parameter values are unknown. We require $t(.95; 8) = 1.860$. From earlier work, we have:

$$\hat{Y}_{55} = 120 \qquad s^2(\hat{Y}_{55}) = .80515$$
$$MSE = 7.5$$

Using (3.37), we obtain:

$$s^2(Y_{55(new)}) = .80515 + 7.5 = 8.30515$$

so that:

$$s(Y_{55(new)}) = 2.88187$$

Hence, the 90 percent prediction interval for $Y_{55(new)}$ is by (3.35):

$$120 - 1.860(2.88187) \le Y_{55(new)} \le 120 + 1.860(2.88187)$$
$$114.6 \le Y_{55(new)} \le 125.4$$

With confidence coefficient .90, we predict that the number of man-hours for the next production run of 55 parts will be somewhere between 114.6 and 125.4.

Comments

1. The 90 percent prediction interval for $Y_{55(new)}$ just obtained is wider than the 90 percent confidence interval for $E(Y_{55})$ obtained on page 75. The reason is that when predicting a new observation, we encounter both the variability in $\hat{Y}_h$ from sample to sample as well as the variation within the probability distribution of Y.

2. Formula (3.37a) indicates that the prediction interval is wider the further X_h is from $\bar{X}$. The reason for this is that the estimate of the mean $\hat{Y}_h$, as noted earlier, is less precise as X_h is located further away from $\bar{X}$.

3. The confidence coefficient for the prediction limits (3.35) refers to the taking of repeated samples based on the same set of X values, and calculating prediction limits for $Y_{h(new)}$ for each sample.

4. Prediction limits lend themselves to statistical control uses. In our example, suppose that the new production run of 55 parts, for which the prediction limits were 114.6 and 125.4 hours, actually required 135 hours. Management here would have an indication that a change in the production process may have occurred, and may wish to initiate a search for the assignable cause.

5. When the sample size is large, the last two terms inside the brackets in (3.37a) are small compared to 1, the first term in the brackets. Also, of course, the t distribution is then approximately normal. Hence, approximate $1 - \alpha$ prediction limits for $Y_{h(new)}$ when n is large are:

$$(3.38) \qquad \hat{Y}_h \pm z(1 - \alpha/2)\sqrt{MSE}$$

6. Prediction limits (3.35) apply for a single prediction based on the sample data. Next, we discuss how to predict the mean of several new observations at a given X_h; and in Chapter 5, we take up how to make several predictions at different X_h values.

7. Prediction intervals resemble confidence intervals. However, they differ conceptually. A confidence interval represents an inference on a parameter, and is an interval which is intended to cover the value of the parameter. A prediction interval, on the other hand, is a statement about the value to be taken by a random variable.

Prediction of mean of m new observations for given X_h

Occasionally, one would like to predict the mean of m new observations on Y for a given level of the independent variable. Suppose the Westwood Company has been asked to bid on a contract that calls for $m = 3$ independent production runs of $X_h = 55$ parts during the next few months. Management would like to predict the mean man-hours per run for these three runs, and then convert this into a prediction of the total man-hours required to fill the contract.

We shall denote the mean value of Y to be predicted as $\bar{Y}_{h(new)}$. It can be shown that the appropriate $1 - \alpha$ prediction limits are:

$$(3.39) \qquad \hat{Y}_h \pm t(1 - \alpha/2; n - 2)s(\bar{Y}_{h(new)})$$

where:

$$(3.39a) \qquad s^2(\bar{Y}_{h(new)}) = s^2(\hat{Y}_h) + \frac{MSE}{m}$$

or equivalently:

$$(3.39b) \qquad s^2(\bar{Y}_{h(new)}) = MSE\left[\frac{1}{m} + \frac{1}{n} + \frac{(X_h - \bar{X})^2}{\Sigma(X_i - \bar{X})^2}\right]$$

Note from (3.39a) that the variance $s^2(\bar{Y}_{h(new)})$ has two components:

1. The variance of the sampling distribution of $\hat{Y}_h$.
2. The variance of the mean of m observations from the probability distribution of Y.

Example. In the Westwood Company example, let us find the 90 percent prediction interval for the mean number of man-hours $\bar{Y}_{h(new)}$ in three new production runs, each for $X_h = 55$ parts. From previous work, we have:

$$\hat{Y}_{55} = 120 \qquad s^2(\hat{Y}_{55}) = .80515$$
$$MSE = 7.5 \qquad t(.95; 8) = 1.860$$

Hence, we obtain:

$$s^2(\bar{Y}_{55(new)}) = .80515 + \frac{7.5}{3} = 3.30515$$

or:

$$s(\bar{Y}_{55(new)}) = 1.81801$$

The prediction interval for the mean man-hours per run then is:

$$120 - 1.860(1.81801) \le \bar{Y}_{55(new)} \le 120 + 1.860(1.81801)$$
$$116.6 \le \bar{Y}_{55(new)} \le 123.4$$

Note that these prediction limits are somewhat narrower than those for predicting the man-hours for a single lot of 55 parts because they involve a prediction of the mean man-hours for three lots.

We obtain the prediction interval for the total number of man-hours in the three production runs by multiplying the prediction limits for $\bar{Y}_{55(new)}$ by three:

$$349.8 = 3(116.6) \le \text{Total man-hours} \le 3(123.4) = 370.2$$

Thus, it can be predicted with 90 percent confidence that between 350 and 370 man-hours will be needed to fill the contract for three lots of 55 parts each.

3.6 CONSIDERATIONS IN APPLYING REGRESSION ANALYSIS

We have now discussed the major uses of regression analysis—to make inferences about the regression parameters, to estimate the mean response for a given X, and to predict a new observation Y for a given X. It remains to make a few cautionary remarks about implementing applications of regression analysis.

1. Frequently, regression analysis is used to make inferences for the future. For instance, the Westwood Company may wish to estimate expected man-hours for given lot sizes for purposes of planning future production. In applications of this type, it is important to remember that the validity of the regression application depends upon whether basic causal conditions in the period ahead will be similar to those in existence during the period upon which the regression analysis is based. This caution applies whether mean responses are to be estimated, new observations predicted, or regression parameters estimated.

2. In predicting new observations on Y, the independent variable X itself often has to be predicted. For instance, we mentioned earlier the prediction of company sales for next year from a demographic projection of the number of

persons 16 years of age or older next year. A prediction of company sales under these circumstances is a conditional prediction, dependent upon the correctness of the population projection. It is easy to forget the conditional nature of this type of prediction.

3. Another caution deals with inferences pertaining to levels of the independent variable which fall outside the range of observations. Unfortunately, this situation frequently occurs in practice. A company which predicts its sales from a regression relation of company sales to disposable personal income will often find the level of disposable personal income of interest (e.g., for the year ahead) to fall beyond the range of past data. If the X level does not fall far beyond this range, one may have reasonable confidence in the application of the regression analysis. On the other hand, if the X level falls far beyond the range of past data, extreme caution should be exercised since one cannot be sure that the regression function which fits the past data is appropriate over the wider range of the independent variable.

4. A statistical test which leads to the conclusion that $\beta_1 \neq 0$ does not establish a cause-and-effect relation between the independent and dependent variables. For instance, with nonexperimental data, both the X and Y variables may be simultaneously influenced by other variables not included in the regression model. Thus, data on grade school children's vocabulary (X) and writing speed (Y) may show a clear linear association, but this could be largely the result of a child's age, amount of education, and similar factors that affect both X and Y. On the other hand, the existence of a regression relation in controlled experiments is often good evidence of a cause-and-effect relation.

5. Finally, we should note again that special problems arise when one wishes to estimate the mean response or predict a new observation for a number of different levels of the independent variable, as is frequently the case. The confidence coefficients for the limits (3.32) for estimating $E(Y)$ and for the prediction limits (3.35) for a new observation apply for a single level of X for a given sample. In Chapter 5, we discuss how to make multiple inferences from a given sample.

3.7 CASE WHEN X IS RANDOM

The normal error model (3.1), which has been used throughout this chapter and will continue to be used, assumes that the X values are known constants. As a consequence of this, the confidence coefficients and risks of errors refer to repeated sampling when the X values are kept the same from sample to sample.

Frequently, it may not be appropriate to consider the X values as known constants. For instance, consider regressing daily bathing suit sales by a department store on mean daily temperature. Surely, the department store cannot control daily temperatures, so it would not be meaningful to think of repeated sampling when the temperature levels are the same from sample to sample.

In this type of situation, it may be preferable to consider both Y and X as random variables. Does this mean then that all of our earlier results are not applicable here? Not at all. It can be shown that all results on estimation, testing, and prediction obtained for model (3.1) still apply if the following conditions hold:

(3.40)   1. The conditional distributions of the Y_i, given X_i, are normal and independent, with conditional means $\beta_0 + \beta_1 X_i$ and conditional variance σ^2.
          2. The X_i are independent random variables, whose probability distribution $g(X_i)$ does not involve the parameters β_0, β_1, σ^2.

These conditions require only that model (3.1) is appropriate for each *conditional* distribution of Y_i, and that the probability distribution of X_i does not involve the regression parameters. If these conditions are met, all earlier results on estimation, testing, and prediction still hold even though the X_i are now random variables. The major modification occurs in the interpretation of confidence coefficients and specified risks of error. When X is random, these refer to repeated sampling of pairs of (X_i, Y_i) values where the X_i values as well as the Y_i values change from sample to sample. Thus, in our bathing suit sales illustration, a confidence coefficient would refer to the proportion of correct interval estimates if repeated samples of n days' sales and temperatures were obtained and the confidence interval calculated for each sample. Another modification occurs in the power of tests which is different when X is a random variable.

3.8 ANALYSIS OF VARIANCE APPROACH TO REGRESSION ANALYSIS

We now have developed the basic regression model and demonstrated its major uses. At this point, we will view the relationships in a regression model in a somewhat different way than before. This new perspective will not enable us to do anything new with the basic regression model, but it will come into its own when we take up more complex regression models and additional types of linear statistical models.

Partitioning of total sum of squares

Basic notions. The analysis of variance approach is based on the partitioning of sums of squares and degrees of freedom associated with the response variable Y. To explain the motivation of this approach, consider again the Westwood Company lot size example. Figure 3.5a shows the man-hours required for the 10 production runs presented earlier in Table 2.1. There is variation in the number of man-hours, as in all statistical data. Indeed, if all observations Y_i were identically equal, in which case $Y_i \equiv \bar{Y}$, there would be no statistical problems. The variation of the Y_i is conventionally measured in terms of the deviations:

(3.41) $$Y_i - \bar{Y}$$

FIGURE 3.5 Partitioning of deviations $Y_i - \bar{Y}$ (Y values not plotted to scale)

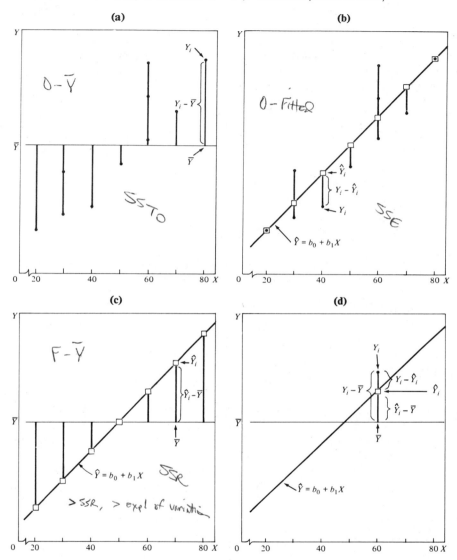

These deviations are shown in Figure 3.5a, and one is labeled explicitly. The measure of total variation, denoted by *SSTO*, is the sum of the squared deviations (3.41):

(3.42)
$$SSTO = \Sigma(Y_i - \bar{Y})^2$$

Here *SSTO* stands for *total sum of squares*. If $SSTO = 0$, all observations are the same. The greater is *SSTO*, the greater is the variation among the Y observations.

When we utilize the regression approach, the variation reflecting the uncertainty in the data is that of the Y observations around the regression line:

$$(3.43) \qquad\qquad Y_i - \hat{Y}_i$$

These deviations are shown in Figure 3.5b. The measure of variation in the data with the regression model is the sum of the squared deviations (3.43), which is the familiar SSE of (2.21):

$$(3.44) \qquad\qquad SSE = \Sigma(Y_i - \hat{Y}_i)^2$$

Again, SSE denotes *error sum of squares*. If $SSE = 0$, all observations fall on the fitted regression line. The larger is SSE, the greater is the variation of the Y observations around the regression line.

For the Westwood Company example, we know from earlier work (Table 3.1) that:

$$SSTO = 13,660$$
$$SSE = 60$$

What accounts for the substantial difference between these two sums of squares? The difference, as we shall show shortly, is another sum of squares:

$$(3.45) \qquad\qquad SSR = \Sigma(\hat{Y}_i - \bar{Y})^2$$

where SSR stands for *regression sum of squares*. Note that SSR is a sum of squared deviations, the deviations being:

$$(3.46) \qquad\qquad \hat{Y}_i - \bar{Y}$$

These deviations are shown in Figure 3.5c. Each deviation is simply the difference between the fitted value on the regression line and the mean of the fitted values $\bar{Y}$. [Recall from (2.18) that the mean of the fitted values $\hat{Y}_i$ is $\bar{Y}$.] If the regression line is horizontal so that $\hat{Y}_i - \bar{Y} \equiv 0$, $SSR = 0$. Otherwise, SSR is positive.

SSR may be considered a measure of the variability of the Y's associated with the regression line. The larger is SSR in relation to $SSTO$, the greater is the effect of the regression relation in accounting for the total variation in the Y observations.

For our lot size example, we have:

$$SSR = SSTO - SSE = 13,660 - 60 = 13,600$$

which indicates that most of the total variability in man-hours is accounted for by the relation between lot size and man-hours.

Formal development of partitioning. Consider the deviation $Y_i - \bar{Y}$, the basic quantity measuring the variation of the observations Y_i. We can decompose this deviation as follows:

(3.47)
$$Y_i - \bar{Y} = \hat{Y}_i - \bar{Y} + Y_i - \hat{Y}_i$$

Total deviation	Deviation of fitted regression value around mean	Deviation around regression line

Thus, the total deviation $Y_i - \bar{Y}$ can be viewed as the sum of two components:

1. The deviation of the fitted value $\hat{Y}_i$ around the mean $\bar{Y}$.
2. The deviation of Y_i around the regression line.

Figure 3.5d shows this decomposition for one of the observations.

It is a remarkable property that the sums of these squared deviations have the same relationship:

(3.48)
$$\Sigma(Y_i - \bar{Y})^2 = \Sigma(\hat{Y}_i - \bar{Y})^2 + \Sigma(Y_i - \hat{Y}_i)^2$$

or, using the notation in (3.42), (3.44), and (3.45):

(3.48a)
$$SSTO = SSR + SSE$$

To prove this basic result in the analysis of variance, we proceed as follows:

$$\Sigma(Y_i - \bar{Y})^2 = \Sigma[(\hat{Y}_i - \bar{Y}) + (Y_i - \hat{Y}_i)]^2$$
$$= \Sigma[(\hat{Y}_i - \bar{Y})^2 + (Y_i - \hat{Y}_i)^2 + 2(\hat{Y}_i - \bar{Y})(Y_i - \hat{Y}_i)]$$
$$= \Sigma(\hat{Y}_i - \bar{Y})^2 + \Sigma(Y_i - \hat{Y}_i)^2 + 2\Sigma(\hat{Y}_i - \bar{Y})(Y_i - \hat{Y}_i)$$

The last term on the right is zero, as can be seen by expanding it out:

$$2\Sigma(\hat{Y}_i - \bar{Y})(Y_i - \hat{Y}_i) = 2\Sigma\hat{Y}_i(Y_i - \hat{Y}_i) - 2\bar{Y}\Sigma(Y_i - \hat{Y}_i)$$

The first summation on the right is zero by (2.20), and the second is zero by (2.17). Hence, (3.48) follows.

Computational formulas. The definitional formulas for SSTO, SSR, and SSE presented above are often not convenient for hand computation. Useful computational formulas for SSTO and SSR, which are algebraically equivalent to the definitional formulas, are:

(3.49)
$$SSTO = \Sigma Y_i^2 - \frac{(\Sigma Y_i)^2}{n} = \Sigma Y_i^2 - n\bar{Y}^2$$

(3.50a)
$$SSR = b_1\left[\Sigma X_i Y_i - \frac{\Sigma X_i \Sigma Y_i}{n}\right]$$
$$= b_1[\Sigma(X_i - \bar{X})(Y_i - \bar{Y})]$$

or:

(3.50b)
$$SSR = b_1^2\Sigma(X_i - \bar{X})^2$$

Computational formulas for SSE were given earlier in (2.24).

Using the results for the Westwood Company example summarized in Table 3.1, we obtain for SSR by (3.50a):

$$SSR = 2.0(6{,}800) = 13{,}600$$

This, of course, is the same result obtained previously by taking the difference $SSTO - SSE$, except sometimes for a slight difference due to rounding effects.

Breakdown of degrees of freedom

Corresponding to the partitioning of the total sum of squares $SSTO$, there is a partitioning of the associated degrees of freedom (abbreviated df). We have $n - 1$ degrees of freedom associated with $SSTO$. One degree of freedom is lost because the deviations $Y_i - \bar{Y}$ are not independent in that they must sum to zero. Equivalently, one degree of freedom is lost because the sample mean $\bar{Y}$ is used to estimate the population mean.

SSE, as noted earlier, has $n - 2$ degrees of freedom associated with it. Two degrees of freedom are lost because the two parameters β_0 and β_1 were estimated in obtaining the fitted values $\hat{Y}_i$.

SSR has one degree of freedom associated with it. There are two parameters in the regression equation, but the deviations $\hat{Y}_i - \bar{Y}$ are not independent because they must sum to zero; hence, one degree of freedom is lost from the possible degrees of freedom.

Note that the degrees of freedom are additive:

$$n - 1 = 1 + (n - 2)$$

For our Westwood Company example, these degrees of freedom are:

$$9 = 1 + 8$$

Mean squares

A sum of squares divided by its associated degrees of freedom is called a *mean square* (abbreviated MS). For instance, an ordinary sample variance is a mean square since a sum of squares, $\Sigma(Y_i - \bar{Y})^2$, is divided by its associated degrees of freedom, $n - 1$. We are interested here in the *regression mean square*, denoted by MSR:

(3.51)
$$MSR = \frac{SSR}{1} = SSR$$

and in the *error mean square, MSE,* defined earlier in (2.22):

(3.52)
$$MSE = \frac{SSE}{n - 2}$$

For our Westwood Company example, we have $SSR = 13{,}600$ and $SSE = 60$.

Hence:

$$MSR = \frac{13,600}{1} = 13,600$$

Also, we obtained earlier:

$$MSE = \frac{60}{8} = 7.5$$

Note

The two mean squares MSR and MSE do not add to $SSTO \div (n - 1) = 13,660 \div 9 = 1,518$. Thus, mean squares are not additive.

Analysis of variance table

Basic table. The breakdowns of the total sum of squares and associated degrees of freedom are displayed in the form of an analysis of variance table (ANOVA table) in Table 3.2. Mean squares of interest also are shown. In addition, there is a column of expected mean squares which will be utilized below. The ANOVA table for our Westwood Company example is shown in Table 3.3.

TABLE 3.2 ANOVA table for simple regression

Source of Variation	SS	df	MS	E(MS)
Regression	$SSR = \sum (\hat{Y}_i - \bar{Y})^2$	1	$MSR = \dfrac{SSR}{1}$	$\sigma^2 + \beta_1^2 \sum (X_i - \bar{X})^2$
Error	$SSE = \sum (Y_i - \hat{Y}_i)^2$	$n - 2$	$MSE = \dfrac{SSE}{n - 2}$	σ^2
Total	$SSTO = \sum (Y_i - \bar{Y})^2$	$n - 1$		

TABLE 3.3 ANOVA table for Westwood Company example

Source of Variation	SS	df	MS
Regression	13,600	1	13,600
Error	60	8	7.5
Total	13,660	9	

Modified table. Sometimes, an ANOVA table showing one additional element of decomposition is utilized. Recall that by (3.49):

$$SSTO = \Sigma(Y_i - \bar{Y})^2 = \Sigma Y_i^2 - n\bar{Y}^2$$

In the modified ANOVA table, the *total uncorrected sum of squares*, denoted by *SSTOU*, is defined as:

(3.53) $$SSTOU = \Sigma Y_i^2$$

and the *correction for the mean sum of squares*, denoted by *SS*(correction for mean), is defined as:

(3.54) $$SS(\text{correction for mean}) = n\bar{Y}^2$$

Table 3.4 shows this modified ANOVA table. The general format is presented in part (a) and the Westwood Company results in part (b). Both types of ANOVA tables are widely used. Ordinarily, we shall utilize the basic type of table.

TABLE 3.4 Modified ANOVA table for simple regression and results for Westwood Company example

(a) General

Source of Variation	SS	df	MS
Regression	$SSR = \Sigma(\hat{Y}_i - \bar{Y})^2$	1	$MSR = \dfrac{SSR}{1}$
Error	$SSE = \Sigma(Y_i - \hat{Y}_i)^2$	$n-2$	$MSE = \dfrac{SSE}{n-2}$
Total	$SSTO = \Sigma(Y_i - \bar{Y})^2$	$n-1$	
Correction for mean	$SS(\text{correction for mean}) = n\bar{Y}^2$	1	
Total, uncorrected	$SSTOU = \Sigma Y_i^2$	n	

(b) Westwood Company Example

Source of Variation	SS	df	MS
Regression	13,600	1	13,600
Error	60	8	7.5
Total	13,660	9	
Correction for mean	121,000	1	
Total, uncorrected	134,660	10	

Expected mean squares

We now find the expected value of each of the mean squares in the ANOVA table so that we can know what quantity each mean square estimates.

We stated earlier that MSE is an unbiased estimator of the error variance σ^2:

(3.55) $$E(MSE) = \sigma^2$$

This follows from theorem (3.11), which states that $SSE/\sigma^2 \sim \chi^2(n-2)$ for model (3.1). Hence, it follows from property (1.37) of the chi-square distribution that:

$$E\left[\frac{SSE}{\sigma^2}\right] = n - 2$$

or that:

$$E\left[\frac{SSE}{n-2}\right] = E(MSE) = \sigma^2$$

To find the expected value of MSR, we begin with (3.50b):

$$SSR = b_1^2 \Sigma(X_i - \bar{X})^2$$

Now by (1.14a), we have:

(3.56) $$\sigma^2(b_1) = E(b_1^2) - [E(b_1)]^2$$

We know from (3.3a) that $E(b_1) = \beta_1$, and from (3.3b) that:

$$\sigma^2(b_1) = \frac{\sigma^2}{\Sigma(X_i - \bar{X})^2}$$

Hence, substituting into (3.56), we obtain:

$$E(b_1^2) = \frac{\sigma^2}{\Sigma(X_i - \bar{X})^2} + \beta_1^2$$

It now follows that:

$$E(SSR) = E(b_1^2)\Sigma(X_i - \bar{X})^2 = \sigma^2 + \beta_1^2\Sigma(X_i - \bar{X})^2$$

Finally, $E(MSR)$ is:

(3.57) $$E(MSR) = E\left(\frac{SSR}{1}\right) = \sigma^2 + \beta_1^2\Sigma(X_i - \bar{X})^2$$

Table 3.2 contains the expected mean squares which we have just derived.

Comments

1. The expectation of MSE is σ^2, whether or not X and Y are linearly related, i.e., whether or not $\beta_1 = 0$.

2. The expectation of MSR is also σ^2 when $\beta_1 = 0$. On the other hand, when $\beta_1 \neq 0$, $E(MSR)$ is greater than σ^2 since the term $\beta_1^2\Sigma(X_i - \bar{X})^2$ in (3.57) then must be positive. Thus, for testing whether or not $\beta_1 = 0$, a comparison of MSR and MSE suggests itself. If MSR and MSE are of the same order of magnitude, this would suggest that $\beta_1 = 0$. On the other hand, if MSR is substantially greater than MSE, this would suggest that $\beta_1 \neq 0$. This indeed is the basic idea underlying the analysis of variance test to be discussed next.

F test of $\beta_1 = 0$ versus $\beta_1 \neq 0$

The general analysis of variance approach provides us with a battery of highly useful tests for regression models (and other linear statistical models). For the simple regression case considered here, the analysis of variance provides us with a test for:

$$(3.58) \qquad \begin{aligned} H_0: \beta_1 &= 0 \\ H_a: \beta_1 &\neq 0 \end{aligned}$$

Test statistic. The test statistic for the analysis of variance approach is denoted by F^*. As just mentioned, it compares MSR and MSE in the following fashion:

$$(3.59) \qquad F^* = \frac{MSR}{MSE}$$

The earlier motivation, based on the expected mean squares in Table 3.2, suggests that large values of F^* support H_a and values of F^* near 1 support H_0. In other words, the appropriate test is an upper-tail one.

Distribution of F^*. In order to be able to construct a statistical decision rule and examine its properties, we need to know the sampling distribution of F^*. We begin by considering the sampling distribution of F^* when H_0 ($\beta_1 = 0$) holds. The famous *Cochran's theorem* will be most helpful in this connection. For our purposes, this theorem can be put as follows:

(3.60)   If all n observations Y_i come from the same normal distribution with mean μ and variance σ^2, and $SSTO$ is decomposed into k sums of squares SS_r, each with degrees of freedom df_r, then the SS_r/σ^2 terms are independent χ^2 variables with df_r degrees of freedom if:

$$\sum_{r=1}^{k} df_r = n - 1$$

Note from Table 3.2 that we have decomposed $SSTO$ into the two sums of squares SSR and SSE, and that their degrees of freedom are additive. Hence:

If $\beta_1 = 0$ so that all Y_i have the same mean $\mu = \beta_0$ and the same variance σ^2, $\dfrac{SSE}{\sigma^2}$ and $\dfrac{SSR}{\sigma^2}$ are independent χ^2 variables.

Now consider the test statistic F^*, which we can write as follows:

$$F^* = \frac{\dfrac{SSR}{\sigma^2}}{1} \div \frac{\dfrac{SSE}{\sigma^2}}{n-2} = \frac{MSR}{MSE}$$

But by Cochran's theorem, we have when H_0 holds:

$$F^* \sim \frac{\chi^2(1)}{1} \div \frac{\chi^2(n-2)}{n-2}$$

where the χ^2 variables are independent. Thus, when H_0 holds, F^* is the ratio of two independent χ^2 variables, each divided by its degrees of freedom. But this is the definition of an F random variable in (1.42).

We have thus established that if H_0 holds, F^* follows the F distribution, specifically the $F(1, n-2)$ distribution.

When H_a holds, it can be shown that F^* follows the noncentral F distribution, a complex distribution that we need not consider further at this time.

Note

SSR and SSE are independent and $SSE/\sigma^2 \sim \chi^2$ even if $\beta_1 \neq 0$. But that both SSR/σ^2 and SSE/σ^2 are χ^2 random variables requires that $\beta_1 = 0$.

Construction of decision rule. Since the test is upper-tailed and F^* is distributed as $F(1, n-2)$ when H_0 holds, the decision rule is as follows when the risk of a Type I error is to be controlled at α:

(3.61) If $F^* \leq F(1 - \alpha; 1, n - 2)$, conclude H_0
        If $F^* > F(1 - \alpha; 1, n - 2)$, conclude H_a

where $F(1 - \alpha; 1, n - 2)$ is the $(1 - \alpha)100$ percentile of the appropriate F distribution.

Example. Using our Westwood Company lot size example again, let us repeat the earlier test on β_1. This time we will use the F test. The alternative conclusions are:

$$H_0: \beta_1 = 0$$
$$H_a: \beta_1 \neq 0$$

As before, let $\alpha = .05$. Since $n = 10$, we require $F(.95; 1, 8)$. We find from Table A–4 in the Appendix that $F(.95; 1, 8) = 5.32$. The decision rule is:

If $F^* \leq 5.32$, conclude H_0
If $F^* > 5.32$, conclude H_a

We have from Table 3.3 that $MSR = 13{,}600$ and $MSE = 7.5$. Hence, F^* is:

$$F^* = \frac{13{,}600}{7.5} = 1{,}813$$

Since $F^* = 1{,}813 > 5.32$, we conclude H_a, that $\beta_1 \neq 0$, or that there is a linear association between man-hours and lot size. This is the same result as when the t test was employed, as it must be according to our discussion below.

The P-value for the test statistic is the probability $P[F(1, 8) > F^* = 1{,}813]$. From Table A–4 we see that this P-value is less than .001 since $F(.999; 1, 8) = 25.4$.

Equivalence of F test and t test. For a given α level, the F test of $\beta_1 = 0$ versus $\beta_1 \neq 0$ is equivalent algebraically to the two-tailed t test. To see this, recall from (3.50b) that:

$$SSR = b_1^2 \Sigma(X_i - \bar{X})^2$$

Thus, we can write:

$$F^* = \frac{SSR \div 1}{SSE \div (n - 2)} = \frac{b_1^2 \Sigma(X_i - \bar{X})^2}{MSE}$$

But since $s^2(b_1) = MSE/\Sigma(X_i - \bar{X})^2$, we obtain:

$$(3.62) \qquad F^* = \frac{b_1^2}{s^2(b_1)} = \left[\frac{b_1}{s(b_1)} \right]^2$$

Now, we know from earlier discussion that the t^* statistic for testing whether or not $\beta_1 = 0$ is by (3.17):

$$t^* = \frac{b_1}{s(b_1)}$$

In squaring, we obtain the expression for F^* in (3.62). Thus:

$$(t^*)^2 = \left[\frac{b_1}{s(b_1)} \right]^2 = F^*$$

In our illustrative problem, we just calculated that $F^* = 1{,}813$. From earlier work, we have: $b_1 = 2.0$, $s(b_1) = .04697$. Thus:

$$(t^*)^2 = \left[\frac{2.0}{.04697} \right]^2 = 1{,}813$$

Corresponding to the relation between t^* and F^*, we have the following relation between the required percentiles of the t and F distributions in the tests: $[t(1 - \alpha/2; n - 2)]^2 = F(1 - \alpha; 1, n - 2)$. In our tests on β_1, these percentiles were: $[t(.975; 8)]^2 = (2.306)^2 = 5.32 = F(.95; 1, 8)$. Remember that the t test is two-tailed while the F test is one-tailed.

Thus, at a given α level, we can use either the t test or the F test for testing $\beta_1 = 0$ versus $\beta_1 \neq 0$. Whenever one test leads to H_0, so will the other, and correspondingly for H_a. The t test, however, is more flexible since it can be used for one-sided alternatives involving $\beta_1(\leq \geq)0$ versus $\beta_1(> <)0$, while the F test cannot.

General linear test. The analysis of variance test of $\beta_1 = 0$ versus $\beta_1 \neq 0$ is an example of a general test of a linear statistical model. We shall briefly explain this general test approach in terms of our simple regression model. We do so at this time because of the generality of the approach and the wide use we shall make of it, and because of the simplicity of understanding the approach in terms of our present problem.

We begin with the model, which in this context is called the *full* or *unrestricted* model. For our simple regression case, the full model is:

$$(3.63) \qquad\qquad Y_i = \beta_0 + \beta_1 X_i + \varepsilon_i \qquad\qquad \text{Full model}$$

We fit this full model by the method of least squares and obtain the error sum of squares SSE. In this context, we shall call this sum of squares $SSE(F)$ to indicate that it measures the variation of the Y_i around the regression line for the full model.

Next, we consider H_0. In this instance, we have:

$$(3.64) \qquad\qquad H_0\colon \beta_1 = 0$$

The model when H_0 holds is called the *reduced* or *restricted* model. Here, it is:

$$(3.65) \qquad\qquad Y_i = \beta_0 + \varepsilon_i \qquad\qquad \text{Reduced model}$$

We fit this reduced model by the method of least squares and obtain the error sum of squares for this reduced model, denoted by $SSE(R)$. When we fit the particular reduced model (3.65), it can be shown that the least squares estimator of β_0 is $\bar{Y}$. Hence, $\hat{Y}_i \equiv \bar{Y}$ and the error sum of squares for this reduced model is:

$$(3.66) \qquad\qquad SSE(R) = \Sigma(Y_i - \bar{Y})^2 = SSTO$$

The logic now is to compare $SSE(F)$ and $SSE(R)$. It can be shown that $SSE(F)$ never is greater than $SSE(R)$:

$$(3.67) \qquad\qquad SSE(F) \leq SSE(R)$$

The reason is that the more parameters there are in the model, the better one can fit the data and the smaller are the deviations around the fitted regression line. If $SSE(F)$ is not much less than $SSE(R)$, using the full model does not account for much more of the variability of the Y_i than the reduced model, in which case the data suggest H_0 holds. To put this another way, if $SSE(F)$ is close to $SSE(R)$, the variation of the observations around the regression line for the full model is almost as great as the variation around the regression line for the reduced model, in which case the added parameters in the full model really do not help to reduce the variation in the Y_i. Thus, a small difference $SSE(R) - SSE(F)$ suggests that H_0 holds. On the other hand, a large difference suggests that H_a holds because the additional parameters in the model do help to reduce substantially the variation of the observations Y_i around the fitted regression line.

The actual test statistic used is a function of $SSE(R) - SSE(F)$, namely:

$$(3.68) \qquad\qquad F^* = \frac{SSE(R) - SSE(F)}{df_R - df_F} \div \frac{SSE(F)}{df_F}$$

which follows the F distribution if H_0 holds. The degrees of freedom df_R and df_F are those associated with the reduced and full model error sums of squares, respectively. Large values of F^* lead to H_a.

For our application, we have:

$$SSE(R) = SSTO \qquad SSE(F) = SSE$$
$$df_R = n - 1 \qquad df_F = n - 2$$

so that we obtain when substituting into (3.68):

$$F^* = \frac{SSTO - SSE}{(n - 1) - (n - 2)} \div \frac{SSE}{n - 2} = \frac{SSR}{1} \div \frac{SSE}{n - 2} = \frac{MSR}{MSE}$$

which is our old test statistic (3.59).

This general approach can be used for highly complex tests of linear statistical models, as well as for simpler tests. The basic steps again are:

1. Fit the full model and obtain the error sum of squares $SSE(F)$.
2. Fit the reduced model under H_0 and obtain the error sum of squares $SSE(R)$.
3. Use the test statistic (3.68).

3.9 DESCRIPTIVE MEASURES OF ASSOCIATION BETWEEN X AND Y IN REGRESSION MODEL

We have discussed the major uses of regression analysis—estimation of parameters and means and prediction of new observations—without mentioning the "degree of linear association" between X and Y, or similar terms. The reason is that the usefulness of estimates or predictions depends upon the width of the interval and the user's needs for precision, which vary from one application to another. Hence, no single descriptive measure of the "degree of linear association" can capture the essential information as to whether a given regression relation is useful in any particular application.

Nevertheless, there are times when the degree of linear association is of interest in its own right. We shall now briefly discuss two descriptive measures that are frequently used in practice to describe the degree of linear association between X and Y.

Coefficient of determination

We saw earlier that $SSTO$ measures the variation in the observations Y_i, or the uncertainty in predicting Y, when no account of the independent variable X is taken. Thus, $SSTO$ is a measure of the uncertainty in predicting Y when X is not considered. Similarly, SSE measures the variation in the Y_i when a regression model utilizing the independent variable X is employed. A natural measure of the effect of X in reducing the variation in Y, i.e., the uncertainty in predicting Y, is therefore:

$$(3.69) \qquad r^2 = \frac{SSTO - SSE}{SSTO} = \frac{SSR}{SSTO} = 1 - \frac{SSE}{SSTO}$$

The measure r^2 is called the *coefficient of determination*. Since $0 \leq SSE \leq SSTO$, it follows that:

(3.70) $$0 \leq r^2 \leq 1$$

We may interpret r^2 as the proportionate reduction of total variation associated with the use of the independent variable X. Thus, the larger is r^2, the more is the total variation of Y reduced by introducing the independent variable X. The limiting values of r^2 occur as follows:

1. If all observations fall on the fitted regression line, $SSE = 0$ and $r^2 = 1$. This case is shown in Figure 3.6a. Here, the independent variable X accounts for all variation in the observations Y_i.

FIGURE 3.6   Scatter plots when $r^2 = 0$ and $r^2 = 1$

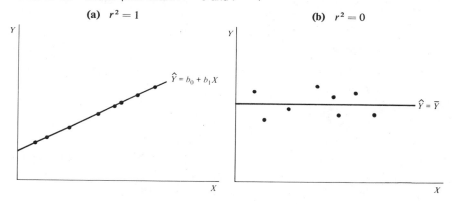

(a) $r^2 = 1$

(b) $r^2 = 0$

2. If the slope of the fitted regression line is $b_1 = 0$ so that $\hat{Y}_i \equiv \bar{Y}$, $SSE = SSTO$ and $r^2 = 0$. This case is shown in Figure 3.6b. Here, there is no linear association between X and Y in the sample data, and the independent variable X is of no help in reducing the variation in the observations Y_i with linear regression.

In practice, r^2 is not likely to be 0 or 1, but rather somewhere in between these limits. The closer it is to 1, the greater is said to be the degree of linear association between X and Y.

Coefficient of correlation

The square root of r^2:

(3.71) $$r = \pm\sqrt{r^2}$$

is called the *coefficient of correlation*. A plus or minus sign is attached to this measure according to whether the slope of the fitted regression line is positive or negative. Thus, the range of r is:

$$(3.72) \qquad\qquad -1 \leq r \leq 1$$

Whereas r^2 indicates the proportional reduction in the variability of Y attained by the use of information about X, the square root, r, does not have such a clear-cut operational interpretation. Nevertheless, there is a tendency to use r instead of r^2 in much applied work.

It is worth noting that since for any r^2 other than 0 or 1, $r^2 < |r|$, r may give the impression of a "closer" relationship between X and Y than does the corresponding r^2. For instance, $r^2 = .10$ indicates that the total variation in Y is reduced by only 10 percent when X is introduced, yet $|r| = .32$ may give an impression of greater linear association between X and Y.

Example

For the Westwood Company example, we obtained $SSTO = 13,660$ and $SSE = 60$. Hence:

$$r^2 = \frac{13,660 - 60}{13,660} = .996$$

Thus, the variation in man-hours is reduced by 99.6 percent when lot size is considered.

The correlation coefficient in this example is:

$$r = +\sqrt{.996} = +.998$$

The plus sign is affixed since b_1 is positive.

Computational formula for r

A direct computational formula for r, which automatically furnishes the proper sign, is:

$$(3.73) \qquad r = \frac{\Sigma(X_i - \bar{X})(Y_i - \bar{Y})}{[\Sigma(X_i - \bar{X})^2 \Sigma(Y_i - \bar{Y})^2]^{1/2}}$$

$$= \frac{\Sigma X_i Y_i - \dfrac{\Sigma X_i \Sigma Y_i}{n}}{\left[\left(\Sigma X_i^2 - \dfrac{(\Sigma X_i)^2}{n}\right)\left(\Sigma Y_i^2 - \dfrac{(\Sigma Y_i)^2}{n}\right)\right]^{1/2}}$$

Comments

1. The following relation between b_1 and r is worth noting:

(3.74)
$$b_1 = \left[\frac{\Sigma(Y_i - \bar{Y})^2}{\Sigma(X_i - \bar{X})^2} \right]^{1/2} r = \left(\frac{s_Y}{s_X} \right) r$$

where $s_Y = [\Sigma(Y_i - \bar{Y})^2/(n - 1)]^{1/2}$ and $s_X = [\Sigma(X_i - \bar{X})^2/(n - 1)]^{1/2}$ are the sample standard deviations for the Y and X observations, respectively. Note that $b_1 = 0$ when $r = 0$, and vice versa. Thus, $r = 0$ implies a horizontal fitted regression line, and vice versa.

2. The value taken by r^2 in a given sample tends to be affected by the spacing of the X observations. This is implied in (3.69). SSE is not affected systematically by the spacing of the X's since for model (3.1), $\sigma^2(Y_i) = \sigma^2$ at all X levels. However, the wider the spacing is of the X's in the sample when $b_1 \neq 0$, the greater will tend to be the spread of the observed Y's around $\bar{Y}$ and hence the greater will be $SSTO$. Consequently, the wider the X's are spaced, the higher will tend to be r^2.

3. The regression sum of squares SSR is often called the "explained variation" in Y. The residual sum of squares SSE is then called the "unexplained variation," and the total sum of squares the "total variation." The coefficient r^2 then is interpreted in terms of the proportion of the total variation in Y which has been "explained" by X. Unfortunately, this terminology frequently is taken literally, hence misunderstood. Remember that in a regression model there is no implication that Y necessarily depends on X in a causal or explanatory sense.

4. A value of r or r^2 relatively close to 1 sometimes is taken as an indication that sufficiently precise inferences on Y can be made from knowledge of X. As mentioned earlier, the usefulness of the regression relation depends upon the width of the confidence or prediction interval and the particular needs for precision, which vary from one application to another. Hence, no single measure is an adequate indicator of the usefulness of the regression relation.

5. Regression models do not contain any parameter to be estimated by r or r^2. These coefficients simply are descriptive measures of the degree of linear association between X and Y in the sample observations which may, or may not, be useful in any one instance. In a later chapter, we discuss correlation models which do contain a parameter for which r is an estimator.

3.10 COMPUTER OUTPUT

Figure 3.7 shows again the computer printout for the Westwood Company case presented in Figure 2.12. We referred to selected items in this printout in Chapter 2. Now, we are in a position to consider the printout as a whole.

The 10 observations on lot size and man-hours are printed at the top. This enables us to verify that the observations were entered into the computer accurately. We have annotated the output in Figure 3.7 in terms of the notation used in this book. The computer package output illustrated in Figure 3.7 does not provide $s(b_0)$, the estimated standard deviation of b_0. However, this estimate can be easily calculated from the data given in the computer output. Note in this connection that the denominator term $\Sigma(X_i - \bar{X})^2$ in (3.21) is equal to $(n - 1)s_X^2$, and that s_X is given in the computer output.

FIGURE 3.7 Segment of computer output for regression run on Westwood Company data (SPSS, Ref. 3.1)

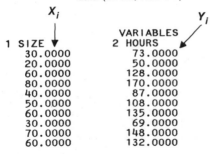

X_i ↓

Y_i ↗↘

```
                        VARIABLES
1 SIZE ▼                2 HOURS
   30.0000                73.0000
   20.0000                50.0000
   60.0000               128.0000
   80.0000               170.0000
   40.0000                87.0000
   50.0000               108.0000
   60.0000               135.0000
   30.0000                69.0000
   70.0000               148.0000
   60.0000               132.0000
```

DEPENDENT VARIABLE..   HOURS

VARIABLE(S) ENTERED ON STEP NUMBER   1..   SIZE

MULTIPLE R   0.99780 ←— r
R SQUARE   0.99561 ←— r^2

STANDARD ERROR   2.73861 ←— $\sqrt{MSE}$
---------------- VARIABLES IN THE EQUATION -----------------
VARIABLE   B   STD ERROR B   F
SIZE   2.000000 ←— b_1 $s(b_1)$ →0.04697   1813.333 ←— F^*
(CONSTANT)   10.00000 ←— b_0

VARIABLE   _   MEAN   STANDARD DEV   CASES
SIZE $\bar{X}$→50.0000 s_X →19.4365   10
HOURS $\bar{Y}$→110.0000 s_Y →38.9587   10 ←— n

ANALYSIS OF VARIANCE   DF   SUM OF SQUARES   MEAN SQUARE
REGRESSION   1. SSR→13600.00000 MSR→13600.00000
RESIDUAL ←— Error   8. SSE→60.00000 MSE→7.50000
```

Computer printouts for regression analysis programs differ substantially in format from one program to another. In addition, differences in the computed results may occur because different program packages do not control roundoff errors equally well. Before using a computer program the first time, it is a good idea to check it on a set of data for which the exact results are known.

# PROBLEMS

**3.1.** A student, working on a summer internship in the economic research office of a large corporation, studied the relation between sales of a product ($Y$, in million dollars) and population ($X$, in million persons) in the firm's 50 marketing districts. Regression model (3.1) was employed. The student first wished to test whether or not a linear association between $Y$ and $X$ existed. Using a time-sharing computer service available to the firm, the student accessed an interactive simple linear
```

regression program and obtained the following information on the regression co-efficients:

Parameter	Estimated Value	95 Percent Confidence Limits	
Intercept	7.43119	−1.18518	16.0476
Slope	.755048	.452886	1.05721

a. The student concluded from these results that there is a linear association between Y and X. Is the conclusion warranted? What is the implied level of significance?

b. Someone questioned the negative lower confidence limit for the intercept, pointing out that dollar sales cannot be negative even if the population in a district is zero. Discuss.

3.2. In a test of the alternatives H_0: $\beta_1 \leq 0$ versus H_a: $\beta_1 > 0$, an analyst concluded H_0. Does this conclusion imply that there is no linear association between X and Y? Explain.

3.3. A member of a student team playing an interactive marketing game received the following computer output when studying the relation between advertising expenditures (X) and sales (Y) for one of the team's products:

$$\text{Estimated regression equation: } \hat{Y} = 350.7 - .18X$$
$$\text{Two-sided } P\text{-value for estimated slope: } .91$$

The student stated: "The message I get here is that the more we spend on advertising this product, the fewer units we sell!" Comment.

3.4. Refer to **Grade point average** Problem 2.15. Some additional results are: $b_0 = -1.700$, $s(b_0) = .7267$, $b_1 = .8399$, $s(b_1) = .144$, $MSE = .1892$.

a. Obtain a 99 percent confidence interval for β_1. Interpret your confidence interval. Does it include zero? Why might the director of admissions be interested in whether the confidence interval includes zero?

b. Test, using the test statistic t^*, whether or not a linear association exists between student's entrance test score (X) and GPA at the end of the freshman year (Y). Use a level of significance of .01. State the alternatives, decision rule, and conclusion.

c. What is the P-value of your test in part (b)? How does it support the conclusion reached in part (b)?

3.5. Refer to **Calculator maintenance** Problem 2.16. Some additional results are: $b_0 = -2.3221$, $s(b_0) = 2.564$, $b_1 = 14.738$, $s(b_1) = .519$, $MSE = 20.086$.

a. Estimate the change in the mean service time when the number of machines serviced increases by one. Use a 90 percent confidence interval. Interpret your confidence interval.

b. Conduct a t test to determine whether or not there is a linear association between X and Y here; control the α risk at .10. State the alternatives, decision rule, and conclusion. What is the P-value of your test?

c. Are your results in parts (a) and (b) consistent? Explain.

d. The manufacturer has suggested that the mean required time should not increase by more than 14 minutes for each additional machine that is serviced on a service call. Conduct a test to decide whether this standard is being satisfied by Tri-City. Control the risk of a Type I error at .05. State the alternatives, decision rule, and conclusion. What is the P-value of the test?

e. Does b_0 give any relevant information here about the "start-up" time on calls—i.e., about the time required before service work is begun on the machines at a customer location?

3.6. Refer to **Airfreight breakage** Problem 2.17.

a. Estimate β_1 with a 95 percent confidence interval. Interpret your interval estimate.

b. Conduct a t test to decide whether or not there is a linear association between number of times a carton is transferred (Y) and number of broken ampules (X). Use a level of significance of .05. State the alternatives, decision rule, and conclusion. What is the P-value of the test?

c. β_0 represents here the mean number of ampules broken when no transfers of the shipment are made—i.e., when $X = 0$. Obtain a 95 percent confidence interval for β_0 and interpret it.

d. A consultant has suggested, based on previous experience, that the mean number of broken ampules should not exceed 9 when no transfers are made. Conduct an appropriate test using $\alpha = .025$. State the alternatives, decision rule, and conclusion. What is the P-value of the test?

e. Obtain the power of your test in part (b) if actually $\beta_1 = 2.0$. Assume $\sigma(b_1) = .50$. Also obtain the power of your test in part (d) if actually $\beta_0 = 11$. Assume $\sigma(b_0) = .75$.

3.7. Refer to **Plastic hardness** Problem 2.18.

a. Estimate the change in the mean hardness when the elapsed time increases by one hour. Use a 99 percent confidence interval. Interpret your interval estimate.

b. The plastic manufacturer has stated that the mean hardness should increase by 2 Brinell units per hour. Conduct a two-sided test to decide whether this standard is being satisfied; use $\alpha = .01$. State the alternatives, decision rule, and conclusion. What is the P-value of the test?

c. Obtain the power of your test in part (b) if the standard actually is being exceeded by .5 Brinell units per hour. Assume $\sigma(b_1) = .16$.

3.8. Refer to Figure 3.7 for the Westwood Company example. A consultant has advised that an increase of one unit in lot size should require an increase of 1.8 in the expected number of man-hours for the given production item.

a. Conduct a test to decide whether or not the increase in the expected number of man-hours in the Westwood Company equals this standard. Use $\alpha = .05$. State the alternatives, decision rule, and conclusion.

b. Obtain the power of your test in part (a) if the consultant's standard actually is being exceeded by .1 hour. Assume $\sigma(b_1) = .05$.

c. Why is $F^* = 1813.333$, given in the printout, not relevant for the test in part (a)?

3.9. Refer to Figure 3.7. A student, noting that $s(b_1)$ is furnished in the printout, asks why $s(\hat{Y}_h)$ is not also given. Discuss.

3.10. For each of the following questions, explain whether a confidence interval for a mean response or a prediction interval for a new observation is appropriate.

a. What will be the humidity level in this greenhouse tomorrow when we set the temperature level at $31°$ C?

   b. How much do families whose disposable income is $23,500 spend, on the average, for meals away from home?

   c. How many kilowatt-hours of electricity will be consumed next month by commercial and industrial users in the Twin Cities service area, given that the index of business activity for the area remains at its present level?

3.11. A person asks if there is a difference between the "mean response at $X = X_h$" and the "mean of m new observations at $X = X_h$." Reply.

3.12. Can $\sigma^2(Y_{h(new)})$ in (3.36) be brought increasingly close to 0 as n becomes large? Is this also the case for $\sigma^2(\hat{Y}_h)$ in (3.28b)? What is the implication of this difference?

3.13. Refer to **Grade point average** Problems 2.15 and 3.4.

   a. Obtain a 95 percent interval estimate of the mean freshman GPA for students whose entrance test score is 4.7. Interpret your confidence interval.

   b. Mary Jones obtained a score of 4.7 on the entrance test. Predict her freshman GPA using a 95 percent prediction interval. Interpret your prediction interval.

   c. Is the prediction interval in part (b) wider than the confidence interval in part (a)? Should it be?

3.14. Refer to **Calculator maintenance** Problems 2.16 and 3.5.

   a. Obtain a 90 percent confidence interval for the mean service time on calls in which six machines are serviced. Interpret your confidence interval.

   b. Obtain a 90 percent prediction interval for the service time on the next call in which six machines are serviced. Is your prediction interval wider than the corresponding confidence interval in part (a)? Should it be?

   c. Suppose that management wishes to estimate the expected service time *per machine* on calls in which six machines are serviced. Obtain an appropriate confidence interval by converting the interval obtained in part (a). Interpret the converted confidence interval.

3.15. Refer to **Airfreight breakage** Problem 2.17.

   a. Because of changes in airline routes, shipments may have to be transferred more frequently than in the past. Estimate the mean breakage for the following numbers of transfers: $X = 2, 4$. Use separate 99 percent confidence intervals. Interpret your results.

   b. The next shipment will entail two transfers. Obtain a 99 percent prediction interval for the number of broken ampules for this shipment. Interpret your prediction interval.

   c. In the next several days, three independent shipments will be made, each entailing two transfers. Obtain a 99 percent prediction interval for the mean number of ampules broken in the three shipments. Convert this interval into a 99 percent prediction interval for the total number of ampules broken in the three shipments.

3.16. Refer to **Plastic hardness** Problem 2.18.

   a. Obtain a 98 percent confidence interval for the mean hardness of molded items with an elapsed time of 60 hours. Interpret your confidence interval.

   b. Obtain a 98 percent prediction interval for the hardness of a newly molded test item with an elapsed time of 60 hours.

   c. Obtain a 98 percent prediction interval for the mean hardness of 10 newly molded test items, each with an elapsed time of 60 hours.

   d. Is the prediction interval in part (c) narrower than the one in part (b)? Should it be?

3.17. An analyst fitted regression model (3.1) and conducted an F test of $\beta_1 = 0$ versus $\beta_1 \neq 0$. The P-value of the test was .033, and the analyst concluded $H_a: \beta_1 \neq 0$. Was the α level used by the analyst greater than or smaller than .033? If the α level had been .01, what would have been the appropriate conclusion?

3.18. For conducting statistical tests concerning the parameter β_1, why is the t test more versatile than the F test?

3.19. When testing whether or not $\beta_1 = 0$, why is the F test a one-sided test even though H_a includes both $\beta_1 < 0$ and $\beta_1 > 0$? [*Hint:* Refer to (3.57).]

3.20. A student asks whether r^2 is a point estimator of any parameter in regression model (3.1). Respond.

3.21. A value of r^2 near 1 is sometimes interpreted to imply that the relation between Y and X is sufficiently close so that suitably precise predictions of Y can be made from knowledge of X. Is this implication a necessary consequence of the definition of r^2?

3.22. Using regression model (3.1) in an engineering safety experiment, a researcher found for the first 10 observations that r^2 was zero. Is it possible that for the complete set of 30 observations r^2 will not be zero? Could r^2 not be zero for the first 10 observations, yet equal zero for all 30 observations? Explain.

3.23. Refer to **Grade point average** Problems 2.15 and 3.4. Some additional calculational results are: $SSE = 3.406$, $SSR = 6.434$.
  a. Set up the ANOVA table.
  b. What is estimated by MSR in your ANOVA table? By MSE? Under what condition do MSR and MSE estimate the same quantity?
  c. Conduct an F test of whether or not $\beta_1 = 0$. Control the α risk at .01. State the alternatives, decision rule, and conclusion.
  d. What is the absolute magnitude of the reduction in the variation of Y when X is introduced into the regression model? What is the relative reduction? What is the name of the latter measure?
  e. Obtain r and attach the appropriate sign.
  f. Which measure, r^2 or r, has the more clear-cut operational interpretation? Explain.

3.24. Refer to **Calculator maintenance** Problems 2.16 and 3.5. Some additional calculational results are: $SSE = 321.396$, $SSR = 16,182.604$.
  a. Set up the basic ANOVA table in the format of Table 3.2. Which elements of your table are additive? Also set up the ANOVA table in the format of Table 3.4a. How do the two tables differ?
  b. Conduct an F test to determine whether or not there is a linear association between time spent and number of machines serviced; use $\alpha = .10$. State the alternatives, decision rule, and conclusion.
  c. By how much, relatively, is the total variation in number of minutes spent on a call reduced when the number of machines serviced is introduced into the analysis? Is this a relatively small or large reduction? What is the name of this measure?
  d. Calculate r and attach the appropriate sign.
  e. Which measure, r or r^2, has the more clear-cut operational interpretation?

3.25. Refer to **Airfreight breakage** Problem 2.17.

    a. Set up the ANOVA table. Which elements are additive?

    b. Conduct an F test to decide whether or not there is a linear association between the number of times a carton is transferred and the number of broken ampules; control the α risk at .05. State the alternatives, decision rule, and conclusion.

    c. Obtain the t^* statistic for the test in part (b) and demonstrate its equivalence to the F^* statistic obtained in part (b).

    d. Calculate r^2 and r. What proportion of the variation in Y is accounted for by introducing X into the regression model?

3.26. Refer to **Plastic hardness** Problem 2.18.

    a. Set up the ANOVA table.

    b. Test by means of an F test whether or not there is a linear association between the hardness of the plastic and the elapsed time. Use $\alpha = .01$. State the alternatives, decision rule, and conclusion.

    c. Plot the deviations $Y_i - \hat{Y}_i$ against X_i on a graph. Plot the deviations $\hat{Y}_i - \bar{Y}$ against X_i on another graph. From your two graphs, does SSE or SSR appear to be the larger component of $SSTO$?

    d. Calculate r^2 and r.

3.27. Refer to **Muscle mass** Problem 2.23.

    a. Conduct a test to decide whether or not there is a negative linear association between amount of muscle mass and age. Control the risk of Type I error at .05. State the alternatives, decision rule, and conclusion. What is the P-value of the test?

    b. The two-sided P-value for b_0 is $0+$. Can it now be concluded that b_0 provides relevant information on the amount of muscle mass at birth for a female child?

    c. Estimate with a 95 percent confidence interval the difference in expected muscle mass for women whose ages differ by one year. Why is it not necessary to know the specific ages to make this estimate?

3.28. Refer to **Muscle mass** Problem 2.23.

    a. Obtain a 95 percent confidence interval for the mean muscle mass for women of age 60. Interpret your confidence interval.

    b. Obtain a 95 percent prediction interval for the muscle mass of a woman whose age is 60. Is the prediction interval relatively precise?

3.29. Refer to **Muscle mass** Problem 2.23.

    a. Plot the deviations $Y_i - \hat{Y}_i$ against X_i on one graph. Plot the deviations $\hat{Y}_i - \bar{Y}$ against X_i on another graph. From your graphs, does SSE or SSR appear to be the larger component of $SSTO$?

    b. Set up the ANOVA table.

    c. Test whether or not $\beta_1 = 0$ using an F test with $\alpha = .10$. State the alternatives, decision rule, and conclusion.

    d. What proportion of the total variation in muscle mass remains "unexplained" when age is introduced into the analysis? Is this proportion relatively small or large?

    e. Obtain r^2 and r.

3.30. Refer to **Robbery rate** Problem 2.24.

    a. Test whether or not there is a linear association between robbery rate and population density using a t test with $\alpha = .01$. State the alternatives, decision rule, and conclusion. What is the P-value of the test?

    b. Test whether or not $\beta_0 = 0$; control the risk of Type I error at .01. State the alternatives, decision rule, and conclusion. Why might there be interest in testing whether or not $\beta_0 = 0$?

    c. Estimate β_1 with a 99 percent confidence interval. Interpret your interval estimate.

3.31. Refer to **Robbery rate** Problem 2.24.

    a. Set up the ANOVA table.

    b. Carry out the test in Problem 3.30a by means of the F test. Show the equivalence of the two test statistics and decision rules. Is the P-value for the F test the same as that for the t test?

    c. By how much is the total variation in robbery rate reduced when population density is introduced into the analysis? Is this a relatively large or small reduction?

    d. Obtain r.

3.32. Refer to **Robbery rate** Problems 2.24 and 3.30. Suppose that the test in Problem 3.30a is to be carried out by means of a general linear test.

    a. State the full and reduced models.

    b. Obtain: (1) $SSE(F)$, (2) $SSE(R)$, (3) df_F, (4) df_R, (5) test statistic F^* for the general linear test, (6) decision rule.

    c. Are the test statistic F^* and the decision rule for the general linear test equivalent to those in Problem 3.30a?

3.33. In empirically developing a cost function from observed data on a complex chemical experiment, an analyst employed regression model (3.1). β_0 was interpreted here as the cost of setting up the experiment. The analyst hypothesized that this cost should be $7.5 thousand and wished to test the hypothesis by means of a general linear test.

    a. Indicate the alternative conclusions for the test.

    b. Specify the full and reduced models.

    c. Without additional information, can you tell what the quantity $df_R - df_F$ in test statistic (3.68) will equal in the analyst's test? Explain.

3.34. Refer to **Grade point average** Problem 2.15.

    a. Would it be more reasonable to consider the X_i as known constants or as random variables here? Explain.

    b. If the X_i were considered to be random variables, would this have any effect on prediction intervals for new applicants? Explain.

3.35. Refer to **Calculator maintenance** Problems 2.16 and 3.5. How would the meaning of the confidence coefficient in Problem 3.5a change if the independent variable were considered a random variable and the conditions in (3.40) were applicable?

EXERCISES

3.36. Show that b_0 as defined in (3.19) is an unbiased estimator of β_0.

3.37. Derive the expression in (3.20b) for the variance of b_0, making use of theorem (3.29). Also explain how variance (3.20b) is a special case of variance (3.28b).

3.38. (Calculus needed.)
  a. Obtain the likelihood function for the sample observations $Y_1, \ldots, Y_n$ given $X_1, \ldots, X_n$, if the conditions in (3.40) apply.
  b. Obtain the maximum likelihood estimators of β_0, β_1, and σ^2. Are the estimators of β_0 and β_1 the same as those in (2.27) when the X_i are fixed?

3.39. Suppose that the normal error regression model (3.1) is applicable except that the error variance is not constant; rather the variance is larger, the larger is X. Does $\beta_1 = 0$ still imply that there is no linear association between X and Y? That there is no association between X and Y? Explain.

3.40. Derive the expression for SSR in (3.50b).

3.41. In a small-scale regression study, five observations on Y were obtained corresponding to $X = 1$, 4, 10, 11, and 14. Assume that $\sigma = .6$, $\beta_0 = 5$, and $\beta_1 = 3$.
  a. What are the expected values of MSR and MSE here?
  b. For purposes of determining whether or not a regression relation exists, would it have been better or worse to have made the five observations at $X = 6$, 7, 8, 9, and 10? Why? Would the same answer apply if the principal purpose were to estimate the mean response for $X = 8$? Discuss.

3.42. The simple linear regression model (3.1) is assumed to be applicable.
  a. When testing H_0: $\beta_1 = 5$ versus H_a: $\beta_1 \neq 5$ by means of a general linear test, what is the reduced model? df_R?
  b. When testing H_0: $\beta_0 = 2$, $\beta_1 = 5$ versus H_a: not both $\beta_0 = 2$ and $\beta_1 = 5$, what is the reduced model? df_R?

PROJECTS

3.43. Refer to the **SMSA** data set and Project 2.38. Using r^2 as the criterion, which independent variable accounts for the largest reduction in the variability in the number of active physicians?

3.44. Refer to the **SMSA** data set and Project 2.39. Obtain a separate interval estimate of β_1 for each region. Use a 90 percent confidence coefficient in each case. Do the regression lines for the different regions appear to have similar slopes?

3.45. Refer to the **SENIC** data set and Project 2.40. Using r^2 as the criterion, which independent variable accounts for the largest reduction in the variability of the average length of stay?

3.46. Refer to the **SENIC** data set and Project 2.41. Obtain a separate interval estimate of β_1 for each region. Use a 95 percent confidence coefficient in each case. Do the regression lines for the different regions appear to have similar slopes?

3.47. Five observations on Y are to be taken when $X = 4$, 8, 12, 16, and 20, respectively. The true regression function is $E(Y) = 20 + 4X$, and the ε_i are independent $N(0, 25)$.
  a. Generate five normal random numbers, with mean 0 and variance 25. Consider these random numbers as the error terms for the five observations at $X = 4$, 8, 12, 16, and 20, and calculate Y_1, Y_2, Y_3, Y_4, and Y_5. Obtain the

least squares estimates b_0 and b_1 when fitting a straight line to the five observations. Also calculate $\hat{Y}_h$ when $X_h = 10$.

b. Repeat part (a) 200 times, generating new random numbers each time.

c. Make a frequency distribution of the 200 estimates b_1. Calculate the mean and standard deviation of the 200 estimates b_1. Are the results consistent with theoretical expectations?

d. For each of the 200 replications, calculate a 95 percent confidence interval for $E(Y_h)$ when $X_h = 10$. What proportion of the 200 confidence intervals include $E(Y_h)$? Is this result consistent with theoretical expectations?

CITED REFERENCE

3.1 Nie, N. H.; C. H. Hull; J. G. Jenkins; K. Steinbrenner; and D. H. Bent. *SPSS: Statistical Package for the Social Sciences*. 2d ed. New York: McGraw-Hill, 1975.

4

Aptness of model and remedial measures

When a regression model, such as the simple linear regression model (3.1), is selected for an application, one can usually not be certain in advance that the model is appropriate for that application. Any one, or several, of the features of the model, such as linearity of the regression function or normality of the error terms, may not be appropriate for the particular data at hand. Hence, it is important to examine the aptness of the model for the data before further analysis based on that model is undertaken. In this chapter, we discuss some simple graphic methods for studying the aptness of a model, as well as some formal statistical tests for doing so. We conclude with a consideration of some techniques whereby the simple regression model (3.1) can be made appropriate when the data do not accord with the conditions of the model.

While the discussion in this chapter is in terms of the aptness of the simple regression model (3.1), the basic principles apply to all statistical models discussed in this book. In later chapters, additional material concerning the aptness of the model and remedial measures will be presented.

4.1 RESIDUALS

A residual e_i, as defined in (2.16), is the difference between the observed value and the fitted value:

$$(4.1) \qquad e_i = Y_i - \hat{Y}_i$$

As such, it may be regarded as the observed error, in distinction to the unknown true error ε_i in the regression model:

$$(4.2) \qquad\qquad \varepsilon_i = Y_i - E(Y_i)$$

For regression model (3.1), the ε_i are assumed to be independent normal random variables, with mean 0 and constant variance σ^2. If the model is appropriate for the data at hand, the observed residuals e_i should then reflect the properties assumed for the ε_i. This is the basic idea underlying *residual analysis*, a highly useful means of examining the aptness of a model.

Properties of residuals

The mean of the n residuals e_i is by (2.17):

$$(4.3) \qquad\qquad \bar{e} = \frac{\Sigma e_i}{n} = 0$$

where $\bar{e}$ denotes the mean of the residuals. Thus, since $\bar{e}$ is always 0, it provides no information as to whether the true errors ε_i have expected value $E(\varepsilon_i) = 0$.

The variance of the n residuals e_i is defined as follows for model (3.1):

$$(4.4) \qquad\qquad \frac{\Sigma(e_i - \bar{e})^2}{n-2} = \frac{\Sigma e_i^2}{n-2} = \frac{SSE}{n-2} = MSE$$

If the model is appropriate, MSE is, as noted earlier, an unbiased estimator of the variance of the error terms σ^2.

Standardized residuals

For analytical convenience, standardized residuals are used at times in residual analysis. Since the standard deviation of the error terms ε_i is σ, which is estimated by $\sqrt{MSE}$, we shall define here the standardized residual as follows:

$$(4.5) \qquad\qquad \frac{e_i - \bar{e}}{\sqrt{MSE}} = \frac{e_i}{\sqrt{MSE}}$$

We shall explain residual analysis mainly in terms of the residuals e_i, but occasionally will employ the standardized residuals.

Nonindependence of residuals

The residuals e_i are not independent random variables because they involve the fitted values $\hat{Y}_i$ which are based on the sample estimates b_0 and b_1. Thus, the residuals are associated with only $n-2$ degrees of freedom. As a result, we know from (2.17) that the sum of the e_i must be 0 and from (2.19) that the products $X_i e_i$ must sum to 0. The same lack of independence holds for the standardized residuals.

When the sample size is large in comparison to the number of parameters in the regression model, the dependency effect among the e_i is relatively unimportant and can be ignored for most purposes.

Departures from model to be studied by residuals

We shall consider the use of residuals for examining six important types of departures from model (3.1), the simple linear regression model with normal errors:

1. The regression function is not linear.
2. The error terms do not have constant variance.
3. The error terms are not independent.
4. The model fits all but one or a few outlier observations.
5. The error terms are not normally distributed.
6. One or several important independent variables have been omitted from the model.

4.2 GRAPHIC ANALYSIS OF RESIDUALS

We take up now some informal ways in which graphs of residuals can be analyzed to provide information on whether any of the six types of departures from the simple linear regression model (3.1) just mentioned are present.

Nonlinearity of regression function

Whether or not a linear regression function is appropriate for the data being analyzed can often be studied from a scatter plot of the data, with the fitted regression function plotted on it. Figure 4.1a contains the data and the fitted regression line for a study of the relation between amount of transit information and bus ridership in eight comparable test cities, where X is the number of bus transit maps distributed free to residents of the city at the beginning of the test period and Y is the increase during the test period in average daily bus ridership during nonpeak hours. The original data and fitted values are given in Table 4.1. The graph suggests strongly that a linear regression function is not appropriate.

Figure 4.1b presents for this same example the residuals e, shown in Table 4.1, plotted against the independent variable X. The lack of fit of a linear regression function is also strongly suggested by the residual plot against X in Figure 4.1b, since the residuals depart from 0 in a systematic fashion. Note that they are negative for smaller X values, positive for medium size X values, and negative again for large X values.

In this case, both Figures 4.1a and 4.1b are effective means of examining the appropriateness of the linearity of the regression function. Figure 4.1b, the residual plot, in general has some advantages over Figure 4.1a, the scatter plot. First, the residual plot can easily be used for examining other facets of the aptness of the model. Second, there are occasions when the scaling of the scatter plot places

FIGURE 4.1 Scatter plot and residual plot for transit example illustrating nonlinear regression function

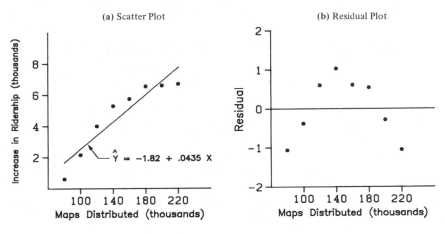

(a) Scatter Plot   (b) Residual Plot

$\hat{Y} = -1.82 + .0435\ X$

TABLE 4.1 Number of maps distributed and increase in ridership—transit example

City i	Increase in Ridership (thousands) Y_i	Maps Distributed (thousands) X_i	Fitted Value $\hat{Y}_i$	Residual $Y_i - \hat{Y}_i = e_i$
1	.60	80	1.66	− 1.06
2	6.70	220	7.75	− 1.05
3	5.30	140	4.27	+ 1.03
4	4.00	120	3.40	+.60
5	6.55	180	6.01	+.54
6	2.15	100	2.53	−.38
7	6.60	200	6.88	−.28
8	5.75	160	5.14	+.61

$$\hat{Y} = -1.82 + .0435X$$

the Y_i observations close to the fitted values $\hat{Y}_i$, for instance, when there is a steep slope. It then becomes more difficult to study the appropriateness of a linear regression function from the scatter plot. A residual plot, on the other hand, can clearly show any systematic pattern in the deviations around the regression line under these conditions.

Figure 4.2a shows a prototype situation of the residual plot against X if the linear model is appropriate. The residuals should tend to fall within a horizontal band centered around 0, displaying no systematic tendencies to be positive and negative.

Figure 4.2b shows a prototype situation of a departure from the linear regression model indicating the need for a curvilinear regression function. Here the

FIGURE 4.2 Prototype residual plots

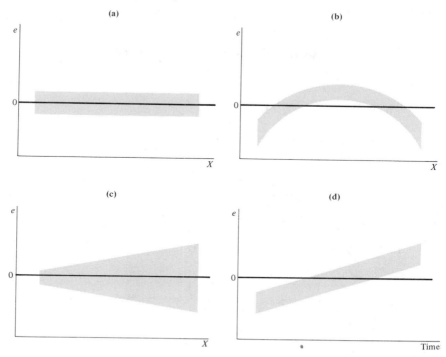

residuals tend to vary in a systematic fashion between being positive and negative. A different type of departure from linearity would, of course, lead to a different picture than the prototype pattern in Figure 4.2b.

Nonconstancy of error variance

A plot of the residuals against the independent variable is not only helpful to study whether a linear regression function is appropriate but also to examine whether the variance of the error terms is constant. For instance, Figure 4.3a shows a residual plot against the independent variable X for an application involving the regression of the diastolic blood pressure of female children (Y) against their age (X). The plot was generated by the BMDP package (Ref. 4.1). The numerical values shown in the graph indicate the number of residuals falling on or near a point. We have added the flared lines to highlight the tendency that the larger X is, the more spread out are the residuals. This suggests that the error variance is larger for older children than for younger ones.

Figure 4.2c shows a prototype picture of a residual plot when the error variance increases with X. In many business, social science, and biological science applications, departures from constancy of the error variance tend to be of the trapezoidal type shown in Figure 4.2c. One can also encounter error variances

FIGURE 4.3 Residual plots for blood pressure example illustrating nonconstant error variance (BMDP2R, Ref. 4.1)

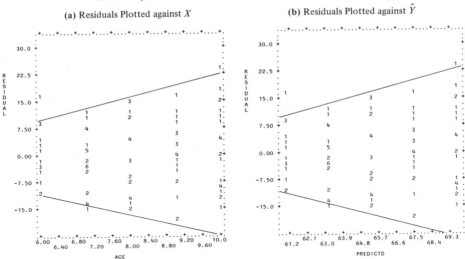

(a) Residuals Plotted against X

(b) Residuals Plotted against $\hat{Y}$

decreasing with increasing levels of the independent variable or varying in some other fashion.

A residual plot against the fitted values $\hat{Y}$ is also an effective means of studying the constancy of the error variance, particularly when the regression function is not linear or when a multiple regression model is employed. Figure 4.3b shows, for the same data as in Figure 4.3a, a plot of the residuals e_i against the fitted values $\hat{Y}_i$ generated by the BMDP package. Note that the horizontal axis is labeled PREDICTD which stands for "predicted," an alternative term often used for "fitted" value. Again we see the prototype pattern of Figure 4.2c, suggesting here that the error variance increases with $\hat{Y}$. Since the relation between blood pressure and age is a positive one, Figure 4.3b also indicates that the error variance increases with X.

Presence of outliers

Outliers are extreme observations. In a residual plot, they are points that lie far beyond the scatter of the remaining residuals, perhaps four or more standard deviations from zero. The residual plot in Figure 4.4 presents standardized residuals and contains one outlier, which is circled. Note that this residual represents an observation almost six standard deviations from the fitted value.

Outliers can create great difficulty. When we encounter one, our first suspicion is that the observation resulted from a mistake or other extraneous effect, and hence should be discarded. A major reason for discarding it is that under the least squares method, a fitted line may be pulled disproportionately toward an outlying observation because the sum of the *squared* deviations is minimized. This could cause a misleading fit if indeed the outlier observation resulted from a

FIGURE 4.4 Residual plot with outlier

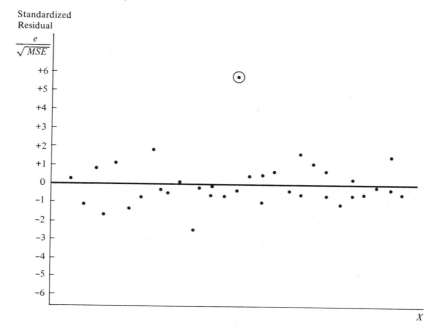

mistake or other extraneous cause. On the other hand, outliers may convey significant information, as when an outlier occurs because of an interaction with another independent variable omitted from the model. A safe rule frequently suggested is to discard an outlier only if there is direct evidence that it represents an error in recording, a miscalculation, a malfunctioning of equipment, or a similar type of circumstance.

Note

When a linear regression model is fitted to a data set with a small number of observations and an outlier is present, the fitted regression may be so distorted by the outlier that the residual plot suggests a lack of fit of the linear regression model in addition to flagging the outlier. Figure 4.5 illustrates this situation. The scatter plot in Figure 4.5a presents a case where all observations except the outlier fall around a straight-line statistical relationship. When a linear regression function is fitted to these data, the outlier causes such a shift in the fitted regression line as to lead to a systematic pattern of deviations from the fitted line for the other observations, as evidenced by the residual plot in Figure 4.5b.

Nonindependence of error terms

Whenever data are obtained in a time sequence, it is a good idea to plot the residuals against time, even though time has not been explicitly incorporated as a variable into the model. The purpose is to see if there is any correlation between the error terms over time. In an experiment to study the relation between the

FIGURE 4.5 Distorting effect on residuals caused by an outlier when remaining data follow linear regression

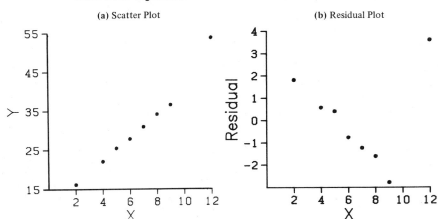

(a) Scatter Plot   (b) Residual Plot

diameter of a weld (X) and the shear strength of the weld (Y), the residual plot against X as shown in Figure 4.6a appears to indicate no departures from the simple regression model, either with respect to linearity or constancy of error variance. When the residuals are plotted in the time order in which the welds were made in Figure 4.6b, however, an evident correlation between the error terms stands out. Negative residuals are associated mainly with the early trials, and positive residuals with the later trials. Apparently, some effect connected with time was present, such as learning by the welder or a gradual change in the welding equipment, so that the shear strength tended to be greater in the later welds on account of this effect.

A prototype of a time-related effect is shown in Figure 4.2d, which portrays a linear time-related effect. It is sometimes useful to view the problem of non-independence of the error terms as one in which an important variable (in this case, time) has been omitted from the model. We shall discuss this type of problem shortly.

When the error terms are independent, we would expect the residuals to fluctuate in a more or less random pattern around the base line 0, such as the scattering shown in Figure 4.7. Lack of randomness can take the form of too much alternation of points around the zero line, or too little alternation. In practice, there is little concern with the former case except in situations where the error term is subject to periodic or offsetting effects in observations at successive levels of X. Too little alternation, in contrast, frequently is a matter of concern, as in the welding example in Figure 4.6b.

Note

When the residuals are plotted against X, as in Figure 4.1b, the scatter may not appear to be random. For this plot, however, the basic problem is probably not lack of independence of the error terms but rather a poorly fitting regression function. This, indeed, is the situation portrayed in Figure 4.1a.

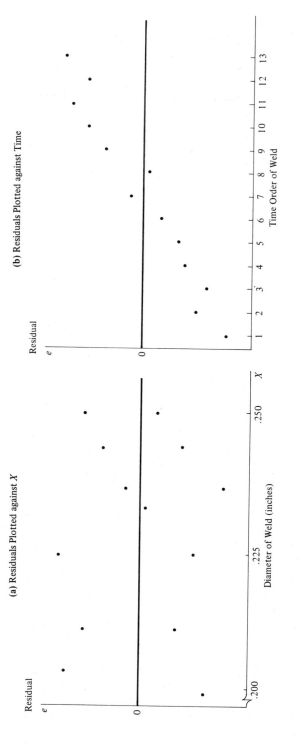

FIGURE 4.6 Residual plots for welding example illustrating nonindependence of error terms

FIGURE 4.7 Residual plot against time suggesting independence of error terms

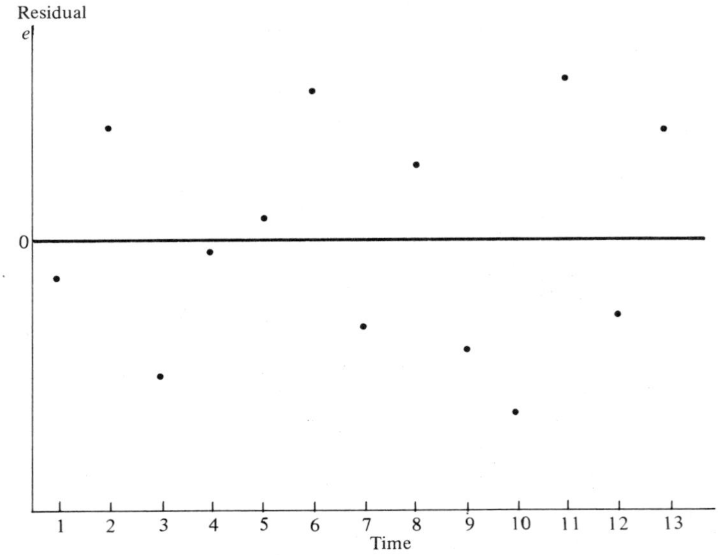

Nonnormality of error terms

As we noted earlier, small departures from normality do not create any serious problems. Major departures, on the other hand, should be of concern. The normality of the error terms can be studied informally by examining the residuals in a variety of graphic ways. One can construct a histogram of the residuals and see if gross departures from normality are shown by it. Another possibility is to determine whether, say, about 68 percent of the standardized residuals $e_i/\sqrt{MSE}$ fall between -1 and $+1$, or about 90 percent between -1.64 and $+1.64$. (If the sample size is small, the corresponding t values would be used.)

Still another possibility is to prepare a normal probability plot of the residuals. Here the residuals are plotted against their expected values when the distribution is normal. A plot which is nearly linear suggests agreement with normality whereas a plot which departs substantially from linearity suggests that the error distribution is not normal.

Table 4.2, column 1, contains the residuals, in ascending order, for a regression study of per capita library usage (Y) and size of city (X) in 10 cities. To find the expected values of the ordered residuals under normality, we utilize the facts that (1) the expected value of the error terms for regression model (3.1) is zero and (2) that the standard deviation of the error terms is estimated by $\sqrt{MSE}$. Statistical theory states that for a normal random variable with mean 0 and estimated standard deviation $\sqrt{MSE}$, the expected value of the ith smallest observation in a random sample of n is given approximately by the following expression:

(4.6)
$$\sqrt{MSE}\left[z\left(\frac{i-.375}{n+.25}\right)\right]$$

where $z(A)$ as usual denotes the $(A)100$ percentile of the standard normal distribution.

TABLE 4.2 Residuals and expected values under normality for library usage example

Ascending Order i	(1) Ordered Residual e_i	(2) Expected Value under Normality
1	−1.33	−1.08
2	−.52	−.70
3	−.27	−.46
4	−.19	−.26
5	−.09	−.08
6	.09	.08
7	.18	.26
8	.44	.46
9	.66	.70
10	1.03	1.08

Squaring the residuals in Table 4.2, summing, and dividing by $n - 2$ we obtain $MSE = .4859$, so $\sqrt{MSE} = .697$. For the smallest residual, we have $i = 1$. Hence, $(i - .375)/(n + .25) = (1 - .375)/(10 + .25) = .061$, and the expected value of the smallest residual under normality is:

$$.697[z(.061)] = .697(-1.55) = -1.08$$

Similarly, the expected value of the second smallest residual under normality is obtained by finding, for $i = 2$, $(i - .375)/(n + .25) = (2 - .375)/(10 + .25) = .159$ so that:

$$.697[z(.159)] = .697(-1.00) = -.70$$

Because of the symmetry of a normal probability distribution, the expected values of the largest and second largest residuals are 1.08 and .70, respectively.

Table 4.2, column 2, contains all 10 expected values under the assumption of normality. Figure 4.8 presents a plot of the residuals against their expected values under normality. This plot is called a *normal probability plot*. Note that the points in Figure 4.8 fall reasonably close to a straight line, suggesting that the error terms are approximately normally distributed.

Many computer packages will prepare normal probability plots at the option of the user. Some of these plots utilize standardized residuals, but this does not affect the basic nature of the plot.

FIGURE 4.8 Normal probability plot of residuals—library usage example

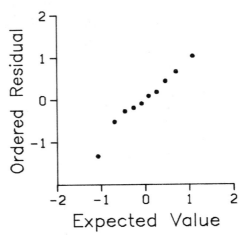

One method of assessing the linearity of the normal probability plot is to calculate the coefficient of correlation (3.73) relating the residuals e_i to their expected values under normality. A high value of the correlation coefficient, say, .90 or more, is indicative of normality. In our library usage example in Table 4.2, the coefficient of correlation is .981, supporting the conclusion of approximate normality for the error terms.

The analysis for model departures with respect to normality is, in many respects, more difficult than that for other types of departures. In the first place, random variation can be particularly mischievous when one studies the nature of a probability distribution unless the sample size is quite large. Even worse, other types of departures can and do affect the distribution of the residuals. For instance, residuals may appear to be not normally distributed because an inappropriate regression function is used or because the error variance is not constant. Hence, it is usually a good strategy to investigate these other types of departures first, before concerning oneself with the normality of the error terms.

Omission of important independent variables

Residuals should be plotted against variables omitted from the model that might have important effects on the response, data being available. The time variable cited earlier in the welding application is an example. The purpose of this additional analysis is to determine whether there are any other key independent variables that could provide important additional descriptive and predictive power to the model.

As another example, in a study to predict output by piece rate workers in an assembling operation, the relation between output (Y) and age (X) of worker was studied for a sample of employees. The plot of the residuals against X is shown in Figure 4.9a, and indicates no ground for suspecting the appropriateness of the linearity of the regression function or the constancy of the error variance.

Machines produced by two companies (A and B) are used in the assembling operation. Residual plots by type of machine were undertaken and are shown in

FIGURE 4.9 Residual plots for productivity example

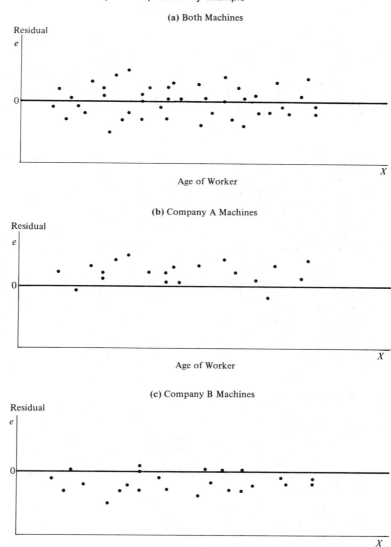

(a) Both Machines

(b) Company A Machines

(c) Company B Machines

Figures 4.9b and c. Note that the residuals for machines made by Company A tend to be positive, while those for machines made by Company B tend to be negative. Thus, type of machine appears to have a definite effect on productivity, and output predictions may turn out to be far superior when this independent variable is added to the model. While this example dealt with a classification variable (type of machine), the residual analysis for an additional quantitative variable is completely analogous. One would simply plot the residuals against the additional independent variable and see whether or not the residuals tend to vary systematically with the level of the additional independent variable.

Note

We do not say that the original model is "wrong" when it can be improved materially by adding one or more independent variables. Only a few of the factors operating on any dependent variable Y in real-world situations can be included explicitly in a regression model. The chief purpose of residual analysis in identifying other important independent variables is therefore to test the adequacy of the model and see whether it could be improved materially by adding one or a few independent variables.

Comments

1. We discussed the model departures one at a time. In actuality, several types of departures may occur together. For instance, a linear regression function may be a poor fit and the variance of the error terms may not be constant. In these cases, the prototype patterns of Figure 4.2 can still be useful, but they would need to be combined into composite patterns.

2. While graphic analysis of residuals is only an informal method of analysis, in many cases it suffices for examining the aptness of a model.

3. The basic approach to residual analysis applies not only to simple linear regression but also to more complex regression and other types of statistical models.

4. Most of the routine work in residual analysis can be handled on computers. Almost all regression programs supply the fitted values and corresponding residuals, and routines are generally available whereby the various types of residual plots can be obtained optionally on printout.

4.3 TESTS INVOLVING RESIDUALS

Graphic analysis of residuals is inherently subjective. Nevertheless, subjective analysis of a variety of interrelated residual plots will frequently reveal difficulties in the model more clearly than particular tests. There are occasions, however, when one wishes to put specific questions to a test. We now review some of the relevant tests briefly, and take up one new type of test.

Most statistical tests require independent observations. As we have seen, however, the residuals are dependent. Fortunately, the dependency becomes quite small for large samples, so that one can usually then ignore it.

Tests for randomness

The runs test is frequently used to test for lack of randomness in the residuals arranged in time order. Another test, specifically designed for lack of randomness in least squares residuals, is the Durbin-Watson test. This test will be discussed in Chapter 13.

Tests for constancy of variance

When a residual plot gives the impression that the variance may be increasing or decreasing in a systematic manner related to X or $E(Y)$, a simple test is to fit separate regression functions to each half of the observations arranged by level of X, calculate error mean squares for each, and test for equality of the error variances by an F test. Another simple test is by means of rank correlation between the absolute value of the residual and the value of the independent variable.

Tests for outliers

A simple test for an outlier observation involves fitting a new regression line to the other $n - 1$ observations. The suspect observation, which was not used in fitting the new line, can now be regarded as a new observation. One can calculate the probability that in n observations, a deviation from the fitted line as great as that of the outlier will be obtained by chance. If this probability is sufficiently small, the outlier can be rejected as not having come from the same population as the other $n - 1$ observations. Otherwise, the outlier is retained.

Many other tests to aid in evaluating outliers have been developed. These are discussed in specialized references such as Reference 4.2 and in statistical journals.

Tests for normality

Goodness of fit tests can be used for examining the normality of the error terms. For instance, the chi-square test or the Kolmogorov-Smirnov test can be employed for testing the normality of the error terms by analyzing the residuals.

Note

The runs test, rank correlation, and goodness of fit tests, mentioned above, are commonly used statistical procedures which are discussed in many basic statistics texts.

4.4 *F* TEST FOR LACK OF FIT

We now take up a formal test for determining whether or not a specified regression function adequately fits the data. This lack of fit test assumes that the observations Y for given X are (1) independent and (2) normally distributed, and (3) the distributions of Y have the same variance σ^2. We illustrate this test for ascertaining whether or not a linear regression function is a good fit for the data.

Replications

The lack of fit test requires repeat observations at one or more X levels. In nonexperimental data, these may occur fortuitously, as when in a productivity study relating workers' output and age, several workers of the same age happen to be included in the study. In an experiment, one can assure by design that there are repeat observations. For instance, in an experiment on the effect of size of salesperson bonus on sales, three salespersons can be offered a particular size of bonus, for each of six bonus sizes, and their sales then observed.

Repeated trials for the same level of the independent variable, of the type described, are called *replications*. The resulting observations are called *replicates*.

Example

In an experiment involving 12 similar but scattered suburban branch offices of a commercial bank, holders of checking accounts at the offices were offered gifts for setting up savings accounts at these same offices. The initial deposit in the new savings account had to be for a specified minimum amount to qualify for the gift. The value of the gift was directly proportional to the specified minimum deposit. Various levels of minimum deposit and related gift values were used in the experiment in order to ascertain the relation between the specified minimum deposit and gift value on the one hand and number of accounts opened at the office on the other. Altogether, six levels of minimum deposit and proportional gift value were used, with two of the branch offices assigned at random to each level. One branch office had a fire during the period and was dropped from the study. Table 4.3 contains the results, where X is the amount of minimum deposit and Y is the number of new savings accounts that were opened and qualified for the gift during the test period.

TABLE 4.3 Data for bank example

Observation i	Size of Minimum Deposit (dollars) X_i	Number of New Accounts Y_i	Observation i	Size of Minimum Deposit (dollars) X_i	Number of New Accounts Y_i
1	125	160	7	75	42
2	100	112	8	175	124
3	200	124	9	125	150
4	75	28	10	200	104
5	150	152	11	100	136
6	175	156			

A linear regression function was fitted in the usual fashion; it is (calculations not shown):

$$\hat{Y} = 50.72251 + .48670X$$

The analysis of variance table also was obtained and is shown in Table 4.4. A scatter plot, together with the fitted regression line, is shown in Figure 4.10. The indications are strong that a linear regression function is inappropriate. To test this formally, we need to perform a decomposition of the error sum of squares *SSE* in Table 4.4 into two components called the pure error and lack of fit components.

TABLE 4.4 ANOVA table for bank example

Source of Variation	SS	df	MS
Regression	$SSR = 5,141.3$	1	$MSR = 5,141.3$
Error	$SSE = 14,741.6$	9	$MSE = 1,638.0$
Total	$SSTO = 19,882.9$	10	

FIGURE 4.10 Scatter plot and fitted regression line—bank example

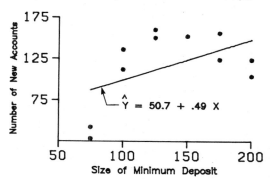

Decomposition of *SSE*

Pure error component.   The basic idea for the first component of *SSE* rests on the fact that there are replications at some levels of *X*. Let us denote the different X levels in the study, whether or not replicated observations are present, as $X_1, \ldots, X_c$. For our example, $c = 6$ since there are six minimum deposit size levels in the study, for five of which there are two observations and for one there is a single observation. We shall let $X_1 = 75$ (the smallest minimum deposit level), $X_2 = 100, \ldots, X_6 = 200$. Further, we shall denote the number of observations for the jth level of X as n_j; for our example, $n_1 = n_2 = n_3 = n_5 = n_6 = 2$ and $n_4 = 1$. Thus, the total number of observations n is given by:

$$(4.7) \qquad n = \sum_{j=1}^{c} n_j$$

If we make no assumption about the nature of the regression function but assume all other elements of model (3.1), we can still estimate the error variance σ^2 because of the repeated observations. Table 4.5 presents the same data as Table 4.3, but in a different arrangement. Table 4.5 also shows the mean of the Y observations for each minimum deposit size. We shall denote the mean of the Y observations when $X = X_j$ by $\bar{Y}_j$. Thus $\bar{Y}_1 = 35$ for the two branches with minimum deposits of $X_1 = \$75$, and so on.

Since the two Y observations for $X_1 = \$75$ come from the same probability distribution, we can estimate the variance of this distribution by calculating the usual sample variance, using the deviations around $\bar{Y}_1 = 35$:

$$\frac{(28 - 35)^2 + (42 - 35)^2}{2 - 1} = 98$$

Likewise, the two observations for $X_2 = \$100$ come from the same probability distribution, so that we can estimate the variance of this distribution by calculating the sample variance:

$$\frac{(112 - 124)^2 + (136 - 124)^2}{2 - 1} = 288$$

Similarly, we can estimate the variance of each of the other distributions except for the one at $X_4 = 150$ where there is only a single observation. Since model (3.1) assumes that all probability distributions of Y have the same variance σ^2, we can combine the results for each of the X levels. The optimum way of combining is to add the numerators:

$$(28 - 35)^2 + (42 - 35)^2 + (112 - 124)^2 + (136 - 124)^2$$
$$+ (160 - 155)^2 + (150 - 155)^2 + (156 - 140)^2 + (124 - 140)^2$$
$$+ (124 - 114)^2 + (104 - 114)^2 = 1{,}148$$

then add the denominators:

$$1 + 1 + 1 + 1 + 1 = 5$$

and finally take the ratio:

$$\frac{1{,}148}{5} = 229.6$$

TABLE 4.5 Data for bank example, arranged by observation number and minimum deposit

Observation	Size of Minimum Deposit (dollars)					
	$X_1 = 75$	$X_2 = 100$	$X_3 = 125$	$X_4 = 150$	$X_5 = 175$	$X_6 = 200$
$i = 1$	28	112	160	152	156	124
$i = 2$	42	136	150		124	104
Mean $\bar{Y}_j$	35	124	155	152	140	114

To generalize, let us denote the ith observation for the jth level of X by Y_{ij}, where $i = 1, \ldots, n_j$; $j = 1, \ldots, c$. For our example (Table 4.5), $Y_{11} = 28$, $Y_{21} = 42$, $Y_{12} = 112$, and so on. First, we calculate the sum of squares of the deviations from the mean at any given level of X. For $X = X_j$, this sum of squares is:

$$(4.8) \qquad \sum_{i=1}^{n_j} (Y_{ij} - \bar{Y}_j)^2$$

We then add these sums of squares over all levels of X and denote this sum of the sums of squares by *SSPE*:

$$(4.9) \qquad SSPE = \sum_{j=1}^{c} \sum_{i=1}^{n_j} (Y_{ij} - \bar{Y}_j)^2$$

Here *SSPE* stands for *pure error sum of squares*. Note that when there is only a single observation at X_j, we have $Y_{1j} = \bar{Y}_j$ so that $Y_{1j} - \bar{Y}_j = 0$. Hence, such X_j levels do not contribute to the pure error sum of squares, as was illustrated by our example.

The degrees of freedom associated with *SSPE* are $n - c$. This is easy to see since there are as usual $n_1 - 1$ degrees of freedom associated with the sum of squares for X_1, $n_2 - 1$ degrees of freedom with the sum of squares for X_2, and so on. The sum of the degrees of freedom is:

$$(4.10) \qquad \sum_{j=1}^{c} (n_j - 1) = \Sigma n_j - c = n - c$$

Again, we see that X_j levels for which $n_j = 1$ do not contribute to the degrees of freedom since $n_j - 1 = 0$ then.

The *pure error mean square MSPE* is given by:

$$(4.11) \qquad MSPE = \frac{SSPE}{n - c}$$

The reason for the term "pure error" is that *MSPE* is an unbiased estimator of the error variance σ^2 no matter what is the nature of the regression function. *MSPE* measures the variability of the distributions of Y without relying on any assumptions about the nature of the regression relation; hence, it is a "pure" measure of the error variance.

Lack of fit component. The second component of *SSE* is:

(4.12) $SSLF = SSE - SSPE$

where *SSLF* denotes *lack of fit sum of squares*. It can be shown that:

(4.12a) $$SSLF = \sum_{j=1}^{c} n_j(\bar{Y}_j - \hat{Y}_j)^2$$

where $\hat{Y}_j$ denotes the fitted value when $X = X_j$. Thus, *SSLF* is a weighted sum of squares (the weights are the sample sizes n_j) of the deviations:

(4.13) $\bar{Y}_j - \hat{Y}_j$

Note that these deviations represent the difference between the mean $\bar{Y}_j$ and the fitted value $\hat{Y}_j$ based on the regression model. The closer the $\bar{Y}_j$ are to the $\hat{Y}_j$, the greater is the evidence that the fitted regression function is a good fit and therefore appropriate. The further the $\bar{Y}_j$ deviate from the $\hat{Y}_j$, the more the indication that the fitted regression function is inappropriate.

Figure 4.11 illustrates for the observation $X_3 = 125$, $Y_{13} = 160$, the partitioning of the error deviation $Y_{13} - \hat{Y}_3 = 48$ into the pure error deviation $Y_{13} - \bar{Y}_3 = 5$ and the lack of fit deviation $\bar{Y}_3 - \hat{Y}_3 = 43$ for testing whether or not a linear regression function is a good fit.

FIGURE 4.11   Illustration of decomposition of $Y_{ij} - \hat{Y}_j$

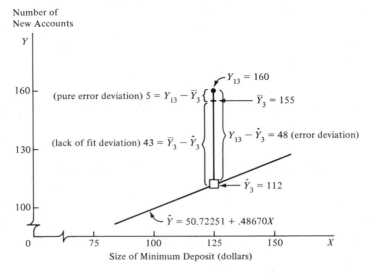

Number of
New Accounts

There are $c - 2$ degrees of freedom associated with *SSLF* when testing for lack of fit of a linear regression function. The reason is that there are c levels of X and two degrees of freedom are lost because two parameters (β_0 and β_1) are estimated in obtaining the fitted values $\hat{Y}_j$. Thus, the lack of fit mean square *MSLF* is:

$$(4.14) \qquad MSLF = \frac{SSLF}{c - 2}$$

For our example, using (4.12) and the earlier results ($SSE = 14{,}741.6$ from Table 4.4, $SSPE = 1{,}148.0$ from p. 126), we obtain:

$$SSLF = 14{,}741.6 - 1{,}148.0 = 13{,}593.6$$

and:

$$MSLF = \frac{13{,}593.6}{6 - 2} = 3{,}398.4$$

Table 4.6a contains the general ANOVA table including the decomposition of *SSE* just explained and the mean squares of interest, and Table 4.6b contains the ANOVA decomposition for our example.

TABLE 4.6 ANOVA table for testing lack of fit of simple linear regression function

(a) General

Source of Variation	SS	df	MS
Regression	SSR	1	MSR
Error	SSE	$n - 2$	MSE
Lack of fit	SSLF	$c - 2$	MSLF
Pure error	SSPE	$n - c$	MSPE
Total	SSTO	$n - 1$	

(b) Bank Example

Source of Variation	SS	df	MS
Regression	$SSR = 5{,}141.3$	1	$MSR = 5{,}141.3$
Error	$SSE = 14{,}741.6$	9	$MSE = 1{,}638.0$
Lack of fit	$SSLF = 13{,}593.6$	4	$MSLF = 3{,}398.4$
Pure error	$SSPE = 1{,}148.0$	5	$MSPE = 229.6$
Total	$SSTO = 19{,}882.9$	10	

F test

Test statistic. For testing lack of fit of the regression function, the appropriate test statistic is:

$$(4.15) \qquad F^* = \frac{MSLF}{MSPE}$$

We noted that $MSPE$ has expectation σ^2 no matter what is the nature of the regression function. It can be shown that for testing lack of fit of a simple linear regression function:

$$(4.16) \qquad E(MSLF) = \sigma^2 + \frac{\Sigma n_j[E(Y_j) - (\beta_0 + \beta_1 X_j)]^2}{c - 2}$$

where $E(Y_j)$ denotes the true mean of the distribution of Y when $X = X_j$ and $\beta_0 + \beta_1 X_j$ is the mean response indicated by the linear regression model. If the regression function is linear, the second term in (4.16) is 0 so that $E(MSLF) = \sigma^2$ then. On the other hand, if the regression function is not linear, $E(Y_j) \neq \beta_0 + \beta_1 X_j$ so that $E(MSLF)$ will be greater than σ^2. Hence, a value of F^* near 1 accords with a linear regression function; large values of F^* indicate that the regression function is not linear.

Decision rule. Since $SSLF$ and $SSPE$ are additive, as are the degrees of freedom, we know from Cochran's theorem that F^* follows the $F(c - 2; n - c)$ distribution if the regression function is linear and all other conditions of model (3.1) hold. To decide between:

$$(4.17) \qquad \begin{aligned} H_0 &: E(Y) = \beta_0 + \beta_1 X \\ H_a &: E(Y) \neq \beta_0 + \beta_1 X \end{aligned}$$

we use the test statistic (4.15). The decision rule to control the risk of a Type I error at α is:

$$(4.18) \qquad \begin{aligned} &\text{If } F^* \leq F(1 - \alpha; c - 2, n - c), \text{ conclude } H_0 \\ &\text{If } F^* > F(1 - \alpha; c - 2, n - c), \text{ conclude } H_a \end{aligned}$$

Example. For our example, the test statistic can be constructed easily from the results in Table 4.6b:

$$F^* = \frac{3,398.4}{229.6} = 14.80$$

If the level of significance is to be $\alpha = .01$, we require $F(.99; 4, 5) = 11.4$. Since $F^* = 14.80 > 11.4$, we conclude H_a, that the regression function is not linear. This, of course, accords with our visual impression from Figure 4.10. To report the P-value for the test statistic, we note that $F^* = 14.80$ lies between $F(.99; 4, 5) = 11.4$ and $F(.995; 4, 5) = 15.6$ and thus the P-value must be between .005 and .01. The exact P-value can be shown to be .006.

Comments

1. As was shown by our example, not all levels of X need have repeat observations for the F test for lack of fit to be applicable. Repeat observations at only one or some levels of X are adequate.

2. The *F* test for lack of fit falls into the framework of a general linear test discussed in Section 3.8. The full model is:

(4.19)   For each X_j, Y is normal with mean $E(Y_j)$ and variance σ^2.

The least squares estimator of $E(Y_j)$ for the full model is $\bar{Y}_j$ so that the error sum of squares is:

(4.20)
$$SSE(F) = \sum_i \sum_j (Y_{ij} - \bar{Y}_j)^2 = SSPE$$

which has associated with it $n - c$ degrees of freedom.
    Since H_0 states:

(4.21)
$$H_0: E(Y) = \beta_0 + \beta_1 X$$

the error sum of squares for the reduced model is:

(4.22)
$$SSE(R) = \sum_i \sum_j (Y_{ij} - \hat{Y}_j)^2 = SSE$$

which has associated with it $n - 2$ degrees of freedom. Substituting into (3.68) and utilizing (4.12), we obtain:

(4.23)
$$\frac{SSE - SSPE}{(n-2) - (n-c)} \div \frac{SSPE}{n-c} = \frac{SSLF}{c-2} \div \frac{SSPE}{n-c} = \frac{MSLF}{MSPE}$$

the same test statistic as in (4.15).

3. Suppose that prior to any analysis of the aptness of the model, we had wished to test whether or not $\beta_1 = 0$ for the data underlying Table 4.4. The test statistic (3.59) would be:

$$F^* = \frac{MSR}{MSE} = \frac{5,141.3}{1,638.0} = 3.14$$

For $\alpha = .10$, $F(.90; 1, 9) = 3.36$, and we would conclude H_0, that $\beta_1 = 0$ or that there is no *linear association* between minimum deposit size (and value of gift) and number of new accounts. A conclusion that there is no *relation* between these variables would be improper, however. Such an inference requires that model (3.1) is appropriate. Here it is not, as we have seen, because the regression function is not linear. There exists indeed a (curvilinear) relation between minimum deposit size and number of new accounts, and testing whether or not $\beta_1 = 0$ under these circumstances has entirely different implications. This illustrates the importance of always examining the aptness of a model before further inferences are drawn.

4. The *F* test approach just explained can be used to test the aptness of other regression functions, not just the simple linear one in (4.17). Only the degrees of freedom for *SSLF* will need be modified. In general, $c - p$ degrees of freedom are associated with *SSLF*, where p is the number of parameters in the regression function. For the test of a simple linear regression function, $p = 2$ because there are two parameters, β_0 and β_1, in the regression function.

5. The alternative H_a in (4.17) includes all regression functions other than a linear one. For instance, it includes a quadratic regression function or a logarithmic one. If H_a is concluded, a study of residuals can be helpful in identifying an appropriate function.

6. Clearly, repeat observations are most valuable whenever we are not certain of the nature of the regression function. If at all possible, provision should be made for some replications.

7. If we conclude that the employed model in H_0 is appropriate, the usual practice is to use the error mean square *MSE* as an estimator of σ^2 in preference to the pure error mean square *MSPE*, since the former contains more degrees of freedom.

8. Observations at the same level of X are genuine repeats only if they involve independent trials with respect to the error term. Suppose that in a regression analysis of the relation between hardness (Y) and amount of carbon (X) in specimens of an alloy, the error term in the model covers, among other things, random errors in the measurement of hardness by the analyst and effects of uncontrolled production factors which vary at random from specimen to specimen and affect hardness. If the analyst takes two readings on the hardness of a specimen, this will not provide genuine replication because the effects of random variation in the production factors are fixed in any given specimen. For genuine replication, different specimens with the same carbon content (X) would have to be measured by the analyst so that *all* the effects covered in the error term could vary at random from one repeated observation to the next.

4.5   REMEDIAL MEASURES

If the simple linear regression model (3.1) is not appropriate for the data at hand, there are two basic choices:

1.   Abandon model (3.1) and search for a more appropriate model.
2.   Use some transformation on the data so that model (3.1) is appropriate for the transformed data.

Each approach has advantages and disadvantages. The first approach may entail a more complex model which may yield better insights, but may also lead into serious difficulties in estimating the parameters. Successful use of transformations, on the other hand, leads to relatively simple methods of estimation and may involve fewer parameters than a complex model, an advantage when the sample size is small. Yet transformations may obscure the fundamental interconnections between the variables, though at other times may illuminate them.

We shall consider the use of transformations in this chapter and the use of more complex models in later chapters. First, we provide a brief overview of remedial measures.

Nonlinearity of regression function

If the regression function is not linear, a direct approach is to modify model (3.1) with respect to the nature of the regression function. For instance, a quadratic regression function might be used:

$$(4.24) \qquad E(Y) = \beta_0 + \beta_1 X + \beta_2 X^2$$

or an exponential regression function:

$$(4.25) \qquad E(Y) = \beta_0 \beta_1^X$$

In Chapter 9, we discuss models where the regression function is a polynomial; and in Chapter 14, we discuss exponential regression functions.

The transformation approach uses a transformation to linearize, at least approximately, a nonlinear regression function. For instance, the transformation:

$$(4.26) \qquad\qquad Y' = \log Y$$

where Y' is the transformed variable, is often useful. We discuss the use of transformations to linearize regression functions in Section 4.6.

Nonconstancy of error variance

If the error variance is not constant but varies in a systematic fashion, a direct approach is to modify the model to allow for this and use the method of *weighted least squares* to obtain the estimators of the parameters. We discuss the use of weighted least squares for this purpose in Section 5.7.

Transformations can also be effective in stabilizing the variance. Some of these are discussed in Section 4.6. For instance, the transformation:

$$(4.27) \qquad\qquad Y' = \sqrt{Y}$$

is useful in a number of applications for stabilizing the variance.

Nonindependence of error terms

If the error terms are correlated, a direct remedial measure is to work with a model which calls for correlated error terms. We discuss such a model in Chapter 13. A simple remedial transformation which is often helpful is to work with first differences, a topic also discussed in Chapter 13.

Nonnormality of error terms

Lack of normality and nonconstant error variances frequently go hand in hand. Fortunately, it is often the case that the same transformation which helps stabilize the variance, such as a logarithmic or a square root transformation, is also helpful in normalizing the error terms. It is therefore desirable that the transformation for stabilizing the error variance be utilized first, and then the residuals studied to see if serious departures from normality are still present. We discuss transformations to achieve normality in Section 4.6.

Omission of important independent variables

When residual analysis indicates that an important independent variable has been omitted from the model, the solution is to modify the model. In Chapter 7 and following chapters of Part II, we discuss multiple regression analysis in which two or more independent variables are utilized.

4.6 TRANSFORMATIONS

We now consider in more detail the use of transformations of one or both of the original variables before carrying out the regression analysis. Simple transformations of either the dependent variable Y or the independent variable X, or of both, are often sufficient to make the simple linear regression model appropriate for the transformed data. We shall illustrate the use of simple transformations by three examples.

Example 1

In columns 1 and 2 of Table 4.7 are presented data on number of days of training (X) and performance score (Y) for 10 sales trainees in a battery of simulated sales situations in an experiment. These observations are shown as a scatter plot in Figure 4.12a. Clearly the regression relation appears to be curvilinear so that the simple linear regression model (3.1) does not seem to be appropriate.

TABLE 4.7 Regression calculations with square root transformation—sales training example

(1) Days of Training X	(2) Performance Score Y	(3) $\sqrt{Y} = Y'$	(4) XY'	(5) X^2
.5	43	6.5574	3.2787	.25
.5	40	6.3246	3.1623	.25
1.0	71	8.4261	8.4261	1.00
1.0	74	8.6023	8.6023	1.00
1.5	107	10.3441	15.5162	2.25
1.5	109	10.4403	15.6605	2.25
2.0	158	12.5698	25.1396	4.00
2.5	209	14.4568	36.1420	6.25
3.0	270	16.4317	49.2951	9.00
3.5	341	18.4662	64.6317	12.25
Total 17.0	1,422	112.6193	229.8545	38.50

In Figure 4.12b, the same data are plotted but the dependent variable has been transformed as follows:

$$Y' = \sqrt{Y}$$

where Y' denotes the transformed variable. Note that the scatter plot now shows a reasonably linear relation and that the variability of the scatter is reasonably constant at the different X levels. Hence, the simple linear regression model (3.1) now appears to be appropriate.

FIGURE 4.12 Scatter plots of original and transformed observations—sales training example

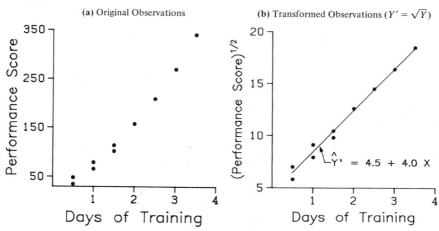

(a) Original Observations

(b) Transformed Observations ($Y' = \sqrt{Y}$)

$\hat{Y}' = 4.5 + 4.0\ X$

Example 2

At times, a curvilinear regression relationship is accompanied by systematic changes in the variability of the error terms and/or by error terms which follow a highly skewed distribution. In columns 1 and 2 of Table 4.8 are presented data on age (X) and plasma level of a polyamine (Y) for 14 healthy children. These data are plotted in Figure 4.13a as a scatter plot. Note the distinct curvilinear regression relationship, as well as the greater extent of scatter for younger children than for older ones.

TABLE 4.8 Regression calculations with logarithmic transformation—plasma levels example

	(1) Age X	(2) Plasma Level Y	(3) $\log_{10} Y = Y'$	(4) XY'	(5) X^2
	0 (newborn)	17.0	1.23045	0	0
	0 (newborn)	11.2	1.04922	0	0
	1	9.2	.96379	.96379	1
	1	12.6	1.10037	1.10037	1
	2	7.4	.86923	1.73846	4
	2	10.5	1.02119	2.04238	4
	3	8.3	.91908	2.75724	9
	3	5.8	.76343	2.29029	9
	4	4.6	.66276	2.65104	16
	4	6.5	.81291	3.25164	16
	5	5.3	.72428	3.62140	25
	5	3.8	.57978	2.89890	25
	6	3.2	.50515	3.03090	36
	6	4.5	.65321	3.91926	36
Total	42	109.9	11.85485	30.26567	182

FIGURE 4.13 Scatter plots of original and transformed observations—plasma levels example

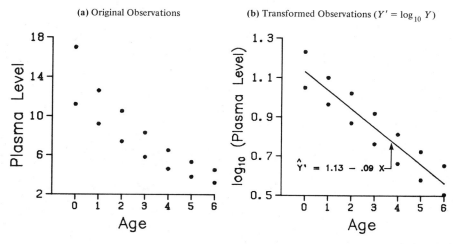

(a) Original Observations (b) Transformed Observations ($Y' = \log_{10} Y$)

In Figure 4.13b, the same data are plotted but with the dependent variable transformed as follows:

$$Y' = \log_{10} Y$$

Note that the transformation not only has led to a reasonably linear regression relation but also that the extent of scatter at the different levels of X has become reasonably constant. Hence, use of the simple linear regression model (3.1) with the transformed data now appears to be appropriate.

Example 3

In Figure 4.14a, we present a scatter plot of data on number of years experience (X) and current hourly earnings (Y) of five employees in a shop that makes hairpieces to order. The data suggest strongly that the regression function is curvilinear. Since the Y values fall within a relatively small range, a transformation on Y is not likely to be effective and we consider a transformation on the independent variable:

$$X' = \frac{1}{X}$$

Figure 4.14b contains a scatter plot with the transformed variable X'. Note that this transformation has been successful since the points tend to fall in a linear pattern.

FIGURE 4.14 Scatter plots of original and transformed observations—hairpiece earnings example

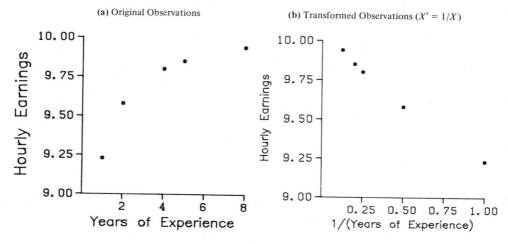

(a) Original Observations

(b) Transformed Observations ($X' = 1/X$)

Useful transformations

For many situations, the few simple transformation types just illustrated suffice to remedy the departures from the simple linear regression model (3.1). These transformations can be applied either to the dependent variable Y, the independent variable X, or occasionally to both variables, as follows:

(4.28a) $\qquad\qquad Y' = \sqrt{Y} \qquad\qquad X' = \sqrt{X}$

(4.28b) $\qquad\qquad Y' = \log_{10} Y \qquad X' = \log_{10} X$

(4.28c) $\qquad\qquad Y' = \dfrac{1}{Y} \qquad\qquad X' = \dfrac{1}{X}$

Figure 4.15 is a guide for the selection of the transformation type. Note that in one case the transformation can be applied either to the Y variable or to the X variable, or to both. Use of an interactive computer package for preparing scatter plots based on the different transformations can be most helpful in deciding on an appropriate transformation.

Comments

1. At times, theoretical or a priori considerations can be utilized to help in choosing an appropriate transformation. For example, when the shape of the scatter in a study of the relation between price of a commodity (X) and quantity demanded (Y) is that in Figure 4.15c, economists may prefer a logarithmic transformation on both Y and X to linearize the relation because the slope of the regression line for the transformed variables then measures the price elasticity of demand. The slope is then commonly interpreted as showing the percent change in quantity demanded per 1 percent change in price, where it is understood that the changes are in opposite directions.

FIGURE 4.15 Potential transformations for different curvilinear patterns

(a) $Y' = \sqrt{Y}$ or log Y or $1/Y$ (b) $X' = \sqrt{X}$ or log X or $1/X$ (c) $Y' = \sqrt{Y}$ or log Y or $1/Y$
and/or
$X' = \sqrt{X}$ or log X or $1/X$

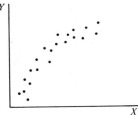

Similarly, scientists may prefer logarithmic transformations of both Y and X when studying the relation between radioactive decay (Y) of a substance and time (X) to linearize a curvilinear relation of the type illustrated in Figure 4.15c because the slope of the regression line for the transformed variables then measures the decay rate.

2. Transformations of X do not affect the variability or shape of the error distribution, while transformations of Y do. Hence, when curvilinearity is accompanied by unequal variability or skewness of the error distributions, the dependent variable needs to be transformed to remedy these additional problems. On the other hand, when curvilinearity is accompanied by constant variability, a transformation of Y may lead to significant unequal variability of the error terms and it is important to check for this possibility by examining residual plots after the transformation.

3. When either variable can be transformed (the case of Figure 4.15c), the variable for which the observations have a wider range should be considered first since a transformation is not likely to be effective when the range of the observations is narrow.

4. After a transformation has been tentatively selected, residual plots and other analyses described earlier need to be employed to ascertain that the simple linear regression model (3.1) is appropriate for the transformed data.

5. When the variance of the error terms is not constant but has a particular relation to the level of the mean response $E(Y)$ for given X, statistical theory indicates an appropriate transformation to stabilize the variance. Three important cases are [where σ_i^2 denotes the error term variance and $E(Y_i)$ denotes the mean response when $X = X_i$]:

(4.29a)   If σ_i^2 is proportional to $E(Y_i)$, use $Y' = \sqrt{Y}$

(4.29b)   If σ_i is proportional to $E(Y_i)$, use $Y' = \log Y$

(4.29c)   If $\sqrt{\sigma_i}$ is proportional to $E(Y_i)$, use $Y' = \dfrac{1}{Y}$

At times, the observed variable Y is a proportion, such as the proportion of families with income X who are planning to purchase a new car next month. An appropriate transformation for this case is:

(4.29d)   If observation is a proportion, use $Y' = 2 \arcsin \sqrt{Y}$

This transformation can be made readily on many calculator models. Also, tables to facilitate this transformation have been prepared, such as the one in Reference 4.3 which incorporates a slight refinement over transformation (4.29d) to improve the variance stabilization.

Regression analysis with transformed data

Once the data have been transformed to make the simple linear regression model (3.1) appropriate, the regression calculations are carried out in the usual fashion with the transformed data. We illustrate these calculations for two earlier examples.

Example 1. Table 4.7 (p. 134) contains for the sales training example the basic calculations required for the least squares estimators b_0 and b_1. Since $Y' = \sqrt{Y}$ now plays the role of Y in all earlier formulas, we obtain:

$$b_1 = \frac{\Sigma X_i Y_i' - \dfrac{\Sigma X_i \Sigma Y_i'}{n}}{\Sigma X_i^2 - \dfrac{(\Sigma X_i)^2}{n}} = \frac{229.8545 - \dfrac{(17)(112.6193)}{10}}{38.5 - \dfrac{(17)^2}{10}} = 4.00017$$

$$b_0 = \frac{1}{n}(\Sigma Y_i' - b_1 \Sigma X_i) = \frac{1}{10}[112.6193 - 4.00017(17)] = 4.46164$$

The fitted regression function is:

$$\hat{Y}' = 4.46164 + 4.00017X$$

where $\hat{Y}'$ is the point estimator of $E(Y')$, the mean of the probability distribution of Y' for given X.

If we wish to obtain the fitted regression equation in the original units, we simply take squares:

$$\hat{Y} = (4.46164 + 4.00017X)^2$$

For example, when $X = 3$, we have:

$$\hat{Y} = [4.46164 + 4.00017(3)]^2 = 271.0$$

Figure 4.16 presents a residual plot for the fitted regression model based on the transformed data. This plot shows no evidence of lack of fit and, in view of the few observations for large X values, no strong evidence of unequal error variances. Hence, the square root transformation of Y appears to have been effective here.

Example 2. Table 4.8 (p. 135) contains the necessary least squares calculations for our plasma levels example where the transformation $Y' = \log_{10} Y$ was found to be effective in linearizing a curvilinear relationship and in stabilizing the

FIGURE 4.16 Residual plot for transformed observations—sales training example

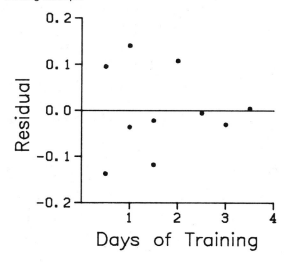

error variance. Since Y' now plays the role of Y in all earlier formulas, we obtain:

$$b_1 = \frac{\Sigma X_i Y_i' - \dfrac{\Sigma X_i \Sigma Y_i'}{n}}{\Sigma X_i^2 - \dfrac{(\Sigma X_i)^2}{n}} = \frac{30.26567 - \dfrac{(42)(11.85485)}{14}}{182 - \dfrac{(42)^2}{14}} = -.094623$$

$$b_0 = \frac{1}{n}(\Sigma Y_i' - b_1 \Sigma X_i) = \frac{1}{14}[11.85485 - (-.094623)(42)] = 1.130644$$

The fitted regression function therefore is:

$$\hat{Y}' = 1.130644 - .094623X$$

where $\hat{Y}'$ is the point estimator of $E(Y')$, the mean of the probability distribution of Y' for given X. This fitted regression function is plotted in Figure 4.13b.

If we wish to obtain the fitted regression equation in the original units, we simply take antilogs:

$$\hat{Y} = \text{antilog}_{10}(1.130644 - .094623X)$$

To find the fitted value $\hat{Y}$ in the original units when $X = 3$, for example, we have:

$$\hat{Y} = \text{antilog}_{10}[1.130644 - .094623(3)] = 7.03$$

Comments

1. We reiterate the importance of checking the model (3.1) assumptions if a transformation on Y is employed. For instance, when the transformation $Y' = \log Y$ is employed with model (3.1), it is assumed that the distribution of $\log Y$ for given X is normal with

constant variance. This needs to be checked after the transformation has been made.

2. When transformed models are employed, the estimators b_0 and b_1 obtained by least squares have the least squares properties with respect to the transformed observations, not the original ones.

PROBLEMS

4.1. Distinguish between: (1) residual and standardized residual, (2) $E(\varepsilon_i) = 0$ and $\bar{e} = 0$, (3) error term and residual.

4.2. Prepare a prototype residual plot for each of the following cases: (1) error variance decreases with X, (2) true regression function is U shaped but a linear regression function is fitted.

4.3. Refer to **Grade point average** Problem 2.15. The fitted values and residuals are:

i:	1	2	3	4	5	6	7	8	9	10
$\hat{Y}_i$:	2.92	2.33	2.25	1.58	2.08	3.51	3.34	2.67	2.25	1.91
e_i:	.18	−.03	.75	.32	.42	.19	.06	−.07	.55	−.31

i:	11	12	13	14	15	16	17	18	19	20
$\hat{Y}_i$:	2.42	2.84	2.50	3.59	2.16	1.91	2.50	3.26	1.74	2.25
e_i:	−.42	.06	−.20	−.39	−.36	−.51	−.50	.54	.46	−.75

a. Plot the residuals e_i against the fitted values $\hat{Y}_i$. What departures from regression model (3.1) can be studied from this plot? What are your findings?

b. Prepare a normal probability plot of the residuals. Also obtain the coefficient of correlation between the ordered residuals and their expected values under normality. Does the normality assumption appear to be reasonable here?

c. Information is given below for each student on two variables not included in the model, namely, intelligence test score (X_2) and high school average (X_3). Prepare additional residual plots to ascertain whether the model can be improved by including either of these variables. What do you conclude?

i:	1	2	3	4	5	6	7	8	9	10
X_2:	105	113	118	107	110	125	115	121	117	111
X_3:	2.9	2.8	3.1	2.4	3.0	2.4	3.5	3.1	3.1	2.9

i:	11	12	13	14	15	16	17	18	19	20
X_2:	123	114	120	132	122	110	119	109	116	108
X_3:	3.2	3.3	3.4	2.6	3.0	2.8	3.3	3.4	2.6	2.7

4.4. Refer to **Calculator maintenance** Problem 2.16. The fitted values and residuals are:

i:	1	2	3	4	5	6	7	8	9
$\hat{Y}_i$:	100.8	86.1	71.4	12.4	71.4	56.6	100.8	41.9	56.6
e_i:	−3.8	−.1	6.6	−2.4	3.6	5.4	.2	−2.9	−3.6

i:	10	11	12	13	14	15	16	17	18
$\hat{Y}_i$:	27.2	115.6	71.4	27.2	71.4	100.8	12.4	56.6	71.4
e_i:	5.8	2.4	−6.4	−2.2	−.4	4.2	4.6	−7.6	−3.4

a. Prepare residual plots of e_i versus $\hat{Y}_i$ and e_i versus X_i. What departures from regression model (3.1) can be studied from this plot? State your findings.

b. Prepare a normal probability plot of the residuals. Also obtain the coefficient of correlation between the ordered residuals and their expected values under normality. Does the normality assumption appear to be tenable here?

c. The observations are given in time order. Plot the residuals against time to ascertain whether or not the error terms are correlated over time. What is your conclusion?

d. Information is given below on two variables not included in the model, namely, mean operational age of machines serviced on the call (X_2, in months) and years of experience of the service person making the call (X_3). Make additional residual plots to ascertain whether the model can be improved by including either or both of these variables. What do you conclude?

i:	1	2	3	4	5	6	7	8	9
X_2:	12	21	38	16	25	32	18	14	12
X_3:	3	6	2	2	3	4	5	2	3

i:	10	11	12	13	14	15	16	17	18
X_2:	35	20	8	15	17	28	29	9	14
X_3:	6	5	3	5	6	3	5	3	6

4.5. Refer to **Airfreight breakage** Problem 2.17.

a. Obtain the residuals e_i and plot them against X_i to ascertain whether any departures from regression model (3.1) are evident. What is your conclusion?

b. Prepare a normal probability plot of the residuals. Also obtain the coefficient of correlation between the ordered residuals and their expected values under normality to ascertain whether the normality assumption is reasonable here. What do you conclude?

4.6. Refer to **Plastic hardness** Problem 2.18.

a. Obtain the residuals e_i and plot them against the fitted values $\hat{Y}_i$ to ascertain whether any departures from regression model (3.1) are evident. State your findings.

b. Prepare a normal probability plot of the residuals. Also obtain the coefficient of correlation between the ordered residuals and their expected values under normality. Does the normality assumption appear to be reasonable here?

4.7. Refer to **Muscle mass** Problem 2.23.

a. Obtain the residuals e_i and plot them against $\hat{Y}_i$ and also against X_i to ascertain whether any departures from regression model (3.1) are evident. State your conclusions.

b. Prepare a normal probability plot of the residuals. Also obtain the coefficient of correlation between the ordered residuals and their expected values under normality to ascertain whether the normality assumption is tenable here. What do you conclude?

c. The observations are given in time order. Plot the residuals against time to see whether or not the error terms are uncorrelated over time. What is your finding?

4.8. Refer to **Robbery rate** Problem 2.24.

a. Obtain the residuals and make a residual plot of e_i versus $\hat{Y}_i$. What does the plot show?

b. Prepare a normal probability plot of the residuals. Also obtain the coefficient of correlation between the ordered residuals and their expected values under normality. What do you conclude?

4.9. Electricity consumption. An economist studying the relation between household electricity consumption (Y) and number of rooms in the home (X) employed linear regression model (3.1) and obtained the following residuals:

i:	1	2	3	4	5	6	7	8	9	10
X_i:	2	3	4	5	6	7	8	9	10	11
e_i:	3.2	2.9	−1.7	−2.0	−2.3	−1.2	−.9	.8	.7	.5

Plot the residuals e_i against X_i. What problem appears to be present here? Might a transformation alleviate this problem?

4.10. Per capita earnings. A sociologist employed linear regression model (3.1) to relate per capita earnings (Y) to average number of years of schooling (X) for 12 cities. The fitted values $\hat{Y}_i$ and the standardized residuals $e_i/\sqrt{MSE}$ follow.

i:	1	2	3	4	5	6	7	8	9	10	11	12
$\hat{Y}_i$:	9.9	9.3	10.2	9.6	10.2	12.4	14.3	9.6	9.2	15.6	11.2	13.1
$e_i/\sqrt{MSE}$:	−1.12	.81	−.76	.43	.65	−.17	1.62	1.79	−.53	−3.78	.74	.32

Plot the standardized residuals against the fitted values. What does the plot suggest?

4.11. Drug concentration. A pharmacologist employed linear regression model (3.1) to study the relation between the concentration of a drug in plasma (Y) and the log-dose of the drug (X). The residuals and log-dose levels follow.

i:	1	2	3	4	5	6	7	8	9
X_i:	−1	0	1	−1	0	1	−1	0	1
e_i:	.5	2.1	−3.4	.3	−1.7	4.2	−.6	2.6	−4.0

Plot the residuals e_i against X_i. What conclusions do you draw from the plot?

4.12. A student states that she doesn't understand why the sum of squares defined in (4.9) is called a pure error sum of squares "since the formula looks like one for an ordinary sum of squares." Explain.

4.13. Refer to **Calculator maintenance** Problem 2.16. Some additional calculational results are: $SSR = 16{,}182.6$, $SSE = 321.4$.

a. In an F test for lack of fit of a linear regression function, what are the alternative conclusions?

b. Perform the test indicated in part (a). Control the risk of Type I error at .05. State the decision rule and conclusion.

c. Does your test in part (b) detect other departures from model (3.1), such as lack of constant variance or lack of normality in the error terms? Could the results of the test be affected by such departures? Discuss.

4.14. Refer to **Plastic hardness** Problem 2.18.

a. Perform an F test to determine whether or not there is lack of fit of a linear regression function. Use a level of significance of .01. State the alternatives, decision rule, and conclusion.

b. Assuming that the number of replications here was limited in advance to four, is there any advantage in conducting these all at the same level of X? Is there any disadvantage?

c. Does the test in part (a) indicate what regression function is appropriate when it leads to the conclusion that the regression function is not linear? How would you proceed?

4.15. Solution concentration. A chemist studied the concentration of a solution (Y) over time (X). Fifteen identical solutions were prepared. The 15 solutions were randomly divided into five sets of three, and the five sets were measured, respectively, after 1, 3, 5, 7, and 9 hours. The results follow.

i:	1	2	3	4	5	6	7	8	9	10	11	12	13	14	15
X_i:	9	9	9	7	7	7	5	5	5	3	3	3	1	1	1
Y_i:	.07	.09	.08	.16	.17	.21	.49	.58	.53	1.22	1.15	1.07	2.84	2.57	3.10

(TIME = X, SOL = Y)

a. Fit a linear regression function.
b. Perform an F test to determine whether or not there is lack of fit of a linear regression function; use $\alpha = .025$. State the alternatives, decision rule, and conclusion.
c. Does the test in part (b) indicate what regression function is appropriate when it leads to the conclusion that lack of fit of a linear regression function exists? Explain.

4.16. Refer to **Solution concentration** Problem 4.15.
a. Prepare a scatter plot of the data. What transformations might you try to achieve linearity?
b. Use transformation $Y' = \log_{10} Y$ and obtain the estimated linear regression function for the transformed data.
c. Plot the estimated regression line and the transformed data. Does the regression line appear to be a good fit to the transformed data?
d. Obtain the residuals and plot them against the fitted values. Also prepare a normal probability plot. What do your plots show?
e. Express the estimated regression equation in the original units.

4.17. Sales growth. A marketing researcher studied annual sales of a product that had been introduced 10 years ago. The data were as follows, where X is the year (coded) and Y is sales in thousands of units:

i:	1	2	3	4	5	6	7	8	9	10
X_i:	0	1	2	3	4	5	6	7	8	9
Y_i:	98	135	162	178	221	232	283	300	374	395

(YR = X, SALE = Y)

a. Prepare a scatter plot of the data. Does a linear relation appear adequate here?
b. Use transformation $Y' = \sqrt{Y}$ and obtain the estimated linear regression function for the transformed data.
c. Plot the estimated regression line and the transformed data. Does the regression line appear to be a good fit to the transformed data?
d. Obtain the residuals and plot them against the fitted values. Also prepare a normal probability plot. What do your plots show?
e. Express the estimated regression equation in the original units.

EXERCISES

4.18. A student fitted a linear regression function for a class assignment. Some results follow.

i:	1	2	3	4	5
Y_i:	35	17	42	28	53
$\hat{Y}_i$:	42	29	32	32	40
e_i:	-7	-12	10	-4	13

The student plotted the residuals e_i against Y_i and found a positive relation. When he plotted the residuals against the fitted values $\hat{Y}_i$, he found no relation. Why is there this difference, and which is the more meaningful plot?

4.19. If the error terms in a regression model are independent $N(0, \sigma^2)$, what can be said about the error terms after transformation $X' = 1/X$ is used? Is the situation the same after transformation $Y' = 1/Y$ is used?

4.20. Using theorems (1.65), (1.36), and (1.37), show that $E(MSPE) = \sigma^2$ for the normal error regression model (3.1).

PROJECTS

4.21. Machine speed. The number of defective items produced by a machine (Y) is known to be linearly related to the speed setting of the machine (X). The data below were collected from recent quality control records.

i:	1	2	3	4	5	6	7	8	9	10	11	12
X_i:	200	400	300	400	200	300	300	400	200	400	200	300
Y_i:	28	75	37	53	22	58	40	96	46	52	30	69

a. Assuming regression model (3.1) is appropriate, obtain the estimated regression function and plot the residuals against X. What does the residual plot show?

b. Calculate the sample variance s^2 of the Y observations for each of the three machine speeds: $X = 200, 300, 400$. What is suggested by these three sample variances about whether or not the true variances at the three X levels are equal?

c. Compute $\bar{Y}/\sqrt{s}$, $\bar{Y}/s$, and $\bar{Y}/s^2$ for each of the three X levels. Suggest an appropriate transformation from (4.29) to stabilize the variance based on your computed ratios.

d. Make the transformation suggested in part (c) and obtain the estimated regression line for the transformed data. Plot the residuals against X. Does it appear from your plot that the purpose of the transformation has been attained?

4.22. Blood pressure. The following data were obtained in a study of the relation between diastolic blood pressure (Y) and age (X) for boys 5 to 13 years old.

i:	1	2	3	4	5	6	7	8
X_i:	5	8	11	7	13	12	12	6
Y_i:	63	67	74	64	75	69	90	60

a. Assuming regression model (3.1) is appropriate, obtain the estimated regression function and plot the residuals e_i against X_i. What does your residual plot show?

b. Omit observation 7 from the data and obtain the estimated regression line based on the remaining seven observations. Compare this estimated regression function to that obtained in part (a). What can you conclude about the effect of observation 7?

c. Using your fitted regression function in part (b), obtain a 99 percent prediction interval for a new Y observation at $X = 12$. Does observation Y_7 fall outside this prediction interval? What is the significance of this?

4.23. Refer to the **SMSA** data set and Project 2.38. For each of the three fitted regression models, obtain the residuals and prepare a residual plot against X and a normal probability plot. Summarize your conclusions. Is linear regression model (3.1) more apt in one case than in the others?

4.24. Refer to the **SMSA** data set and Project 2.39. For each geographic region, obtain the residuals and prepare a residual plot against X and a normal probability plot. Do the four regions appear to have similar error variances? What other conclusions do you draw from your plots?

4.25. Refer to the **SENIC** data set and Project 2.40.

a. For each of the three fitted regression models, obtain the residuals and prepare a residual plot against X and a normal probability plot. Summarize your conclusions. Is linear regression model (3.1) more apt in one case than in the others?

b. Obtain the fitted regression line for the relation between length of stay and infection risk after deleting observations 47 ($X_{47} = 6.5$, $Y_{47} = 19.56$) and 112 ($X_{112} = 5.9$, $Y_{112} = 17.94$). From this fitted regression line obtain separate 95 percent prediction intervals for new Y observations at $X = 6.5$ and $X = 5.9$, respectively. Do observations Y_{47} and Y_{112} fall outside these prediction intervals? Discuss the significance of this.

4.26. Refer to the **SENIC** data set and Project 2.41. For each geographic region, obtain the residuals and prepare a residual plot against X and a normal probability plot. Do the four regions appear to have similar error variances? What other conclusions do you draw from your plots?

CITED REFERENCES

4.1 Dixon, W. J., and M. B. Brown, eds. *BMDP-81, Biomedical Computer Programs, P-Series*. Berkeley, Calif.: University of California Press, 1981.

4.2 Barnett, Vic, and Toby Lewis. *Outliers in Statistical Data*. New York: John Wiley & Sons, 1978.

4.3 Owen, Donald B. *Handbook of Statistical Tables*. Reading, Mass.: Addison-Wesley Publishing, 1962.

5

Simultaneous inferences and other topics in regression analysis—I

In this chapter, we take up a variety of topics in simple regression analysis. Several of the topics pertain to the problem of how to make simultaneous inferences from the same set of sample observations.

5.1   JOINT ESTIMATION OF β_0 AND β_1

Need for joint estimation

A market research analyst conducted a study of the relation between level of advertising (X) and sales (Y), in which there was no advertising $(X = 0)$ for some observations while for other observations the level of advertising was varied. The scatter plot suggested a linear regression in the range of the advertising expenditures levels studied. The analyst wished to draw inferences about both the intercept β_0 and the slope β_1. One means of doing this is by constructing a joint confidence region for β_0 and β_1 so that with confidence level $1 - \alpha$ both β_0 and β_1 are contained in this region.

Joint confidence region

A joint $1 - \alpha$ confidence region for β_0 and β_1 is illustrated in Figure 5.1. It can be shown that such a region is given by:

(5.1) $$\frac{n(b_0 - \beta_0)^2 + 2(\Sigma X_i)(b_0 - \beta_0)(b_1 - \beta_1) + (\Sigma X_i^2)(b_1 - \beta_1)^2}{2MSE}$$

$$\leq F(1 - \alpha; 2, n - 2)$$

The confidence coefficient $1 - \alpha$ indicates that with repeated sampling, the confidence region (5.1) will contain both β_0 and β_1 in $(1 - \alpha)100$ percent of the cases. The confidence region consists of all points (β_0, β_1) which satisfy the inequality (5.1). The boundary of the confidence region is obtained from the equality in (5.1):

(5.2) $$\frac{n(b_0 - \beta_0)^2 + 2(\Sigma X_i)(b_0 - \beta_0)(b_1 - \beta_1) + (\Sigma X_i^2)(b_1 - \beta_1)^2}{2MSE}$$

$$= F(1 - \alpha; 2, n - 2)$$

The boundary is an ellipse centered around the point (b_0, b_1), as illustrated in Figure 5.1. We explain now by an example how the boundary is calculated.

Example. Let us return to the Westwood Company lot size example of the previous chapters. Suppose that we require a joint confidence region for β_0 and

FIGURE 5.1 Elliptical joint 90 percent confidence region for β_0 and β_1—Westwood Company example

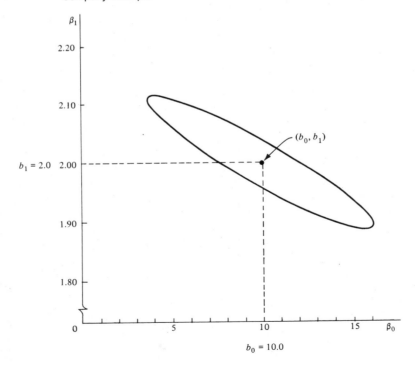

β_1 with .90 confidence coefficient. Then $1 - \alpha = .90$ and $F(.90; 2, 8) = 3.11$. From previous work (Table 3.1), we have: $b_0 = 10.0$, $b_1 = 2.0$, $MSE = 7.5$, $\Sigma X_i = n\bar{X} = 500$, $\Sigma X_i^2 = 28{,}400$. Substitution into (5.2) gives:

$$\frac{10(10.0 - \beta_0)^2 + 2(500)(10.0 - \beta_0)(2.0 - \beta_1) + 28{,}400(2.0 - \beta_1)^2}{2(7.5)} = 3.11$$

Boundary points are calculated by assigning a value to either β_0 or β_1 and finding corresponding values of the other unknown. For example, let $\beta_0 = 10.0$. The quantity $(10.0 - \beta_0)$ then equals 0, and the above expression reduces to:

$$\frac{28{,}400(2.0 - \beta_1)^2}{2(7.5)} = 3.11$$

Manipulation gives:

$$\beta_1^2 - 4.0\beta_1 + 3.998357 = 0$$

This quadratic equation has two roots, 1.95947 and 2.04053 (see the following Comment 3 for a brief review of the solution of a quadratic equation). Hence, two boundary points are $(10.0, 1.96)$ and $(10.0, 2.04)$.

Additional boundary points are found in the same manner, by assigning a value to either β_0 or β_1 and solving for the other. The points can be plotted on a graph and connected to form the boundary of the elliptical confidence region, as shown in Figure 5.1. The region is always centered around (b_0, b_1). In our example, it is centered around $(b_0, b_1) = (10.0, 2.0)$. The joint confidence region indicates that β_1 is somewhere between 1.88 and 2.12, and that β_0 would be in the neighborhood of 4.0 if β_1 were near the upper limit and about 16.0 if β_1 were near its lower limit. Note carefully the interrelation between the estimates for β_0 and β_1: the larger is β_1, the smaller would be β_0, and vice versa. This interrelation is the result of the tilted position of the ellipse, with the major axis being sloped negatively.

Comments

1. The joint confidence region can be used directly for testing. To illustrate this use, suppose an industrial engineer working for the Westwood Company theorized that the regression function should have an intercept of 13.0 and a slope of 2.10. Since the point $(13.0, 2.10)$ does not fall in the joint confidence region, we would conclude at the $\alpha = .10$ level of significance that either $\beta_0 \neq 13.0$ or $\beta_1 \neq 2.10$ or both. Note from Figure 5.1 that the engineer may be correct with respect to either the intercept or the slope, but that this particular *combination* is not supported by the data.

2. The tilt of the ellipse is a function of the covariance between b_0 and b_1. It can be shown that:

$$(5.3) \qquad\qquad \sigma(b_0, b_1) = -\bar{X}\sigma^2(b_1)$$

Both the tilt of the ellipse and the covariance indicate the degree to which the point estimates of β_0 and β_1 obtained from the same sample are likely to err in a similar or an opposite direction because of sampling error. A positive covariance indicates a tendency for the values of b_0 and b_1 to be jointly too high or jointly too low, while a negative covariance means that the joint errors tend to be in opposite directions.

In our example, $\bar{X} = 50$; hence, the covariance is negative and the ellipse's major axis is sloped negatively. This implies that the estimators b_0 and b_1 tend to err in opposite directions. We expect this intuitively. Since the observed points (X_i, Y_i) fall in the first quadrant, we anticipate that if the slope of the fitted regression line is too steep (b_1 overestimates β_1), the intercept is most likely to be too low (b_0 underestimates β_0), and vice versa.

When the independent variable is $X_i - \bar{X}$, we know that b_0^* and b_1 are independent and hence have zero covariance. Thus, the ellipse will in this case have axes parallel to the axes of the graph so that there is no tilt in the confidence region.

3. The roots of a quadratic equation of the form $ax^2 + bx + c = 0$ are given by:

$$x = \frac{-b \pm \sqrt{b^2 - 4ac}}{2a}$$

In our earlier example, the quadratic equation was $\beta_1^2 - 4.0\beta_1 + 3.998357 = 0$, so that $a = 1$, $b = -4.0$, and $c = 3.998357$. Hence:

$$\beta_1 = \frac{-(-4.0) \pm \sqrt{(-4.0)^2 - 4(1)(3.998357)}}{2(1)}$$

$$= 1.95947 \text{ and } 2.04053$$

Bonferroni joint confidence intervals

The procedure for developing a joint confidence region for β_0 and β_1 is somewhat cumbersome. Also, in multiple regression, where additional parameters are involved in the model, the joint confidence region involves three or more dimensions, which is difficult to visualize.

Hence, it is often preferable to construct separate confidence intervals for each parameter. For instance, by the methods of Chapter 3, the market research analyst in our earlier example could construct separate 95 percent confidence intervals for β_0 and β_1. The difficulty is that these would not provide 95 percent confidence that the conclusions for *both* β_0 and β_1 are correct. If the inferences were independent, the probability of both being correct would be $(.95)^2$, or only .9025. The inferences are not, however, independent, coming as they do from the same set of sample data, which makes the determination of the probability of both inferences being correct much more difficult.

Analysis of data frequently requires a series of estimates (or tests) where the analyst would like to have an assurance about the correctness of the entire set of estimates (or tests). We shall call the set of estimates of interest the *family* of estimates. In our illustration, the *family* consists of the estimates of β_0 and β_1. We then distinguish between a *statement confidence coefficient* and a *family confidence coefficient*. The former is the familiar type of confidence coefficient discussed earlier, which indicates the proportion of correct estimates that are obtained when repeated samples are selected and the specified confidence interval is calculated for each sample. A family confidence coefficient, on the other hand, indicates the proportion of correct families of estimates when repeated samples are selected and the specified confidence intervals for the entire family are calculated for each sample.

To illustrate the meaning of a family confidence coefficient further, let us return to the joint estimation of β_0 and β_1. A family confidence coefficient of, say, .95 would indicate for this situation that if repeated samples are selected and interval estimates for both β_0 and β_1 are calculated for each sample by specified procedures, 95 percent of the samples would lead to a family of estimates where *both* confidence intervals are correct. For 5 percent of the samples, either one or both of the interval estimates would be incorrect.

Clearly, a procedure which provides a family confidence coefficient is often highly desirable since it permits the analyst to weave the separate results together into an integrated set of conclusions, with an assurance that the entire set of estimates is correct. The Bonferroni method of developing joint confidence intervals with a specified family confidence coefficient is a very simple one: each statement confidence coefficient is adjusted to be higher than $1 - \alpha$ so that the family confidence coefficient is $1 - \alpha$. The method is a general one which can be applied in many cases, as we shall see, not just for the joint estimation of β_0 and β_1. Here, we explain the Bonferroni method as it applies for estimating β_0 and β_1 jointly.

Development of joint confidence intervals. We start with ordinary confidence limits for β_0 and β_1 with statement confidence coefficients $1 - \alpha$ each. These are:

$$b_0 \pm t(1 - \alpha/2; n - 2)s(b_0)$$
$$b_1 \pm t(1 - \alpha/2; n - 2)s(b_1)$$

We then ask what is the probability that both sets of limits are correct. Let A_1 denote the event that the first confidence interval does not cover β_0 and A_2 denote the event that the second confidence interval does not cover β_1. We know:

$$P(A_1) = \alpha \qquad P(A_2) = \alpha$$

Probability theorem (1.6) states:

$$P(A_1 \cup A_2) = P(A_1) + P(A_2) - P(A_1 \cap A_2)$$

and hence:

(5.4) $$1 - P(A_1 \cup A_2) = 1 - P(A_1) - P(A_2) + P(A_1 \cap A_2)$$

Now by probability theorems (1.9) and (1.10), we have:

$$1 - P(A_1 \cup A_2) = P(\overline{A_1 \cup A_2}) = P(\overline{A}_1 \cap \overline{A}_2)$$

$P(\overline{A}_1 \cap \overline{A}_2)$ is the probability that both confidence intervals are correct. We thus have from (5.4):

(5.5) $$P(\overline{A}_1 \cap \overline{A}_2) = 1 - P(A_1) - P(A_2) + P(A_1 \cap A_2)$$

Since $P(A_1 \cap A_2) \geq 0$, we obtain from (5.5) the Bonferroni inequality:

(5.6) $$P(\overline{A}_1 \cap \overline{A}_2) \geq 1 - P(A_1) - P(A_2)$$

which for our situation is:

(5.6a)
$$P(\bar{A}_1 \cap \bar{A}_2) \geq 1 - \alpha - \alpha = 1 - 2\alpha$$

Thus, if β_0 and β_1 are separately estimated with, say, 95 percent confidence intervals, the Bonferroni inequality guarantees us a family confidence coefficient of at least 90 percent that both intervals based on the same sample are correct.

We can easily use the Bonferroni inequality (5.6a) to obtain a family confidence coefficient of at least $1 - \alpha$ for estimating β_0 and β_1. We do this by estimating β_0 and β_1 separately with statement confidence coefficients of $1 - \alpha/2$ each. Thus, the $1 - \alpha$ family confidence limits for β_0 and β_1, often called a *confidence set*, are by the Bonferroni procedure:

(5.7)
$$b_0 \pm Bs(b_0)$$
$$b_1 \pm Bs(b_1)$$

where:

(5.7a)
$$B = t(1 - \alpha/4; n - 2)$$

Note that a statement confidence coefficient of $1 - \alpha/2$ requires the $(1 - \alpha/4)100$ percentile of the t distribution for a two-sided confidence interval.

Example. For the Westwood Company lot size application, 90 percent family confidence intervals for β_0 and β_1 require $B = t(1 - .10/4; 8) = t(.975; 8) = 2.306$. We have from before:

$$b_0 = 10.0 \qquad s(b_0) = 2.50294$$
$$b_1 = 2.0 \qquad s(b_1) = .04697$$

Hence, the two pairs of confidence limits are $10.0 \pm 2.306(2.50294)$ and $2.0 \pm 2.306(.04697)$, and the joint confidence intervals are:

$$4.2282 \leq \beta_0 \leq 15.7718$$
$$1.8917 \leq \beta_1 \leq 2.1083$$

Thus, we conclude that β_0 is between 4.23 and 15.77 *and* β_1 is between 1.89 and 2.11. The family confidence coefficient is at least .90 that the procedure leads to correct pairs of interval estimates.

Comments

1. We reiterate that the Bonferroni $1 - \alpha$ family confidence coefficient is actually a lower bound on the true (but often unknown) family confidence coefficient. To the extent that incorrect interval estimates of β_0 and β_1 tend to pair up in the family (particularly when the covariance between b_0 and b_1 is large), the families of statements will tend to be correct more than $(1 - \alpha)100$ percent of the time.

2. The Bonferroni inequality (5.6a) can easily be extended to g simultaneous confidence intervals with family confidence coefficient $1 - \alpha$:

(5.8)
$$P\left(\bigcap_{i=1}^{g} \bar{A}_i\right) \geq 1 - g\alpha$$

Thus, if g interval estimates are desired with a family confidence coefficient $1 - \alpha$, constructing each interval estimate with statement confidence coefficient $1 - \alpha/g$ will suffice.

3. For a given family confidence coefficient, the larger the number of confidence intervals in the family, the greater becomes the multiple B, which may make some or all of the confidence intervals too wide to be helpful. The Bonferroni technique is ordinarily most useful when the number of simultaneous estimates is not too large.

4. It is not necessary with the Bonferroni procedure that the confidence intervals have the same statement confidence coefficient. Different statement confidence coefficients can be used, depending on the importance of each estimate. For instance, in our earlier illustration β_0 might be estimated with a 92 percent confidence interval and β_1 with a 98 percent confidence interval. The family confidence coefficient by (5.6) will still be at least 90 percent.

Comparison of two approaches.   Figure 5.2 contains the joint 90 percent confidence region by the Bonferroni approach for our example. Note that the region is a rectangle since the Bonferroni approach does not utilize the existing relationship between b_0 and b_1. The rectangle is centered at (b_0, b_1). Also shown in Figure 5.2 is the elliptical 90 percent joint confidence region which we obtained earlier, which is also centered at (b_0, b_1). The elliptical region is more efficient in that it covers fewer (β_0, β_1) points for the same confidence coeffi-

FIGURE 5.2   Bonferroni and elliptical joint 90 percent confidence regions for β_0 and β_1—Westwood Company example

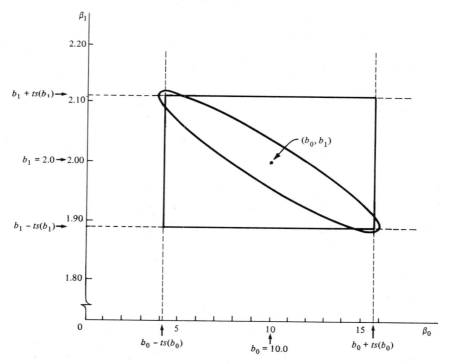

cient. Nevertheless, the Bonferroni approach can be highly useful in many cases. In multiple regression applications, in particular, the Bonferroni method comes into its own because of the ease of obtaining and interpreting the joint estimates.

5.2 CONFIDENCE BAND FOR REGRESSION LINE

At times we would like to obtain a confidence band for the regression line $E(Y) = \beta_0 + \beta_1 X$ so that we can see in what region the entire regression line lies. This differs from estimating $E(Y_h) = \beta_0 + \beta_1 X_h$ for a particular value of X_h by an interval estimate, which we took up in Section 3.4.

To obtain a confidence band for the entire regression line, we essentially need to consider the regression lines for all possible (β_0, β_1) combinations in the elliptical joint confidence region (5.1) for β_0 and β_1. For our Westwood Company example, three possible (β_0, β_1) combinations (Figure 5.1) and their corresponding regression lines are:

β_0	β_1	$E(Y) = \beta_0 + \beta_1 X$
4.00	2.11	$E(Y) = 4.00 + 2.11X$
9.00	2.00	$E(Y) = 9.00 + 2.00X$
16.00	1.89	$E(Y) = 16.00 + 1.89X$

Figure 5.3 contains a plot of these three possible regression lines. As additional lines for other possible (β_0, β_1) combinations in the confidence region in Figure 5.1 are plotted, we will fill out a confidence band for the regression line. The boundaries of the confidence band are sketched in Figure 5.3 by the broken lines, which are hyperbolas.

Working and Hotelling derived the formula for the $1 - \alpha$ hyperbolic confidence band for the regression line. At any level X_h, the two boundary values of the confidence band are:

$$(5.9) \qquad \hat{Y}_h \pm W s(\hat{Y}_h)$$

where:

$$(5.9a) \qquad W^2 = 2F(1 - \alpha; 2, n - 2)$$

and $\hat{Y}_h$ and $s(\hat{Y}_h)$ are defined in (3.27) and (3.30), respectively.

Example

For our Westwood Company example, suppose that we wish to set up the 90 percent confidence band for the regression line. We developed earlier:

$$\hat{Y}_h = 10.0 + 2.0X_h \qquad \text{(p. 41)}$$

$$s^2(\hat{Y}_h) = 7.5 \left[\frac{1}{10} + \frac{(X_h - 50)^2}{3,400} \right] \qquad \text{(p. 75)}$$

We need $F(.90; 2, 8) = 3.11$. Hence, we obtain:

$$W^2 = 2(3.11) = 6.22 \quad \text{or} \quad W = 2.494$$

FIGURE 5.3 Plot of three possible regression lines for joint confidence region in Figure 5.1 (Y values not plotted to scale)

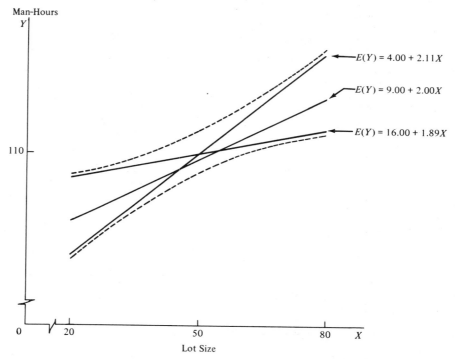

Let us find the boundary points of the confidence band at $X_h = 55$:

$$\hat{Y}_{55} = 10.0 + 2.0(55) = 120.0$$

$$s^2(\hat{Y}_{55}) = 7.5\left[\frac{1}{10} + \frac{(55 - 50)^2}{3,400}\right] = .80515$$

or:

$$s(\hat{Y}_{55}) = .89730$$

Hence, the 90 percent boundary points of the confidence band for the regression line at $X_h = 55$ are $120.0 \pm 2.494(.89730)$ or:

$$117.8 \leq \beta_0 + \beta_1 X_h \leq 122.2$$

In similar fashion, the boundary points at a number of other values of X_h can be developed and the boundary curves then sketched in. This has been done in Figure 5.4, which contains the 90 percent confidence band for the regression line for the Westwood Company example.

FIGURE 5.4 90 percent confidence band for regression line—Westwood Company example

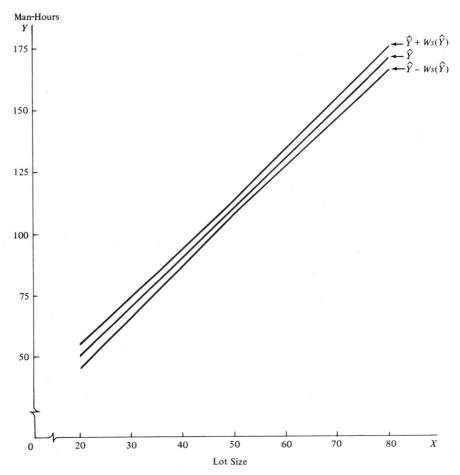

Man-Hours

Lot Size

Comments

1. The boundary points (5.9) of the confidence band for the regression line are of exactly the same form as the confidence limits for a single mean response $E(Y_h)$ in (3.32) except that the t multiple has been replaced by the W multiple (named after Working).

2. If we had wished to estimate a single mean response with a 90 percent confidence coefficient, the required t value would have been $t(.95; 8) = 1.860$. Note that this t multiple, though smaller than the multiple $W = 2.494$ for the entire regression line, does not differ by any major extent. This is typically the case, so that the boundary points of the regression line band usually are not very much further apart than the confidence limits for a single mean response $E(Y_h)$ at a given X_h value. With the somewhat wider limits for the entire regression line, one is able to draw conclusions about any and all mean responses

for the entire regression line and not just about the mean response at a given X level. One use of this broader base for inferences will be explained in the next section.

3. The confidence band (5.9) applies to the entire regression line over all real-numbered values of X from $-\infty$ to $+\infty$. The confidence coefficient indicates the percent of time the estimating procedure will yield a band which covers the entire line, in a long series of samples in which the X observations are kept at the same level as in the sample actually taken.

In applications, the confidence band is ignored for that part of the regression line which is not of interest in the problem at hand. In the Westwood Company example, for instance, negative lot sizes would be ignored. The confidence coefficient for a limited segment of the band of interest is somewhat higher than $1 - \alpha$, so that $1 - \alpha$ serves then as a lower bound to the confidence coefficient.

4. Research continues on confidence banding of the regression line. An alternative procedure to the Working-Hotelling one, for instance, gives a confidence band of uniform width over a finite interval on the X axis centered around $\bar{X}$ (Ref. 5.1).

5.3 SIMULTANEOUS ESTIMATION OF MEAN RESPONSES

Often one would like to estimate the mean responses at a number of X levels from the same sample data. The Westwood Company, for instance, may wish to estimate the mean number of man-hours for lots of 30, 55, and 80 parts. We already know how to do this for any one level of X with given statement confidence coefficient. Now we shall discuss two approaches for simultaneous estimation of mean responses with a family confidence coefficient, so that there is a known assurance of all estimates of mean responses being correct. The two approaches are the Working-Hotelling approach and the Bonferroni approach.

The reason for concern with a family confidence coefficient is that separate interval estimates of $E(Y_h)$ at various levels X_h need not all be correct or all be incorrect, even though they are all based on the same sample data and fitted regression line. The combination of sampling errors in b_0 and b_1 may be such that the interval estimates of $E(Y_h)$ will be correct over some range of X levels and incorrect elsewhere.

Working-Hotelling approach

Since the Working-Hotelling confidence band for the entire regression line holds for all values of X, it certainly must hold for selected levels of the independent variable. Hence, to obtain with the Working-Hotelling approach a family of interval estimates of mean responses at different levels of X, with a $1 - \alpha$ family confidence coefficient, we simply use formula (5.9) repetitively to calculate boundary points of the confidence band at the various X levels being considered. These boundary points serve then as the family confidence limits for the interval estimates of the mean responses of interest.

Example. For the Westwood Company lot size example, suppose that we require a family of estimates of the mean number of man-hours at the following levels of lot size: 30, 55, 80. The family confidence coefficient is to be .90. We obtained earlier $\hat{Y}_h$ and $s(\hat{Y}_h)$ for $X_h = 55$, and found $W = 2.494$. In similar fashion, we can obtain the needed results for the other lot sizes. We summarize them here, without showing the calculations:

X_h	$\hat{Y}_h$	$s(\hat{Y}_h)$	$Ws(\hat{Y}_h)$
30	70.0	1.27764	3.1864
55	120.0	.89730	2.2379
80	170.0	1.65387	4.1248

Thus, the boundary points of the regression line band at $X_h = 30, 55$, and 80 are:

$$66.8 = 70.0 - 3.1864 \le E(Y_{30}) \le 70.0 + 3.1864 = 73.2$$
$$117.8 = 120.0 - 2.2379 \le E(Y_{55}) \le 120.0 + 2.2379 = 122.2$$
$$165.9 = 170.0 - 4.1248 \le E(Y_{80}) \le 170.0 + 4.1248 = 174.1$$

With family confidence coefficient .90, we conclude that the mean number of man-hours for lots of 30 parts is between 66.8 and 73.2, for lots of 55 parts is between 117.8 and 122.2, and for lots of 80 parts is between 165.9 and 174.1. The family confidence coefficient .90 provides assurance that the procedure leads to all correct estimates in the family of estimates.

Bonferroni approach

The Bonferroni approach, discussed earlier for simultaneous estimation of β_0 and β_1, is a completely general approach. To construct a family of confidence intervals for mean responses at different X levels, we calculate the usual confidence limits for a single mean response $E(Y_h)$, given in (3.32), and adjust the statement confidence coefficient to yield the specified family confidence coefficient.

If $E(Y_h)$ is to be estimated for g levels of X, with a family confidence coefficient of $1 - \alpha$, the Bonferroni confidence limits are:

(5.10)
$$\hat{Y}_h \pm Bs(\hat{Y}_h)$$

where:

(5.10a) $B = t(1 - \alpha/2g; n - 2)$

g is the number of confidence intervals in the family

Example. The estimates of the mean man-hours for lot sizes of 30, 55, and 80 parts with a family confidence coefficient of .90 by the Bonferroni approach require the same data as the Working-Hotelling approach presented above. In addition, we require $B = t(1 - .10/2(3); 8) = t(.983; 8)$. By linear interpolation, we obtain $t(.983; 8) = 2.56$ (see the following Comment 4).

We thus obtain the confidence intervals, with a 90 percent family confidence coefficient:

$$66.7 = 70.0 - 2.56(1.27764) \leq E(Y_{30}) \leq 70.0 + 2.56(1.27764) = 73.3$$
$$117.7 = 120.0 - 2.56(.89730) \leq E(Y_{55}) \leq 120.0 + 2.56(.89730) = 122.3$$
$$165.8 = 170.0 - 2.56(1.65387) \leq E(Y_{80}) \leq 170.0 + 2.56(1.65387) = 174.2$$

Comments

1. In this instance the Working-Hotelling confidence limits are slightly tighter than the Bonferroni limits. In other cases where the number of statements is small, the Bonferroni limits may be tighter. For larger families, the Working-Hotelling confidence limits will always be the tighter, since W in (5.9a) stays the same for any number of statements in the family whereas B in (5.10a) becomes larger as the number of statements increases. In practice, once the family confidence coefficient has been decided upon, one can calculate the W and B multiples to determine which procedure leads to tighter confidence limits.

2. Both the Working-Hotelling and Bonferroni approaches to multiple estimation of mean responses provide lower bounds to the actual family confidence coefficient. The reason why the Working-Hotelling approach furnishes a lower bound is that the confidence coefficient $1 - \alpha$ actually applies to the entire line from $-\infty$ to $+\infty$.

3. Sometimes it is not known in advance for which levels of the independent variable to estimate the mean response. That is determined as the analysis proceeds. In such cases, it is better to use the Working-Hotelling approach.

4. To obtain an untabled percentile of the t distribution, linear interpolation in Table A–2 ordinarily will give a reasonably close approximation as long as the degrees of freedom are not minimal. In our illustration of the Bonferroni method, we required $t(.983; 8)$. From Table A–2, we know that:

$$t(.980; 8) = 2.449 \qquad t(.985; 8) = 2.634$$

Linear interpolation therefore gives:

$$t(.983; 8) = 2.449 + \left(\frac{.983 - .980}{.985 - .980} \right)(2.634 - 2.449) = 2.56$$

5.4 SIMULTANEOUS PREDICTION INTERVALS FOR NEW OBSERVATIONS

Now we consider the simultaneous prediction of g new observations on Y in g independent trials at g different levels of X. To illustrate this type of application, let us suppose the Westwood Company plans to produce the next three lots in sizes of 30, 55, and 80 parts, and wishes to predict the man-hours for each of these lots with a family confidence coefficient of .95.

Two procedures will be considered here, the Scheffé procedure and the Bonferroni procedure. Both utilize the same type of limits as for predicting a single observation, given in (3.35), and only the multiple of the estimated standard deviation is changed. The Scheffé procedure uses the F distribution, while the Bonferroni procedure uses the t distribution. The simultaneous prediction limits for g predictions with the Scheffé procedure with family confidence coefficient $1 - \alpha$ are:

(5.11) $$\hat{Y}_h \pm Ss(Y_{h(new)})$$

where:

(5.11a) $$S^2 = gF(1 - \alpha; g, n - 2)$$

With the Bonferroni procedure, the $1 - \alpha$ simultaneous prediction limits are:

(5.12) $$\hat{Y}_h \pm Bs(Y_{h(\text{new})})$$

where:

(5.12a) $$B = t(1 - \alpha/2g; n - 2)$$

We can evaluate the S and B multiples to see which procedure provides tighter prediction limits. For our example, we have:

$$S^2 = 3F(.95; 3, 8) = 3(4.07) = 12.21 \quad \text{or} \quad S = 3.49$$
$$B = t(1 - .05/2(3); 8) = t(.992; 8) = 3.04$$

so that the Bonferroni method will be used here. From earlier results, we obtain (calculations not shown):

X_h	$\hat{Y}_h$	$s(Y_{h(\text{new})})$	$Bs(Y_{h(\text{new})})$
30	70.0	3.02198	9.18682
55	120.0	2.88187	8.76088
80	170.0	3.19926	9.72575

and the simultaneous prediction limits are:

$$60.8 = 70.0 - 9.18682 \le Y_{30(\text{new})} \le 70.0 + 9.18682 = 79.2$$
$$111.2 = 120.0 - 8.76088 \le Y_{55(\text{new})} \le 120.0 + 8.76088 = 128.8$$
$$160.3 = 170.0 - 9.72575 \le Y_{80(\text{new})} \le 170.0 + 9.72575 = 179.7$$

With family confidence coefficient at least .95, we can predict that the man-hours for the next three production runs all will be within the above limits.

Comments

1. Simultaneous prediction intervals for g new observations on Y at g different levels of X with a $1 - \alpha$ family confidence coefficient are wider than the corresponding single prediction intervals of (3.35). When the number of simultaneous predictions is not large, however, the difference in the width is only moderate. For instance, a single 95 percent prediction interval for our example would have utilized the t multiple $t(.975; 8) = 2.306$, which is only moderately smaller than the multiple $B = 3.04$ for three simultaneous predictions.

2. Note that both B and S become larger as g increases. This contrasts with simultaneous estimation of mean responses where B becomes larger but not W. When g is large, both the B and S multiples may become so large that the prediction intervals will be too wide to be useful. Other simultaneous estimation techniques could then be considered, as discussed in Reference 5.2.

5.5 REGRESSION THROUGH THE ORIGIN

Sometimes the regression line is known to go through the origin at $(0, 0)$. This occurs, for instance, when X is units of output and Y is variable cost, so Y is zero by definition when X is zero. Another example is where X is the number of brands of cigarettes stocked in a supermarket in an experiment (including some supermarkets with no brands stocked) and Y is the volume of cigarette sales in the supermarket. The normal error model for these cases is the same as the general model (3.1) except $\beta_0 = 0$:

$$(5.13) \qquad Y_i = \beta_1 X_i + \varepsilon_i$$

where:

$\quad\quad \beta_1$ is a parameter

$\quad\quad X_i$ are known constants

$\quad\quad \varepsilon_i$ are independent $N(0, \sigma^2)$

The regression function for model (5.13) is:

$$(5.14) \qquad E(Y) = \beta_1 X$$

The least squares estimator of β_1 is obtained by minimizing:

$$(5.15) \qquad Q = \Sigma(Y_i - \beta_1 X_i)^2$$

with respect to β_1. The resulting normal equation is:

$$(5.16) \qquad \Sigma X_i(Y_i - b_1 X_i) = 0$$

leading to the point estimator:

$$(5.17) \qquad b_1 = \frac{\Sigma X_i Y_i}{\Sigma X_i^2}$$

b_1 as given in (5.17) is also the maximum likelihood estimator.

An unbiased estimator of $E(Y)$ is:

$$(5.18) \qquad \hat{Y} = b_1 X$$

Also, an unbiased estimator of σ^2 is:

$$(5.19) \qquad MSE = \frac{\Sigma(Y_i - \hat{Y}_i)^2}{n - 1} = \frac{\Sigma(Y_i - b_1 X_i)^2}{n - 1}$$

The reason for the denominator $n - 1$ is that only one degree of freedom is lost in estimating the single parameter of the regression equation (5.14).

Confidence intervals for β_1, $E(Y_h)$, and a new observation $Y_{h(new)}$ are shown in Table 5.1. Note that the t multiple has $n - 1$ degrees of freedom here, the

TABLE 5.1 Confidence limits for regression through origin

Estimate of—	Estimated Variance		Confidence Limits
β_1	$s^2(b_1) = \dfrac{MSE}{\Sigma X_i^2}$	(5.20)	$b_1 \pm ts(b_1)$
$E(Y_h)$	$s^2(\hat{Y}_h) = \dfrac{X_h^2 MSE}{\Sigma X_i^2}$	(5.21)	$\hat{Y}_h \pm ts(\hat{Y}_h)$
$Y_{h(new)}$	$s^2(Y_{h(new)}) = MSE\left[1 + \dfrac{X_h^2}{\Sigma X_i^2}\right]$	(5.22)	$\hat{Y}_h \pm ts(Y_{h(new)})$
		where:	
			$t = t(1 - \alpha/2; n - 1)$

degrees of freedom associated with *MSE*. The results in Table 5.1 are derived in analogous fashion to the earlier results for our general model (3.1). Whereas for the general case, we encounter terms $(X_i - \bar{X})^2$ or $(X_h - \bar{X})^2$, here we find X_i^2 and X_h^2 because of the regression through the origin.

Example

The Charles Plumbing Supplies Company operates 12 warehouses. In an attempt to tighten procedures for planning and control, a consultant studied the relation between number of work units performed (X) and total variable labor cost (Y) in the warehouses during a test period. The data are given in Table 5.2, and the observations are shown as a scatter plot in Figure 5.5.

Model (5.13) for regression through the origin was employed since Y involves variable costs only and the other conditions of the model appeared to be satisfied as well. From Table 5.2, we have $\Sigma X_i Y_i = 894,714$ and $\Sigma X_i^2 = 190,963$. Hence:

$$b_1 = \frac{\Sigma X_i Y_i}{\Sigma X_i^2} = \frac{894,714}{190,963} = 4.68527$$

and the estimated regression function is:

$$\hat{Y} = 4.68527X$$

The fitted regression line is plotted in Figure 5.5.

To illustrate inferences for regression through the origin, suppose an interval estimate of β_1 is desired with a 95 percent confidence coefficient. We obtain (calculations not shown):

$$MSE = \frac{\Sigma(Y_i - b_1 X_i)^2}{n - 1} = \frac{2,457.66}{11} = 223.42$$

TABLE 5.2 Data for regression through origin—warehouse example

Warehouse *i*	Work Units Performed X_i	Variable Labor Cost (dollars) Y_i	$X_i Y_i$	X_i^2
1	20	114	2,280	400
2	196	921	180,516	38,416
3	115	560	64,400	13,225
4	50	245	12,250	2,500
5	122	575	70,150	14,884
6	100	475	47,500	10,000
7	33	138	4,554	1,089
8	154	727	111,958	23,716
9	80	375	30,000	6,400
10	147	670	98,490	21,609
11	182	828	150,696	33,124
12	160	762	121,920	25,600
Total	1,359	6,390	894,714	190,963

FIGURE 5.5 Scatter plot and fitted regression through origin—warehouse example

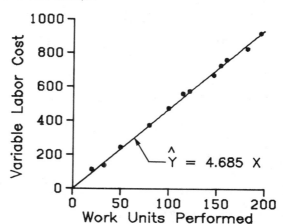

From Table 5.2, we have $\Sigma X_i^2 = 190{,}963$. Hence:

$$s^2(b_1) = \frac{MSE}{\Sigma X_i^2} = \frac{223.42}{190{,}963} = .0011700 \quad \text{or} \quad s(b_1) = .034205$$

For a 95 percent confidence coefficient, we require $t(.975; 11) = 2.201$. The confidence limits, by (5.20) in Table 5.1, are $b_1 \pm ts(b_1)$ or $4.68527 \pm 2.201(.034205)$. The 95 percent confidence interval for β_1 therefore is:

$$4.61 \leq \beta_1 \leq 4.76$$

Thus, with 95 percent confidence, it is estimated that the mean of the distribution of total variable labor costs increases by somewhere between \$4.61 and \$4.76 for each additional work unit performed.

Comments

1. In linear regression through the origin, there is no property of the form $\Sigma(Y_i - b_1 X_i) = \Sigma e_i = 0$. Consequently, the residuals usually will not sum to zero here. The only property comes from the normal equation (5.16), namely, $\Sigma X_i e_i = 0$.

2. In interval estimation of $E(Y_h)$ or $Y_{h(new)}$, note that the intervals (5.21) and (5.22) in Table 5.1 widen, the further X_h is from the origin. The reason is that the value of the true regression function is known precisely at the origin, so that the effect of the sampling error in the slope b_1 becomes increasingly important the farther X_h is from the origin.

3. Since only one regression parameter, β_1, must be estimated for the regression function (5.14), simultaneous estimation methods are not required to make a family of statements about several mean responses. For a given confidence coefficient $1 - \alpha$, formula (5.21) can be used repetitively with the given sample results to generate a family of statements for which the family confidence coefficient is still $1 - \alpha$.

4. Like any other model, model (5.13) should be evaluated for aptness. Even when it is known that the regression function must go through the origin, the function might not be

linear or the variance of the error terms might not be constant. Often one cannot be sure in advance that the regression function goes through the origin, and it is then safe practice to use the general model (3.1). If the regression does go through the origin, b_0 will differ from 0 only by a small sampling error, and unless the sample size is very small, use of model (3.1) has no disadvantages of any consequence. If the regression does not go through the origin, use of the general model (3.1) will avoid potentially serious difficulties resulting from forcing the regression through the origin when this is not appropriate.

5.6 EFFECT OF MEASUREMENT ERRORS

In our discussion of the regression model up to this point, we have not explicitly considered the presence of measurement errors in either X or Y. We now examine briefly the effect of measurement errors.

Measurement errors in Y

If random measurement errors are present in the dependent variable Y, no new problems are created if these errors are uncorrelated and not biased (positive and negative measurement errors tend to cancel out). Consider, for example, a study of the relation between the time required to complete a task (Y) and the complexity of the task (X). The time to complete the task may not be measured accurately because the person operating the stopwatch may not do so at the precise instants called for. As long as such measurement errors are of a random nature, uncorrelated, and not biased, these measurement errors can simply be absorbed in the model error term ε. The model error term reflects the composite effects of a large number of factors not considered in the model, one of which simply would be random errors due to inaccuracy in the process of measuring Y.

Measurement errors in X

Unfortunately, a different situation holds if the independent variable X is known only with measurement error. Frequently, to be sure, X is known without measurement error, as when the independent variable is price of a product, number of variables in an optimization problem, or wage rate for a class of employees. At other times, however, measurement errors may enter the value observed for the independent variable, for instance, when it is pressure, temperature, production line speed, or person's age.

We shall use the latter illustration in our development of the nature of the problem. Suppose we are regressing employees' piecework earnings on age. Let X_i denote the true age of the ith employee and X_i^* the age given by the employee on his or her employment record. Needless to say, the two are not always the same. We define the measurement error δ_i as follows:

$$(5.23) \qquad \delta_i = X_i^* - X_i$$

The regression model we would like to study is:

$$(5.24) \qquad Y_i = \beta_0 + \beta_1 X_i + \varepsilon_i$$

Since, however, we only observe X_i^*, model (5.24) becomes:

$$(5.25) \qquad Y_i = \beta_0 + \beta_1(X_i^* - \delta_i) + \varepsilon_i$$

where we make use of (5.23) in replacing X_i. We can rewrite (5.25) as follows:

$$(5.26) \qquad Y_i = \beta_0 + \beta_1 X_i^* + (\varepsilon_i - \beta_1\delta_i)$$

Model (5.26) may appear like an ordinary regression model, with independent variable X^* and error term $\varepsilon - \beta_1\delta$, but it is not. The independent variable observation X_i^* is a random variable, which, as we shall see, is correlated with the error term $\varepsilon_i - \beta_1\delta_i$. Theorem (3.40) for the case of random independent variables requires that the error term be independent of the independent variable. Hence, the standard regression results are not applicable for model (5.26).

Intuitively, we know that $\varepsilon_i - \beta_1\delta_i$ is not independent of X_i^* since (5.23) constrains $X_i^* - \delta_i$ to equal X_i. To determine the dependence formally, let us assume:

$$(5.27a) \qquad E(\delta_i) = 0$$

$$(5.27b) \qquad E(\varepsilon_i) = 0$$

$$(5.27c) \qquad E(\delta_i\varepsilon_i) = 0$$

Note that (5.27a) implies that $E(X_i^*) = E(X_i + \delta_i) = X_i$, and that (5.27c) assumes the measurement error δ_i is not correlated with the model error ε_i because by (1.19a) we have $\sigma(\delta_i, \varepsilon_i) = E(\delta_i\varepsilon_i)$ since $E(\delta_i) = E(\varepsilon_i) = 0$ by (5.27a) and (5.27b). We now wish to find the covariance:

$$\begin{aligned}
\sigma(X_i^*, \varepsilon_i - \beta_1\delta_i) &= E\{[X_i^* - E(X_i^*)][(\varepsilon_i - \beta_1\delta_i) - E(\varepsilon_i - \beta_1\delta_i)]\} \\
&= E[(X_i^* - X_i)(\varepsilon_i - \beta_1\delta_i)] \\
&= E[\delta_i(\varepsilon_i - \beta_1\delta_i)] \\
&= E(\delta_i\varepsilon_i - \beta_1\delta_i^2)
\end{aligned}$$

Now $E(\delta_i\varepsilon_i) = 0$ by (5.27c), and $E(\delta_i^2) = \sigma^2(\delta_i)$ by (1.14a) because $E(\delta_i) = 0$ by (5.27a). We therefore obtain:

$$(5.28) \qquad \sigma(X_i^*, \varepsilon_i - \beta_1\delta_i) = -\beta_1\sigma^2(\delta_i)$$

This covariance is not zero if there is a linear regression relation between X and Y.

If standard least squares procedures are applied to model (5.26), the estimators b_0 and b_1 are biased and also lack the property of consistency. Great difficulties are encountered in developing unbiased estimators when there are measurement errors in X. One approach is to impose severe conditions on the problem—for example, to make fairly strong assumptions about the properties of the distributions of δ_i, the covariance of δ_i and ε_j, and so on. Another approach is to use additional variables which are known to be related to the true value of X but not with the errors of measurement δ. Such variables are called "instrumental" variables because they are used as an instrument in studying the relation between X and Y. Instrumental variables make it possible to obtain consistent estimators of the regression parameters.

Discussions of possible approaches and further references will be found in specialized works such as Reference 5.3 and in statistical journals.

Note

It may be asked what is the distinction between the case when X is a random variable, considered in Chapter 3, and the case when X is subject to random measurement errors, and why are there special problems with the latter. When X is a random variable, it is not under the control of the analyst and will vary at random from trial to trial, as when X is the number of persons entering a store in a day. If this random variable X is not subject to measurement errors, however, it can be accurately ascertained for a given trial. Thus, if there are no measurement errors in counting the number of persons entering a store in a day, the analyst has accurate information to study the relation between number of persons entering the store and sales, even though the levels of number of customers which actually occur cannot be controlled. If, on the other hand, measurement errors are present in the number of persons entering the store, a distorted picture of the relation between number of persons and sales occurs because the sales observations will frequently be matched against an incorrect number of customers. This distorting effect of measurement errors is present whether X is fixed or random.

Berkson model

There is one situation where measurement errors in X are no problem. This case was first noted by Berkson (Ref. 5.4). Frequently, the independent variable in an experiment is set at a target value. For instance, in an experiment on the effect of temperature on typist productivity, the temperature may be set at target levels of 68°F, 70°F, and so on, according to the temperature control on the thermostat. The observed temperature X_i^* is fixed here, while the actual temperature X_i is a random variable since the thermostat may not be completely accurate. Similar situations exist when water pressure is set according to a gauge, or employees of specified ages according to their employment records are selected for a study.

In all of these cases, the observation X_i^* is a fixed quantity, while the unobserved true value X_i is a random variable. The measurement error is, as before:

$$(5.29) \qquad \delta_i = X_i^* - X_i$$

Here, however, there is no constraint on the relation between X_i^* and δ_i, since X_i^* is a fixed quantity. Again, we assume that $E(\delta_i) = 0$.

Model (5.26), which we obtained when replacing X_i by $X_i^* - \delta_i$, is still applicable for the Berkson case:

$$(5.30) \qquad Y_i = \beta_0 + \beta_1 X_i^* + (\varepsilon_i - \beta_1 \delta_i)$$

The expected value of the error term, $E(\varepsilon_i - \beta_1 \delta_i)$, is zero as before, since $E(\varepsilon_i) = 0$ and $E(\delta_i) = 0$. Further, $\varepsilon_i - \beta_1 \delta_i$ is now independent of X_i^*, since X_i^*

is a constant for the Berkson case. Hence, the conditions of an ordinary regression model are met:

1. The error terms have expectation 0.
2. The independent variable is a constant, and hence the error terms are independent of it.

Thus, standard least squares procedures can be applied for the Berkson case without modification, and the estimators b_0 and b_1 will be unbiased. If we can make the standard normality and constant variance assumptions for the errors $\varepsilon_i - \beta_1\delta_i$, the usual tests and interval estimates can be utilized.

5.7 WEIGHTED LEAST SQUARES

General approach

The least squares criterion (2.8):

$$Q = \sum_{i=1}^{n} (Y_i - \beta_0 - \beta_1 X_i)^2$$

weights each observation equally. There are times, however, when some observations should receive greater weight and others smaller weight. The weighted least squares criterion for simple linear regression is:

(5.31),
$$Q_w = \sum_{i=1}^{n} w_i(Y_i - \beta_0 - \beta_1 X_i)^2$$

where w_i is the given weight of the ith observation. Minimizing Q_w with respect to β_0 and β_1 leads to the normal equations:

(5.32)
$$\Sigma w_i Y_i = b_0 \Sigma w_i + b_1 \Sigma w_i X_i$$
$$\Sigma w_i X_i Y_i = b_0 \Sigma w_i X_i + b_1 \Sigma w_i X_i^2$$

In turn, these can be solved for the weighted least squares estimators b_0 and b_1:

(5.33a)
$$b_1 = \frac{\Sigma w_i X_i Y_i - \dfrac{\Sigma w_i X_i \Sigma w_i Y_i}{\Sigma w_i}}{\Sigma w_i X_i^2 - \dfrac{(\Sigma w_i X_i)^2}{\Sigma w_i}}$$

(5.33b)
$$b_0 = \frac{\Sigma w_i Y_i - b_1 \Sigma w_i X_i}{\Sigma w_i}$$

Note that if all weights are equal so w_i is identically equal to a constant, the normal equations (5.32) for weighted least squares reduce to the ones for unweighted least squares in (2.9) and the weighted least squares estimators (5.33) reduce to the ones for unweighted least squares in (2.10).

Weighted least squares should be used when the error term variance is not constant for all observations. The weight for an observation should then be the reciprocal of the observation's error term variance:

$$(5.34) \qquad w_i = \frac{1}{\sigma_i^2}$$

where σ_i^2 is the variance of the error term for the ith observation. Thus, observations whose error terms are subject to large variation receive less weight and observations whose error terms are subject to small variation receive more weight.

Unfortunately, the error term variances σ_i^2 are usually unknown. However, when the error term variance varies with the level of the independent variable in a systematic fashion, this relation can be exploited. For instance, if the error term variance σ_i^2 is proportional to X_i^2 so that $\sigma_i^2 = kX_i^2$ where k is a proportionality factor, the weights w_i would be:

$$(5.35) \qquad w_i = \frac{1}{kX_i^2}$$

Since the proportionality constant k drops out of the normal equations (5.32), we can simply use the weights:

$$(5.35a) \qquad w_i = \frac{1}{X_i^2}$$

This particular weight relation of the error term variance being proportional to X^2 is frequently encountered in business, economic, and biological applications, as in studies of savings regressed on income.

We illustrate these concepts by an example.

Example

The Nielsen Construction Company studied the relation between the size of a bid in million dollars (X) and the cost to the firm of preparing the bid in thousand dollars (Y), for 12 recent bids. The data are presented in Table 5.3, columns 1 and 2, and are shown in a scatter plot in Figure 5.6. The scatter plot strongly suggests that the error variance increases with X. An analyst after conducting a preliminary residual analysis concluded that the error variance is approximately proportional to X^2. (See the following Comment 4 for a discussion of how to study the relation between the error variance and the level of X.) The analyst, therefore, decided to employ the regression model:

$$(5.36) \qquad Y_i = \beta_0 + \beta_1 X_i + \varepsilon_i \qquad \text{where } \sigma_i^2 = kX_i^2$$

Column 3 in Table 5.3 contains the weights $w_i = 1/X_i^2$, and columns 4–6 contain the needed calculations. The least squares estimators (5.33a) and (5.33b) are, using the data in Table 5.3 and noting that $w_i X_i^2 \equiv 1$ here so that $\Sigma w_i X_i^2 = n = 12$:

TABLE 5.3 Regression calculations for weighted least squares—bid preparation example

(1) X_i	(2) Y_i	(3) $w_i = \dfrac{1}{X_i^2}$	(4) $w_i X_i = \dfrac{1}{X_i}$	(5) $w_i Y_i$	(6) $w_i X_i Y_i = \dfrac{Y_i}{X_i}$
2.13	15.5	.220415	.46948	3.41643	7.27700
1.21	11.1	.683013	.82645	7.58144	9.17355
11.00	62.6	.008264	.09091	.51733	5.69091
6.00	35.4	.027778	.16667	.98334	5.90000
5.60	24.9	.031888	.17857	.79401	4.44643
6.91	28.1	.020943	.14472	.58850	4.06657
2.97	15.0	.113367	.33670	1.70051	5.05051
3.35	23.2	.089107	.29851	2.06728	6.92537
10.39	42.0	.009263	.09625	.38905	4.04235
1.10	10.0	.826446	.90909	8.26446	9.09091
4.36	20.0	.052605	.22936	1.05210	4.58716
8.00	47.5	.015625	.12500	.74219	5.93750
Total 63.02	335.3	2.098714	3.87171	28.09664	72.18826

$$b_1 = \frac{72.18826 - \dfrac{(3.87171)(28.09664)}{2.098714}}{12 - \dfrac{(3.87171)^2}{2.098714}} = 4.19057$$

$$b_0 = \frac{28.09664 - 4.19057(3.87171)}{2.098714} = 5.65678$$

Hence, the fitted regression line is:

$$\hat{Y} = 5.6568 + 4.1906X$$

FIGURE 5.6 Scatter plot and fitted weighted regression line—bid preparation example

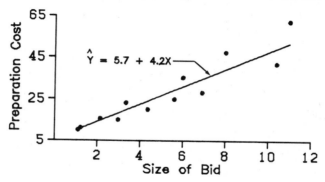

The fitted regression line is shown in Figure 5.6 and appears to be a reasonably good fit to the data.

If we had fitted a straight line by unweighted least squares to the original data in our example, we would have obtained the regression function:

$$\hat{Y} = 4.2289 + 4.5153X$$

This differs from the weighted least squares results, as will generally be the case. The reason is that the weights $w_i = 1/X_i^2$ give more emphasis to observations for smaller X (for which the distribution of Y has a smaller variance) and less emphasis to observations for larger X (for which the distribution of Y has a larger variance).

While the estimates obtained by unweighted least squares are unbiased, as are the estimates obtained by weighted least squares, the unweighted least squares estimates are subject to greater sampling variation. In our example, the estimated standard deviations of the regression coefficients for the two methods are:

Unweighted Least Squares	Weighted Least Squares
$s(b_0) = 3.2517$	$s(b_0) = .9652$
$s(b_1) = .5285$	$s(b_1) = .4037$

Comments

1. The condition of the error variance not being constant over all observations is called *heteroscedasticity,* in contrast to the condition of equal error variances, called *homoscedasticity.*

2. When heteroscedasticity prevails but the other conditions of model (3.1) are met, the estimators b_0 and b_1 obtained by ordinary least squares procedures are still unbiased and consistent, but they are no longer minimum variance unbiased estimators, as illustrated in the previous example.

3. Heteroscedasticity is inherent when the response in regression analysis follows a distribution in which the variance is functionally related to the mean. (Significant nonnormality in Y is encountered as well in most such cases.) Consider, in this connection, a regression analysis where X is the speed of a machine which puts a plastic coating on cable and Y is the number of blemishes in the coating per thousand feet drum of cable. If Y is Poisson distributed with a mean which increases as X increases, the distributions of Y cannot have constant variance at all levels of X since the variance of a Poisson variable equals the mean, which is increasing with X.

4. Replicate observations at several of the X levels are very desirable for obtaining information about the nature of the relation of the error variance to X. When replicates are not available, one may divide the residuals into, say, three or four groups of approximately equal size, according to ascending or descending order of X, calculate the variance of the residuals for each group, and examine the relation of these variances to various functions of X, such as $\sqrt{X}$, X, and X^2. This rough analysis frequently will be an adequate guide to decide about the approximate relation between the error variance and the level of X.

Weighted least squares by means of transformations

Many regression packages allow the user to perform a weighted least squares analysis as an option, with the user specifying the weights. When this option is not available, weighted least squares estimators can still be obtained by employing unweighted least squares on specially transformed observations.

To illustrate that weighted least squares is equivalent to unweighted least squares of specially transformed data, we consider again the weighted least squares criterion (5.31) for the case $\sigma_i^2 = kX_i^2$ where $w_i = 1/X_i^2$:

$$Q_w = \Sigma w_i(Y_i - \beta_0 - \beta_1 X_i)^2$$
$$= \Sigma \frac{1}{X_i^2}(Y_i - \beta_0 - \beta_1 X_i)^2$$

so that:

(5.37)
$$Q_w = \Sigma \left(\frac{Y_i}{X_i} - \frac{\beta_0}{X_i} - \beta_1 \right)^2$$

We shall use the following notation:

(5.38)
$$Y_i' = \frac{Y_i}{X_i} \qquad \beta_0' = \beta_1$$

$$X_i' = \frac{1}{X_i} \qquad \beta_1' = \beta_0$$

We can then express (5.37) as follows:

(5.39)
$$Q_w = \Sigma(Y_i' - \beta_0' - \beta_1' X_i')^2$$

which is the unweighted least squares criterion for the observations X_i' and Y_i' transformed according to (5.38).

The error variances for the model with the transformed variables are now constant. This can be seen by dividing the terms in the original model:

$$Y_i = \beta_0 + \beta_1 X_i + \varepsilon_i$$

by $\sqrt{w_i} = 1/X_i$:

$$\frac{Y_i}{X_i} = \frac{\beta_0}{X_i} + \beta_1 + \frac{\varepsilon_i}{X_i}$$

so that we obtain, using the notation in (5.38):

(5.40)
$$Y_i' = \beta_0' + \beta_1' X_i' + \varepsilon_i'$$

where:

(5.40a)
$$\varepsilon_i' = \frac{\varepsilon_i}{X_i}$$

Note that (5.40) is the simple linear regression model using the transformed variables. The variances of the error terms ε_i' in model (5.40) are now constant, as can be seen using (1.15b):

$$(5.41) \qquad \sigma^2(\varepsilon_i') = \sigma^2\left(\frac{\varepsilon_i}{X_i}\right) = \frac{1}{X_i^2}\sigma^2(\varepsilon_i) = \frac{1}{X_i^2}(kX_i^2) = k$$

Hence, using unweighted least squares for the dependent variable $Y' = Y/X$ and the independent variable $X' = 1/X$ will yield the same results as a weighted least squares analysis with weights $w_i = 1/X_i^2$. Once the unweighted least squares estimators b_0' and b_1' are obtained for the transformed variables, they are related to the weighted least squares estimators for the original variables using (5.38), so that $b_0' = b_1$ and $b_1' = b_0$.

Appropriate transformations of the observations for other variance relationships, e.g., $\sigma_i^2 = kX_i$, can be found in the same fashion as explained here for $\sigma_i^2 = kX_i^2$.

5.8 INVERSE PREDICTIONS

At times, a regression model of Y on X is used to make a prediction of the value of X which gave rise to a new observation Y. This is known as an *inverse prediction*. We illustrate inverse predictions by two examples:

1. A trade association analyst has regressed the selling price of a product (Y) on its cost (X) for the 15 member firms of the association. The selling price $Y_{h(new)}$ for another firm not belonging to the trade association is known, and it is desired to estimate the cost $X_{h(new)}$ for this firm.

2. A regression analysis of the decrease in cholesterol level (Y) against dosage of a new drug (X) has been conducted, based on observations for 50 patients. A physician is treating a new patient for whom the cholesterol level should decrease by $Y_{h(new)}$. It is desired to estimate the appropriate dosage level $X_{h(new)}$ to be administered to bring about the desired cholesterol level decrease $Y_{h(new)}$.

In inverse predictions, model (3.1) is assumed as before:

$$(5.42) \qquad Y_i = \beta_0 + \beta_1 X_i + \varepsilon_i$$

The estimated regression function based on n observations is obtained as usual:

$$(5.43) \qquad \hat{Y} = b_0 + b_1 X$$

A new observation $Y_{h(new)}$ becomes available, and it is desired to estimate the level $X_{h(new)}$ which gave rise to this new observation. A natural point estimator is obtained by solving (5.43) for X, given $Y_{h(new)}$:

$$(5.44) \qquad \hat{X}_{h(new)} = \frac{Y_{h(new)} - b_0}{b_1} \qquad b_1 \neq 0$$

where $\hat{X}_{h(new)}$ denotes the point estimator of the new level $X_{h(new)}$. Figure 5.7

contains a representation of this point estimator for an example to be discussed shortly. $\hat{X}_{h(new)}$ is, indeed, the maximum likelihood estimator of $X_{h(new)}$ for regression model (3.1).

It can be shown that approximate $1 - \alpha$ confidence limits for $X_{h(new)}$ are:

(5.45) $$\hat{X}_{h(new)} \pm t(1 - \alpha/2; n - 2)s(\hat{X}_{h(new)})$$

where:

(5.45a) $$s^2(\hat{X}_{h(new)}) = \frac{MSE}{b_1^2}\left[1 + \frac{1}{n} + \frac{(\hat{X}_{h(new)} - \bar{X})^2}{\Sigma(X_i - \bar{X})^2}\right]$$

Example

A medical researcher in an experiment employed a new method for measuring low concentration of galactose (sugar) in the blood (Y) on 12 samples containing known concentrations (X). Altogether, four concentration levels were used in the experiment. Linear regression model (3.1) was fitted with the following results:

$$n = 12 \qquad b_0 = -.100 \qquad b_1 = 1.017 \qquad MSE = .0272$$
$$s(b_1) = .0142 \qquad \bar{X} = 5.500 \qquad \bar{Y} = 5.492 \qquad \Sigma(X_i - \bar{X})^2 = 135$$
$$\hat{Y} = -.100 + 1.017X$$

The data and the estimated regression line are plotted in Figure 5.7.

FIGURE 5.7 Scatter plot and fitted regression line—calibration example

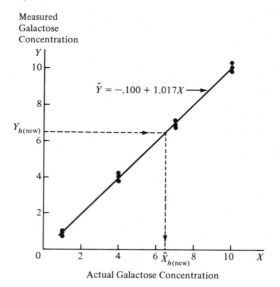

Actual Galactose Concentration

The researcher first wished to make sure that there is a linear association between the two variables. A test of $H_0: \beta_1 = 0$ versus $H_a: \beta_1 \neq 0$ utilizing the test statistic $b_1/s(b_1) = 1.017/.0142 = 71.6$ was conducted for $\alpha = .05$. Since $t(.975; 10) = 2.228$, $|t^*| = 71.6 > 2.228$ and it was concluded that $\beta_1 \neq 0$, or that a linear association exists between the measured concentration and the actual concentration.

The researcher now wishes to use the regression relation for a new patient for whom the new measurement procedure yielded $Y_{h(new)} = 6.52$. It is desired to estimate the actual concentration $X_{h(new)}$ for this patient by means of a 95 percent confidence interval.

Using (5.44) and (5.45a), we obtain:

$$\hat{X}_{h(new)} = \frac{6.52 - (-.100)}{1.017} = 6.509$$

$$s^2(\hat{X}_{h(new)}) = \frac{.0272}{(1.017)^2}\left[1 + \frac{1}{12} + \frac{(6.509 - 5.500)^2}{135}\right] = .0287$$

so that $s(\hat{X}_{h(new)}) = \sqrt{.0287} = .1694$. We require $t(.975; 10) = 2.228$, and using (5.45), we obtain the confidence limits $6.509 \pm 2.228(.1694)$. Hence, the 95 percent confidence interval is:

$$6.13 \leq X_{h(new)} \leq 6.89$$

Thus, it can be concluded with 95 percent confidence that the actual galactose concentration is between 6.13 and 6.89. This is approximately a ± 6 percent error, which was considered reasonable by the researcher.

Comments

1. The inverse prediction problem is also known as a *calibration problem* since it is applicable when inexpensive, quick, and approximate measurements (Y) are related to precise, often expensive, and time-consuming measurements (X) based on n observations. The resulting regression model is then used to estimate for a new approximate measurement $Y_{h(new)}$ what is the precise measurement $X_{h(new)}$. We illustrated this use in our previous example.

2. The approximate confidence interval (5.45) is appropriate if the quantity:

(5.46)
$$\frac{[t(1 - \alpha/2; n - 2)]^2 MSE}{b_1^2 \Sigma(X_i - \bar{X})^2}$$

is small, say less than .1. For our example, this quantity is:

$$\frac{(2.228)^2(.0272)}{(1.017)^2(135)} = .00097$$

so that the approximate confidence interval is appropriate.

3. Simultaneous prediction intervals based on g different new observed measurements $Y_{h(new)}$, with a $1 - \alpha$ family confidence coefficient, are easily obtained using either the Bonferroni or the Scheffé procedures discussed in Section 5.4. The value of $t(1 - \alpha/2; n - 2)$ in (5.45) is replaced by either $B = t(1 - \alpha/2g; n - 2)$ or $S = [gF(1 - \alpha; g, n - 2)]^{1/2}$.

5.9 CHOICE OF X LEVELS

When regression data are obtained by experiment, the levels of X at which observations on Y are to be taken are under the control of the experimenter. Among other things, the experimenter will have to consider:

1. How many levels of X should be investigated?
2. What shall the two extreme levels be?
3. How shall the other levels of X, if any, be spaced?
4. How many observations should be taken at each level of X?

There is no single answer to these questions, since different purposes of the regression analysis lead to different answers. The possible objectives in regression analysis are varied, as we have noted earlier. The main objective may be to estimate the slope of the regression line, or in some cases to estimate the intercept. In many cases, the main objective is to predict one or more new observations or to estimate one or more mean responses. When the regression function is curvilinear, the main objective may be to locate the maximum or minimum mean response. At still other times, the main purpose is to determine the nature of the regression function.

To illustrate how the purpose affects the design, consider the variances of b_0, b_1, $\hat{Y}_h$, and $Y_{h(new)}$ which were developed earlier:

$$(5.47) \qquad \sigma^2(b_0) = \sigma^2 \frac{\Sigma X_i^2}{n\Sigma(X_i - \bar{X})^2} = \sigma^2 \left[\frac{1}{n} + \frac{\bar{X}^2}{\Sigma(X_i - \bar{X})^2} \right]$$

$$(5.48) \qquad \sigma^2(b_1) = \frac{\sigma^2}{\Sigma(X_i - \bar{X})^2}$$

$$(5.49) \qquad \sigma^2(\hat{Y}_h) = \sigma^2 \left[\frac{1}{n} + \frac{(X_h - \bar{X})^2}{\Sigma(X_i - \bar{X})^2} \right]$$

$$(5.50) \qquad \sigma^2(Y_{h(new)}) = \sigma^2 \left[1 + \frac{1}{n} + \frac{(X_h - \bar{X})^2}{\Sigma(X_i - \bar{X})^2} \right]$$

The variance of the slope is minimized if $\Sigma(X_i - \bar{X})^2$ is maximized. This is accomplished by using two levels of X, at the two extremes for the scope of the model, and placing half of the observations at each of the two levels. Of course, if one were not sure of the linearity of the regression function, one would be hesitant to use only two levels since they would provide no information about possible departures from linearity.

If the main purpose is to estimate β_0, the number and placement of levels does not matter as long as $\bar{X} = 0$. On the other hand, to estimate the mean response or predict a new observation at X_h, it is best to use X levels so that $\bar{X} = X_h$. If a number of mean responses are to be estimated or a number of new observations are to be predicted, it would be best to spread out the X levels such that $\bar{X}$ is in the center of the X_h levels of interest.

Although the number and spacing of X levels depends very much on the major purpose of the regression analysis, some general advice can be given, at least to be used as a point of departure. D. R. Cox, a well-known statistician, suggests as follows:

> Use two levels when the object is primarily to examine whether or not . . . (the independent variable) . . . has an effect and in which direction that effect is. Use three levels whenever a description of the response curve by its slope and curvature is likely to be adequate; this should cover most cases. Use four levels if further examination of the shape of the response curve is important. Use more than four levels when it is required to estimate the detailed shape of the response curve, or when the curve is expected to rise to an asymptotic value, or in general to show features not adequately described by slope and curvature. Except in these last cases it is generally satisfactory to use equally spaced levels with equal numbers of observations per level [Ref. 5.5].

PROBLEMS

5.1. When joint confidence intervals for β_0 and β_1 are developed by the Bonferroni method with a family confidence coefficient of 90 percent, does this imply that 10 percent of the time the confidence intervals for β_0 will be incorrect? That 5 percent of the time the confidence interval for β_0 will be incorrect and 5 percent of the time that for β_1 will be incorrect? Discuss.

5.2. Refer to Problem 3.1. Suppose the student combines the two confidence intervals into a confidence set. What can you say about the family confidence coefficient for this set?

5.3. Refer to **Calculator maintenance** Problems 2.16 and 3.5.
    a.   Will b_0 and b_1 tend to err in the same direction or in opposite directions here? What does this imply about the tilt of the elliptical joint confidence region for β_0 and β_1 here?
    b.   Obtain a few boundary points of the joint confidence region for β_0 and β_1 and plot the boundary. Use a 95 percent confidence coefficient. Interpret your joint confidence region.
    c.   A consultant has suggested that β_0 should be zero and β_1 should equal 14.0. Does your joint confidence region support this view?
    d.   Obtain Bonferroni joint confidence intervals for β_0 and β_1 using a 95 percent family confidence coefficient. Do these intervals support the view of the consultant in part (c)?

5.4. Refer to **Airfreight breakage** Problem 2.17.
    a.   Will b_0 and b_1 tend to err in the same direction or in opposite directions here? What does this imply about the tilt of the elliptical joint confidence region for β_0 and β_1 here?
    b.   Obtain a few boundary points of the joint confidence region for β_0 and β_1 and plot the boundary. Use a 99 percent confidence coefficient. Interpret your joint confidence region.
    c.   Obtain Bonferroni joint confidence intervals for β_0 and β_1 using a 99 percent family confidence coefficient. Do the Bonferroni joint confidence intervals

provide substantially less precise information than the joint confidence region in part (b)?

5.5. Refer to **Plastic hardness** Problem 2.18.

a. Obtain Bonferroni joint confidence intervals for β_0 and β_1 using a 90 percent family confidence coefficient. Interpret your confidence intervals.

b. Are b_0 and b_1 positively or negatively correlated here? Is this reflected in your joint confidence intervals in part (a)?

c. What is the meaning of the family confidence coefficient in part (a)?

5.6. Refer to **Muscle mass** Problem 2.23.

a. Obtain Bonferroni joint confidence intervals for β_0 and β_1 using a 99 percent family confidence coefficient. Interpret your confidence intervals.

b. Will b_0 and b_1 tend to err in the same direction or in opposite directions here? Explain.

c. A researcher has suggested that β_0 should equal approximately 160 and that β_1 should be between -1.9 and -1.5. Do the joint confidence intervals in part (a) support this expectation?

5.7. Refer to **Airfreight breakage** Problem 2.17. Obtain selected boundary points for the 90 percent confidence band for the true regression line. Plot the confidence band, together with the estimated regression line. How precisely do we estimate the location of the true regression line here?

5.8. Refer to **Plastic hardness** Problem 2.18. Obtain selected boundary points for the 95 percent confidence band for the true regression line. Plot the confidence band, together with the estimated regression line. How precisely do we estimate the location of the true regression line here?

5.9. Refer to **Calculator maintenance** Problems 2.16 and 3.5.

a. Estimate the expected number of minutes spent when there are 3, 5, and 7 machines to be serviced, respectively. Use interval estimates with a 90 percent family confidence coefficient based on the Working-Hotelling approach.

b. Two service calls for preventive maintenance are scheduled in which the numbers of machines to be serviced are 4 and 7, respectively. A family of prediction intervals for the times to be spent on these calls is desired with a 90 percent family confidence coefficient. Which procedure, Scheffé or Bonferroni, will provide tighter prediction limits here?

c. Obtain the family of prediction limits required in part (b) using the more efficient procedure.

5.10. Refer to **Airfreight breakage** Problem 2.17.

a. It is desired to obtain interval estimates of the mean number of broken ampules when there are 0, 1, and 2 transfers for the shipment using a 95 percent family confidence coefficient. Obtain the desired confidence intervals using the Working-Hotelling approach.

b. Are the confidence intervals obtained in part (a) more efficient than Bonferroni intervals here? Explain.

c. The next three shipments will make 0, 1, and 2 transfers, respectively. Obtain prediction limits for the number of broken ampules for each of these three shipments using the Scheffé procedure and a 95 percent family confidence coefficient.

    d. Would the Bonferroni procedure have been more efficient in developing the prediction intervals in part (c)? Explain.

5.11. Refer to **Plastic hardness** Problem 2.18.

    a. Management wishes to obtain interval estimates of the mean hardness when the elapsed time is 40, 50, and 60 hours, respectively. Calculate the desired confidence intervals using the Bonferroni procedure and a 90 percent family confidence coefficient. What is the meaning of the family confidence coefficient here?

    b. Is the Bonferroni procedure employed in part (a) the most efficient one that could be employed here? Explain.

    c. The next two test items will be measured after 30 and 40 hours of elapsed time, respectively. Predict the hardness for each of these two items using the most efficient procedure and a 90 percent family confidence coefficient.

5.12. Refer to **Muscle mass** Problem 2.23.

    a. The nutritionist is particularly interested in the mean muscle mass for women aged 45, 55, and 65. Obtain joint confidence intervals for the means of interest using the Working-Hotelling procedure and a 95 percent family confidence coefficient.

    b. Is the Working-Hotelling approach the most efficient one to be employed in part (a)? Explain.

    c. Three additional women aged 48, 59, and 74 have contacted the nutritionist. Predict the muscle mass for each of these three women using the Bonferroni approach and a 95 percent family confidence coefficient.

    d. Subsequently, the nutritionist wishes to predict the muscle mass for a fourth woman aged 64, with a family confidence coefficient of 95 percent for the four predictions. Will the three prediction intervals in part (c) have to be recalculated? Would this also be true if the Scheffé method had been used in constructing the prediction intervals?

5.13. A behavioral scientist stated recently: "I am never sure whether the regression line goes through the origin. Hence, I will not use such a model." Comment.

5.14. **Typographical errors.** Shown below are the number of galleys of type set (X) and the dollar cost of correcting typographical errors (Y) in a random sample of recent printing orders handled by a firm specializing in technical printing. Since Y involves variable costs only, an analyst wished to determine whether the regression through the origin model (5.13) is apt for studying the relation between the two variables.

i:	1	2	3	4	5	6	7	8	9	10	11	12
X_i:	7	12	10	10	14	25	30	25	18	10	4	6
Y_i:	128	213	191	178	250	446	540	457	324	177	75	107

    a. Fit model (5.13) and state the estimated regression function.

    b. Plot the estimated regression function and the data. Does a linear regression function through the origin appear to provide a good fit here? Comment.

    c. In estimating costs of handling prospective orders, management has used a standard of $17.50 per galley for the cost of correcting typographical errors. Test whether or not this standard should be revised; use $\alpha = .02$. State the alternatives, decision rule, and conclusion.

d. Obtain a prediction interval for the correction cost on a forthcoming job involving 10 galleys. Use a confidence coefficient of 98 percent.

5.15. Refer to **Typographical errors** Problem 5.14. Conduct a formal test for lack of fit of linear regression through the origin; use $\alpha = .01$. State the alternatives, decision rule, and conclusion.

5.16. Refer to **Grade point average** Problem 2.15. Assume that linear regression through the origin model (5.13) is appropriate.
   a. Fit model (5.13) and state the estimated regression function.
   b. Estimate β_1 with a 95 percent confidence interval. Interpret your interval estimate.
   c. Estimate the mean freshman GPA for students whose entrance test score is 5.7. Use a 95 percent confidence interval.

5.17. Refer to **Grade point average** Problem 5.16.
   a. Plot the fitted regression line and the data. Does the linear regression function through the origin appear to be a good fit here?
   b. Conduct a formal test for lack of fit of linear regression through the origin; use $\alpha = .005$. State the alternatives, decision rule, and conclusion. What is the P-value of the test?

5.18. Refer to **Calculator maintenance** Problem 2.16. Assume that linear regression through the origin model (5.13) is appropriate.
   a. Obtain the estimated regression function.
   b. Estimate β_1 with a 90 percent confidence interval. Interpret your interval estimate.
   c. Predict the service time on a new call in which six machines are to be serviced. Use a 90 percent prediction interval.

5.19. Refer to **Calculator maintenance** Problem 5.18.
   a. Plot the fitted regression line and the data. Does the linear regression function through the origin appear to be a good fit here?
   b. Conduct a formal test for lack of fit of linear regression through the origin; use $\alpha = .01$. State the alternatives, decision rule, and conclusion. What is the P-value of the test?

5.20. Refer to **Plastic hardness** Problem 2.18. Suppose that errors arise in X because the laboratory technician is instructed to measure the hardness of the ith specimen (Y_i) at a prerecorded elapsed time (X_i), but the timing is imperfect so the true elapsed time varies at random from the prerecorded elapsed time. Will ordinary least squares estimates be biased here? Discuss.

5.21. **Computer-assisted learning.** Data from a study of computer-assisted learning by 12 students, showing the total number of responses in completing a lesson (X) and the cost of computer time $(Y$, in cents), follow.

i:	1	2	3	4	5	6	7	8	9	10	11	12
X_i:	16	14	22	10	14	17	10	13	19	12	18	11
Y_i:	77	70	85	50	62	70	52	63	88	57	81	54

   a. Fit a linear regression function by ordinary least squares, obtain the residuals, and prepare a plot of the residuals against X. What does the residual plot suggest?

b. Assume that $\sigma_i^2 = kX_i^2$, and use weighted least squares to fit the linear regression function. Are the regression coefficients similar to those obtained in part (a) with ordinary least squares? How do the standard deviations of the regression coefficients compare?

c. What transformation of the variables, if used with ordinary least squares, would have yielded the same regression coefficients as weighted least squares here?

5.22. Refer to **Machine speed** Problem 4.21.

a. Verify that the relation $\sigma_i = kX_i$ holds approximately here.

b. Use weighted least squares to fit the linear regression function.

c. Obtain an interval estimate of β_1 with a 95 percent confidence coefficient.

d. What transformation of the variables when used with ordinary least squares would have yielded the same regression coefficients as weighted least squares here?

5.23. Refer to **Grade point average** Problems 2.15 and 3.4. A new student earned a grade point average of 3.4 in the freshman year.

a. Obtain a 90 percent confidence interval for the student's entrance test score. Interpret your confidence interval.

b. Is criterion (5.46) met as to the appropriateness of the approximate confidence interval?

5.24. Refer to **Plastic hardness** Problem 2.18. The measurement of a new test item showed 298 Brinell units of hardness.

a. Obtain a 99 percent confidence interval for the elapsed time before the hardness was measured. Interpret your confidence interval.

b. Is criterion (5.46) met as to the appropriateness of the approximate confidence interval?

EXERCISES

5.25. What simplification takes place in the boundary of the joint confidence region (5.2) for β_0 and β_1 if the independent variable is so coded that $\bar{X} = 0$? What happens to the shape of the boundary?

5.26. If the independent variable is so coded that $\bar{X} = 0$ and the normal error model (3.1) applies, are b_0 and b_1 independent? Are the confidence intervals for β_0 and β_1 then independent?

5.27. Derive an extension of the Bonferroni inequality (5.6a) for the case of three statements, each with statement confidence coefficient $1 - \alpha$.

5.28. Show that for the fitted least squares regression line through the origin (5.18), $\Sigma X_i e_i = 0$.

5.29. Show that $\hat{Y}$ as defined in (5.18) for linear regression through the origin is an unbiased estimator of $E(Y)$.

5.30. Derive the formula for $s^2(\hat{Y}_h)$ given in Table 5.1 for linear regression through the origin.

5.31. (Calculus needed.) Derive the weighted least squares normal equations for fitting a linear regression function when the variance of the error terms is proportional to

X—i.e., $\sigma^2(\varepsilon_i) = kX_i$. Check your results by making appropriate substitutions in (5.32).

5.32. Express the weighted least squares estimator b_1 in (5.33a) in terms of the deviations $X_i - \bar{X}_w$ and $Y_i - \bar{Y}_w$, where $\bar{X}_w$ and $\bar{Y}_w$ are weighted means.

PROJECTS

5.33. Refer to the **SMSA** data set and Project 2.38. Consider the regression relation of number of active physicians to total population.
  a. Obtain a joint confidence region for β_0 and β_1 and plot it; use a 95 percent confidence coefficient.
  b. Plot the 95 percent confidence band for the regression line. Does it provide a fairly precise indication of the location of the regression line?

5.34. Refer to the **SENIC** data set and Project 2.40. Consider the regression relation of length of stay to infection risk.
  a. Obtain a joint confidence region for β_0 and β_1 and plot it; use a 90 percent confidence coefficient.
  b. Plot the 90 percent confidence band for the regression line. Does it provide a fairly precise indication of the location of the regression line?

5.35. Five observations on Y are to be taken when $X = 10, 20, 30, 40,$ and 50, respectively. The true regression function is $E(Y) = 20 + 10X$. The error terms are independent and normally distributed, with $E(\varepsilon_i) = 0$ and $\sigma^2(\varepsilon_i) = .8X_i$.
  a. Generate a random Y observation for each X level, and calculate both the ordinary and weighted least squares estimates of the regression coefficient β_1 in linear regression model (3.1).
  b. Repeat part (a) 200 times, generating new random numbers each time.
  c. Calculate the mean and variance of the 200 ordinary least squares estimates of β_1, and do the same for the 200 weighted least squares estimates.
  d. Do both the ordinary least squares and weighted least squares estimators appear to be unbiased? Explain. Which estimator appears to be more precise here? Comment.

CITED REFERENCES

5.1 Gafarian, A. V. "Confidence Bands in Straight Line Regression." *Journal of the American Statistical Association* 59 (1964), pp. 182–213.
5.2 Miller, Rupert G., Jr. *Simultaneous Statistical Inference.* 2d ed. New York: Springer-Verlag, 1981, pp. 114–16.
5.3 Johnston, J. *Econometric Methods.* 2d ed. New York: McGraw-Hill, 1971.
5.4 Berkson, J. "Are There Two Regressions?" *Journal of the American Statistical Association* 45 (1950), pp. 164–80.
5.5 Cox, D. R. *Planning of Experiments.* New York: John Wiley & Sons, 1958, pp. 141–42.

PART II

General regression and correlation analysis

6

Matrix approach to
simple regression analysis

Matrix algebra is widely used for mathematical and statistical analysis. The matrix approach is practically a necessity in multiple regression analysis, since it permits extensive systems of equations and large arrays of data to be denoted compactly and operated upon efficiently.

In this chapter, we first take up a brief introduction to matrix algebra. (A fuller treatment of matrix algebra may be found in specialized texts such as Reference 6.1.) Then we apply matrix methods to the simple linear regression model discussed in Part I. While matrix algebra is not really required for simple regression with one independent variable, the application of matrix methods to this case will provide a transition to multiple regression which will be taken up in succeeding chapters.

Readers who are familiar with matrix algebra may wish to scan the introductory parts of this chapter and focus upon the parts dealing with the use of matrix methods in regression analysis.

6.1 MATRICES

Definition of matrix

A matrix is a rectangular array of elements arranged in rows and columns. An example of a matrix is:

$$
\begin{array}{cc}
 & \text{Column} \quad \text{Column} \\
 & 1 \qquad\quad 2 \\
\begin{array}{c} \text{Row 1} \\ \text{Row 2} \\ \text{Row 3} \end{array}
& \begin{bmatrix} 6{,}000 & 23 \\ 13{,}000 & 47 \\ 11{,}000 & 35 \end{bmatrix}
\end{array}
$$

The *elements* of this particular matrix are numbers representing income (column 1) and age (column 2) of three persons. The elements are arranged by row (person) and column (characteristic of person). Thus, the element in the first row and first column (6,000) represents the income of the first person. The element in the first row and second column (23) represents the age of the first person. The *dimension* of the matrix is 3×2, i.e., 3 rows by 2 columns. If we wanted to present income and age for 1,000 persons in a matrix with the same format as the one earlier, we would require a $1{,}000 \times 2$ matrix.

Other examples of matrices are:

$$
\begin{bmatrix} 1 & 0 \\ 5 & 10 \end{bmatrix} \qquad
\begin{bmatrix} 4 & 7 & 12 & 16 \\ 3 & 15 & 9 & 8 \end{bmatrix}
$$

These two matrices have dimensions of 2×2 and 2×4, respectively. Note that we always specify the number of rows first and then the number of columns in giving the dimension of a matrix.

As in ordinary algebra, we may use symbols to identify the elements of a matrix:

$$
\begin{array}{c}
\qquad\quad j = 1 \quad\ j = 2 \quad\ j = 3 \\
\begin{array}{c} i = 1 \\ i = 2 \end{array}
\begin{bmatrix} a_{11} & a_{12} & a_{13} \\ a_{21} & a_{22} & a_{23} \end{bmatrix}
\end{array}
$$

Note that the first subscript identifies the row number and the second the column number. We shall use the general notation a_{ij} for the element in the ith row and the jth column. In our above example, $i = 1, 2$ and $j = 1, 2, 3$.

A matrix may be denoted by a symbol such as $\mathbf{A}$, $\mathbf{X}$, or $\mathbf{Z}$. The symbol is in **boldface** to identify that it refers to a matrix. Thus, we might define for the above matrix:

$$
\mathbf{A} = \begin{bmatrix} a_{11} & a_{12} & a_{13} \\ a_{21} & a_{22} & a_{23} \end{bmatrix}
$$

Reference to the matrix $\mathbf{A}$ then implies reference to the 2×3 array just given.

Another notation for the matrix $\mathbf{A}$ just given is:

$$
\mathbf{A} = [a_{ij}] \qquad i = 1, 2; j = 1, 2, 3
$$

which avoids the need for writing out all elements of the matrix by stating only the general element. This notation can only be used, of course, when the elements of a matrix are symbols.

To summarize, a matrix with r rows and c columns will be represented either in full:

$$A = \begin{bmatrix} a_{11} & a_{12} & \cdots & a_{1j} & \cdots & a_{1c} \\ a_{21} & a_{22} & \cdots & a_{2j} & \cdots & a_{2c} \\ \cdot & \cdot & & \cdot & & \cdot \\ \cdot & \cdot & & \cdot & & \cdot \\ \cdot & \cdot & & \cdot & & \cdot \\ a_{i1} & a_{i2} & \cdots & a_{ij} & \cdots & a_{ic} \\ \cdot & \cdot & & \cdot & & \cdot \\ \cdot & \cdot & & \cdot & & \cdot \\ \cdot & \cdot & & \cdot & & \cdot \\ a_{r1} & a_{r2} & \cdots & a_{rj} & \cdots & a_{rc} \end{bmatrix}$$

(6.1)

or in the abbreviated form:

(6.2) $$A = [a_{ij}] \qquad i = 1, \ldots, r; j = 1, \ldots, c$$

or simply by a boldface symbol, such as A.

Comments

1. Do not think of a matrix as a number. It is a set of elements arranged in an array. Only when the matrix has dimension 1×1 is there a single number in the matrix, in which case one *can* think of it interchangeably as either a matrix or a number.

2. The following is *not* a matrix:

$$\begin{bmatrix} & 14 & \\ & 8 & \\ 10 & & 15 \\ 9 & & 16 \end{bmatrix}$$

since the numbers are not arranged in columns and rows.

Square matrix

A matrix is said to be square if the number of rows equals the number of columns. Two examples are:

$$\begin{bmatrix} 4 & 7 \\ 3 & 9 \end{bmatrix} \qquad \begin{bmatrix} a_{11} & a_{12} & a_{13} \\ a_{21} & a_{22} & a_{23} \\ a_{31} & a_{32} & a_{33} \end{bmatrix}$$

Vector

A matrix containing only one column is called a *column vector* or simply a *vector*. Two examples are:

$$A = \begin{bmatrix} 4 \\ 7 \\ 10 \end{bmatrix} \qquad C = \begin{bmatrix} c_1 \\ c_2 \\ c_3 \\ c_4 \\ c_5 \end{bmatrix}$$

The vector A is a 3×1 matrix, and the vector C is a 5×1 matrix.

A matrix containing only one row is called a *row vector*. Two examples are:

$$\mathbf{B}' = [15 \quad 25 \quad 50] \qquad \mathbf{F}' = [f_1 \quad f_2]$$

We use the prime symbol for row vectors for reasons to be seen shortly. Note that the row vector $\mathbf{B}'$ is a 1×3 matrix and the row vector $\mathbf{F}'$ is a 1×2 matrix.

A single subscript suffices to identify the elements of a vector.

Transpose

The transpose of a matrix $\mathbf{A}$ is another matrix, denoted by $\mathbf{A}'$, that is obtained by interchanging corresponding columns and rows of the matrix $\mathbf{A}$.

For example, if:

$$\underset{3 \times 2}{\mathbf{A}} = \begin{bmatrix} 2 & 5 \\ 7 & 10 \\ 3 & 4 \end{bmatrix}$$

then the transpose $\mathbf{A}'$ is:

$$\underset{2 \times 3}{\mathbf{A}'} = \begin{bmatrix} 2 & 7 & 3 \\ 5 & 10 & 4 \end{bmatrix}$$

Note that the first column of $\mathbf{A}$ is the first row of $\mathbf{A}'$, and similarly the second column of $\mathbf{A}$ is the second row of $\mathbf{A}'$. Correspondingly, the first row of $\mathbf{A}$ has become the first column of $\mathbf{A}'$, and so on. Note that the dimension of $\mathbf{A}$, indicated under the symbol $\mathbf{A}$, becomes reversed for the dimension of $\mathbf{A}'$.

As another example, consider:

$$\underset{3 \times 1}{\mathbf{C}} = \begin{bmatrix} 4 \\ 7 \\ 10 \end{bmatrix} \qquad \underset{1 \times 3}{\mathbf{C}'} = [4 \quad 7 \quad 10]$$

Thus, the transpose of a column vector is a row vector, and vice versa. This is the reason why we used the symbol $\mathbf{B}'$ earlier to identify a row vector, since it may be thought of as the transpose of a column vector $\mathbf{B}$.

In general, we have:

$$\underset{r \times c}{\mathbf{A}} = \begin{bmatrix} a_{11} & \cdots & a_{1c} \\ \cdot & & \cdot \\ \cdot & & \cdot \\ \cdot & & \cdot \\ a_{r1} & \cdots & a_{rc} \end{bmatrix} = [a_{ij}] \qquad i = 1, \ldots, r; j = 1, \ldots, c$$

$$\overset{\text{Row}}{\underset{\text{index}}{\uparrow}} \overset{\text{Column}}{\underset{\text{index}}{\nwarrow}}$$

(6.3)

$$\underset{c \times r}{\mathbf{A}'} = \begin{bmatrix} a_{11} & \cdots & a_{r1} \\ \cdot & & \cdot \\ \cdot & & \cdot \\ \cdot & & \cdot \\ a_{1c} & \cdots & a_{rc} \end{bmatrix} = [a_{ji}] \qquad j = 1, \ldots, c; i = 1, \ldots, r$$

$$\overset{\text{Row}}{\underset{\text{index}}{\uparrow}} \overset{\text{Column}}{\underset{\text{index}}{\nwarrow}}$$

Thus, the element in the *i*th row and *j*th column in **A** is found in the *j*th row and *i*th column in **A'**.

Equality of matrices

Two matrices **A** and **B** are said to be equal if they have the same dimension and if all corresponding elements are equal. Conversely, if two matrices are equal, their corresponding elements are equal. For example, if:

$$\mathbf{A} = \begin{bmatrix} a_1 \\ a_2 \\ a_3 \end{bmatrix} \qquad \mathbf{B} = \begin{bmatrix} 4 \\ 7 \\ 3 \end{bmatrix}$$

then **A** = **B** implies:

$$a_1 = 4$$
$$a_2 = 7$$
$$a_3 = 3$$

Similarly, if:

$$\mathbf{A} = \begin{bmatrix} a_{11} & a_{12} \\ a_{21} & a_{22} \\ a_{31} & a_{32} \end{bmatrix} \qquad \mathbf{B} = \begin{bmatrix} 17 & 2 \\ 14 & 5 \\ 13 & 9 \end{bmatrix}$$

then **A** = **B** implies:

$$a_{11} = 17 \qquad a_{12} = 2$$
$$a_{21} = 14 \qquad a_{22} = 5$$
$$a_{31} = 13 \qquad a_{32} = 9$$

Regression examples

In regression analysis, one basic matrix is the vector **Y**, consisting of the *n* observations on the dependent variable:

$$(6.4) \qquad \underset{n \times 1}{\mathbf{Y}} = \begin{bmatrix} Y_1 \\ Y_2 \\ \cdot \\ \cdot \\ \cdot \\ Y_n \end{bmatrix}$$

Note that the transpose **Y'** is the row vector:

$$(6.5) \qquad \underset{1 \times n}{\mathbf{Y'}} = [Y_1 \quad Y_2 \quad \cdots \quad Y_n]$$

Another basic matrix in regression analysis is the **X** matrix, which is defined as follows for simple regression analysis:

(6.6)
$$\mathop{\mathbf{X}}_{n \times 2} = \begin{bmatrix} 1 & X_1 \\ 1 & X_2 \\ \cdot & \cdot \\ \cdot & \cdot \\ \cdot & \cdot \\ 1 & X_n \end{bmatrix}$$

The matrix $\mathbf{X}$ consists of a column of 1's and a column containing the n values of the independent variable X. Note that the transpose of $\mathbf{X}$ is:

(6.7)
$$\mathop{\mathbf{X}'}_{2 \times n} = \begin{bmatrix} 1 & 1 & \cdots & 1 \\ X_1 & X_2 & \cdots & X_n \end{bmatrix}$$

For the Westwood Company lot size example, the $\mathbf{Y}$ and $\mathbf{X}$ matrices are (Table 2.1):

$$\mathbf{Y} = \begin{bmatrix} 73 \\ 50 \\ \cdot \\ \cdot \\ 132 \end{bmatrix} \qquad \mathbf{X} = \begin{bmatrix} 1 & 30 \\ 1 & 20 \\ \cdot & \cdot \\ \cdot & \cdot \\ 1 & 60 \end{bmatrix}$$

6.2 MATRIX ADDITION AND SUBTRACTION

Adding or subtracting two matrices requires that they have the same dimension. The sum, or difference, of two matrices is another matrix whose elements each consists of the sum, or difference, of the corresponding elements of the two matrices. Suppose:

$$\mathop{\mathbf{A}}_{3 \times 2} = \begin{bmatrix} 1 & 4 \\ 2 & 5 \\ 3 & 6 \end{bmatrix} \qquad \mathop{\mathbf{B}}_{3 \times 2} = \begin{bmatrix} 1 & 2 \\ 2 & 3 \\ 3 & 4 \end{bmatrix}$$

then:

$$\mathop{\mathbf{A} + \mathbf{B}}_{3 \times 2} = \begin{bmatrix} 1 + 1 & 4 + 2 \\ 2 + 2 & 5 + 3 \\ 3 + 3 & 6 + 4 \end{bmatrix} = \begin{bmatrix} 2 & 6 \\ 4 & 8 \\ 6 & 10 \end{bmatrix}$$

Similarly:

$$\mathop{\mathbf{A} - \mathbf{B}}_{3 \times 2} = \begin{bmatrix} 1 - 1 & 4 - 2 \\ 2 - 2 & 5 - 3 \\ 3 - 3 & 6 - 4 \end{bmatrix} = \begin{bmatrix} 0 & 2 \\ 0 & 2 \\ 0 & 2 \end{bmatrix}$$

In general, if:

$$\mathop{\mathbf{A}}_{r \times c} = [a_{ij}] \qquad \mathop{\mathbf{B}}_{r \times c} = [b_{ij}] \qquad i = 1, \ldots, r; j = 1, \ldots, c$$

then:

(6.8) $$\mathbf{A} + \mathbf{B} = [a_{ij} + b_{ij}] \quad \text{and} \quad \mathbf{A} - \mathbf{B} = [a_{ij} - b_{ij}]$$
$$\underset{r \times c}{} \qquad\qquad\qquad \underset{r \times c}{}$$

Formula (6.8) generalizes in an obvious way to addition and subtraction of more than two matrices. Note also that $\mathbf{A} + \mathbf{B} = \mathbf{B} + \mathbf{A}$, as in ordinary algebra.

Regression example

The regression model:

$$Y_i = E(Y_i) + \varepsilon_i \quad i = 1, \ldots, n$$

can be written compactly in matrix notation. First, let us define the vector of mean responses:

(6.9)
$$\mathbf{E(Y)} = \begin{bmatrix} E(Y_1) \\ E(Y_2) \\ \cdot \\ \cdot \\ \cdot \\ E(Y_n) \end{bmatrix}$$
$$\underset{n \times 1}{}$$

and the vector of the error terms:

(6.10)
$$\boldsymbol{\varepsilon} = \begin{bmatrix} \varepsilon_1 \\ \varepsilon_2 \\ \cdot \\ \cdot \\ \cdot \\ \varepsilon_n \end{bmatrix}$$
$$\underset{n \times 1}{}$$

Recalling the definition of the $\mathbf{Y}$ observation vector (6.4), we can write the regression model as follows:

$$\mathbf{Y} = \mathbf{E(Y)} + \boldsymbol{\varepsilon}$$

because:

$$\begin{bmatrix} Y_1 \\ Y_2 \\ \cdot \\ \cdot \\ \cdot \\ Y_n \end{bmatrix} = \begin{bmatrix} E(Y_1) \\ E(Y_2) \\ \cdot \\ \cdot \\ \cdot \\ E(Y_n) \end{bmatrix} + \begin{bmatrix} \varepsilon_1 \\ \varepsilon_2 \\ \cdot \\ \cdot \\ \cdot \\ \varepsilon_n \end{bmatrix} = \begin{bmatrix} E(Y_1) + \varepsilon_1 \\ E(Y_2) + \varepsilon_2 \\ \cdot \\ \cdot \\ \cdot \\ E(Y_n) + \varepsilon_n \end{bmatrix}$$

Thus, the observations vector $\mathbf{Y}$ equals the sum of two vectors, a vector containing the expected values and another containing the error terms.

6.3 MATRIX MULTIPLICATION

Multiplication of a matrix by a scalar

A scalar is an ordinary number or a symbol representing a number. We frequently encounter multiplication of a matrix by a scalar. In this, every element of the matrix is multiplied by the scalar. For example, suppose the matrix $\mathbf{A}$ is given by:

$$\mathbf{A} = \begin{bmatrix} 2 & 7 \\ 9 & 3 \end{bmatrix}$$

Then $4\mathbf{A}$, where 4 is the scalar, equals:

$$4\mathbf{A} = 4\begin{bmatrix} 2 & 7 \\ 9 & 3 \end{bmatrix} = \begin{bmatrix} 8 & 28 \\ 36 & 12 \end{bmatrix}$$

Similarly, $\lambda\mathbf{A}$ equals:

$$\lambda\mathbf{A} = \lambda\begin{bmatrix} 2 & 7 \\ 9 & 3 \end{bmatrix} = \begin{bmatrix} 2\lambda & 7\lambda \\ 9\lambda & 3\lambda \end{bmatrix}$$

where λ denotes the scalar.

If every element of a matrix has a common factor, this factor can be taken outside the matrix and treated as a scalar. For example:

$$\begin{bmatrix} 9 & 27 \\ 15 & 18 \end{bmatrix} = 3\begin{bmatrix} 3 & 9 \\ 5 & 6 \end{bmatrix}$$

Similarly:

$$\begin{bmatrix} \dfrac{5}{\lambda} & \dfrac{2}{\lambda} \\ \dfrac{3}{\lambda} & \dfrac{8}{\lambda} \end{bmatrix} = \dfrac{1}{\lambda}\begin{bmatrix} 5 & 2 \\ 3 & 8 \end{bmatrix}$$

In general, if $\mathbf{A} = [a_{ij}]$ and λ is a scalar, we have:

(6.11) $$\lambda\mathbf{A} = \mathbf{A}\lambda = [\lambda a_{ij}]$$

Multiplication of a matrix by a matrix

Multiplication of a matrix by a matrix may appear somewhat complicated at first, but a little practice will make it into a routine operation.

Consider the two matrices:

$$\mathbf{A}_{2\times2} = \begin{bmatrix} 2 & 5 \\ 4 & 1 \end{bmatrix} \qquad \mathbf{B}_{2\times2} = \begin{bmatrix} 4 & 6 \\ 5 & 8 \end{bmatrix}$$

The product $\mathbf{AB}$ will be a 2×2 matrix whose elements are obtained by finding the cross products of rows of $\mathbf{A}$ with columns of $\mathbf{B}$ and summing the cross products. For instance, to find the element in the first row and first column of the product **AB**, we work with the first row of **A** and the first column of **B**, as follows:

$$
\begin{array}{ccc}
\mathbf{A} & \mathbf{B} & \mathbf{AB} \\
\end{array}
$$

Row 1 $\begin{bmatrix} \boxed{2} & 5 \\ 4 & 1 \end{bmatrix}$ $\begin{bmatrix} \boxed{4} & 6 \\ \boxed{5} & 8 \end{bmatrix}$ Row 1 $\begin{bmatrix} 33 \\ \end{bmatrix}$

Row 2 Column Column Column
 1 2 1

We take the cross products and sum:

$$2(4) + 5(5) = 33$$

The number 33 is the element in the first row and first column of the matrix **AB**.
To find the element in the first row and second column of **AB**, we work with the first row of **A** and the second column of **B**:

$$
\begin{array}{ccc}
\mathbf{A} & \mathbf{B} & \mathbf{AB} \\
\end{array}
$$

Row 1 $\begin{bmatrix} \boxed{2} & 5 \\ 4 & 1 \end{bmatrix}$ $\begin{bmatrix} 4 & \boxed{6} \\ 5 & \boxed{8} \end{bmatrix}$ Row 1 $\begin{bmatrix} 33 & 52 \\ & \end{bmatrix}$

 Row 2 Column Column Column Column
 1 2 1 2

The sum of the cross products is:

$$2(6) + 5(8) = 52$$

Continuing this process, we find the product **AB** to be:

$$
\underset{2\times2}{\mathbf{AB}} = \begin{bmatrix} 2 & 5 \\ 4 & 1 \end{bmatrix} \begin{bmatrix} 4 & 6 \\ 5 & 8 \end{bmatrix} = \begin{bmatrix} 33 & 52 \\ 21 & 32 \end{bmatrix}
$$

Let us consider another example:

$$
\underset{2\times3}{\mathbf{A}} = \begin{bmatrix} 1 & 3 & 4 \\ 0 & 5 & 8 \end{bmatrix} \qquad \underset{3\times1}{\mathbf{B}} = \begin{bmatrix} 3 \\ 5 \\ 2 \end{bmatrix}
$$

$$
\underset{2\times1}{\mathbf{AB}} = \begin{bmatrix} 1 & 3 & 4 \\ 0 & 5 & 8 \end{bmatrix} \begin{bmatrix} 3 \\ 5 \\ 2 \end{bmatrix} = \begin{bmatrix} 26 \\ 41 \end{bmatrix}
$$

When obtaining the product **AB**, we say that **A** is *postmultiplied* by **B** or **B** is *premultiplied* by **A**. The reason for this precise terminology is that multiplication rules for ordinary algebra do not apply to matrix algebra. In ordinary algebra, $xy = yx$. In matrix algebra, $\mathbf{AB} \neq \mathbf{BA}$ usually. In fact, even though the product **AB** may be defined, the product **BA** may not be defined at all.

In general, the product $\mathbf{AB}$ is only defined when the number of columns in $\mathbf{A}$ equals the number of rows in $\mathbf{B}$ so that there will be corresponding terms in the cross products. Thus, in our previous two examples, we had:

$$
\underset{\substack{\text{Dimension} \\ \text{of} \\ \text{product}}}{\underset{2\times2}{\mathbf{A}}\overset{\text{Equal}}{\swarrow}\underset{2\times2}{\mathbf{B}}} = \underset{2\times2}{\mathbf{AB}}
\qquad
\underset{\substack{\text{Dimension} \\ \text{of} \\ \text{product}}}{\underset{2\times3}{\mathbf{A}}\overset{\text{Equal}}{\swarrow}\underset{3\times1}{\mathbf{B}}} = \underset{2\times1}{\mathbf{AB}}
$$

Note that the dimension of the product $\mathbf{AB}$ is given by the number of rows in $\mathbf{A}$ and the number of columns in $\mathbf{B}$. Note also that in the second case the product $\mathbf{BA}$ would not be defined since the number of columns in $\mathbf{B}$ is not equal to the number of rows in $\mathbf{A}$:

$$
\underset{3\times1}{\mathbf{B}}\overset{\text{Unequal}}{\swarrow}\underset{2\times3}{\mathbf{A}}
$$

Here is another example of matrix multiplication:

$$
\mathbf{AB} = \begin{bmatrix} a_{11} & a_{12} & a_{13} \\ a_{21} & a_{22} & a_{23} \end{bmatrix} \begin{bmatrix} b_{11} & b_{12} \\ b_{21} & b_{22} \\ b_{31} & b_{32} \end{bmatrix}
$$

$$
= \begin{bmatrix} a_{11}b_{11} + a_{12}b_{21} + a_{13}b_{31} & a_{11}b_{12} + a_{12}b_{22} + a_{13}b_{32} \\ a_{21}b_{11} + a_{22}b_{21} + a_{23}b_{31} & a_{21}b_{12} + a_{22}b_{22} + a_{23}b_{32} \end{bmatrix}
$$

In general, if $\mathbf{A}$ has dimension $r \times c$ and $\mathbf{B}$ has dimension $c \times s$, the product $\mathbf{AB}$ is a matrix of dimension $r \times s$ whose element in the ith row and jth column is:

$$
\sum_{k=1}^{c} a_{ik}b_{kj}
$$

so that:

(6.12)
$$
\underset{r\times s}{\mathbf{AB}} = \left[\sum_{k=1}^{c} a_{ik}b_{kj} \right] \qquad i = 1,\ldots,r; j = 1,\ldots,s
$$

Thus, in the foregoing example, the element in the first row and second column of the product $\mathbf{AB}$ is:

$$
\sum_{k=1}^{3} a_{1k}b_{k2} = a_{11}b_{12} + a_{12}b_{22} + a_{13}b_{32}
$$

as indeed we found by taking the cross products of the elements in the first row of $\mathbf{A}$ and second column of $\mathbf{B}$ and summing.

Additional examples

1.

$$\begin{bmatrix} 4 & 2 \\ 5 & 8 \end{bmatrix} \begin{bmatrix} a_1 \\ a_2 \end{bmatrix} = \begin{bmatrix} 4a_1 + 2a_2 \\ 5a_1 + 8a_2 \end{bmatrix}$$

2.

$$[2 \quad 3 \quad 5] \begin{bmatrix} 2 \\ 3 \\ 5 \end{bmatrix} = [2^2 + 3^2 + 5^2] = [38]$$

Here, the product is a 1×1 matrix, which is equivalent to a scalar. Thus, the matrix product here equals the number 38.

3.

$$\begin{bmatrix} 1 & X_1 \\ 1 & X_2 \\ 1 & X_3 \end{bmatrix} \begin{bmatrix} \beta_0 \\ \beta_1 \end{bmatrix} = \begin{bmatrix} \beta_0 + \beta_1 X_1 \\ \beta_0 + \beta_1 X_2 \\ \beta_0 + \beta_1 X_3 \end{bmatrix}$$

Regression examples. Let us define the vector $\boldsymbol{\beta}$ of the regression coefficients as follows:

(6.13)
$$\underset{2 \times 1}{\boldsymbol{\beta}} = \begin{bmatrix} \beta_0 \\ \beta_1 \end{bmatrix}$$

Then the product $\mathbf{X}\boldsymbol{\beta}$, where $\mathbf{X}$ is defined in (6.6), is an $n \times 1$ matrix:

(6.14)
$$\underset{n \times 1}{\mathbf{X}\boldsymbol{\beta}} = \begin{bmatrix} 1 & X_1 \\ 1 & X_2 \\ \cdot & \cdot \\ \cdot & \cdot \\ \cdot & \cdot \\ 1 & X_n \end{bmatrix} \begin{bmatrix} \beta_0 \\ \beta_1 \end{bmatrix} = \begin{bmatrix} \beta_0 + \beta_1 X_1 \\ \beta_0 + \beta_1 X_2 \\ \cdot \\ \cdot \\ \cdot \\ \beta_0 + \beta_1 X_n \end{bmatrix}$$

Since $\beta_0 + \beta_1 X_i = E(Y_i)$, we see that $\mathbf{X}\boldsymbol{\beta}$ is the vector of expected values $E(Y_i)$ for the simple linear regression model, i.e., $\mathbf{E(Y)} = \mathbf{X}\boldsymbol{\beta}$, where $\mathbf{E(Y)}$ is defined in (6.9).

Another product frequently needed is $\mathbf{Y'Y}$, where $\mathbf{Y}$ is the vector of observations on the dependent variable as defined in (6.4):

(6.15)
$$\underset{1 \times 1}{\mathbf{Y'Y}} = [Y_1 \quad Y_2 \quad \cdots \quad Y_n] \begin{bmatrix} Y_1 \\ Y_2 \\ \cdot \\ \cdot \\ \cdot \\ Y_n \end{bmatrix} = [Y_1^2 + Y_2^2 + \cdots + Y_n^2] = [\Sigma Y_i^2]$$

Note that $\mathbf{Y'Y}$ is a 1×1 matrix, or a scalar. We thus have a compact way of writing a sum of squares: $\mathbf{Y'Y} = \Sigma Y_i^2$.

We also will need $\mathbf{X}'\mathbf{X}$, which is a 2×2 matrix:

$$
\underset{2\times 2}{\mathbf{X}'\mathbf{X}} = \begin{bmatrix} 1 & 1 & \cdots & 1 \\ X_1 & X_2 & \cdots & X_n \end{bmatrix} \begin{bmatrix} 1 & X_1 \\ 1 & X_2 \\ \vdots & \vdots \\ 1 & X_n \end{bmatrix} = \begin{bmatrix} n & \Sigma X_i \\ \Sigma X_i & \Sigma X_i^2 \end{bmatrix}
$$

(6.16)

and $\mathbf{X}'\mathbf{Y}$, which is a 2×1 matrix:

$$
\underset{2\times 1}{\mathbf{X}'\mathbf{Y}} = \begin{bmatrix} 1 & 1 & \cdots & 1 \\ X_1 & X_2 & \cdots & X_n \end{bmatrix} \begin{bmatrix} Y_1 \\ Y_2 \\ \vdots \\ Y_n \end{bmatrix} = \begin{bmatrix} \Sigma Y_i \\ \Sigma X_i Y_i \end{bmatrix}
$$

(6.17)

6.4 SPECIAL TYPES OF MATRICES

Certain special types of matrices arise regularly in regression analysis. We shall consider the most important of these.

Symmetric matrix

If $\mathbf{A} = \mathbf{A}'$, $\mathbf{A}$ is said to be symmetric. Thus, $\mathbf{A}$ below is symmetric:

$$
\mathbf{A} = \begin{bmatrix} 1 & 4 & 6 \\ 4 & 2 & 5 \\ 6 & 5 & 3 \end{bmatrix} \qquad \mathbf{A}' = \begin{bmatrix} 1 & 4 & 6 \\ 4 & 2 & 5 \\ 6 & 5 & 3 \end{bmatrix}
$$

Clearly, a symmetric matrix necessarily is square. Symmetric matrices arise typically in regression analysis when we premultiply a matrix, say, $\mathbf{X}$, by its transpose, $\mathbf{X}'$. The resulting matrix, $\mathbf{X}'\mathbf{X}$, is symmetric, as can readily be seen from (6.16).

Diagonal matrix

A diagonal matrix is a square matrix whose off-diagonal elements are all zeros, such as:

$$
\mathbf{A} = \begin{bmatrix} a_1 & 0 & 0 \\ 0 & a_2 & 0 \\ 0 & 0 & a_3 \end{bmatrix} \qquad \mathbf{B} = \begin{bmatrix} 4 & 0 & 0 & 0 \\ 0 & 1 & 0 & 0 \\ 0 & 0 & 10 & 0 \\ 0 & 0 & 0 & 5 \end{bmatrix}
$$

We will often not show all zeros for a diagonal matrix, presenting it in the form:

$$\mathbf{A} = \begin{bmatrix} a_1 & & 0 \\ & a_2 & \\ 0 & & a_3 \end{bmatrix} \qquad \mathbf{B} = \begin{bmatrix} 4 & & \\ & 1 & 0 \\ & & 10 \\ 0 & & & 5 \end{bmatrix}$$

Two important types of diagonal matrices are the identity matrix and the scalar matrix.

Identity matrix. The identity matrix or unit matrix is denoted by $\mathbf{I}$. It is a diagonal matrix whose elements on the main diagonal are all 1's. Premultiplying or postmultiplying any $r \times r$ matrix $\mathbf{A}$ by the $r \times r$ identity matrix $\mathbf{I}$ leaves $\mathbf{A}$ unchanged. For example:

$$\mathbf{IA} = \begin{bmatrix} 1 & 0 & 0 \\ 0 & 1 & 0 \\ 0 & 0 & 1 \end{bmatrix} \begin{bmatrix} a_{11} & a_{12} & a_{13} \\ a_{21} & a_{22} & a_{23} \\ a_{31} & a_{32} & a_{33} \end{bmatrix} = \begin{bmatrix} a_{11} & a_{12} & a_{13} \\ a_{21} & a_{22} & a_{23} \\ a_{31} & a_{32} & a_{33} \end{bmatrix}$$

Similarly, we have:

$$\mathbf{AI} = \begin{bmatrix} a_{11} & a_{12} & a_{13} \\ a_{21} & a_{22} & a_{23} \\ a_{31} & a_{32} & a_{33} \end{bmatrix} \begin{bmatrix} 1 & 0 & 0 \\ 0 & 1 & 0 \\ 0 & 0 & 1 \end{bmatrix} = \begin{bmatrix} a_{11} & a_{12} & a_{13} \\ a_{21} & a_{22} & a_{23} \\ a_{31} & a_{32} & a_{33} \end{bmatrix}$$

Note that the identity matrix $\mathbf{I}$ therefore corresponds to the number 1 in ordinary algebra, since we have there that $1 \cdot x = x \cdot 1 = x$.

In general, we have for any $r \times r$ matrix $\mathbf{A}$:

(6.18) $$\mathbf{AI} = \mathbf{IA} = \mathbf{A}$$

Thus, the identity matrix can be inserted or dropped from a matrix expression whenever it is convenient to do so.

Scalar matrix. A scalar matrix is a diagonal matrix whose main-diagonal elements are the same. Two examples of scalar matrices are:

$$\begin{bmatrix} 2 & 0 \\ 0 & 2 \end{bmatrix} \qquad \begin{bmatrix} \lambda & 0 & 0 \\ 0 & \lambda & 0 \\ 0 & 0 & \lambda \end{bmatrix}$$

A scalar matrix can be expressed $\lambda\mathbf{I}$, where λ is the scalar. For instance:

$$\begin{bmatrix} 2 & 0 \\ 0 & 2 \end{bmatrix} = 2 \begin{bmatrix} 1 & 0 \\ 0 & 1 \end{bmatrix} = 2\mathbf{I}$$

$$\begin{bmatrix} \lambda & 0 & 0 \\ 0 & \lambda & 0 \\ 0 & 0 & \lambda \end{bmatrix} = \lambda \begin{bmatrix} 1 & 0 & 0 \\ 0 & 1 & 0 \\ 0 & 0 & 1 \end{bmatrix} = \lambda\mathbf{I}$$

Multiplying an $r \times r$ matrix $\mathbf{A}$ by the $r \times r$ scalar matrix $\lambda \mathbf{I}$ is equivalent to multiplying $\mathbf{A}$ by the scalar λ.

Vector and matrix with all elements 1

A column vector with all elements 1 will be denoted by $\mathbf{1}$:

$$(6.19) \qquad \underset{r \times 1}{\mathbf{1}} = \begin{bmatrix} 1 \\ 1 \\ \cdot \\ \cdot \\ \cdot \\ 1 \end{bmatrix}$$

and a square matrix with all elements 1 will be denoted by $\mathbf{J}$:

$$(6.20) \qquad \underset{r \times r}{\mathbf{J}} = \begin{bmatrix} 1 & \cdots & 1 \\ 1 & & 1 \\ \cdot & & \cdot \\ \cdot & & \cdot \\ \cdot & & \cdot \\ 1 & \cdots & 1 \end{bmatrix}$$

For instance, we have:

$$\underset{3 \times 1}{\mathbf{1}} = \begin{bmatrix} 1 \\ 1 \\ 1 \end{bmatrix} \qquad \underset{3 \times 3}{\mathbf{J}} = \begin{bmatrix} 1 & 1 & 1 \\ 1 & 1 & 1 \\ 1 & 1 & 1 \end{bmatrix}$$

Note that for an $n \times 1$ vector $\mathbf{1}$ we obtain:

$$\underset{1 \times 1}{\mathbf{1}'\mathbf{1}} = [1 \quad \cdots \quad 1] \begin{bmatrix} 1 \\ \cdot \\ \cdot \\ \cdot \\ 1 \end{bmatrix} = [n] = n$$

and:

$$\underset{n \times n}{\mathbf{1}\mathbf{1}'} = \begin{bmatrix} 1 \\ \cdot \\ \cdot \\ \cdot \\ 1 \end{bmatrix} [1 \quad \cdots \quad 1] = \begin{bmatrix} 1 & \cdots & 1 \\ \cdot & & \cdot \\ \cdot & & \cdot \\ \cdot & & \cdot \\ 1 & \cdots & 1 \end{bmatrix} = \underset{n \times n}{\mathbf{J}}$$

Zero vector

A zero vector is a column vector containing only zeros. It will be denoted by $\mathbf{0}$:

(6.21)

$$\mathbf{0}_{r \times 1} = \begin{bmatrix} 0 \\ 0 \\ \cdot \\ \cdot \\ \cdot \\ 0 \end{bmatrix}$$

For example, we have:

$$\mathbf{0}_{3 \times 1} = \begin{bmatrix} 0 \\ 0 \\ 0 \end{bmatrix}$$

6.5 LINEAR DEPENDENCE AND RANK OF MATRIX

Linear dependence

Consider the following matrix:

$$\mathbf{A} = \begin{bmatrix} 1 & 2 & 5 & 1 \\ 2 & 2 & 10 & 6 \\ 3 & 4 & 15 & 1 \end{bmatrix}$$

Let us think now of the columns of this matrix as vectors. Thus, we view $\mathbf{A}$ as being made up of four column vectors. It happens here that the columns are interrelated in a special manner. Note that the third column vector is a multiple of the first column vector:

$$\begin{bmatrix} 5 \\ 10 \\ 15 \end{bmatrix} = 5 \begin{bmatrix} 1 \\ 2 \\ 3 \end{bmatrix}$$

We say that the columns of $\mathbf{A}$ are linearly dependent. They contain redundant information, so to speak, since one column can be obtained as a linear combination of the others.

We define a set of column vectors to be linearly dependent if one vector can be expressed as a linear combination of the others. If no vector in the set can be so expressed, we define the set of vectors to be linearly independent. A more general, though equivalent, definition for the c column vectors $\mathbf{C}_1, \ldots, \mathbf{C}_c$ in an $r \times c$ matrix is:

(6.22)   When c scalars $\lambda_1, \ldots, \lambda_c$, not all zero, can be found such that:

$$\lambda_1 \mathbf{C}_1 + \lambda_2 \mathbf{C}_2 + \cdots + \lambda_c \mathbf{C}_c = \mathbf{0}$$

where $\mathbf{0}$ denotes the zero column vector, the c column vectors are linearly dependent. If the only set of scalars for which the equality holds is $\lambda_1 = 0, \ldots, \lambda_c = 0$, the set of c column vectors is linearly independent.

To illustrate for our example, $\lambda_1 = 5$, $\lambda_2 = 0$, $\lambda_3 = -1$, $\lambda_4 = 0$ leads to:

$$5\begin{bmatrix} 1 \\ 2 \\ 3 \end{bmatrix} + 0\begin{bmatrix} 2 \\ 2 \\ 4 \end{bmatrix} - 1\begin{bmatrix} 5 \\ 10 \\ 15 \end{bmatrix} + 0\begin{bmatrix} 1 \\ 6 \\ 1 \end{bmatrix} = \begin{bmatrix} 0 \\ 0 \\ 0 \end{bmatrix}$$

Hence, the column vectors are linearly dependent. Note that some of the $\lambda_j = 0$ here. It is only required for linear dependence that not all λ_j are zero.

Rank of a matrix

The rank of a matrix is defined to be the maximum number of linearly independent columns in the matrix. We know that the rank of $\mathbf{A}$ in our earlier example cannot be 4, since the four columns are linearly dependent. We can, however, find 3 columns (1, 2, and 4) which are linearly independent. There are no scalars λ_1, λ_2, λ_4 such that $\lambda_1\mathbf{C}_1 + \lambda_2\mathbf{C}_2 + \lambda_4\mathbf{C}_4 = \mathbf{0}$ other than $\lambda_1 = \lambda_2 = \lambda_4 = 0$. Thus, the rank of $\mathbf{A}$ in our example is 3.

The rank of a matrix is unique and can equivalently be defined as the maximum number of linearly independent rows. It follows that the rank of an $r \times c$ matrix cannot exceed $\min(r, c)$, the minimum of the two values r and c.

6.6 INVERSE OF A MATRIX

In ordinary algebra, the inverse of a number is its reciprocal. Thus, the inverse of 6 is $\frac{1}{6}$. A number multiplied by its inverse always equals 1:

$$6 \cdot \tfrac{1}{6} = 1$$

$$x \cdot \frac{1}{x} = x \cdot x^{-1} = x^{-1} \cdot x = 1$$

In matrix algebra, the inverse of a matrix $\mathbf{A}$ is another matrix, denoted by $\mathbf{A}^{-1}$, such that:

(6.23) $$\mathbf{A}^{-1}\mathbf{A} = \mathbf{A}\mathbf{A}^{-1} = \mathbf{I}$$

where $\mathbf{I}$ is the identity matrix. Thus, again, the identity matrix $\mathbf{I}$ plays the same role as the number 1 in ordinary algebra. An inverse of a matrix is defined only for square matrices. Even so, many square matrices do not have an inverse. If a square matrix does have an inverse, the inverse is unique.

Examples

1. The inverse of the matrix:

$$\mathbf{A} = \begin{bmatrix} 2 & 4 \\ 3 & 1 \end{bmatrix}$$

is:

$$\mathbf{A}^{-1} = \begin{bmatrix} -.1 & .4 \\ .3 & -.2 \end{bmatrix}$$

since:

$$\mathbf{A}^{-1}\mathbf{A} = \begin{bmatrix} -.1 & .4 \\ .3 & -.2 \end{bmatrix} \begin{bmatrix} 2 & 4 \\ 3 & 1 \end{bmatrix} = \begin{bmatrix} 1 & 0 \\ 0 & 1 \end{bmatrix}$$

or:

$$\mathbf{A}\mathbf{A}^{-1} = \begin{bmatrix} 2 & 4 \\ 3 & 1 \end{bmatrix} \begin{bmatrix} -.1 & .4 \\ .3 & -.2 \end{bmatrix} = \begin{bmatrix} 1 & 0 \\ 0 & 1 \end{bmatrix}$$

2. The inverse of the matrix:

$$\mathbf{A} = \begin{bmatrix} 3 & 0 & 0 \\ 0 & 4 & 0 \\ 0 & 0 & 2 \end{bmatrix}$$

is:

$$\mathbf{A}^{-1} = \begin{bmatrix} \frac{1}{3} & 0 & 0 \\ 0 & \frac{1}{4} & 0 \\ 0 & 0 & \frac{1}{2} \end{bmatrix}$$

since:

$$\mathbf{A}^{-1}\mathbf{A} = \begin{bmatrix} \frac{1}{3} & 0 & 0 \\ 0 & \frac{1}{4} & 0 \\ 0 & 0 & \frac{1}{2} \end{bmatrix} \begin{bmatrix} 3 & 0 & 0 \\ 0 & 4 & 0 \\ 0 & 0 & 2 \end{bmatrix} = \begin{bmatrix} 1 & 0 & 0 \\ 0 & 1 & 0 \\ 0 & 0 & 1 \end{bmatrix}$$

Note that the inverse of a diagonal matrix is a diagonal matrix consisting simply of the reciprocals of the elements on the diagonal.

Finding the inverse   *need to do by hand?*

Up to this point, the inverse of a matrix $\mathbf{A}$ has been given, and we have only checked to make sure it is the inverse by seeing whether or not $\mathbf{A}^{-1}\mathbf{A} = \mathbf{I}$. But how does one find the inverse, and when does it exist?

An inverse of a square $r \times r$ matrix exists if the rank of the matrix is r. Such a matrix is said to be *nonsingular*. An $r \times r$ matrix with rank less than r is said to be *singular*, and does not have an inverse.

Finding the inverse of a matrix can often require a tremendous amount of computing. We shall take the approach in this book that the inverse of a 2 × 2 matrix and a 3 × 3 matrix can be calculated by hand. For any larger matrix, one ordinarily uses a computer or a programmable calculator to find the inverse, unless the matrix is of a special form such as a diagonal matrix. It can be shown that the inverses for 2 × 2 and 3 × 3 matrices are as follows:

1. If:

$$\mathbf{A} = \begin{bmatrix} a & b \\ c & d \end{bmatrix}$$

then:

(6.24)
$$\mathbf{A}^{-1} = \begin{bmatrix} a & b \\ c & d \end{bmatrix}^{-1} = \begin{bmatrix} \dfrac{d}{D} & \dfrac{-b}{D} \\ \dfrac{-c}{D} & \dfrac{a}{D} \end{bmatrix}$$

where:

$$D = ad - bc$$

D is called the *determinant* of the matrix $\mathbf{A}$. If $\mathbf{A}$ were singular, its determinant would equal zero and no inverse of $\mathbf{A}$ would exist.

2. If:

$$\mathbf{B} = \begin{bmatrix} a & b & c \\ d & e & f \\ g & h & k \end{bmatrix}$$

then:

(6.25)
$$\mathbf{B}^{-1} = \begin{bmatrix} a & b & c \\ d & e & f \\ g & h & k \end{bmatrix}^{-1} = \begin{bmatrix} A & B & C \\ D & E & F \\ G & H & K \end{bmatrix}$$

where:

$$A = (ek - fh)/Z \qquad B = -(bk - ch)/Z \qquad C = (bf - ce)/Z$$
$$D = -(dk - fg)/Z \qquad E = (ak - cg)/Z \qquad F = -(af - cd)/Z$$
$$G = (dh - eg)/Z \qquad H = -(ah - bg)/Z \qquad K = (ae - bd)/Z$$

and:

$$Z = a(ek - fh) - b(dk - fg) + c(dh - eg)$$

Z is called the determinant of the matrix $\mathbf{B}$.

Let us use (6.24) to find the inverse of:

$$\mathbf{A} = \begin{bmatrix} 2 & 4 \\ 3 & 1 \end{bmatrix}$$

We have:

$$a = 2 \qquad b = 4$$
$$c = 3 \qquad d = 1$$
$$D = ad - bc = 2(1) - 4(3) = -10$$

Hence:

$$\mathbf{A}^{-1} = \begin{bmatrix} \dfrac{1}{-10} & \dfrac{-4}{-10} \\[2mm] \dfrac{-3}{-10} & \dfrac{2}{-10} \end{bmatrix} = \begin{bmatrix} -.1 & .4 \\ .3 & -.2 \end{bmatrix}$$

as was given in an earlier example.

When an inverse $\mathbf{A}^{-1}$ has been obtained, either by hand calculations or from a computer run, it is usually wise to compute $\mathbf{A}^{-1}\mathbf{A}$ to check whether the product equals the identity matrix, allowing for minor rounding departures from 0 and 1.

Regression example

The principal inverse matrix encountered in regression analysis is the inverse of the matrix $\mathbf{X}'\mathbf{X}$ in (6.16):

$$\mathbf{X}'\mathbf{X} = \begin{bmatrix} n & \Sigma X_i \\ \Sigma X_i & \Sigma X_i^2 \end{bmatrix}$$

Using rule (6.24), we have:

$$\begin{aligned} a &= n & b &= \Sigma X_i \\ c &= \Sigma X_i & d &= \Sigma X_i^2 \end{aligned}$$

so that:

$$D = n\Sigma X_i^2 - (\Sigma X_i)(\Sigma X_i) = n\left(\Sigma X_i^2 - \frac{(\Sigma X_i)^2}{n} \right) = n\Sigma(X_i - \bar{X})^2$$

Hence:

(6.26)
$$(\mathbf{X}'\mathbf{X})^{-1} = \begin{bmatrix} \dfrac{\Sigma X_i^2}{n\Sigma(X_i - \bar{X})^2} & \dfrac{-\Sigma X_i}{n\Sigma(X_i - \bar{X})^2} \\[4mm] \dfrac{-\Sigma X_i}{n\Sigma(X_i - \bar{X})^2} & \dfrac{n}{n\Sigma(X_i - \bar{X})^2} \end{bmatrix}$$

Since $\Sigma X_i = n\bar{X}$, we can simplify (6.26):

(6.27)
$$(\mathbf{X}'\mathbf{X})^{-1} = \begin{bmatrix} \dfrac{\Sigma X_i^2}{n\Sigma(X_i - \bar{X})^2} & \dfrac{-\bar{X}}{\Sigma(X_i - \bar{X})^2} \\[4mm] \dfrac{-\bar{X}}{\Sigma(X_i - \bar{X})^2} & \dfrac{1}{\Sigma(X_i - \bar{X})^2} \end{bmatrix}$$

Uses of inverse matrix

In ordinary algebra, we solve an equation of the type:

$$5y = 20$$

by multiplying both sides of the equation by the inverse of 5, namely:

$$\tfrac{1}{5}(5y) = \tfrac{1}{5}(20)$$

We obtain:

$$y = \tfrac{1}{5}(20) = 4$$

In matrix algebra, if we have an equation:

$$\mathbf{AY} = \mathbf{C}$$

we correspondingly premultiply both sides by $\mathbf{A}^{-1}$, assuming $\mathbf{A}$ has an inverse, and obtain:

$$\mathbf{A}^{-1}\mathbf{AY} = \mathbf{A}^{-1}\mathbf{C}$$

Since $\mathbf{A}^{-1}\mathbf{AY} = \mathbf{IY} = \mathbf{Y}$, we obtain:

$$\mathbf{Y} = \mathbf{A}^{-1}\mathbf{C}$$

To illustrate this use, suppose we have two simultaneous equations:

$$2y_1 + 4y_2 = 20$$
$$3y_1 + y_2 = 10$$

which can be written as follows in matrix notation:

$$\begin{bmatrix} 2 & 4 \\ 3 & 1 \end{bmatrix} \begin{bmatrix} y_1 \\ y_2 \end{bmatrix} = \begin{bmatrix} 20 \\ 10 \end{bmatrix}$$

The solution of these equations then is:

$$\begin{bmatrix} y_1 \\ y_2 \end{bmatrix} = \begin{bmatrix} 2 & 4 \\ 3 & 1 \end{bmatrix}^{-1} \begin{bmatrix} 20 \\ 10 \end{bmatrix}$$

Earlier we found the required inverse, so we obtain:

$$\begin{bmatrix} y_1 \\ y_2 \end{bmatrix} = \begin{bmatrix} -.1 & .4 \\ .3 & -.2 \end{bmatrix} \begin{bmatrix} 20 \\ 10 \end{bmatrix} = \begin{bmatrix} 2 \\ 4 \end{bmatrix}$$

Hence, $y_1 = 2$ and $y_2 = 4$ satisfy these two equations.

6.7 SOME BASIC THEOREMS FOR MATRICES

We list here, without proof, some basic theorems for matrices which we will utilize in later work.

(6.28) $$\mathbf{A} + \mathbf{B} = \mathbf{B} + \mathbf{A}$$

(6.29) $$(\mathbf{A} + \mathbf{B}) + \mathbf{C} = \mathbf{A} + (\mathbf{B} + \mathbf{C})$$

(6.30) $$(\mathbf{AB})\mathbf{C} = \mathbf{A}(\mathbf{BC})$$

$$(6.31) \qquad \mathbf{C(A + B)} = \mathbf{CA} + \mathbf{CB}$$
$$(6.32) \qquad \lambda\mathbf{(A + B)} = \lambda\mathbf{A} + \lambda\mathbf{B}$$
$$(6.33) \qquad \mathbf{(A')'} = \mathbf{A}$$
$$(6.34) \qquad \mathbf{(A + B)'} = \mathbf{A'} + \mathbf{B'}$$
$$(6.35) \qquad \mathbf{(AB)'} = \mathbf{B'A'}$$
$$(6.36) \qquad \mathbf{(ABC)'} = \mathbf{C'B'A'}$$
$$(6.37) \qquad \mathbf{(AB)}^{-1} = \mathbf{B}^{-1}\mathbf{A}^{-1}$$
$$(6.38) \qquad \mathbf{(ABC)}^{-1} = \mathbf{C}^{-1}\mathbf{B}^{-1}\mathbf{A}^{-1}$$
$$(6.39) \qquad \mathbf{(A}^{-1})^{-1} = \mathbf{A}$$
$$(6.40) \qquad \mathbf{(A')}^{-1} = \mathbf{(A}^{-1})'$$

6.8   RANDOM VECTORS AND MATRICES

A random vector or a random matrix contains elements which are random variables. Thus, the observation vector $\mathbf{Y}$ in (6.4) is a random vector since the Y_i elements are random variables.

Expectation of random vector or matrix

Suppose we have $n = 3$ observations and are concerned with the observation vector:

$$\mathbf{Y} = \begin{bmatrix} Y_1 \\ Y_2 \\ Y_3 \end{bmatrix}$$

The expected value of $\mathbf{Y}$ is a vector, denoted by $\mathbf{E(Y)}$, which is defined as follows:

$$\mathbf{E(Y)} = \begin{bmatrix} E(Y_1) \\ E(Y_2) \\ E(Y_3) \end{bmatrix}$$

Thus, the expected value of a random vector is a vector whose elements are the expected values of the random variables which are the elements of the random vector. Similarly, the expectation of a random matrix is a matrix whose elements are the expected values of the corresponding random variables in the original matrix. We encountered a vector of expected values earlier in (6.9).

In general, for a random vector $\mathbf{Y}$ the expectation is:

$$(6.41) \qquad \underset{n \times 1}{\mathbf{E(Y)}} = [E(Y_i)] \qquad i = 1, \ldots, n$$

and for a random matrix $\mathbf{Y}$ with dimension $n \times p$, the expectation is:

$$(6.42) \qquad \underset{n \times p}{\mathbf{E(Y)}} = [E(Y_{ij})] \qquad i = 1, \ldots, n; j = 1, \ldots, p$$

Regression example. Suppose the number of observations in a regression application is $n = 3$. The three error terms ε_1, ε_2, ε_3 each have expectation zero. For the error vector:

$$\boldsymbol{\varepsilon} = \begin{bmatrix} \varepsilon_1 \\ \varepsilon_2 \\ \varepsilon_3 \end{bmatrix}$$

we have:

$$E(\boldsymbol{\varepsilon}) = \mathbf{0}$$

since:

$$E(\boldsymbol{\varepsilon}) = \begin{bmatrix} E(\varepsilon_1) \\ E(\varepsilon_2) \\ E(\varepsilon_3) \end{bmatrix} = \begin{bmatrix} 0 \\ 0 \\ 0 \end{bmatrix} = \mathbf{0}$$

Variance-covariance matrix of a random vector

Consider again the random vector $\mathbf{Y}$ consisting of three observations Y_1, Y_2, Y_3. Each random variable has a variance, $\sigma^2(Y_i)$, and any two random variables have a covariance, $\sigma(Y_i, Y_j)$. We can assemble these in a matrix called the *variance-covariance matrix of* $\mathbf{Y}$, denoted by $\boldsymbol{\sigma}^2(\mathbf{Y})$:

$$(6.43) \qquad \boldsymbol{\sigma}^2(\mathbf{Y}) = \begin{bmatrix} \sigma^2(Y_1) & \sigma(Y_1, Y_2) & \sigma(Y_1, Y_3) \\ \sigma(Y_2, Y_1) & \sigma^2(Y_2) & \sigma(Y_2, Y_3) \\ \sigma(Y_3, Y_1) & \sigma(Y_3, Y_2) & \sigma^2(Y_3) \end{bmatrix}$$

Note that the variances are on the main diagonal and the covariance $\sigma(Y_i, Y_j)$ is found in the ith row and jth column of the matrix. Thus, $\sigma(Y_2, Y_1)$ is found in the second row, first column, and $\sigma(Y_1, Y_2)$ is found in the first row, second column. Remember, of course, that $\sigma(Y_2, Y_1) = \sigma(Y_1, Y_2)$. Since in general $\sigma(Y_i, Y_j) = \sigma(Y_j, Y_i)$ for $i \neq j$, $\boldsymbol{\sigma}^2(\mathbf{Y})$ is a symmetric matrix.

It follows readily that:

$$(6.44) \qquad \boldsymbol{\sigma}^2(\mathbf{Y}) = E\{[\mathbf{Y} - E(\mathbf{Y})] \ [\mathbf{Y} - E(\mathbf{Y})]'\}$$

For our illustration, we have:

$$\boldsymbol{\sigma}^2(\mathbf{Y}) = E \begin{bmatrix} Y_1 - E(Y_1) \\ Y_2 - E(Y_2) \\ Y_3 - E(Y_3) \end{bmatrix} [Y_1 - E(Y_1) \quad Y_2 - E(Y_2) \quad Y_3 - E(Y_3)]$$

Multiplying the two matrices and then taking expectations, we obtain:

Location in Product	Term	Expected Value
Row 1, column 1	$[Y_1 - E(Y_1)]^2$	$\sigma^2(Y_1)$
Row 1, column 2	$[Y_1 - E(Y_1)][Y_2 - E(Y_2)]$	$\sigma(Y_1, Y_2)$
Row 1, column 3	$[Y_1 - E(Y_1)][Y_3 - E(Y_3)]$	$\sigma(Y_1, Y_3)$
Row 2, column 1	$[Y_2 - E(Y_2)][Y_1 - E(Y_1)]$	$\sigma(Y_2, Y_1)$
etc.	etc.	etc.

This, of course, leads to the variance-covariance matrix in (6.43). Remember the definitions of variance and covariance in (1.14) and (1.19), respectively, when taking expectations.

To generalize, the variance-covariance matrix for an $n \times 1$ random vector $\mathbf{Y}$ is:

$$(6.45) \qquad \underset{n \times n}{\boldsymbol{\sigma}^2(\mathbf{Y})} = \begin{bmatrix} \sigma^2(Y_1) & \sigma(Y_1, Y_2) & \cdots & \sigma(Y_1, Y_n) \\ \sigma(Y_2, Y_1) & \sigma^2(Y_2) & \cdots & \sigma(Y_2, Y_n) \\ \cdot & \cdot & & \cdot \\ \cdot & \cdot & & \cdot \\ \cdot & \cdot & & \cdot \\ \sigma(Y_n, Y_1) & \sigma(Y_n, Y_2) & \cdots & \sigma^2(Y_n) \end{bmatrix}$$

Note again that $\boldsymbol{\sigma}^2(\mathbf{Y})$ is a symmetric matrix.

Regression example. Let us return to the example based on $n = 3$ observations. Suppose that the three error terms have constant variance, $\sigma^2(\varepsilon_i) = \sigma^2$, and are uncorrelated so that $\sigma(\varepsilon_i, \varepsilon_j) = 0$ for $i \neq j$. We can then write the variance-covariance matrix for the random vector $\boldsymbol{\varepsilon}$ of the previous example as follows:

$$\boldsymbol{\sigma}^2(\boldsymbol{\varepsilon}) = \sigma^2 \mathbf{I}$$

since:

$$\sigma^2 \mathbf{I} = \sigma^2 \begin{bmatrix} 1 & 0 & 0 \\ 0 & 1 & 0 \\ 0 & 0 & 1 \end{bmatrix} = \begin{bmatrix} \sigma^2 & 0 & 0 \\ 0 & \sigma^2 & 0 \\ 0 & 0 & \sigma^2 \end{bmatrix}$$

Note that all variances are σ^2 and all covariances are zero.

Some basic theorems

Frequently, we shall encounter a random vector $\mathbf{W}$ which is obtained by premultiplying the random vector $\mathbf{Y}$ by a constant matrix $\mathbf{A}$ (a matrix whose elements are fixed):

$$(6.46) \qquad\qquad\qquad \mathbf{W} = \mathbf{AY}$$

Some basic theorems for this case are:

$$(6.47) \qquad\qquad\qquad \mathbf{E}(\mathbf{A}) = \mathbf{A}$$
$$(6.48) \qquad\qquad\qquad \mathbf{E}(\mathbf{W}) = \mathbf{E}(\mathbf{AY}) = \mathbf{AE}(\mathbf{Y})$$
$$(6.49) \qquad\qquad\qquad \boldsymbol{\sigma}^2(\mathbf{W}) = \boldsymbol{\sigma}^2(\mathbf{AY}) = \mathbf{A}[\boldsymbol{\sigma}^2(\mathbf{Y})]\mathbf{A}'$$

where $\boldsymbol{\sigma}^2(\mathbf{Y})$ is the variance-covariance matrix of $\mathbf{Y}$.

Example. As a simple illustration of the use of these theorems, consider:

$$\underset{\underset{2 \times 1}{\mathbf{W}}}{\begin{bmatrix} W_1 \\ W_2 \end{bmatrix}} = \underset{\underset{2 \times 2}{\mathbf{A}}}{\begin{bmatrix} 1 & -1 \\ 1 & 1 \end{bmatrix}} \underset{\underset{2 \times 1}{\mathbf{Y}}}{\begin{bmatrix} Y_1 \\ Y_2 \end{bmatrix}} = \begin{bmatrix} Y_1 - Y_2 \\ Y_1 + Y_2 \end{bmatrix}$$

We then have by (6.48):

$$\mathbf{E(W)} = \begin{bmatrix} 1 & -1 \\ 1 & 1 \end{bmatrix} \begin{bmatrix} E(Y_1) \\ E(Y_2) \end{bmatrix} = \begin{bmatrix} E(Y_1) - E(Y_2) \\ E(Y_1) + E(Y_2) \end{bmatrix}$$

and by (6.49):

$$\boldsymbol{\sigma}^2\mathbf{(W)} = \begin{bmatrix} 1 & -1 \\ 1 & 1 \end{bmatrix} \begin{bmatrix} \sigma^2(Y_1) & \sigma(Y_1, Y_2) \\ \sigma(Y_2, Y_1) & \sigma^2(Y_2) \end{bmatrix} \begin{bmatrix} 1 & 1 \\ -1 & 1 \end{bmatrix}$$

$$= \begin{bmatrix} \sigma^2(Y_1) + \sigma^2(Y_2) - 2\sigma(Y_1, Y_2) & \sigma^2(Y_1) - \sigma^2(Y_2) \\ \sigma^2(Y_1) - \sigma^2(Y_2) & \sigma^2(Y_1) + \sigma^2(Y_2) + 2\sigma(Y_1, Y_2) \end{bmatrix}$$

Thus:

$$\sigma^2(W_1) = \sigma^2(Y_1 - Y_2) = \sigma^2(Y_1) + \sigma^2(Y_2) - 2\sigma(Y_1, Y_2)$$
$$\sigma^2(W_2) = \sigma^2(Y_1 + Y_2) = \sigma^2(Y_1) + \sigma^2(Y_2) + 2\sigma(Y_1, Y_2)$$
$$\sigma(W_1, W_2) = \sigma(Y_1 - Y_2, Y_1 + Y_2) = \sigma^2(Y_1) - \sigma^2(Y_2)$$

6.9 SIMPLE LINEAR REGRESSION MODEL IN MATRIX TERMS

We are now ready to develop simple linear regression in matrix terms. Remember again that we will not present any new results, but shall only state in matrix terms the results obtained earlier. We shall begin with the regression model (3.1):

(6.50) $$Y_i = \beta_0 + \beta_1 X_i + \varepsilon_i \qquad i = 1, \ldots, n$$

This implies:

$$Y_1 = \beta_0 + \beta_1 X_1 + \varepsilon_1$$
$$Y_2 = \beta_0 + \beta_1 X_2 + \varepsilon_2$$

(6.51)

$$\cdot$$
$$\cdot$$
$$\cdot$$

$$Y_n = \beta_0 + \beta_1 X_n + \varepsilon_n$$

We defined earlier the observation vector $\mathbf{Y}$ in (6.4), the $\mathbf{X}$ matrix in (6.6), the $\boldsymbol{\varepsilon}$ vector in (6.10), and the $\boldsymbol{\beta}$ vector in (6.13). Let us repeat these definitions:

(6.52) $$\mathbf{Y} = \begin{bmatrix} Y_1 \\ Y_2 \\ \cdot \\ \cdot \\ \cdot \\ Y_n \end{bmatrix} \qquad \mathbf{X} = \begin{bmatrix} 1 & X_1 \\ 1 & X_2 \\ \cdot & \cdot \\ \cdot & \cdot \\ \cdot & \cdot \\ 1 & X_n \end{bmatrix} \qquad \boldsymbol{\beta} = \begin{bmatrix} \beta_0 \\ \beta_1 \end{bmatrix} \qquad \boldsymbol{\varepsilon} = \begin{bmatrix} \varepsilon_1 \\ \varepsilon_2 \\ \cdot \\ \cdot \\ \cdot \\ \varepsilon_n \end{bmatrix}$$

Now we can write (6.51) in matrix terms compactly as follows:

(6.53)
$$\underset{n \times 1}{\mathbf{Y}} = \underset{n \times 2}{\mathbf{X}} \; \underset{2 \times 1}{\boldsymbol{\beta}} + \underset{n \times 1}{\boldsymbol{\varepsilon}}$$

since:

$$
\begin{bmatrix} Y_1 \\ Y_2 \\ \cdot \\ \cdot \\ \cdot \\ Y_n \end{bmatrix} = \begin{bmatrix} 1 & X_1 \\ 1 & X_2 \\ \cdot & \cdot \\ \cdot & \cdot \\ \cdot & \cdot \\ 1 & X_n \end{bmatrix} \begin{bmatrix} \beta_0 \\ \beta_1 \end{bmatrix} + \begin{bmatrix} \varepsilon_1 \\ \varepsilon_2 \\ \cdot \\ \cdot \\ \cdot \\ \varepsilon_n \end{bmatrix} = \begin{bmatrix} \beta_0 + \beta_1 X_1 + \varepsilon_1 \\ \beta_0 + \beta_1 X_2 + \varepsilon_2 \\ \cdot \\ \cdot \\ \cdot \\ \beta_0 + \beta_1 X_n + \varepsilon_n \end{bmatrix}
$$

The column of 1's in the **X** matrix may be viewed as consisting of the dummy variable $X_0 \equiv 1$ in the alternative regression model (2.5):

$$ Y_i = \beta_0 X_0 + \beta_1 X_i + \varepsilon_i \qquad \text{where } X_0 \equiv 1 $$

Thus, the **X** matrix may be considered to contain a column vector of the dummy variable X_0 and another column vector consisting of the independent variable observations X_i.

With respect to the error terms, model (3.1) assumes that $E(\varepsilon_i) = 0$, $\sigma^2(\varepsilon_i) = \sigma^2$, and that the ε_i are independent normal random variables. The condition $E(\varepsilon_i) = 0$ in matrix terms is:

(6.54) $$ E(\boldsymbol{\varepsilon}) = \mathbf{0} $$

since (6.54) states:

$$
\mathbf{E} \begin{bmatrix} \varepsilon_1 \\ \varepsilon_2 \\ \cdot \\ \cdot \\ \cdot \\ \varepsilon_n \end{bmatrix} = \begin{bmatrix} E(\varepsilon_1) \\ E(\varepsilon_2) \\ \cdot \\ \cdot \\ \cdot \\ E(\varepsilon_n) \end{bmatrix} = \begin{bmatrix} 0 \\ 0 \\ \cdot \\ \cdot \\ \cdot \\ 0 \end{bmatrix}
$$

The condition that the error terms have constant variance σ^2 and that all covariances $\sigma(\varepsilon_i, \varepsilon_j)$ for $i \neq j$ are zero (since the ε_i are independent) is expressed in matrix terms through the variance-covariance matrix:

(6.55) $$ \underset{n \times n}{\boldsymbol{\sigma}^2(\boldsymbol{\varepsilon})} = \underset{n \times n}{\sigma^2 \mathbf{I}} $$

since (6.55) states:

$$
\boldsymbol{\sigma}^2(\boldsymbol{\varepsilon}) = \sigma^2 \begin{bmatrix} 1 & 0 & 0 & \cdots & 0 \\ 0 & 1 & 0 & \cdots & 0 \\ \cdot & \cdot & \cdot & & \cdot \\ \cdot & \cdot & \cdot & & \cdot \\ 0 & 0 & 0 & \cdots & 1 \end{bmatrix} = \begin{bmatrix} \sigma^2 & 0 & 0 & \cdots & 0 \\ 0 & \sigma^2 & 0 & \cdots & 0 \\ \cdot & \cdot & \cdot & & \cdot \\ \cdot & \cdot & \cdot & & \cdot \\ 0 & 0 & 0 & \cdots & \sigma^2 \end{bmatrix}
$$

Thus, the normal error model (3.1) in matrix terms is:

(6.56) $$ \mathbf{Y} = \mathbf{X}\boldsymbol{\beta} + \boldsymbol{\varepsilon} $$

where:

$\boldsymbol{\varepsilon}$ is a vector of independent normal random variables with $\mathbf{E}(\boldsymbol{\varepsilon}) = \mathbf{0}$ and $\boldsymbol{\sigma}^2(\boldsymbol{\varepsilon}) = \sigma^2 \mathbf{I}$

6.10 LEAST SQUARES ESTIMATION OF REGRESSION PARAMETERS

Normal equations

The normal equations (2.9):

$$nb_0 + b_1 \Sigma X_i = \Sigma Y_i$$

(6.57)

$$b_0 \Sigma X_i + b_1 \Sigma X_i^2 = \Sigma X_i Y_i$$

in matrix terms are:

(6.58)
$$\mathbf{X}'\mathbf{X}\mathbf{b} = \mathbf{X}'\mathbf{Y}$$

where $\mathbf{b}$ is the vector of the least squares regression coefficients:

(6.58a)
$$\underset{2 \times 1}{\mathbf{b}} = \begin{bmatrix} b_0 \\ b_1 \end{bmatrix}$$

To see this, recall that we obtained $\mathbf{X}'\mathbf{X}$ in (6.16) and $\mathbf{X}'\mathbf{Y}$ in (6.17). Equation (6.58) thus states:

$$\begin{bmatrix} n & \Sigma X_i \\ \Sigma X_i & \Sigma X_i^2 \end{bmatrix} \begin{bmatrix} b_0 \\ b_1 \end{bmatrix} = \begin{bmatrix} \Sigma Y_i \\ \Sigma X_i Y_i \end{bmatrix}$$

or:

$$\begin{bmatrix} nb_0 + b_1 \Sigma X_i \\ b_0 \Sigma X_i + b_1 \Sigma X_i^2 \end{bmatrix} = \begin{bmatrix} \Sigma Y_i \\ \Sigma X_i Y_i \end{bmatrix}$$

These are precisely the normal equations in (6.57).

Estimated regression coefficients

To obtain the estimated regression coefficients from the normal equations:

$$\mathbf{X}'\mathbf{X}\mathbf{b} = \mathbf{X}'\mathbf{Y}$$

by matrix methods, we premultiply both sides by the inverse of $\mathbf{X}'\mathbf{X}$ (we assume this exists):

$$(\mathbf{X}'\mathbf{X})^{-1}\mathbf{X}'\mathbf{X}\mathbf{b} = (\mathbf{X}'\mathbf{X})^{-1}\mathbf{X}'\mathbf{Y}$$

so that we find, since $(\mathbf{X}'\mathbf{X})^{-1}\mathbf{X}'\mathbf{X} = \mathbf{I}$ and $\mathbf{I}\mathbf{b} = \mathbf{b}$:

(6.59)
$$\mathbf{b} = (\mathbf{X}'\mathbf{X})^{-1}\mathbf{X}'\mathbf{Y}$$

The estimators b_0 and b_1 in $\mathbf{b}$ are the same as those given earlier in (2.10a) and (2.10b). We shall demonstrate this by an example.

Example. Let us find the estimated regression coefficients for the Westwood Company lot size example by matrix methods. From earlier work, we have (Table 2.2):

$$n = 10 \qquad \Sigma Y_i = 1{,}100 \qquad \Sigma X_i = 500 \qquad \Sigma X_i^2 = 28{,}400$$
$$\Sigma X_i Y_i = 61{,}800$$

Let us now use (6.26) to evaluate $(\mathbf{X'X})^{-1}$. We have:

$$n\Sigma(X_i - \bar{X})^2 = n\left[\Sigma X_i^2 - \frac{(\Sigma X_i)^2}{n}\right] = 10\left[28{,}400 - \frac{(500)^2}{10}\right] = 34{,}000$$

Therefore:

$$(\mathbf{X'X})^{-1} = \begin{bmatrix} \dfrac{\Sigma X_i^2}{n\Sigma(X_i - \bar{X})^2} & \dfrac{-\Sigma X_i}{n\Sigma(X_i - \bar{X})^2} \\ \dfrac{-\Sigma X_i}{n\Sigma(X_i - \bar{X})^2} & \dfrac{n}{n\Sigma(X_i - \bar{X})^2} \end{bmatrix} = \begin{bmatrix} \dfrac{28{,}400}{34{,}000} & \dfrac{-500}{34{,}000} \\ \dfrac{-500}{34{,}000} & \dfrac{10}{34{,}000} \end{bmatrix}$$

$$= \begin{bmatrix} .83529412 & -.01470588 \\ -.01470588 & .00029412 \end{bmatrix}$$

We also wish to make use of (6.17) to evaluate $\mathbf{X'Y}$:

$$\mathbf{X'Y} = \begin{bmatrix} \Sigma Y_i \\ \Sigma X_i Y_i \end{bmatrix} = \begin{bmatrix} 1{,}100 \\ 61{,}800 \end{bmatrix}$$

Hence, by (6.59):

$$\mathbf{b} = \begin{bmatrix} b_0 \\ b_1 \end{bmatrix} = (\mathbf{X'X})^{-1}\mathbf{X'Y} = \begin{bmatrix} .83529412 & -.01470588 \\ -.01470588 & .00029412 \end{bmatrix} \begin{bmatrix} 1{,}100 \\ 61{,}800 \end{bmatrix}$$

$$= \begin{bmatrix} 10.0 \\ 2.0 \end{bmatrix}$$

or $b_0 = 10.0$ and $b_1 = 2.0$. This agrees with the results in Chapter 2. Any difference would have been due to rounding errors.

To reduce the effect of rounding errors when obtaining the vector $\mathbf{b}$ by hand calculations, it is often desirable to move the constant in the denominator of the elements of $(\mathbf{X'X})^{-1}$ outside the matrix, and do the division as the last step. For our example, this would lead to:

$$(\mathbf{X'X})^{-1} = \frac{1}{n\Sigma(X_i - \bar{X})^2}\begin{bmatrix} \Sigma X_i^2 & -\Sigma X_i \\ -\Sigma X_i & n \end{bmatrix}$$

$$= \frac{1}{34{,}000}\begin{bmatrix} 28{,}400 & -500 \\ -500 & 10 \end{bmatrix}$$

$$\mathbf{b} = \frac{1}{34{,}000}\begin{bmatrix} 28{,}400 & -500 \\ -500 & 10 \end{bmatrix}\begin{bmatrix} 1{,}100 \\ 61{,}800 \end{bmatrix}$$

$$= \frac{1}{34{,}000}\begin{bmatrix} 340{,}000 \\ 68{,}000 \end{bmatrix} = \begin{bmatrix} 10.0 \\ 2.0 \end{bmatrix}$$

In this instance, the two methods of calculation lead to identical results. Often, however, postponing division by $n\Sigma(X_i - \bar{X})^2$ until the end yields more accurate results.

Comments

1. To derive the normal equations by the method of least squares, we minimize the quantity:

$$Q = \Sigma[Y_i - (\beta_0 + \beta_1 X_i)]^2$$

In matrix notation:

(6.60) $$Q = (\mathbf{Y} - \mathbf{X\beta})'(\mathbf{Y} - \mathbf{X\beta})$$

Expanding out, we obtain:

$$Q = \mathbf{Y'Y} - \mathbf{\beta'X'Y} - \mathbf{Y'X\beta} + \mathbf{\beta'X'X\beta}$$

since $(\mathbf{X\beta})' = \mathbf{\beta'X'}$ by (6.35). Note now that $\mathbf{Y'X\beta}$ is 1×1, hence is equal to its transpose, which according to (6.36) is $\mathbf{\beta'X'Y}$. Thus, we find:

(6.61) $$Q = \mathbf{Y'Y} - 2\mathbf{\beta'X'Y} + \mathbf{\beta'X'X\beta}$$

To find the value of $\mathbf{\beta}$ which minimizes Q, we differentiate with respect to β_0 and β_1. Let:

(6.62) $$\frac{\partial}{\partial\mathbf{\beta}}(Q) = \begin{bmatrix} \dfrac{\partial Q}{\partial \beta_0} \\[2mm] \dfrac{\partial Q}{\partial \beta_1} \end{bmatrix}$$

Then it follows that:

(6.63) $$\frac{\partial}{\partial\mathbf{\beta}}(Q) = -2\mathbf{X'Y} + 2\mathbf{X'X\beta}$$

Equating to zero and substituting $\mathbf{b}$ for $\mathbf{\beta}$ gives the matrix form of the least squares normal equations:

$$\mathbf{X'Xb} = \mathbf{X'Y}$$

2. A comparison of the normal equations and $\mathbf{X'X}$ shows that whenever the columns of $\mathbf{X'X}$ are linearly dependent, the normal equations will be linearly dependent also. No unique solutions can be obtained for b_0 and b_1 in that case. Fortunately, in most regression applications, the columns of $\mathbf{X'X}$ are linearly independent, leading to unique solutions for b_0 and b_1.

6.11 ANALYSIS OF VARIANCE RESULTS

Fitted values and residuals

Let the vector of the fitted values $\hat{Y}_i$ be denoted by $\hat{\mathbf{Y}}$:

$$(6.64) \qquad \hat{\mathbf{Y}}_{n \times 1} = \begin{bmatrix} \hat{Y}_1 \\ \hat{Y}_2 \\ \cdot \\ \cdot \\ \cdot \\ \hat{Y}_n \end{bmatrix}$$

and the vector of the residuals $e_i = Y_i - \hat{Y}_i$ be denoted by $\mathbf{e}$:

$$(6.65) \qquad \mathbf{e}_{n \times 1} = \begin{bmatrix} e_1 \\ e_2 \\ \cdot \\ \cdot \\ \cdot \\ e_n \end{bmatrix}$$

In matrix notation, we then have:

$$(6.66) \qquad \hat{\mathbf{Y}}_{n \times 1} = \mathbf{X}_{n \times 2} \mathbf{b}_{2 \times 1}$$

because:

$$\begin{bmatrix} \hat{Y}_1 \\ \hat{Y}_2 \\ \cdot \\ \cdot \\ \hat{Y}_n \end{bmatrix} = \begin{bmatrix} 1 & X_1 \\ 1 & X_2 \\ \cdot & \cdot \\ \cdot & \cdot \\ 1 & X_n \end{bmatrix} \begin{bmatrix} b_0 \\ b_1 \end{bmatrix} = \begin{bmatrix} b_0 + b_1 X_1 \\ b_0 + b_1 X_2 \\ \cdot \\ \cdot \\ b_0 + b_1 X_n \end{bmatrix}$$

Similarly:

$$(6.67) \qquad \mathbf{e}_{n \times 1} = \mathbf{Y}_{n \times 1} - \hat{\mathbf{Y}}_{n \times 1} = \mathbf{Y}_{n \times 1} - \mathbf{X} \mathbf{b}_{n \times 1}$$

Sums of squares

To see how the sums of squares are expressed in matrix notation, we begin with $SSTO$. We know from (3.49) that:

$$(6.68) \qquad SSTO = \Sigma Y_i^2 - n\bar{Y}^2 = \Sigma Y_i^2 - \frac{(\Sigma Y_i)^2}{n}$$

We also know from (6.15) that:

$$\mathbf{Y}'\mathbf{Y} = \Sigma Y_i^2$$

The subtraction term $n\bar{Y}^2 = (\Sigma Y_i)^2/n$ in matrix form uses $\mathbf{1}$, the vector of 1's defined in (6.19), as follows:

$$(6.69) \qquad \frac{(\Sigma Y_i)^2}{n} = \left(\frac{1}{n}\right)\mathbf{Y}'\mathbf{1}\mathbf{1}'\mathbf{Y} \; = \; n\bar{Y}^2$$

For instance, if $n = 2$, we have:

$$\left(\frac{1}{n}\right)[Y_1 \quad Y_2]\begin{bmatrix} 1 \\ 1 \end{bmatrix}[1 \quad 1]\begin{bmatrix} Y_1 \\ Y_2 \end{bmatrix} = \left(\frac{1}{n}\right)(\Sigma Y_i)(\Sigma Y_i) = \frac{(\Sigma Y_i)^2}{n}$$

Hence, it follows that:

(6.70a) $$SSTO = \mathbf{Y'Y} - \left(\frac{1}{n}\right)\mathbf{Y'11'Y}$$

Just as ΣY_i^2 is represented by $\mathbf{Y'Y}$ in matrix terms, so $SSE = \Sigma e_i^2 = \Sigma(Y_i - \hat{Y}_i)^2$ can be represented as follows:

(6.70b) $$SSE = \mathbf{e'e} = (\mathbf{Y} - \mathbf{Xb})'(\mathbf{Y} - \mathbf{Xb})$$

which can be shown to equal:

(6.70c) $$SSE = \mathbf{Y'Y} - \mathbf{b'X'Y}$$

Finally, it can be shown that:

(6.70d) $$SSR = \mathbf{b'X'Y} - \left(\frac{1}{n}\right)\mathbf{Y'11'Y}$$

Example. Let us find SSE for the Westwood Company lot size example by matrix methods, using (6.70c). We know from earlier results:

$$\mathbf{Y'Y} = \Sigma Y_i^2 = 134{,}660$$

We also know from earlier:

$$\mathbf{b} = \begin{bmatrix} 10.0 \\ 2.0 \end{bmatrix} \qquad \mathbf{X'Y} = \begin{bmatrix} 1{,}100 \\ 61{,}800 \end{bmatrix}$$

Hence:

$$\mathbf{b'X'Y} = [10.0 \quad 2.0]\begin{bmatrix} 1{,}100 \\ 61{,}800 \end{bmatrix} = 134{,}600$$

and:

$$SSE = \mathbf{Y'Y} - \mathbf{b'X'Y} = 134{,}660 - 134{,}600 = 60$$

which is the same result as that obtained in Chapter 2. Any difference would have been due to rounding errors.

Similarly, we can find SSR using (6.70d):

$$SSR = \mathbf{b'X'Y} - \left(\frac{1}{n}\right)\mathbf{Y'11'Y}$$
$$= 134{,}600 - 10(110)^2 = 13{,}600$$

since the subtraction term in SSR equals $n\bar{Y}^2$, and $\bar{Y} = 110$ for the Westwood Company example.

Note

To illustrate the derivation of the sums of squares expressions in matrix notation, consider SSE:

$$SSE = \mathbf{e'e} = (\mathbf{Y} - \mathbf{Xb})'(\mathbf{Y} - \mathbf{Xb}) = \mathbf{Y'Y} - 2\mathbf{b'X'Y} + \mathbf{b'X'Xb}$$

In substituting for the right-most $\mathbf{b}$ we obtain by (6.59):

$$SSE = \mathbf{Y'Y} - 2\mathbf{b'X'Y} + \mathbf{b'X'X(X'X)^{-1}X'Y}$$
$$= \mathbf{Y'Y} - 2\mathbf{b'X'Y} + \mathbf{b'IX'Y}$$

In dropping $\mathbf{I}$ and subtracting, we obtain the result in (6.70c):

$$SSE = \mathbf{Y'Y} - \mathbf{b'X'Y}$$

Sums of squares as quadratic forms

The ANOVA sums of squares can be shown to be *quadratic forms*. An example of a quadratic form of the observations Y_i when $n = 2$ is:

(6.71) $$5Y_1^2 + 6Y_1Y_2 + 4Y_2^2$$

Note that this expression is a second-degree polynomial containing terms involving the squares of the observations and the cross product. We can express (6.71) in matrix terms as follows:

(6.71a) $$[Y_1 \quad Y_2] \begin{bmatrix} 5 & 3 \\ 3 & 4 \end{bmatrix} \begin{bmatrix} Y_1 \\ Y_2 \end{bmatrix} = \mathbf{Y'AY}$$

where $\mathbf{A}$ is a symmetric matrix of coefficients.

In general, a quadratic form is defined as:

(6.72) $$\underset{1 \times 1}{\mathbf{Y'AY}} = \sum_{i=1}^{n} \sum_{j=1}^{n} a_{ij}Y_iY_j \qquad \text{where } a_{ij} = a_{ji}$$

$\mathbf{A}$ is a symmetric $n \times n$ matrix and is called the *matrix of the quadratic form.*

The ANOVA sums of squares $SSTO$, SSR, and SSE are all quadratic forms. To see this, we need to express the matrix forms for these sums of squares in (6.70) still more compactly. We do this by noting that:

(6.73) $$\underset{n \times 1}{\mathbf{1}} \quad \underset{1 \times n}{\mathbf{1'}} = \underset{n \times n}{\mathbf{J}}$$

where $\mathbf{J}$ is the $n \times n$ matrix all of whose elements are 1's, as defined in (6.20). Also, the transpose of $\mathbf{b}$ in (6.59) can be obtained using (6.36) and (6.33):

(6.74) $$\mathbf{b'} = [(\mathbf{X'X})^{-1}\mathbf{X'Y}]' = \mathbf{Y'X(X'X)^{-1}}$$

by noting that $(\mathbf{X'X})^{-1}$ is a symmetric matrix so that it equals its transpose. Hence:

(6.75a) $$SSTO = \mathbf{Y'}\left[\mathbf{I} - \left(\frac{1}{n}\right)\mathbf{J}\right]\mathbf{Y}$$

$$(6.75b) \qquad SSR = \mathbf{Y}' \left[\mathbf{X}(\mathbf{X}'\mathbf{X})^{-1}\mathbf{X}' - \left(\frac{1}{n} \right) \mathbf{J} \right] \mathbf{Y}$$

$$(6.75c) \qquad SSE = \mathbf{Y}'[\mathbf{I} - \mathbf{X}(\mathbf{X}'\mathbf{X})^{-1}\mathbf{X}']\mathbf{Y}$$

Each of these sums of squares can now be seen to be of the form $\mathbf{Y}'\mathbf{A}\mathbf{Y}$. It can be shown that the three $\mathbf{A}$ matrices:

$$(6.76a) \qquad \mathbf{I} - \left(\frac{1}{n} \right) \mathbf{J}$$

$$(6.76b) \qquad \mathbf{X}(\mathbf{X}'\mathbf{X})^{-1}\mathbf{X}' - \left(\frac{1}{n} \right) \mathbf{J}$$

$$(6.76c) \qquad \mathbf{I} - \mathbf{X}(\mathbf{X}'\mathbf{X})^{-1}\mathbf{X}'$$

are symmetric. Hence, $SSTO$, SSR, and SSE are quadratic forms, with the matrices of the quadratic forms given in (6.76). Quadratic forms play an important role in statistics because all sums of squares in the analysis of variance for linear statistical models can be expressed as quadratic forms.

6.12 INFERENCES IN REGRESSION ANALYSIS

As we saw in earlier chapters, all interval estimates are of the following form: point estimator plus and minus a certain number of estimated standard deviations of the point estimator. Similarly, all tests require the point estimator and the estimated standard deviation of the point estimator or, in the case of analysis of variance tests, various sums of squares. Matrix algebra is of principal help in inference making when obtaining the estimated standard deviations and sums of squares. We have already given the matrix equivalents of the sums of squares for the analysis of variance. Hence, we focus here chiefly on the matrix expressions for the estimated standard deviations of point estimators of interest.

Regression coefficients

The variance-covariance matrix of $\mathbf{b}$:

$$(6.77) \qquad \boldsymbol{\sigma}^2(\mathbf{b}) = \begin{bmatrix} \sigma^2(b_0) & \sigma(b_0, b_1) \\ \sigma(b_1, b_0) & \sigma^2(b_1) \end{bmatrix}$$

is:

$$(6.78) \qquad \underset{2\times2}{\boldsymbol{\sigma}^2(\mathbf{b})} = \sigma^2(\mathbf{X}'\mathbf{X})^{-1}$$

or, using (6.27):

$$(6.78a) \qquad \boldsymbol{\sigma}^2(\mathbf{b}) = \begin{bmatrix} \dfrac{\sigma^2 \Sigma X_i^2}{n\Sigma(X_i - \bar{X})^2} & \dfrac{-\bar{X}\sigma^2}{\Sigma(X_i - \bar{X})^2} \\ \dfrac{-\bar{X}\sigma^2}{\Sigma(X_i - \bar{X})^2} & \dfrac{\sigma^2}{\Sigma(X_i - \bar{X})^2} \end{bmatrix}$$

When MSE is substituted for σ^2 in (6.78a) we have:

$$(6.79) \qquad s^2(\mathbf{b}) = MSE(\mathbf{X}'\mathbf{X})^{-1}_{2\times 2} = \begin{bmatrix} \dfrac{MSE\Sigma X_i^2}{n\Sigma(X_i - \bar{X})^2} & \dfrac{-\bar{X}MSE}{\Sigma(X_i - \bar{X})^2} \\ \dfrac{-\bar{X}MSE}{\Sigma(X_i - \bar{X})^2} & \dfrac{MSE}{\Sigma(X_i - \bar{X})^2} \end{bmatrix}$$

where $s^2(\mathbf{b})$ is the estimated variance-covariance matrix of $\mathbf{b}$. In (6.78a), you will recognize the variances of b_0 (3.20b) and b_1 (3.3b) and the covariance of b_0 and b_1 (5.3). Likewise, the estimated variances in (6.79) are familiar from earlier chapters.

Joint confidence region for β_0 and β_1

The boundary for the joint confidence region for β_0 and β_1, given in (5.2), is expressed in matrix terms as follows:

$$(6.80) \qquad \frac{(\mathbf{b} - \boldsymbol{\beta})'\mathbf{X}'\mathbf{X}(\mathbf{b} - \boldsymbol{\beta})}{2MSE} = F(1 - \alpha; 2, n - 2)$$

Mean response

To estimate the mean response at X_h, let us define the vector:

$$(6.81) \qquad \mathbf{X}_h_{2\times 1} = \begin{bmatrix} 1 \\ X_h \end{bmatrix} \quad \text{or} \quad \mathbf{X}'_h = [1 \quad X_h]$$

The fitted value in matrix notation then is:

$$(6.82) \qquad \hat{Y}_h = \mathbf{X}'_h\mathbf{b}_{1\times 1}$$

since:

$$\mathbf{X}'_h\mathbf{b} = [1 \quad X_h] \begin{bmatrix} b_0 \\ b_1 \end{bmatrix} = [b_0 + b_1X_h] = [\hat{Y}_h] = \hat{Y}_h$$

Note that $\mathbf{X}'_h\mathbf{b}$ is a 1×1 matrix; hence, we can write the final result as a scalar.

The variance of $\hat{Y}_h$, given earlier in (3.28b), is in matrix notation:

$$(6.83) \qquad \sigma^2(\hat{Y}_h) = \sigma^2\mathbf{X}'_h(\mathbf{X}'\mathbf{X})^{-1}\mathbf{X}_h = \mathbf{X}'_h\sigma^2(\mathbf{b})\mathbf{X}_h$$

where $\sigma^2(\mathbf{b})$ is the variance-covariance matrix of the regression coefficients in (6.78). Note, therefore, that $\sigma^2(\hat{Y}_h)$ is a function of the variances $\sigma^2(b_0)$ and $\sigma^2(b_1)$ and of the covariance $\sigma(b_0, b_1)$.

The estimated variance of $\hat{Y}_h$, given earlier in (3.30), is in matrix notation:

$$(6.84) \qquad s^2(\hat{Y}_h) = MSE(\mathbf{X}'_h(\mathbf{X}'\mathbf{X})^{-1}\mathbf{X}_h) = \mathbf{X}'_h s^2(\mathbf{b})\mathbf{X}_h$$

where $s^2(\mathbf{b})$ is the estimated variance-covariance matrix of the regression coefficients in (6.79).

Prediction of new observation

The estimated variance $s^2(Y_{h(\text{new})})$, given earlier in (3.37), is in matrix notation:

$$(6.85) \qquad s^2(Y_{h(\text{new})}) = MSE + s^2(\hat{Y}_h) = MSE + \mathbf{X}_h'\mathbf{s}^2(\mathbf{b})\mathbf{X}_h$$

$$= MSE(1 + \mathbf{X}_h'(\mathbf{X}'\mathbf{X})^{-1}\mathbf{X}_h)$$

Examples

1. We wish to find $s^2(b_0)$ and $s^2(b_1)$ for the Westwood Company lot size example by matrix methods. We found earlier that $MSE = 7.5$ and:

$$(\mathbf{X}'\mathbf{X})^{-1} = \begin{bmatrix} .83529412 & -.01470588 \\ -.01470588 & .00029412 \end{bmatrix}$$

Hence, by (6.79):

$$\mathbf{s}^2(\mathbf{b}) = MSE(\mathbf{X}'\mathbf{X})^{-1} = 7.5 \begin{bmatrix} .83529412 & -.01470588 \\ -.01470588 & .00029412 \end{bmatrix}$$

$$= \begin{bmatrix} 6.264706 & -.1102941 \\ -.1102941 & .0022059 \end{bmatrix}$$

Thus, $s^2(b_0) = 6.26471$ and $s^2(b_1) = .002206$. These are the same as the results obtained in Chapter 3.

Note how simple it is to find the estimated variances of the regression coefficients as soon as $(\mathbf{X}'\mathbf{X})^{-1}$ has been obtained. This inverse is needed in the first place to find the regression coefficients, so that practically no extra work is required to obtain their estimated variances.

2. We wish to find $s^2(\hat{Y}_h)$ for the Westwood Company example when $X_h = 55$. We define:

$$\mathbf{X}_h' = \begin{bmatrix} 1 & 55 \end{bmatrix}$$

and obtain by (6.84):

$$s^2(\hat{Y}_{55}) = \mathbf{X}_h'\mathbf{s}^2(\mathbf{b})\mathbf{X}_h$$

$$= \begin{bmatrix} 1 & 55 \end{bmatrix} \begin{bmatrix} 6.264706 & -.1102941 \\ -.1102941 & .0022059 \end{bmatrix} \begin{bmatrix} 1 \\ 55 \end{bmatrix} = .80520$$

This is the same result as that obtained in Chapter 3, except for a minor difference due to rounding.

Comments

1. To illustrate a derivation in matrix terms, let us find the variance-covariance matrix of $\mathbf{b}$. Recall that:

$$\mathbf{b} = (\mathbf{X}'\mathbf{X})^{-1}\mathbf{X}'\mathbf{Y} = \mathbf{A}\mathbf{Y}$$

where $\mathbf{A}$ is a constant matrix:

$$\mathbf{A} = (\mathbf{X}'\mathbf{X})^{-1}\mathbf{X}'$$

Hence by (6.49), we have:

$$\sigma^2(\mathbf{b}) = \mathbf{A}[\sigma^2(\mathbf{Y})]\mathbf{A}'$$

Now $\sigma^2(\mathbf{Y}) = \sigma^2\mathbf{I}$. Further, it follows from (6.74) that:

$$\mathbf{A}' = \mathbf{X}(\mathbf{X}'\mathbf{X})^{-1}$$

We find therefore:

$$\begin{aligned}
\sigma^2(\mathbf{b}) &= (\mathbf{X}'\mathbf{X})^{-1}\mathbf{X}'\sigma^2\mathbf{I}\mathbf{X}(\mathbf{X}'\mathbf{X})^{-1} \\
&= \sigma^2(\mathbf{X}'\mathbf{X})^{-1}\mathbf{X}'\mathbf{X}(\mathbf{X}'\mathbf{X})^{-1} \\
&= \sigma^2(\mathbf{X}'\mathbf{X})^{-1}\mathbf{I} \\
&= \sigma^2(\mathbf{X}'\mathbf{X})^{-1}
\end{aligned}$$

2. Since $\hat{Y}_h = \mathbf{X}_h'\mathbf{b}$, it follows at once from (6.49) that:

$$\sigma^2(\hat{Y}_h) = \mathbf{X}_h'[\sigma^2(\mathbf{b})]\mathbf{X}_h$$

Hence:

$$\sigma^2(\hat{Y}_h) = [1 \quad X_h] \begin{bmatrix} \sigma^2(b_0) & \sigma(b_0, b_1) \\ \sigma(b_1, b_0) & \sigma^2(b_1) \end{bmatrix} \begin{bmatrix} 1 \\ X_h \end{bmatrix}$$

or:

$$(6.86) \qquad \sigma^2(\hat{Y}_h) = \sigma^2(b_0) + 2X_h\sigma(b_0, b_1) + X_h^2\sigma^2(b_1)$$

Using the results from (6.78a), we obtain:

$$\sigma^2(\hat{Y}_h) = \frac{\sigma^2\Sigma X_i^2}{n\Sigma(X_i - \bar{X})^2} + \frac{2X_h(-\bar{X})\sigma^2}{\Sigma(X_i - \bar{X})^2} + \frac{X_h^2\sigma^2}{\Sigma(X_i - \bar{X})^2}$$

which reduces to the familiar expression:

$$(6.87) \qquad \sigma^2(\hat{Y}_h) = \sigma^2\left[\frac{1}{n} + \frac{(X_h - \bar{X})^2}{\Sigma(X_i - \bar{X})^2}\right]$$

Thus, we see explicitly that the variance expression in (6.87) contains contributions from $\sigma^2(b_0)$, $\sigma^2(b_1)$, and $\sigma(b_0, b_1)$, which it must according to theorem (1.25b) since $\hat{Y}_h$ is a linear combination of b_0 and b_1:

$$\hat{Y}_h = b_0 + b_1X_h$$

3. We do not show the results in matrix terms for other types of inferences, such as simultaneous prediction of several new observations on Y at different X_h levels, since these are based on results we have developed.

6.13 WEIGHTED LEAST SQUARES

The regression results for weighted least squares can be stated compactly with matrix algebra. Let the matrix $\mathbf{W}$ be a diagonal matrix containing the weights w_i.

$$\text{(6.88)} \qquad \underset{n\times n}{\mathbf{W}} = \begin{bmatrix} w_1 & & & \\ & w_2 & & 0 \\ & & \cdot & \\ & & & \cdot \\ & & & & \cdot \\ 0 & & & w_n \end{bmatrix}$$

The weighted least squares normal equations (5.32) can then be expressed as follows:

$$\text{(6.89)} \qquad \mathbf{X'WXb} = \mathbf{X'WY}$$

and the weighted least squares estimators are:

$$\text{(6.90)} \qquad \underset{2\times 1}{\mathbf{b}} = (\mathbf{X'WX})^{-1}\mathbf{X'WY}$$

Note that if $\mathbf{W} = \mathbf{I}$, as it would for unweighted least squares, (6.90) reduces to the unweighted estimators (6.59).

Other results for weighted least squares bear a similar relation to the earlier results for unweighted least squares. For instance, when the error term variances σ_i^2 are not equal, the weights w_i are chosen to be inversely proportional to σ_i^2, so that $\sigma_i^2 = \sigma^2/w_i$. The variance-covariance matrix of the weighted least squares estimators, then, is:

$$\text{(6.91)} \qquad \underset{2\times 2}{\sigma^2(\mathbf{b})} = \sigma^2(\mathbf{X'WX})^{-1}$$

and the estimated variance-covariance matrix is:

$$\text{(6.92)} \qquad \underset{2\times 2}{s^2(\mathbf{b})} = MSE_w(\mathbf{X'WX})^{-1}$$

where MSE_w is based on the weighted squared deviations:

$$\text{(6.92a)} \qquad MSE_w = \frac{\Sigma w_i(Y_i - \hat{Y}_i)^2}{n-2}$$

6.14 RESIDUALS

For later analysis of residuals, it will be useful to recognize that each residual e_i can be expressed as a linear combination of the observations Y_i. It can be shown that the vector of the residuals $\mathbf{e}$, defined in (6.65), equals:

$$\text{(6.93)} \qquad \underset{n\times 1}{\mathbf{e}} = (\underset{n\times n}{\mathbf{I}} - \underset{n\times n}{\mathbf{H}})\underset{n\times 1}{\mathbf{Y}}$$

where:

$$\text{(6.93a)} \qquad \underset{n\times n}{\mathbf{H}} = \mathbf{X}(\mathbf{X'X})^{-1}\mathbf{X'}$$

Note from (6.76c) that the matrix $\mathbf{I} - \mathbf{H}$ is the matrix of the quadratic form (6.75c) for $SSE = \Sigma e_i^2$.

The square $n \times n$ matrix $\mathbf{H}$ is called the *hat matrix* and plays an important role in regression analysis, as we shall see in Chapter 11 when we consider whether or not regression results are unduly influenced by one or a few observations. The matrix $\mathbf{I} - \mathbf{H}$ is symmetric and has the special property (called idempotency):

$$(6.94) \qquad (\mathbf{I} - \mathbf{H})(\mathbf{I} - \mathbf{H}) = \mathbf{I} - \mathbf{H}$$

In general, a matrix $\mathbf{M}$ is said to be idempotent if $\mathbf{MM} = \mathbf{M}$.

It can be shown that the variance-covariance matrix of the vector of residuals $\mathbf{e}$ also involves the matrix $\mathbf{I} - \mathbf{H}$:

$$(6.95) \qquad \sigma^2(\mathbf{e}) = \sigma^2(\mathbf{I} - \mathbf{H})$$

and is estimated by:

$$(6.96) \qquad s^2(\mathbf{e}) = MSE(\mathbf{I} - \mathbf{H})$$

Note

The variance-covariance matrix of $\mathbf{e}$ can be derived by means of (6.49). Since $\mathbf{e} = (\mathbf{I} - \mathbf{H})\mathbf{Y}$, we obtain:

$$\sigma^2(\mathbf{e}) = (\mathbf{I} - \mathbf{H})\sigma^2(\mathbf{Y})(\mathbf{I} - \mathbf{H})'$$

Now $\sigma^2(\mathbf{Y}) = \sigma^2(\boldsymbol{\varepsilon}) = \sigma^2\mathbf{I}$ for the normal error model according to (6.55). Also, $(\mathbf{I} - \mathbf{H})' = \mathbf{I} - \mathbf{H}$ because of the symmetry of the matrix. Hence:

$$\sigma^2(\mathbf{e}) = \sigma^2(\mathbf{I} - \mathbf{H})\mathbf{I}(\mathbf{I} - \mathbf{H})$$
$$= \sigma^2(\mathbf{I} - \mathbf{H})(\mathbf{I} - \mathbf{H})$$

In view of property (6.94), we obtain formula (6.95):

$$\sigma^2(\mathbf{e}) = \sigma^2(\mathbf{I} - \mathbf{H})$$

PROBLEMS

6.1. For the matrices below, obtain: (1) $\mathbf{A} + \mathbf{B}$, (2) $\mathbf{A} - \mathbf{B}$, (3) $\mathbf{AC}$, (4) $\mathbf{AB}'$, (5) $\mathbf{B}'\mathbf{A}$.

$$\mathbf{A} = \begin{bmatrix} 1 & 4 \\ 2 & 6 \\ 3 & 8 \end{bmatrix} \qquad \mathbf{B} = \begin{bmatrix} 1 & 3 \\ 1 & 4 \\ 2 & 5 \end{bmatrix} \qquad \mathbf{C} = \begin{bmatrix} 3 & 8 & 1 \\ 5 & 4 & 0 \end{bmatrix}$$

State the dimension of each resulting matrix.

6.2. For the matrices below, obtain: (1) $\mathbf{A} + \mathbf{C}$, (2) $\mathbf{A} - \mathbf{C}$, (3) $\mathbf{B}'\mathbf{A}$, (4) $\mathbf{AC}'$, (5) $\mathbf{C}'\mathbf{A}$.

$$\mathbf{A} = \begin{bmatrix} 2 & 1 \\ 3 & 5 \\ 5 & 7 \\ 4 & 8 \end{bmatrix} \qquad \mathbf{B} = \begin{bmatrix} 6 \\ 9 \\ 3 \\ 1 \end{bmatrix} \qquad \mathbf{C} = \begin{bmatrix} 3 & 8 \\ 8 & 6 \\ 5 & 1 \\ 2 & 4 \end{bmatrix}$$

State the dimension of each resulting matrix.

6.3. Show how the following expressions are written in terms of matrices: (1) $Y_i - \hat{Y}_i$ $= e_i$, (2) $\Sigma X_i e_i = 0$. Assume $i = 1, \ldots, 4$.

6.4. **Flavor deterioration.** The results shown below were obtained in a small-scale experiment to study the relation between °F of storage temperature (X) and number of weeks before flavor deterioration of a food product begins to occur (Y).

i:	1	2	3	4	5
X_i:	+8	+4	0	−4	−8
Y_i:	7.8	9.0	10.2	11.0	11.7

Assume that the first-order regression model (3.1) is applicable. Using matrix methods, find: (1) $\mathbf{Y'Y}$, (2) $\mathbf{X'X}$, (3) $\mathbf{X'Y}$.

6.5. **Consumer finance.** The data below show for a consumer finance company operating in six cities, the number of competing loan companies operating in the city (X) and the number per thousand of the company's loans made in that city that are currently delinquent (Y):

i:	1	2	3	4	5	6
X_i:	4	1	2	3	3	4
Y_i:	16	5	10	15	13	22

Assume that the first-order regression model (3.1) is applicable. Using matrix methods, find: (1) $\mathbf{Y'Y}$, (2) $\mathbf{X'X}$, (3) $\mathbf{X'Y}$.

6.6. Refer to **Airfreight breakage** Problem 2.17. Using matrix methods, find: (1) $\mathbf{Y'Y}$, (2) $\mathbf{X'X}$, (3) $\mathbf{X'Y}$.

6.7. Refer to **Plastic hardness** Problem 2.18. Using matrix methods, find: (1) $\mathbf{Y'Y}$, (2) $\mathbf{X'X}$, (3) $\mathbf{X'Y}$.

6.8. Let **B** be defined as follows:

$$\mathbf{B} = \begin{bmatrix} 1 & 5 & 0 \\ 1 & 0 & 5 \\ 1 & 0 & 5 \end{bmatrix}$$

   a. Are the column vectors of **B** linearly dependent?
   b. What is the rank of **B**?
   c. What must be the determinant of **B**?

6.9. Let **A** be defined as follows:

$$\mathbf{A} = \begin{bmatrix} 0 & 1 & 8 \\ 0 & 3 & 1 \\ 0 & 5 & 5 \end{bmatrix}$$

   a. Are the column vectors of **A** linearly dependent?
   b. Restate definition (6.22) in terms of row vectors. Are the row vectors of **A** linearly dependent?
   c. What is the rank of **A**?
   d. Calculate the determinant of **A**.

6.10. Find the inverse of each of the following matrices:

$$\mathbf{A} = \begin{bmatrix} 2 & 4 \\ 3 & 1 \end{bmatrix} \qquad \mathbf{B} = \begin{bmatrix} 4 & 3 & 2 \\ 6 & 5 & 10 \\ 10 & 1 & 6 \end{bmatrix}$$

Check in each case that the resulting matrix is indeed the inverse.

6.11. Find the inverse of the following matrix:

$$A = \begin{bmatrix} 5 & 1 & 3 \\ 4 & 0 & 5 \\ 1 & 9 & 6 \end{bmatrix}$$

Check that the resulting matrix is indeed the inverse.

6.12. Refer to **Flavor deterioration** Problem 6.4. Find $(X'X)^{-1}$.

6.13. Refer to **Consumer finance** Problem 6.5. Find $(X'X)^{-1}$.

6.14. Consider the simultaneous equations:

$$4y_1 + 7y_2 = 25$$
$$2y_1 + 3y_2 = 12$$

a. Write these equations in matrix notation.
b. Using matrix methods, find the solutions for y_1 and y_2.

6.15. Consider the simultaneous equations:

$$5y_1 + 2y_2 = 8$$
$$23y_1 + 7y_2 = 28$$

a. Write these equations in matrix notation.
b. Using matrix methods, find the solutions for y_1 and y_2.

6.16. Consider the estimated linear regression function in the form of (2.15). Write expressions in this form for the fitted values $\hat{Y}_i$ in matrix terms for $i = 1, \ldots, 5$.

6.17. Consider the following functions of the random variables Y_1, Y_2, and Y_3:

$$W_1 = Y_1 + Y_2 + Y_3$$
$$W_2 = Y_1 - Y_2$$
$$W_3 = Y_1 - Y_2 - Y_3$$

a. State the above in matrix notation.
b. Find the expectation of the random vector W.
c. Find the variance-covariance matrix of W.

6.18. Consider the following functions of the random variables Y_1, Y_2, Y_3, and Y_4:

$$W_1 = \frac{1}{4}(Y_1 + Y_2 + Y_3 + Y_4)$$

$$W_2 = \frac{1}{2}(Y_1 + Y_2) - \frac{1}{2}(Y_3 + Y_4)$$

a. State the above in matrix notation.
b. Find the expectation of the random vector W.
c. Find the variance-covariance matrix of W.

6.19. Find the matrix A of the quadratic form:

$$3Y_1^2 + 10Y_1Y_2 + 17Y_2^2$$

6.20. Find the matrix A of the quadratic form:

$$7Y_1^2 - 8Y_1Y_2 + 8Y_2^2$$

6.21. For the matrix:

$$A = \begin{bmatrix} 5 & 2 \\ 2 & 1 \end{bmatrix}$$

find the quadratic form of the observations Y_1 and Y_2.

6.22. For the matrix:

$$A = \begin{bmatrix} 1 & 0 & 4 \\ 0 & 3 & 0 \\ 4 & 0 & 9 \end{bmatrix}$$

find the quadratic form of the observations Y_1, Y_2, and Y_3.

6.23. Refer to **Flavor deterioration** Problems 6.4 and 6.12.
  a. Using matrix methods, obtain the following: (1) vector of estimated regression coefficients, (2) vector of residuals, (3) SSR, (4) SSE, (5) estimated variance-covariance matrix of **b**, (6) point estimate of $E(Y_h)$ when $X_h = -6$, (7) estimated variance of $\hat{Y}_h$ when $X_h = -6$.
  b. What simplifications arose from the spacing of the X levels in the experiment?
  c. Using matrix methods, obtain the numerator of the left term in (6.80).

6.24. Refer to **Consumer finance** Problems 6.5 and 6.13.
  a. Using matrix methods, obtain the following: (1) vector of estimated regression coefficients, (2) vector of residuals, (3) SSR, (4) SSE, (5) estimated variance-covariance matrix of **b**, (6) point estimate of $E(Y_h)$ when $X_h = 4$, (7) estimated variance of $Y_{h(\text{new})}$ when $X_h = 4$.
  b. From your estimated variance-covariance matrix in part (a5), obtain the following: (1) $s(b_0, b_1)$; (2) $s^2(b_0)$; (3) $s(b_1)$.

6.25. Refer to **Airfreight breakage** Problems 2.17 and 6.6.
  a. Using matrix methods, obtain the following: (1) $(X'X)^{-1}$, (2) **b**, (3) **e**, (4) SSR, (5) SSE, (6) $s^2(\mathbf{b})$, (7) $\hat{Y}_h$ when $X_h = 2$, (8) $s^2(\hat{Y}_h)$ when $X_h = 2$.
  b. From part (a6), obtain the following: (1) $s^2(b_1)$; (2) $s(b_0, b_1)$; (3) $s(b_0)$.

6.26. Refer to **Plastic hardness** Problems 2.18 and 6.7.
  a. Using matrix methods, obtain the following: (1) $(X'X)^{-1}$, (2) **b**, (3) $\hat{Y}$, (4) SSR, (5) SSE, (6) $s^2(\mathbf{b})$, (7) $s^2(Y_{h(\text{new})})$ when $X_h = 30$.
  b. From part (a6), obtain the following: (1) $s^2(b_0)$; (2) $s(b_0, b_1)$; (3) $s(b_1)$.

EXERCISES

6.27. Refer to regression model (5.13). Set up the expectation vector for $\boldsymbol{\varepsilon}$. Assume that $i = 1, \ldots, 4$.

6.28. Consider model (5.13) for regression through the origin and the estimator b_1 given in (5.17). Obtain (5.17) by utilizing (6.59) with **X** suitably defined.

6.29. Consider the least squares estimator **b** given in (6.59). Using matrix methods, show that **b** is an unbiased estimator.

6.30. Show that $\hat{Y}_h$ in (6.82) can be expressed in matrix terms as $\mathbf{b'X}_h$.

6.31. Refer to regression model (5.36). Set up the variance-covariance matrix for the error terms when $i = 1, \ldots, 4$. Assume $\sigma(\varepsilon_i, \varepsilon_j) = 0$ for $i \neq j$.

6.32. Derive the variance-covariance matrix $\boldsymbol{\sigma}^2(\mathbf{b})$ in (6.91) for the weighted least squares estimators when the variance-covariance matrix of the observation Y_i is $\sigma^2 \mathbf{W}^{-1}$, where $\mathbf{W}$ is given in (6.88).

6.33. a. Obtain an expression for $\hat{\mathbf{Y}}$ in terms of the $\mathbf{H}$ matrix defined in (6.93a). [*Hint:* Use (6.93).]

  b. Obtain an expression for the variance-covariance matrix of the fitted values $\hat{Y}_i$, $i = 1, \ldots, n$, in terms of the hat matrix.

CITED REFERENCE

6.1 Graybill, Franklin A. *Introduction to Matrices with Applications in Statistics*. Belmont, Calif.: Wadsworth, 1969.

7

Multiple regression—I

Multiple regression analysis is one of the most widely used of all statistical tools. In this chapter, we first discuss a variety of multiple regression models. Then we present the basic statistical results for multiple regression in matrix form. Since the matrix expressions for multiple regression are the same as for simple regression, we state the results without much discussion. We then give an example, illustrating a variety of inferences in multiple regression analysis. Finally, we take up some additional facets of multiple regression analysis.

7.1 MULTIPLE REGRESSION MODELS

Need for several independent variables

When we first introduced regression analysis in Chapter 2, we spoke of regression models containing a number of independent variables. We mentioned a regression model where the dependent variable was direct operating cost for a branch office of a consumer finance chain, and four independent variables were considered, including average number of loans outstanding at the branch and total number of new loan applications processed by the branch. We also mentioned a tractor purchase study where the response variable was volume of tractor purchases in a sales territory, and the nine independent variables included number of farms in the territory and quantity of crop production in the territory. In addition, we mentioned a study of short children where the response variable was the peak plasma growth hormone level, and the 14 independent variables included sex, age, and various body measurements. In all these examples, one independent variable in the model would have provided an inadequate description since a number of key independent variables affect the response variable in important and distinctive ways. Furthermore, in situations of this type, one will frequently find that predictions of the response variable based on a model containing only a single independent variable are too imprecise to be useful. A more complex model, containing additional independent variables, typically is more helpful in providing sufficiently precise predictions of the response variable.

In each of the examples mentioned, the analysis is based on observational data because some or all of the independent variables are not susceptible to direct control. Multiple regression analysis is also highly useful in experimental situations where the experimenter can control the independent variables. An experimenter typically will wish to investigate a number of independent variables simultaneously because almost always more than one key independent variable influences the response. For example, in a study on productivity of work crews, the experimenter may wish to control both the size of the crew and the level of bonus pay. Similarly, in a study on responsiveness to a drug, the experimenter may wish to control both the dose of the drug and the body surface area of the subject.

First-order model with two independent variables

When there are two independent variables X_1 and X_2, the model:

$$(7.1) \qquad Y_i = \beta_0 + \beta_1 X_{i1} + \beta_2 X_{i2} + \varepsilon_i$$

is called a first-order model with two independent variables. A first-order model, it will be recalled from Chapter 2, is linear in the parameters and linear in the independent variables. Y_i denotes as usual the response in the ith trial, and X_{i1} and X_{i2} are the values of the two independent variables in the ith trial. The parameters of the model are β_0, β_1, and β_2, and the error term is ε_i.

Assuming that $E(\varepsilon_i) = 0$, the regression function for model (7.1) is:

$$(7.2) \qquad E(Y) = \beta_0 + \beta_1 X_1 + \beta_2 X_2$$

Analogous to simple linear regression, where the regression function $E(Y) = \beta_0 + \beta_1 X$ is a line, the regression function (7.2) is a plane. Figure 7.1 contains a representation of a portion of the response plane:

$$(7.3) \qquad E(Y) = 20.0 + .95X_1 - .50X_2$$

Note that a point on the response plane (7.3) corresponds to the mean response $E(Y)$ at the given combination of levels of X_1 and X_2.

Figure 7.1 also shows a series of observations Y_i corresponding to given levels of the two independent variables (X_{i1}, X_{i2}). Note that each vertical rule in Figure 7.1 represents the difference between Y_i and the mean $E(Y_i)$ of the probability distribution for (X_{i1}, X_{i2}) on the response plane. Hence, the vertical distance from Y_i to the response plane represents the error term $\varepsilon_i = Y_i - E(Y_i)$.

Frequently the regression function in multiple regression is called a *regression*

FIGURE 7.1 Example of response surface—a response plane with observations scattered about it

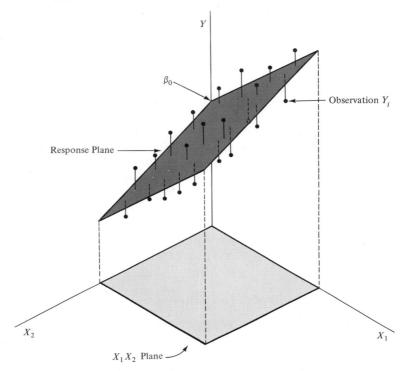

surface or a *response surface*. In Figure 7.1, the response surface is just a simple plane, but in other cases the response surface may be complex in nature.

Meaning of regression coefficients. Let us now consider the meaning of the regression parameters in the multiple regression function (7.2). The parameter β_0 is the Y intercept of the regression plane. If the scope of the model includes $X_1 = 0$, $X_2 = 0$, β_0 gives the mean response at $X_1 = 0$, $X_2 = 0$. Otherwise, β_0 does not have any particular meaning as a separate term in the regression model.

The parameter β_1 indicates the change in the mean response per unit increase in X_1 when X_2 is held constant. Likewise, β_2 indicates the change in the mean response per unit increase in X_2 when X_1 is held constant. To see this for our example, suppose X_2 is held at the level $X_2 = 20$. The regression function (7.3) now is:

$$(7.4) \quad E(Y) = 20.0 + .95X_1 - .50(20) = (20.0 - 10.0) + .95X_1$$
$$= 10.0 + .95X_1$$

Note that for $X_2 = 20$, the response function is a straight line with slope .95. The same is true for any other value of X_2; only the intercept of the response function differs. Hence, $\beta_1 = .95$ indicates that the mean response increases by .95 with a unit increase in X_1 when X_2 is constant, no matter what the level of X_2. More loosely speaking, we state that β_1 indicates the change in $E(Y)$ with a unit increase in X_1 when X_2 is held constant.

Similarly, $\beta_2 = -.50$ in model (7.3) indicates that the mean response decreases by .50 with a unit increase in X_2 when X_1 is held constant.

When the effect of X_1 on the mean response does not depend on the level of X_2, and correspondingly the effect of X_2 does not depend on the level of X_1, the two independent variables are said to have *additive effects* or *not to interact*. Thus, the first-order model (7.1) is designed for independent variables whose effects on the mean response are additive or do not interact.

The parameters β_1 and β_2 are frequently called *partial regression coefficients* because they reflect the partial effect of one independent variable when the other independent variable is included in the model and is held constant.

Example. Suppose that the response surface in (7.3) pertains to urban full-service stations of a major oil company and shows the effect of variety and adequacy of services (X_1) and average time taken to reach car (X_2) on the ratio of actual gallonage of gasoline sold to potential gallonage (Y), where X_1 is expressed as an index with $100 =$ average, X_2 is in seconds, and Y is stated as a percent. Increasing the index of adequacy of services by one point while holding average time to reach car constant leads to an increase of .95 percent point in the expected ratio of actual to potential gallonage. If the index of adequacy of services is held constant and the average time to reach car is increased by one second, the expected ratio of actual to potential gallonage decreases by .50 percent point.

Comments

1. A regression model for which the response surface is a plane can be used either in its own right when it is appropriate, or as an approximation to a more complex response surface. Many complex response surfaces can be approximated well by a plane for limited ranges of X_1 and X_2.

2. We can readily establish the meaning of β_1 and β_2 by calculus, taking partial derivatives of the response surface (7.2) with respect to X_1 and X_2 in turn:

$$\frac{\partial E(Y)}{\partial X_1} = \beta_1 \qquad \frac{\partial E(Y)}{\partial X_2} = \beta_2$$

The partial derivatives measure the rate of change in $E(Y)$ with respect to one independent variable when the other is held constant.

First-order model with more than two independent variables

We consider now the case where there are $p - 1$ independent variables $X_1, \ldots, X_{p-1}$. The model:

(7.5) $$Y_i = \beta_0 + \beta_1 X_{i1} + \beta_2 X_{i2} + \cdots + \beta_{p-1} X_{i,p-1} + \varepsilon_i$$

is called a first-order model with $p - 1$ independent variables. It can also be written:

(7.5a)
$$Y_i = \beta_0 + \sum_{k=1}^{p-1} \beta_k X_{ik} + \varepsilon_i$$

or, if we let $X_{i0} \equiv 1$, it can be written as:

(7.5b)
$$Y_i = \sum_{k=0}^{p-1} \beta_k X_{ik} + \varepsilon_i \qquad \text{where } X_{i0} \equiv 1$$

Assuming that $E(\varepsilon_i) = 0$, the response function for model (7.5) is:

(7.6)
$$E(Y) = \beta_0 + \beta_1 X_{i1} + \beta_2 X_{i2} + \cdots + \beta_{p-1} X_{i,p-1}$$

This response function is a *hyperplane*, which is a plane in more than two dimensions. It is no longer possible to picture this response surface, as we were able to do in Figure 7.1 for the case of two independent variables. Nevertheless, the meaning of the parameters is analogous to the two independent variables case. The parameter β_k indicates the change in the mean response $E(Y)$ with a unit increase in the independent variable X_k, when all other independent variables X_1, X_2, etc., included in the model are held constant. Note again that the effect of any independent variable on the mean response is the same for model (7.5), no matter what are the levels at which the other independent variables are held. Hence, the first-order model (7.5) is designed for independent variables whose effects on the mean response are additive and therefore do not interact.

Note

If $p - 1 = 1$, model (7.5) reduces to:

$$Y_i = \beta_0 + \beta_1 X_{i1} + \varepsilon_i$$

which is the simple linear regression model considered in earlier chapters.

General linear regression model

In general, the variables $X_1, \ldots, X_{p-1}$ in a regression model do not have to represent different independent variables, as we shall shortly see. We therefore define the general linear regression model, with normal error terms, simply in terms of X variables:

(7.7)
$$Y_i = \beta_0 + \beta_1 X_{i1} + \beta_2 X_{i2} + \cdots + \beta_{p-1} X_{i,p-1} + \varepsilon_i$$

where:

$\beta_0, \beta_1, \ldots, \beta_{p-1}$ are parameters
$X_{i1}, \ldots, X_{i,p-1}$ are known constants
ε_i are independent $N(0, \sigma^2)$
$i = 1, \ldots, n$

If we let $X_{i0} \equiv 1$, model (7.7) can be written as follows:

(7.7a)
$$Y_i = \beta_0 X_{i0} + \beta_1 X_{i1} + \beta_2 X_{i2} + \cdots + \beta_{p-1} X_{i,p-1} + \varepsilon_i$$
$$\text{where } X_{i0} \equiv 1$$

or:

(7.7b) $$Y_i = \sum_{k=0}^{p-1} \beta_k X_{ik} + \varepsilon_i \qquad \text{where } X_{i0} \equiv 1$$

The response function for model (7.7) is, since $E(\varepsilon_i) = 0$:

(7.8) $$E(Y) = \beta_0 + \beta_1 X_1 + \beta_2 X_2 + \cdots + \beta_{p-1} X_{p-1}$$

Thus, the general linear regression model implies that the observations Y_i are independent normal variables, with mean $E(Y_i)$ as given by (7.8) and with constant variance σ^2.

This general linear model encompasses a vast variety of situations. We shall consider a few of these now:

$p - 1$ independent variables. When $X_1, \ldots, X_{p-1}$ represent $p - 1$ different independent variables, the general linear model (7.7) is, as we have seen, a first-order model in which there are no interacting effects between the independent variables.

Polynomial regression. Consider the curvilinear regression model with one independent variable:

(7.9) $$Y_i = \beta_0 + \beta_1 X_i + \beta_2 X_i^2 + \varepsilon_i$$

If we let $X_{i1} = X_i$ and $X_{i2} = X_i^2$, we can write (7.9) as follows:

$$Y_i = \beta_0 + \beta_1 X_{i1} + \beta_2 X_{i2} + \varepsilon_i$$

so that model (7.9) is a particular case of the general linear regression model. While (7.9) illustrates a curvilinear model where the response function is quadratic, models with higher degree polynomial response functions are also particular cases of the general linear regression model.

Transformed variables. Consider the model:

(7.10) $$\log Y_i = \beta_0 + \beta_1 X_{i1} + \beta_2 X_{i2} + \beta_3 X_{i3} + \varepsilon_i$$

Here, the response surface is a highly complex one, yet model (7.10) can be treated as a general linear regression model. If we let $Y_i' = \log Y_i$, we can write model (7.10) as follows:

$$Y_i' = \beta_0 + \beta_1 X_{i1} + \beta_2 X_{i2} + \beta_3 X_{i3} + \varepsilon_i$$

which is in the form of the general linear regression model. The dependent variable just happens to be measured as the logarithm of Y.

Many models can be transformed into general linear regression models. Thus, the model:

(7.11) $$Y_i = \frac{1}{\beta_0 + \beta_1 X_{i1} + \beta_2 X_{i2} + \varepsilon_i}$$

can be transformed to a general linear regression model by letting $Y_i' = 1/Y_i$. We then have:

$$Y_i' = \beta_0 + \beta_1 X_{i1} + \beta_2 X_{i2} + \varepsilon_i$$

Interaction effects. Consider the model in two independent variables X_1 and X_2:

(7.12) $$Y_i = \beta_0 + \beta_1 X_{i1} + \beta_2 X_{i2} + \beta_3 X_{i1} X_{i2} + \varepsilon_i$$

The meaning of β_1 and β_2 here is not the same as that given earlier because of the cross-product term $\beta_3 X_{i1} X_{i2}$. It can be shown that the change in the mean response with a unit increase in X_1 when X_2 is held constant is:

(7.13) $$\beta_1 + \beta_3 X_2$$

Similarly, the change in the mean response with a unit change in X_2 when X_1 is held constant is:

(7.14) $$\beta_2 + \beta_3 X_1$$

Hence, in model (7.12) both the effect of X_1 for given level of X_2 and the effect of X_2 for given level of X_1 depend on the level of the other independent variable.

In Figure 7.2, we illustrate the effect of the cross-product term in model (7.12). In Figure 7.2a, we consider a response function without a cross-product term:

$$E(Y) = 10 + 2X_1 + 5X_2$$

and show there the response function $E(Y)$ when $X_2 = 1$ and when $X_2 = 3$. Note that the mean response increases by the amount $\beta_1 = 2$ with a unit increase of X_1, whether $X_2 = 1$ or $X_2 = 3$.

In Figure 7.2b, we consider the same response function but with the cross-product term $.5X_1 X_2$ added:

$$E(Y) = 10 + 2X_1 + 5X_2 + .5X_1 X_2$$

and show the response function $E(Y)$ when $X_2 = 1$ and when $X_2 = 3$. Note that the slope of the response function when plotted against X_1 now differs for $X_2 = 1$ and $X_2 = 3$. The slope of the response function when $X_2 = 1$ is by (7.13):

$$\beta_1 + \beta_3 X_2 = 2 + .5(1) = 2.5$$

and when $X_2 = 3$, the slope is:

$$\beta_1 + \beta_3 X_2 = 2 + .5(3) = 3.5$$

Hence, β_1 in model (7.12) containing a cross-product term no longer indicates the change in the mean response for a unit increase in X_1 for any given X_2 level. That effect in this model depends on the level of X_2. Model (7.12) with the cross-product term is therefore designed for independent variables whose effects on the dependent variable *interact*. The cross-product term $\beta_3 X_{i1} X_{i2}$ is called an *interaction term*. While the mean response in model (7.12) when X_2 is constant is

FIGURE 7.2 Effect of cross-product term in response function with two independent variables

(a)

$E(Y) = 10 + 2X_1 + 5X_2$
Slope is $\beta_1 = 2$ for all X_2

(b)

$E(Y) = 10 + 2X_1 + 5X_2 + .5X_1X_2$
Slope is 2.5 when $X_2 = 1$ but is 3.5 when $X_2 = 3$

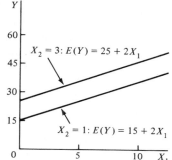

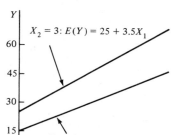

still a linear function of X_1, now both the intercept and the slope of the response function change as the level at which X_2 is held constant is varied. The same holds when the mean response is regarded as a function of X_2, with X_1 constant.

Despite these complexities of model (7.12), it can still be regarded as a general linear regression model. Let $X_{i3} = X_{i1}X_{i2}$. We can then write (7.12) as follows:

$$Y_i = \beta_0 + \beta_1 X_{i1} + \beta_2 X_{i2} + \beta_3 X_{i3} + \varepsilon_i$$

which is in the form of the general linear regression model.

Note

To derive (7.13) and (7.14), we differentiate:

$$E(Y) = \beta_0 + \beta_1 X_1 + \beta_2 X_2 + \beta_3 X_1 X_2$$

with respect to X_1 and X_2, respectively:

$$\frac{\partial E(Y)}{\partial X_1} = \beta_1 + \beta_3 X_2 \qquad \frac{\partial E(Y)}{\partial X_2} = \beta_2 + \beta_3 X_1$$

Combination of cases. A regression model may combine a number of the elements we have just noted and still can be treated as a general linear regression model. Consider a model with two independent variables, each in quadratic form, with an interaction term:

(7.15) $\quad Y_i = \beta_0 + \beta_1 X_{i1} + \beta_2 X_{i1}^2 + \beta_3 X_{i2} + \beta_4 X_{i2}^2 + \beta_5 X_{i1}X_{i2} + \varepsilon_i$

Let us define:

$$Z_{i1} = X_{i1} \qquad Z_{i2} = X_{i1}^2 \qquad Z_{i3} = X_{i2} \qquad Z_{i4} = X_{i2}^2 \qquad Z_{i5} = X_{i1}X_{i2}$$

We can then write model (7.15) as follows:

$$Y_i = \beta_0 + \beta_1 Z_{i1} + \beta_2 Z_{i2} + \beta_3 Z_{i3} + \beta_4 Z_{i4} + \beta_5 Z_{i5} + \varepsilon_i$$

which is in the form of the general linear regression model.

Comments

1. It should be clear from the various examples that the general linear regression model (7.7) is not restricted to linear response surfaces. The term *linear model* refers to the fact that (7.7) is linear in the parameters, not to the shape of the response surface.

2. Figure 7.3 illustrates some complex response surfaces that may be encountered when there are two independent variables.

FIGURE 7.3 Additional examples of response functions

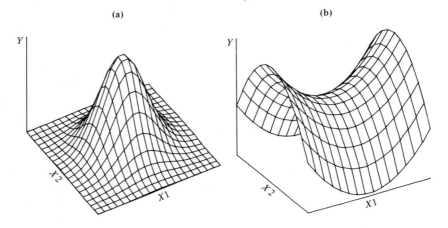

Interactions and nature of response surface

We introduced the concept of interacting independent variables earlier, and now shall illustrate further how the response surface differs when two independent variables do not interact and when they do interact.

Figure 7.4a contains a representation of a response surface in which the two independent variables (mean season temperature, amount of rainfall) do not interact on the dependent variable (corn yield). The absence of interactions can be seen by considering the corn yield curves for given mean season temperatures as a function of rainfall. These curves all have the same shape and differ only by a constant. Thus, each ordinate of the corn yield curve when the mean temperature is 70° is a constant number of units higher than the corresponding ordinate for the corn yield curve when the mean temperature is 78°.

Equivalently, one can note the absence of interactions by considering the corn yield curves for given amounts of rainfall as a function of temperature. Again, these curves are the same in shape and differ only by a constant.

FIGURE 7.4 Response surfaces for additive and interacting independent variables

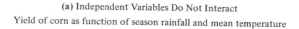

(a) Independent Variables Do Not Interact

Yield of corn as function of season rainfall and mean temperature

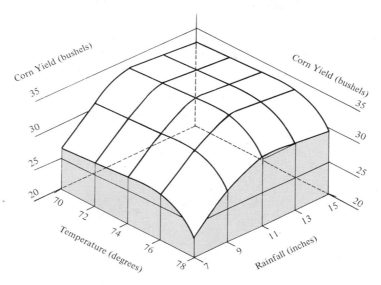

Absence of interactions therefore implies that the mean response $E(Y)$ can be expressed in the form:

$$(7.16) \qquad E(Y) = f_1(X_1) + f_2(X_2)$$

where f_1 and f_2 can be any functions, not necessarily simple ones.

Figure 7.4b illustrates a case where the two independent variables (age, percent of normal weight) interact on the dependent variable (mortality ratio). Here, the shape of the mortality ratio curve as a function of percent of normal weight varies for different ages. For men 22 years old, both underweight and overweight persons have higher mortality rates than normal (normal = 100) for that age. On the other hand, for men 52 years old, the mortality rate is above normal for that age for overweight persons but not for underweight persons. Similarly, the mortality ratio curves as a function of age vary in shape for different weights.

We can illustrate the difference in the shape of the response function when the two independent variables do and do not interact in yet another way, namely, by representing the response surface by means of a contour diagram. Such a diagram shows, for a number of different response levels, the various combinations of the two independent variables which yield the same level of response. Figure 7.5a shows a contour diagram for the response surface portrayed in Figure 7.1:

FIGURE 7.4 *(concluded)*

(b) Independent Variables Interact

Mortality ratio for men as function of age and percent of normal weight

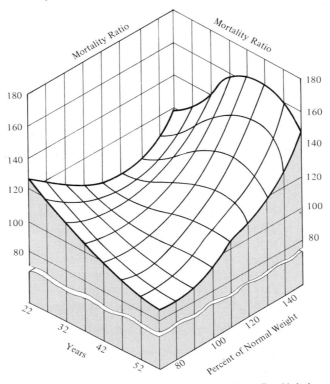

Source: Reprinted, with permission, from M. Ezekiel and K. A. Fox, *Methods of Correlation and Regression Analysis*, 3d ed. (New York: John Wiley & Sons, 1959), pp. 349–50.

$$E(Y) = 20.0 + .95X_1 - .50X_2$$

Note that the independent variables do not interact in this response function and that the contour lines are parallel. Figure 7.5b shows a contour diagram for the response function:

$$E(Y) = 5X_1 + 7X_2 + 3X_1X_2$$

where the two independent variables interact and the contour curves are not parallel.

In general, additive or noninteracting independent variables lead to parallel contour curves while interacting independent variables lead to nonparallel contour curves.

FIGURE 7.5   Response contour diagrams

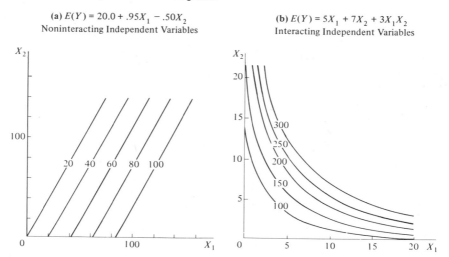

(a) $E(Y) = 20.0 + .95X_1 - .50X_2$
Noninteracting Independent Variables

(b) $E(Y) = 5X_1 + 7X_2 + 3X_1X_2$
Interacting Independent Variables

7.2   GENERAL LINEAR REGRESSION MODEL IN MATRIX TERMS

We shall now present the principal results for the general linear regression model (7.7) in matrix terms. This model, as we have noted, encompasses a wide variety of particular cases. The results to be presented are applicable to all of these.

It is a remarkable property of matrix algebra that the results for the general linear regression model (7.7) appear exactly the same in matrix notation as those for the simple linear regression model (6.56). Only the degrees of freedom and other constants related to the number of independent variables and the dimensions of some matrices will be different. Hence, we shall be able to present the results very concisely.

The matrix notation, to be sure, may hide enormous computational complexities. The inverse of a 10×10 matrix $\mathbf{A}$ requires tremendous amounts of computation, yet is simply represented as $\mathbf{A}^{-1}$. Our reason for emphasizing matrix algebra is that it indicates the essential conceptual steps in the solution. The actual computations will in all but the very simplest cases be done by programmable calculator or computer. Hence, it does not matter for us whether $(\mathbf{X}'\mathbf{X})^{-1}$ represents finding the inverse of a 2×2 or a 10×10 matrix. The important point is to know what the inverse of the matrix represents.

To express the general linear regression model (7.7):

$$Y_i = \beta_0 + \beta_1 X_{i1} + \beta_2 X_{i2} + \cdots + \beta_{p-1} X_{i,p-1} + \varepsilon_i$$

in matrix terms, we need to define the following matrices:

(7.17a)   (7.17b)

$$\mathbf{Y}_{n\times 1} = \begin{bmatrix} Y_1 \\ Y_2 \\ \cdot \\ \cdot \\ Y_n \end{bmatrix} \qquad \mathbf{X}_{n\times p} = \begin{bmatrix} 1 & X_{11} & X_{12} & \cdots & X_{1,p-1} \\ 1 & X_{21} & X_{22} & \cdots & X_{2,p-1} \\ \cdot & \cdot & \cdot & & \cdot \\ \cdot & \cdot & \cdot & & \cdot \\ \cdot & \cdot & \cdot & & \cdot \\ 1 & X_{n1} & X_{n2} & \cdots & X_{n,p-1} \end{bmatrix}$$

(7.17)

(7.17c)   (7.17d)

$$\mathbf{\beta}_{p\times 1} = \begin{bmatrix} \beta_0 \\ \beta_1 \\ \cdot \\ \cdot \\ \cdot \\ \beta_{p-1} \end{bmatrix} \qquad \mathbf{\varepsilon}_{n\times 1} = \begin{bmatrix} \varepsilon_1 \\ \varepsilon_2 \\ \cdot \\ \cdot \\ \cdot \\ \varepsilon_n \end{bmatrix}$$

Note that the $\mathbf{Y}$ and $\mathbf{\varepsilon}$ vectors are the same as for simple regression. The $\mathbf{\beta}$ vector contains additional regression parameters, and the $\mathbf{X}$ matrix contains a column of 1's as well as a column of the n values for each of the $p - 1$ X variables in the regression model. The row subscript for each element X_{ik} in the $\mathbf{X}$ matrix identifies the trial, and the column subscript identifies the X variable.

In matrix terms, the general linear regression model (7.7) is:

(7.18)
$$\underset{n\times 1}{\mathbf{Y}} = \underset{n\times p}{\mathbf{X}} \underset{p\times 1}{\mathbf{\beta}} + \underset{n\times 1}{\mathbf{\varepsilon}}$$

where:

        $\mathbf{Y}$ is a vector of observations
        $\mathbf{\beta}$ is a vector of parameters
        $\mathbf{X}$ is a matrix of constants
        $\mathbf{\varepsilon}$ is a vector of independent normal random variables with expectation $E(\mathbf{\varepsilon}) = \mathbf{0}$ and variance-covariance matrix $\sigma^2(\mathbf{\varepsilon}) = \sigma^2\mathbf{I}$

Consequently, the random vector $\mathbf{Y}$ has expectation:

(7.18a) $E(\mathbf{Y}) = \mathbf{X}\mathbf{\beta}$

and the variance-covariance matrix of $\mathbf{Y}$ is:

(7.18b) $\sigma^2(\mathbf{Y}) = \sigma^2\mathbf{I}$

7.3 LEAST SQUARES ESTIMATORS

Let us denote the vector of estimated regression coefficients $b_0, b_1, \ldots, b_{p-1}$ as $\mathbf{b}$:

$$(7.19) \qquad \mathop{\mathbf{b}}_{p \times 1} = \begin{bmatrix} b_0 \\ b_1 \\ b_2 \\ \cdot \\ \cdot \\ \cdot \\ b_{p-1} \end{bmatrix}$$

The least squares normal equations for the general linear regression model (7.18) are:

$$(7.20) \qquad \mathop{(\mathbf{X'X})}_{p \times p} \mathop{\mathbf{b}}_{p \times 1} = \mathop{\mathbf{X'}}_{p \times n} \mathop{\mathbf{Y}}_{n \times 1}$$

and the least squares estimators are:

$$(7.21) \qquad \mathop{\mathbf{b}}_{p \times 1} = \mathop{(\mathbf{X'X})^{-1}}_{p \times p} \mathop{\mathbf{X'Y}}_{p \times 1}$$

For model (7.18), these least squares estimators are also maximum likelihood estimators and have all the properties mentioned in Chapter 2: they are unbiased, minimum variance unbiased, consistent, and sufficient.

7.4 ANALYSIS OF VARIANCE RESULTS

Let the vector of the fitted values $\hat{Y}_i$ be denoted by $\hat{\mathbf{Y}}$ and the vector of the residual terms $e_i = Y_i - \hat{Y}_i$ be denoted by $\mathbf{e}$:

$$(7.22) \qquad (7.22a) \quad \mathop{\hat{\mathbf{Y}}}_{n \times 1} = \begin{bmatrix} \hat{Y}_1 \\ \hat{Y}_2 \\ \cdot \\ \cdot \\ \cdot \\ \hat{Y}_n \end{bmatrix} \qquad (7.22b) \quad \mathop{\mathbf{e}}_{n \times 1} = \begin{bmatrix} e_1 \\ e_2 \\ \cdot \\ \cdot \\ \cdot \\ e_n \end{bmatrix}$$

The fitted values are represented by:

$$(7.23) \qquad \hat{\mathbf{Y}} = \mathbf{Xb}$$

and the residual terms by:

$$(7.24) \qquad \mathbf{e} = \mathbf{Y} - \hat{\mathbf{Y}} = \mathbf{Y} - \mathbf{Xb}$$

Sums of squares and mean squares

The sums of squares for the analysis of variance are:

$$(7.25) \qquad SSTO = \mathbf{Y'Y} - \left(\frac{1}{n}\right)\mathbf{Y'11'Y}$$

$$(7.26) \qquad SSR = \mathbf{b'X'Y} - \left(\frac{1}{n}\right)\mathbf{Y'11'Y}$$

$$(7.27) \qquad SSE = \mathbf{e'e} = (\mathbf{Y} - \mathbf{Xb})'(\mathbf{Y} - \mathbf{Xb}) = \mathbf{Y'Y} - \mathbf{b'X'Y}$$

where 1 is an $n \times 1$ vector of 1's as defined in (6.19).

SSTO, as usual, has $n - 1$ degrees of freedom associated with it. SSE has $n - p$ degrees of freedom associated with it since p parameters need to be estimated in the regression function for model (7.18). Finally, SSR has $p - 1$ degrees of freedom associated with it, representing the number of X variables $X_1, \ldots, X_{p-1}$.

Table 7.1 shows these analysis of variance results, as well as the mean squares MSR and MSE:

$$(7.28) \qquad MSR = \frac{SSR}{p - 1}$$

$$(7.29) \qquad MSE = \frac{SSE}{n - p}$$

The expectation of MSE is σ^2, as for simple regression. The expectation of MSR is σ^2 plus a quantity that is nonnegative. For instance, when $p - 1 = 2$, we have:

$$E(MSR) = \sigma^2 + [\beta_1^2 \Sigma(X_{i1} - \bar{X}_1)^2 + \beta_2^2 \Sigma(X_{i2} - \bar{X}_2)^2 \\ + 2\beta_1\beta_2\Sigma(X_{i1} - \bar{X}_1)(X_{i2} - \bar{X}_2)]/2$$

Note that if both β_1 and β_2 equal zero, $E(MSR) = \sigma^2$. Otherwise $E(MSR) > \sigma^2$.

TABLE 7.1 ANOVA table for general linear regression model (7.18)

Source of Variation	SS	df	MS
Regression	$SSR = \mathbf{b'X'Y} - \left(\dfrac{1}{n}\right)\mathbf{Y'11'Y}$	$p - 1$	$MSR = \dfrac{SSR}{p - 1}$
Error	$SSE = \mathbf{Y'Y} - \mathbf{b'X'Y}$	$n - p$	$MSE = \dfrac{SSE}{n - p}$
Total	$SSTO = \mathbf{Y'Y} - \left(\dfrac{1}{n}\right)\mathbf{Y'11'Y}$	$n - 1$	

F test for regression relation

To test whether there is a regression relation between the dependent variable Y and the set of X variables $X_1, \ldots, X_{p-1}$, i.e., to choose between the alternatives:

$$(7.30a) \qquad \begin{aligned} H_0&: \beta_1 = \beta_2 = \cdots = \beta_{p-1} = 0 \\ H_a&: \text{not all } \beta_k \ (k = 1, \ldots, p - 1) \text{ equal } 0 \end{aligned}$$

we use the test statistic:

$$(7.30b) \qquad F^* = \frac{MSR}{MSE}$$

The decision rule to control the Type I error at α is:

$$(7.30c) \qquad \begin{array}{l} \text{If } F^* \le F(1 - \alpha; p - 1, n - p), \text{ conclude } H_0 \\ \text{If } F^* > F(1 - \alpha; p - 1, n - p), \text{ conclude } H_a \end{array}$$

The existence of a regression relation by itself does not of course assure that useful predictions can be made by using it.

Note that when $p - 1 = 1$, this test reduces to the F test in (3.61) for testing in simple linear regression whether or not $\beta_1 = 0$.

Coefficient of multiple determination

The coefficient of multiple determination, denoted by R^2, is defined as follows:

$$(7.31) \qquad R^2 = \frac{SSR}{SSTO} = 1 - \frac{SSE}{SSTO}$$

It measures the proportionate reduction of total variation in Y associated with the use of the set of X variables $X_1, \ldots, X_{p-1}$. The coefficient of multiple determination R^2 reduces to the coefficient of determination r^2 in (3.69) for simple linear regression when $p - 1 = 1$, i.e., when one independent variable is in model (7.18). Just as for r^2, we have:

$$(7.32) \qquad 0 \le R^2 \le 1$$

R^2 assumes the value 0 when all $b_k = 0$ ($k = 1, \ldots, p - 1$). R^2 takes on the value 1 when all observations fall directly on the fitted response surface, i.e., when $Y_i = \hat{Y}_i$ for all i.

Comments

1. To distinguish between the coefficients of determination for simple and multiple regression, we shall from now on call r^2 the *coefficient of simple determination*.

. 2. It can be shown that the coefficient of multiple determination R^2 can be viewed as a coefficient of simple determination r^2 between the responses Y_i and the fitted values $\hat{Y}_i$.

3. A large R^2 does not necessarily imply that the fitted model is a useful one. For instance, observations may have been taken at only a few levels of the independent variables. Despite a high R^2 in this case, the fitted model may not be useful because most predictions would require extrapolations outside the region of observations. Again, even though R^2 is large, MSE may still be too large for inferences to be useful in a case where high precision is required.

4. Adding more independent variables to the model can only increase R^2 and never reduce it, because SSE can never become larger with more independent variables and $SSTO$ is always the same for a given set of responses. Since R^2 often can be made large by including a large number of independent variables, it is sometimes suggested that a modified measure be used which recognizes the number of independent variables in the model. This *adjusted coefficient of multiple determination*, denoted by R_a^2, is defined:

$$(7.33) \qquad R_a^2 = 1 - \left(\frac{n - 1}{n - p} \right) \frac{SSE}{SSTO}$$

This adjusted coefficient of multiple determination may actually become smaller when another independent variable is introduced into the model, because the decrease in *SSE* may be more than offset by the loss of a degree of freedom in the denominator $n - p$.

Coefficient of multiple correlation

The coefficient of multiple correlation R is the positive square root of R^2:

$$(7.34) \qquad\qquad R = \sqrt{R^2}$$

It equals in absolute value the correlation coefficient r in (3.71) for simple correlation when $p - 1 = 1$, i.e., when there is one independent variable in model (7.18).

Note

From now on, we shall call r the *coefficient of simple correlation* to distinguish it from the coefficient of multiple correlation.

7.5 INFERENCES ABOUT REGRESSION PARAMETERS

The least squares estimators in **b** are unbiased:

$$(7.35) \qquad\qquad E(\mathbf{b}) = \boldsymbol{\beta}$$

The variance-covariance matrix $\boldsymbol{\sigma}^2(\mathbf{b})$:

$$(7.36) \quad \boldsymbol{\sigma}^2(\mathbf{b}) = \begin{bmatrix} \sigma^2(b_0) & \sigma(b_0, b_1) & \cdots & \sigma(b_0, b_{p-1}) \\ \sigma(b_1, b_0) & \sigma^2(b_1) & \cdots & \sigma(b_1, b_{p-1}) \\ \vdots & \vdots & & \vdots \\ \sigma(b_{p-1}, b_0) & \sigma(b_{p-1}, b_1) & \cdots & \sigma^2(b_{p-1}) \end{bmatrix}$$

is given by:

$$(7.37) \qquad\qquad \underset{p \times p}{\boldsymbol{\sigma}^2(\mathbf{b})} = \sigma^2(\mathbf{X'X})^{-1}$$

The estimated variance-covariance matrix $\mathbf{s}^2(\mathbf{b})$:

$$(7.38) \quad \mathbf{s}^2(\mathbf{b}) = \begin{bmatrix} s^2(b_0) & s(b_0, b_1) & \cdots & s(b_0, b_{p-1}) \\ s(b_1, b_0) & s^2(b_1) & \cdots & s(b_1, b_{p-1}) \\ \vdots & \vdots & & \vdots \\ s(b_{p-1}, b_0) & s(b_{p-1}, b_1) & \cdots & s^2(b_{p-1}) \end{bmatrix}$$

is given by:

$$(7.39) \qquad\qquad \underset{p \times p}{\mathbf{s}^2(\mathbf{b})} = MSE(\mathbf{X'X})^{-1}$$

From $\mathbf{s}^2(\mathbf{b})$, one can obtain $s^2(b_0)$, $s^2(b_1)$ or whatever other variance is needed, or any needed covariances.

Interval estimation of β_k

For the normal error model (7.18), we have:

(7.40) $$\frac{b_k - \beta_k}{s(b_k)} \sim t(n - p) \qquad k = 0, 1, \ldots, p - 1$$

Hence, the confidence limits for β_k with $1 - \alpha$ confidence coefficient are:

(7.41) $$b_k \pm t(1 - \alpha/2; n - p)s(b_k)$$

Tests for β_k

Tests for β_k are set up in the usual fashion. To test:

(7.42a) $$H_0: \beta_k = 0$$
$$H_a: \beta_k \neq 0$$

we may use the test statistic:

(7.42b) $$t^* = \frac{b_k}{s(b_k)}$$

and the decision rule:

(7.42c) If $|t^*| \leq t(1 - \alpha/2; n - p)$, conclude H_0
Otherwise conclude H_a

The power of the t test can be obtained as explained in Chapter 3, with the degrees of freedom modified to $n - p$.

As with simple regression, the test whether or not $\beta_k = 0$ in multiple regression models can also be conducted by means of an F test. We discuss this test in Chapter 8.

Joint inferences

The boundary of the joint confidence region for all p of the β_k regression parameters $(k = 0, 1, \ldots, p - 1)$ with confidence coefficient $1 - \alpha$ is:

(7.43) $$\frac{(\mathbf{b} - \boldsymbol{\beta})'\mathbf{X}'\mathbf{X}(\mathbf{b} - \boldsymbol{\beta})}{pMSE} = F(1 - \alpha; p, n - p)$$

The region defined by this boundary is generally difficult to obtain and interpret.

The Bonferroni joint confidence intervals, on the other hand, are easy to obtain and interpret. If g parameters are to be estimated jointly (where $g \leq p$), the confidence limits with family confidence coefficient $1 - \alpha$ are:

(7.44) $$b_k \pm Bs(b_k)$$

where:

(7.44a) $$B = t(1 - \alpha/2g; n - p)$$

In Section 8.4, we discuss tests concerning a subset of the regression parameters.

7.6 INFERENCES ABOUT MEAN RESPONSE

Interval estimation of $E(Y_h)$

For given values of $X_1, \ldots, X_{p-1}$, denoted by $X_{h1}, \ldots, X_{h,p-1}$, the mean response is denoted by $E(Y_h)$. We define the vector $\mathbf{X}_h$:

$$(7.45) \qquad \mathbf{X}_h = \begin{bmatrix} 1 \\ X_{h1} \\ X_{h2} \\ \cdot \\ \cdot \\ \cdot \\ X_{h,p-1} \end{bmatrix}$$

so that the mean response to be estimated is:

$$(7.46) \qquad E(Y_h) = \mathbf{X}'_h \boldsymbol{\beta}$$

The estimated mean response corresponding to $\mathbf{X}_h$, denoted by $\hat{Y}_h$, is:

$$(7.47) \qquad \hat{Y}_h = \mathbf{X}'_h \mathbf{b}$$

This estimator is unbiased:

$$(7.48) \qquad E(\hat{Y}_h) = \mathbf{X}'_h \boldsymbol{\beta} = E(Y_h)$$

and its variance is:

$$(7.49) \qquad \sigma^2(\hat{Y}_h) = \sigma^2 \mathbf{X}'_h (\mathbf{X}'\mathbf{X})^{-1} \mathbf{X}_h = \mathbf{X}'_h \sigma^2(\mathbf{b}) \mathbf{X}_h$$

Note that the variance $\sigma^2(\hat{Y}_h)$ is a function of the variances $\sigma^2(b_k)$ of the regression coefficients and of the covariances $\sigma(b_k, b_{k'})$ between pairs of regression coefficients, just as in simple linear regression. The estimated variance $s^2(\hat{Y}_h)$ is given by:

$$(7.50) \qquad s^2(\hat{Y}_h) = MSE(\mathbf{X}'_h (\mathbf{X}'\mathbf{X})^{-1} \mathbf{X}_h) = \mathbf{X}'_h s^2(\mathbf{b}) \mathbf{X}_h$$

The $1 - \alpha$ confidence limits for $E(Y_h)$ are:

$$(7.51) \qquad \hat{Y}_h \pm t(1 - \alpha/2; n - p) s(\hat{Y}_h)$$

Confidence region for regression surface

The $1 - \alpha$ confidence region for the entire regression surface is an extension of the Working-Hotelling confidence band for the regression line when there is one independent variable. Boundary points of the confidence region at $\mathbf{X}_h$ are:

$$(7.52) \qquad \hat{Y}_h \pm W s(\hat{Y}_h)$$

where:

$$(7.52a) \qquad W^2 = pF(1 - \alpha; p, n - p)$$

The confidence coefficient is $1 - \alpha$ that the region contains the entire regression surface over all combinations of real-numbered values of the X variables.

Simultaneous confidence intervals for several mean responses

When it is desired to estimate a number of mean responses $E(Y_h)$ corresponding to different $\mathbf{X}_h$ vectors, one can employ two basic approaches:

1. Use the Working-Hotelling type confidence region bounds from (7.52) for the several $\mathbf{X}_h$ vectors of interest. Since these bounds cover the mean responses for all possible $\mathbf{X}_h$ vectors with confidence coefficient $1 - \alpha$, they will cover the mean responses for selected $\mathbf{X}_h$ vectors with confidence coefficient greater than $1 - \alpha$.

2. Use Bonferroni simultaneous confidence intervals. When g statements are to be made with family confidence coefficient $1 - \alpha$, the Bonferroni confidence limits are:

$$(7.53) \qquad \hat{Y}_h \pm Bs(\hat{Y}_h)$$

where:

$$(7.53a) \qquad B = t(1 - \alpha/2g; n - p)$$

For any particular application, one should compare W and B to see which procedure will lead to narrower confidence intervals. If the $\mathbf{X}_h$ levels are not specified in advance but are determined as the analysis proceeds, it is better to use the Working-Hotelling type limits (7.52).

F test for lack of fit

To test whether the response function:

$$(7.54) \qquad E(Y) = \beta_0 + \beta_1 X_1 + \cdots + \beta_{p-1} X_{p-1}$$

is an appropriate response surface for the data at hand requires repeat observations, as for simple regression analysis. Repeat observations in multiple regression are replicate observations on Y corresponding to levels of each of the X variables which are constant from trial to trial. Thus, with two independent variables repeat observations require that X_1 and X_2 each remain at given levels from trial to trial.

The procedures described in Chapter 4 for the F test for lack of fit are applicable to multiple regression. Once the ANOVA table, shown in Table 7.1, has been obtained, SSE is decomposed into pure error and lack of fit components. The pure error sum of squares $SSPE$ is obtained by first calculating for each replicate group the sum of squared deviations of the Y observations around the group mean, where a replicate group has the same values for the X variables. Suppose there are c replicate groups with distinct sets of levels for the X variables, and let the mean of the Y observations for the jth group be denoted by $\bar{Y}_j$. Then the sum of squares for the jth group is given by (4.8), and the pure error sum of squares is the sum of these sums of squares, as given by (4.9). The lack of fit sum of squares $SSLF$ equals the difference $SSE - SSPE$, as indicated by (4.12).

The number of degrees of freedom associated with $SSPE$ is $n - c$, and the number of degrees of freedom associated with $SSLF$ is $(n - p) - (n - c) = c - p$.

The F test is conducted as described in Chapter 4, but with the degrees of freedom modified to those just stated.

7.7 PREDICTIONS OF NEW OBSERVATIONS

Prediction of new observation $Y_{h(\text{new})}$

The prediction limits with $1 - \alpha$ confidence coefficient for a new observation $Y_{h(\text{new})}$ corresponding to X_h, the specified values of the X variables, are:

$$(7.55) \qquad \hat{Y}_h \pm t(1 - \alpha/2; n - p)s(Y_{h(\text{new})})$$

where:

$$(7.55a) \qquad s^2(Y_{h(\text{new})}) = MSE + s^2(\hat{Y}_h) = MSE + X_h's^2(\mathbf{b})X_h$$
$$= MSE(1 + X_h'(\mathbf{X'X})^{-1}X_h)$$

Prediction of mean of m new observations at X_h

When m new observations are to be selected at X_h and their mean $\bar{Y}_{h(\text{new})}$ is to be predicted, the $1 - \alpha$ prediction limits are:

$$(7.56) \qquad \hat{Y}_h \pm t(1 - \alpha/2; n - p)s(\bar{Y}_{h(\text{new})})$$

where:

$$(7.56a) \qquad s^2(\bar{Y}_{h(\text{new})}) = \frac{MSE}{m} + s^2(\hat{Y}_h) = \frac{MSE}{m} + X_h's^2(\mathbf{b})X_h$$
$$= MSE\left(\frac{1}{m} + X_h'(\mathbf{X'X})^{-1}X_h\right)$$

Predictions of g new observations

Simultaneous prediction limits for g new observations at g different levels of X_h with family confidence coefficient $1 - \alpha$ are given by:

$$(7.57) \qquad \hat{Y}_h \pm Ss(Y_{h(\text{new})})$$

where:

$$(7.57a) \qquad S^2 = gF(1 - \alpha; g, n - p)$$

and $s^2(Y_{h(\text{new})})$ is given by (7.55a).

Alternatively, the Bonferroni simultaneous prediction limits can be used. For g predictions with a $1 - \alpha$ family confidence coefficient, they are:

$$(7.58) \qquad \hat{Y}_h \pm Bs(Y_{h(\text{new})})$$

where:

(7.58a) $$B = t(1 - \alpha/2g; n - p)$$

A comparison of S and B in advance of any particular use will indicate which procedure will lead to narrower prediction intervals.

7.8 AN EXAMPLE—MULTIPLE REGRESSION WITH TWO INDEPENDENT VARIABLES

In this section, we shall develop a multiple regression application with two independent variables. We shall illustrate a number of different types of inferences which might be made for this application but will not take up every possible type of inference.

Setting

The Zarthan Company sells a special skin cream through fashion stores exclusively. It operates in 15 marketing districts and is interested in predicting district sales. Table 7.2 contains data on sales by district, as well as district data on target population and per capita discretionary income. Sales are to be treated as the dependent variable Y, and target population and per capita discretionary income as independent variables X_1 and X_2, respectively, in an exploration of the feasibility of predicting district sales from target population and per capita discretionary income. The first-order model:

(7.59) $$Y_i = \beta_0 + \beta_1 X_{i1} + \beta_2 X_{i2} + \varepsilon_i$$

with normal error terms is expected to be appropriate.

TABLE 7.2 Basic data—Zarthan Company example

District i	Sales (gross of jars; 1 gross = 12 dozen) Y_i	Target Population (thousands of persons) X_{i1}	Per Capita Discretionary Income (dollars) X_{i2}
1	162	274	2,450
2	120	180	3,254
3	223	375	3,802
4	131	205	2,838
5	67	86	2,347
6	169	265	3,782
7	81	98	3,008
8	192	330	2,450
9	116	195	2,137
10	55	53	2,560
11	252	430	4,020
12	232	372	4,427
13	144	236	2,660
14	103	157	2,088
15	212	370	2,605

Basic calculations

The **Y** and **X** matrices for the Zarthan Company illustration are shown in Table 7.3. We shall require:

1.

$$\mathbf{X'X} = \begin{bmatrix} 1 & 1 & \cdots & 1 \\ 274 & 180 & \cdots & 370 \\ 2,450 & 3,254 & \cdots & 2,605 \end{bmatrix} \begin{bmatrix} 1 & 274 & 2,450 \\ 1 & 180 & 3,254 \\ \vdots & \vdots & \vdots \\ 1 & 370 & 2,605 \end{bmatrix}$$

which yields:

(7.60)
$$\mathbf{X'X} = \begin{bmatrix} 15 & 3,626 & 44,428 \\ 3,626 & 1,067,614 & 11,419,181 \\ 44,428 & 11,419,181 & 139,063,428 \end{bmatrix}$$

2.

$$\mathbf{X'Y} = \begin{bmatrix} 1 & 1 & \cdots & 1 \\ 274 & 180 & \cdots & 370 \\ 2,450 & 3,254 & \cdots & 2,605 \end{bmatrix} \begin{bmatrix} 162 \\ 120 \\ \vdots \\ 212 \end{bmatrix}$$

which yields:

(7.61)
$$\mathbf{X'Y} = \begin{bmatrix} 2,259 \\ 647,107 \\ 7,096,619 \end{bmatrix}$$

TABLE 7.3 **Y** and **X** matrices—Zarthan Company example

$$\mathbf{Y} = \begin{bmatrix} 162 \\ 120 \\ 223 \\ 131 \\ 67 \\ 169 \\ 81 \\ 192 \\ 116 \\ 55 \\ 252 \\ 232 \\ 144 \\ 103 \\ 212 \end{bmatrix} \quad \mathbf{X} = \begin{bmatrix} 1 & 274 & 2,450 \\ 1 & 180 & 3,254 \\ 1 & 375 & 3,802 \\ 1 & 205 & 2,838 \\ 1 & 86 & 2,347 \\ 1 & 265 & 3,782 \\ 1 & 98 & 3,008 \\ 1 & 330 & 2,450 \\ 1 & 195 & 2,137 \\ 1 & 53 & 2,560 \\ 1 & 430 & 4,020 \\ 1 & 372 & 4,427 \\ 1 & 236 & 2,660 \\ 1 & 157 & 2,088 \\ 1 & 370 & 2,605 \end{bmatrix}$$

3.

$$(\mathbf{X'X})^{-1} = \begin{bmatrix} 15 & 3{,}626 & 44{,}428 \\ 3{,}626 & 1{,}067{,}614 & 11{,}419{,}181 \\ 44{,}428 & 11{,}419{,}181 & 139{,}063{,}428 \end{bmatrix}^{-1}$$

Using (6.25), we define:

$$a = 15 \qquad b = 3{,}626 \qquad c = 44{,}428$$
$$d = 3{,}626 \qquad e = 1{,}067{,}614 \qquad f = 11{,}419{,}181$$
$$g = 44{,}428 \qquad h = 11{,}419{,}181 \qquad k = 139{,}063{,}428$$

so that:

$$Z = 14{,}497{,}044{,}060{,}000$$
$$A = 1.246348416$$
$$B = .0002129664176$$

and so on. We obtain:

(7.62) $(\mathbf{X'X})^{-1} =$

$$\begin{bmatrix} 1.2463484 & 2.1296642E - 4 & -4.1567125E - 4 \\ 2.1296642E - 4 & 7.7329030E - 6 & -7.0302518E - 7 \\ -4.1567125E - 4 & -7.0302518E - 7 & 1.9771851E - 7 \end{bmatrix}$$

Note that some of the results in the $(\mathbf{X'X})^{-1}$ matrix are given in the E format, where, say, $E - 4$ stands for $10^{-4} = 1/10^4$. Thus, $2.1296642E - 4$ stands for .00021296642.

Algebraic equivalents. Note that $\mathbf{X'X}$ for the first-order model (7.59) with two independent variables is:

$$\mathbf{X'X} = \begin{bmatrix} 1 & 1 & \cdots & 1 \\ X_{11} & X_{21} & \cdots & X_{n1} \\ X_{12} & X_{22} & \cdots & X_{n2} \end{bmatrix} \begin{bmatrix} 1 & X_{11} & X_{12} \\ 1 & X_{21} & X_{22} \\ \cdot & \cdot & \cdot \\ \cdot & \cdot & \cdot \\ \cdot & \cdot & \cdot \\ 1 & X_{n1} & X_{n2} \end{bmatrix}$$

or:

(7.63) $$\mathbf{X'X} = \begin{bmatrix} n & \Sigma X_{i1} & \Sigma X_{i2} \\ \Sigma X_{i1} & \Sigma X_{i1}^2 & \Sigma X_{i1}X_{i2} \\ \Sigma X_{i2} & \Sigma X_{i2}X_{i1} & \Sigma X_{i2}^2 \end{bmatrix}$$

Thus, for our example:

$$n = 15$$
$$\Sigma X_{i1} = 274 + 180 + \cdots = 3{,}626$$
$$\Sigma X_{i1}X_{i2} = 274(2{,}450) + 180(3{,}254) + \cdots = 11{,}419{,}181$$
$$\text{etc.}$$

These elements are found in (7.60).

Also note that $\mathbf{X'Y}$ for the first-order model with two independent variables is:

$$(7.64) \qquad \mathbf{X'Y} = \begin{bmatrix} 1 & 1 & \cdots & 1 \\ X_{11} & X_{21} & \cdots & X_{n1} \\ X_{12} & X_{22} & \cdots & X_{n2} \end{bmatrix} \begin{bmatrix} Y_1 \\ Y_2 \\ \vdots \\ Y_n \end{bmatrix} = \begin{bmatrix} \Sigma Y_i \\ \Sigma X_{i1} Y_i \\ \Sigma X_{i2} Y_i \end{bmatrix}$$

For our example, we have:

$$\Sigma Y_i = 162 + 120 + \cdots = 2,259$$
$$\Sigma X_{i1} Y_i = 274(162) + 180(120) + \cdots = 647,107$$
$$\Sigma X_{i2} Y_i = 2,450(162) + 3,254(120) + \cdots = 7,096,619$$

These are the elements found in (7.61).

Estimated regression function

The least squares estimates $\mathbf{b}$ are readily obtained by (7.21), given our basic calculations in (7.61) and (7.62):

$$\mathbf{b} = (\mathbf{X'X})^{-1}\mathbf{X'Y}$$

$$= \begin{bmatrix} 1.2463484 & 2.1296642E-4 & -4.1567125E-4 \\ 2.1296642E-4 & 7.7329030E-6 & -7.0302518E-7 \\ -4.1567125E-4 & -7.0302518E-7 & 1.9771851E-7 \end{bmatrix}$$

$$\times \begin{bmatrix} 2,259 \\ 647,107 \\ 7,096,619 \end{bmatrix}$$

$$= \begin{bmatrix} 3.4526127900 \\ .4960049761 \\ .009199080867 \end{bmatrix}$$

Thus:

$$\begin{bmatrix} b_0 \\ b_1 \\ b_2 \end{bmatrix} = \begin{bmatrix} 3.4526127900 \\ .4960049761 \\ .009199080867 \end{bmatrix}$$

and the estimated regression function is:

$$\hat{Y} = 3.453 + .496X_1 + .00920X_2$$

This estimated regression function indicates that mean sales are expected to increase by .496 gross when the target population increases by one thousand, holding per capita discretionary income constant, and that mean sales are expected to increase by .0092 gross when per capita discretionary income increases by one dollar, holding population constant.

Algebraic version of normal equations.   The normal equations in algebraic form for the case of two independent variables can be obtained readily from (7.63) and (7.64). We have:

$$(\mathbf{X'X})\mathbf{b} = \mathbf{X'Y}$$

$$\begin{bmatrix} n & \Sigma X_{i1} & \Sigma X_{i2} \\ \Sigma X_{i1} & \Sigma X_{i1}^2 & \Sigma X_{i1}X_{i2} \\ \Sigma X_{i2} & \Sigma X_{i2}X_{i1} & \Sigma X_{i2}^2 \end{bmatrix} \begin{bmatrix} b_0 \\ b_1 \\ b_2 \end{bmatrix} = \begin{bmatrix} \Sigma Y_i \\ \Sigma X_{i1}Y_i \\ \Sigma X_{i2}Y_i \end{bmatrix}$$

from which we obtain the normal equations:

$$(7.65) \quad \begin{aligned} \Sigma Y_i &= nb_0 & + b_1\Sigma X_{i1} & + b_2\Sigma X_{i2} \\ \Sigma X_{i1}Y_i &= b_0\Sigma X_{i1} + b_1\Sigma X_{i1}^2 & + b_2\Sigma X_{i1}X_{i2} \\ \Sigma X_{i2}Y_i &= b_0\Sigma X_{i2} + b_1\Sigma X_{i1}X_{i2} + b_2\Sigma X_{i2}^2 \end{aligned}$$

Analysis of aptness of model

To examine the aptness of regression model (7.59) with independent variables X_1 and X_2 for the data at hand, we require the fitted values $\hat{Y}_i$ and the residuals $e_i = Y_i - \hat{Y}_i$. We obtain by (7.23):

$$\hat{\mathbf{Y}} = \mathbf{Xb}$$

$$\begin{bmatrix} \hat{Y}_1 \\ \hat{Y}_2 \\ \cdot \\ \cdot \\ \cdot \\ \hat{Y}_{15} \end{bmatrix} = \begin{bmatrix} 1 & 274 & 2,450 \\ 1 & 180 & 3,254 \\ \cdot & \cdot & \cdot \\ \cdot & \cdot & \cdot \\ \cdot & \cdot & \cdot \\ 1 & 370 & 2,605 \end{bmatrix} \begin{bmatrix} 3.4526127900 \\ .4960049761 \\ .009199080867 \end{bmatrix} = \begin{bmatrix} 161.896 \\ 122.667 \\ \cdot \\ \cdot \\ \cdot \\ 210.938 \end{bmatrix}$$

Further, by (7.24) we find:

$$\mathbf{e} = \mathbf{Y} - \hat{\mathbf{Y}}$$

$$\begin{bmatrix} e_1 \\ e_2 \\ \cdot \\ \cdot \\ \cdot \\ e_{15} \end{bmatrix} = \begin{bmatrix} 162 \\ 120 \\ \cdot \\ \cdot \\ \cdot \\ 212 \end{bmatrix} - \begin{bmatrix} 161.896 \\ 122.667 \\ \cdot \\ \cdot \\ \cdot \\ 210.938 \end{bmatrix} = \begin{bmatrix} .104 \\ -2.667 \\ \cdot \\ \cdot \\ \cdot \\ 1.062 \end{bmatrix}$$

Figures 7.6, 7.7, and 7.8 contain plots of the residuals e_i against the fitted values $\hat{Y}_i$, against X_{i1}, and against X_{i2}, respectively. These plots were generated by the BMDP computer package. There are no suggestions in any of these plots that systematic deviations from the fitted response plane are present, nor that the error variance varies either with the level of $\hat{Y}$ or with the levels of X_1 or X_2. We do not show a normal probability plot, but it does not indicate any major departure from normality. Hence, model (7.59) appears to be apt for this application.

FIGURE 7.6 Residual plot against $\hat{Y}$—Zarthan Company example

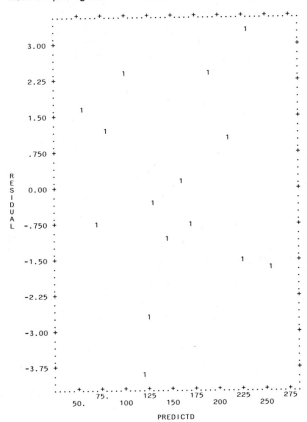

Analysis of variance

To test whether sales are related to population and per capita discretionary income, we construct the ANOVA table in Table 7.4 (p. 255). The basic quantities needed are:

$$\mathbf{Y'Y} = [162 \quad 120 \quad \cdots \quad 212]\begin{bmatrix} 162 \\ 120 \\ \cdot \\ \cdot \\ \cdot \\ 212 \end{bmatrix}$$

$$= (162)^2 + (120)^2 + \cdots + (212)^2$$

$$= 394,107.000$$

FIGURE 7.7 Residual plot against X_1—Zarthan Company example

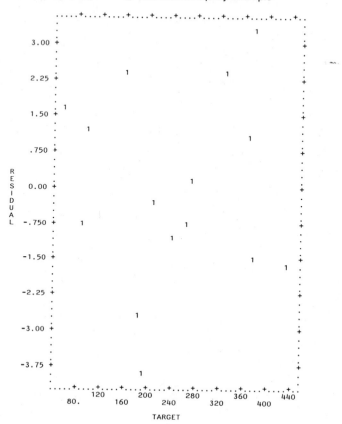

$$\left(\frac{1}{n}\right)\mathbf{Y'11'Y} = \frac{1}{15}[162 \quad 120 \quad \cdots \quad 212]\begin{bmatrix}1\\1\\\cdot\\\cdot\\1\end{bmatrix}[1 \quad 1 \quad \cdots \quad 1]\begin{bmatrix}162\\120\\\cdot\\\cdot\\212\end{bmatrix}$$

$$= \frac{1}{n}(\Sigma Y_i)(\Sigma Y_i) = \frac{(2,259)^2}{15} = 340,205.400$$

Thus:

$$SSTO = \mathbf{Y'Y} - \left(\frac{1}{n}\right)\mathbf{Y'11'Y} = 394,107.000 - 340,205.400 = 53,901.600$$

FIGURE 7.8 Residual plot against X_2—Zarthan Company example

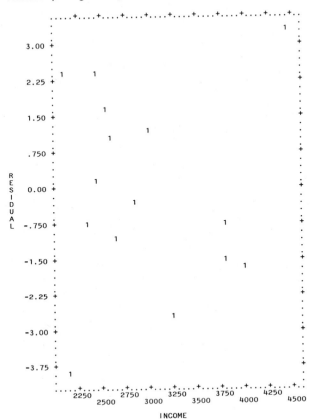

and using our result in (7.61):

$$SSE = \mathbf{Y'Y} - \mathbf{b'X'Y}$$
$$= 394{,}107.000 - [3.4526127900 \quad .4960049761 \quad .009199080867]$$

$$\times \begin{bmatrix} 2{,}259 \\ 647{,}107 \\ 7{,}096{,}619 \end{bmatrix}$$

$$= 394{,}107.000 - 394{,}050.116 = 56.884$$

Finally, we obtain by subtraction:

$$SSR = SSTO - SSE = 53{,}901.600 - 56.884 = 53{,}844.716$$

The degrees of freedom and mean squares are entered in Table 7.4. Note that three regression parameters had to be estimated; hence, $15 - 3 = 12$ degrees of

TABLE 7.4 ANOVA table—Zarthan Company example

Source of Variation	SS	df	MS
Regression	$SSR = 53,844.716$	2	$MSR = 26,922.358$
Error	$SSE = 56.884$	12	$MSE = 4.740$
Total	$SSTO = 53,901.600$	14	

freedom are associated with SSE. Also, the number of degrees of freedom associated with SSR are two—the number of X variables in the model.

Test of regression relation. To test whether sales are related to population and per capita discretionary income:

$$H_0: \beta_1 = 0 \text{ and } \beta_2 = 0$$
$$H_a: \text{not both } \beta_1 \text{ and } \beta_2 \text{ equal } 0$$

we use test statistic (7.30b):

$$F^* = \frac{MSR}{MSE} = \frac{26,922.358}{4.740} = 5,680$$

Assuming α is to be .05, we require $F(.95; 2, 12) = 3.89$. Since $F^* = 5,680 > 3.89$, we conclude H_a, that sales are related to population and per capita discretionary income. Whether this relation is useful for making predictions of sales or estimates of mean sales still remains to be seen.

The P-value for this test is less than .001 since we note from Table A–4 that $F(.999; 2, 12) = 13.0$. In fact, it can be shown that the P-value is $0+$.

Coefficient of multiple determination. For our example, we have by (7.31):

$$R^2 = \frac{SSR}{SSTO} = \frac{53,844.716}{53,901.600} = .9989$$

Thus, when the two independent variables population and per capita discretionary income are considered, the variation in sales is reduced by 99.9 percent.

Algebraic expression for SSE. The error sum of squares for the case of two independent variables in algebraic terms is:

$$SSE = \mathbf{Y'Y} - \mathbf{b'X'Y} = \Sigma Y_i^2 - [b_0 \quad b_1 \quad b_2] \begin{bmatrix} \Sigma Y_i \\ \Sigma X_{i1} Y_i \\ \Sigma X_{i2} Y_i \end{bmatrix}$$

or:

(7.66) $$SSE = \Sigma Y_i^2 - b_0 \Sigma Y_i - b_1 \Sigma X_{i1} Y_i - b_2 \Sigma X_{i2} Y_i$$

Note how this expression is a straightforward extension of (2.24a) for the case of one independent variable.

Estimation of regression parameters

The Zarthan Company is not interested in the parameter β_0 since it falls far outside the scope of the model. It is desired to estimate β_1 and β_2 jointly with a family confidence coefficient .90. We shall use the simultaneous Bonferroni confidence limits in (7.44), since these are easy to develop and interpret.

First, we need the estimated variance-covariance matrix $s^2(\mathbf{b})$:

$$s^2(\mathbf{b}) = MSE(\mathbf{X'X})^{-1}$$

MSE is given in Table 7.4, and $(\mathbf{X'X})^{-1}$ was obtained in (7.62). Hence:

(7.67)

$$s^2(\mathbf{b}) = 4.7403$$

$$\times \begin{bmatrix} 1.2463484 & 2.1296642E-4 & -4.1567125E-4 \\ 2.1296642E-4 & 7.7329030E-6 & -7.0302518E-7 \\ -4.1567125E-4 & -7.0302518E-7 & 1.9771851E-7 \end{bmatrix}$$

$$= \begin{bmatrix} 5.9081 & 1.0095E-3 & -1.9704E-3 \\ 1.0095E-3 & 3.6656E-5 & -3.3326E-6 \\ -1.9704E-3 & -3.3326E-6 & 9.3725E-7 \end{bmatrix}$$

The two elements we require are:

$$s^2(b_1) = .000036656 \quad \text{or} \quad s(b_1) = .006054$$
$$s^2(b_2) = .00000093725 \quad \text{or} \quad s(b_2) = .0009681$$

Next, we require for $g = 2$ simultaneous estimates:

$$B = t(1 - .10/2(2); 12) = t(.975; 12) = 2.179$$

Now we are ready to obtain the two simultaneous confidence intervals:

$$.4960 - 2.179(.006054) \le \beta_1 \le .4960 + 2.179(.006054)$$

or:

$$.483 \le \beta_1 \le .509$$

$$.009199 - 2.179(.0009681) \le \beta_2 \le .009199 + 2.179(.0009681)$$

or:

$$.0071 \le \beta_2 \le .0113$$

With family confidence coefficient .90, we conclude that β_1 falls between .483 and .509 and that β_2 falls between .0071 and .0113.

Note that the simultaneous confidence intervals suggest that both β_1 and β_2

are positive, which is in accord with theoretical expectations that sales should increase with either higher target population or higher per capita discretionary income, the other variable being held constant.

Estimation of mean response

Suppose the Zarthan Company would like to estimate expected (mean) sales in a district with target population $X_{h1} = 220$ thousand persons and per capita discretionary income $X_{h2} = 2,500$ dollars. We define:

$$\mathbf{X}_h = \begin{bmatrix} 1 \\ 220 \\ 2,500 \end{bmatrix}$$

The point estimate of mean sales is by (7.47):

$$\hat{Y}_h = \mathbf{X}_h'\mathbf{b} = \begin{bmatrix} 1 & 220 & 2,500 \end{bmatrix} \begin{bmatrix} 3.4526 \\ .4960 \\ .009199 \end{bmatrix} = 135.57$$

The estimated variance by (7.50) and using the results in (7.67) is:

$$s^2(\hat{Y}_h) = \mathbf{X}_h's^2(\mathbf{b})\mathbf{X}_h$$

$$= \begin{bmatrix} 1 & 220 & 2,500 \end{bmatrix}$$

$$\times \begin{bmatrix} 5.9081 & 1.0095E - 3 & -1.9704E - 3 \\ 1.0095E - 3 & 3.6656E - 5 & -3.3326E - 6 \\ -1.9704E - 3 & -3.3326E - 6 & 9.3725E - 7 \end{bmatrix} \begin{bmatrix} 1 \\ 220 \\ 2,500 \end{bmatrix}$$

$$= .46638$$

or:

$$s(\hat{Y}_h) = .68292$$

Assume that the confidence coefficient for the interval estimate of $E(Y_h)$ is to be .95. We then need $t(.975; 12) = 2.179$, and obtain by (7.51):

$$135.57 - 2.179(.68292) \leq E(Y_h) \leq 135.57 + 2.179(.68292)$$

or:

$$134.1 \leq E(Y_h) \leq 137.1$$

Thus, with confidence coefficient .95, we estimate that mean sales in a district with target population of 220 thousand and per capita discretionary income of $2,500 are somewhere between 134.1 and 137.1 gross.

Algebraic version of estimated variance $s^2(\hat{Y}_h)$.   Since by (7.50):

$$s^2(\hat{Y}_h) = \mathbf{X}_h's^2(\mathbf{b})\mathbf{X}_h$$

it follows for the case of two independent variables:

$$(7.68) \quad s^2(\hat{Y}_h) = s^2(b_0) + X_{h1}^2 s^2(b_1) + X_{h2}^2 s^2(b_2) + 2X_{h1}s(b_0, b_1)$$
$$+ 2X_{h2}s(b_0, b_2) + 2X_{h1}X_{h2}s(b_1, b_2)$$

When we substitute in (7.68), utilizing the estimated variances and covariances from (7.67), we obtain the same result as before, namely, $s^2(\hat{Y}_h) = .46638$.

Prediction limits for new observations

Suppose the Zarthan Company would like to predict sales in two districts. The two districts have the following characteristics:

	District A	District B
X_{h1}	220	375
X_{h2}	2,500	3,500

To determine which simultaneous prediction intervals are best here, we shall find S as given in (7.57a) and B as given in (7.58a) for $g = 2$, assuming the family confidence coefficient is to be .90:

$$S^2 = 2F(.90; 2, 12) = 2(2.81) = 5.62$$

or:

$$S = 2.37$$

and

$$B = t(1 - .10/2(2); 12) = t(.975; 12) = 2.179$$

Hence, the Bonferroni limits are more efficient here.

For district A, we shall use the results we found when estimating mean sales, since the levels of the independent variables are the same as before. We have from earlier:

$$\hat{Y}_A = 135.57 \qquad s^2(\hat{Y}_A) = .46638 \qquad MSE = 4.7403$$

Hence, by (7.55a):

$$s^2(Y_{A(new)}) = MSE + s^2(\hat{Y}_A) = 4.7403 + .46638 = 5.20668$$

or:

$$s(Y_{A(new)}) = 2.28182$$

In similar fashion, we obtain:

$$\hat{Y}_B = 221.65 \qquad s(Y_{B(new)}) = 2.34536$$

We found before that $B = 2.179$. Hence, the simultaneous Bonferroni prediction intervals with family confidence coefficient .90 are by (7.58):

$$135.57 - 2.179(2.28182) \le Y_{A(new)} \le 135.57 + 2.179(2.28182)$$

or:

$$130.6 \leq Y_{A(new)} \leq 140.5$$

$$221.65 - 2.179(2.34536) \leq Y_{B(new)} \leq 221.65 + 2.179(2.34536)$$

or:

$$216.5 \leq Y_{B(new)} \leq 226.8$$

With family confidence coefficient .90, we predict that sales in the two districts will be within the indicated limits. The Zarthan Company considers these prediction limits sufficiently precise, and hence useful.

Computer printout

Figure 7.9 contains an illustrative computer printout for the Zarthan Company example, obtained by using the GLM (general linear model) program of the SAS (Statistical Analysis System) computer package (Ref. 7.1). Regression analysis printouts differ in format from one computer program to another, as may be seen by comparing the output in Figure 7.9 with other outputs presented in earlier chapters. However, the basic information presented in the different outputs is essentially the same for the major statistical regression packages.

We have annotated the output in Figure 7.9 to tie in with the notation of this book. The first two blocks of information contain intermediate regression analysis results in matrix form, specifically the $\mathbf{X'X}$ and $(\mathbf{X'X})^{-1}$ matrices. The label "intercept" in these matrices refers to $X_{i0} \equiv 1$ in the alternative regression model (7.7b).

The next block presents information about the estimated regression coefficients b_k. Shown, in turn, are the estimates b_k, the test statistics $t_k^* = b_k/s(b_k)$ for testing whether or not $\beta_k = 0$, the two-sided P-values for the test statistics, and the estimated standard deviations $s(b_k)$.

The fourth block contains ANOVA information: the ANOVA table, the F^* value for the test of whether or not a regression relation exists, the P-value for this test, $\sqrt{MSE}$, and R^2.

The final block shows the observed values Y_i, the fitted values $\hat{Y}_i$, and the residuals e_i.

Because of rounding, some results in Figure 7.9 do not coincide precisely with the corresponding results given earlier. In this connection, it should be noted that different computer regression packages may lead to somewhat different results because final results are rounded to different extents, and even more importantly because rounding errors are not handled equally well by all packages. Particularly when there are a number of independent variables, some of which are highly correlated, rounding errors can be a serious source of difficulty. It is a wise policy to investigate a computer regression package before using it, for instance, by comparing its output for a test problem against results known to be accurate.

FIGURE 7.9 Computer printout for Zarthan Company example (SAS, Ref. 7.1)

THE X'X MATRIX

	INTERCEPT	TARGTP	INCOME
INTERCEPT ← X_0	15.00	3626.00	44428.00
TARGTP ← X_1	3626.00	1067614.00	11419181.00
INCOME ← X_2	44428.00	11419181.00	139063428.00

X'X INVERSE MATRIX

	INTERCEPT	TARGTP	INCOME
INTERCEPT	1.24634842	0.00021297	-0.00041567
TARGTP	0.00021297	0.00000773	-0.00000070
INCOME	-0.00041567	-0.00000070	0.00000020

PARAMETER	ESTIMATE	T FOR H0: PARAMETER=0	PR > ITI	STD ERROR OF ESTIMATE
INTERCEPT	3.45261279	1.42	0.1809	2.43065049
TARGTP	0.49600498	81.92	0.0001	0.00605444
INCOME	0.00919908	9.50	0.0001	0.00096811

b_k $t_k^* = b_k/s(b_k)$ Two-sided P-value $s(b_k)$

SOURCE	DF	SUM OF SQUARES	MEAN SQUARE	F VALUE
MODEL	2	SSR → 53844.71643444	MSR → 26922.35821722	F^* → 5679.47
ERROR	12	SSE → 56.88356556	MSE → 4.74029713	
CORRECTED TOTAL	14	53901.60000000		

One-sided P-value

SSTO ↑

PR > F   R-SQUARE

→ 0.0001   0.998945 ← R^2

STD DEV

2.17722234 ← $\sqrt{MSE}$

OBSERVATION	Y_i OBSERVED VALUE	$\hat{Y}_i$ PREDICTED VALUE	e_i RESIDUAL
1	162.00000000	161.89572437	0.10427563
2	120.00000000	122.66731763	-2.66731763
3	223.00000000	224.42938429	-1.42938429
4	131.00000000	131.24062439	-0.24062439
5	67.00000000	67.69928353	-0.69928353
6	169.00000000	169.68485530	-0.68485530
7	81.00000000	79.73193570	1.26806430
8	192.00000000	189.67200303	2.32799697
9	116.00000000	119.83201895	-3.83201895
10	55.00000000	53.29052354	1.70947646
11	252.00000000	253.71505760	-1.71505760
12	232.00000000	228.69079490	3.30920510
13	144.00000000	144.97934226	-0.97934226
14	103.00000000	100.53307489	2.46692511
15	212.00000000	210.93805961	1.06194039

Caution about hidden extrapolations

Before concluding this illustration of multiple regression analysis, we should caution again about making estimates or predictions outside the scope of the model. The danger, of course, is that the model may not be appropriate when extended outside the region of the observations. In multiple regression, it is particularly easy to lose track of this region since the levels of $X_1, \ldots, X_{p-1}$ *jointly* define the region. Thus, one cannot merely look at the ranges of each independent variable. Consider Figure 7.10, where the shaded region is the region of observations for a multiple regression application with two independent variables. The circled dot is within the ranges of the independent variables X_1 and X_2 individually, yet is well outside the joint region of observations.

FIGURE 7.10 Region of observations on X_1 and X_2 jointly, compared with ranges of X_1 and X_2 individually

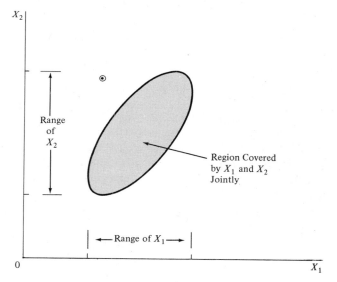

7.9 STANDARDIZED REGRESSION COEFFICIENTS

Standardized regression coefficients have been proposed to facilitate comparisons between regression coefficients. Ordinarily, it is difficult to compare regression coefficients because of differences in the units involved. We cite two examples.

1. When considering the fitted response function:

$$\hat{Y} = 200 + 20{,}000X_1 + .2X_2$$

one may be tempted to conclude that X_1 is the only important independent variable, and that X_2 has little effect on the dependent variable Y. A little reflection should make one wary of this conclusion. The reason is that we do not know the units involved. Suppose the units are:

Y in dollars
X_1 in thousand dollars
X_2 in cents

In that event, the effect on the mean response of a \$1,000 increase in X_1 when X_2 is constant would be exactly the same as the effect of a \$1,000 increase in X_2 when X_1 is constant, despite the difference in the regression coefficients.

2. In our Zarthan Company example, we cannot make any comparison between b_1 and b_2 because b_1 is in units of one gross of jars per thousand persons while b_2 is in units of one gross of jars per dollar of per capita discretionary income.

Standardized regression coefficients, also sometimes called *beta coefficients*, are defined as follows:

$$(7.69) \quad B_k = b_k \left[\frac{s_k}{s_Y} \right] = b_k \left[\frac{\dfrac{\sum_i (X_{ik} - \bar{X}_k)^2}{n-1}}{\dfrac{\sum_i (Y_i - \bar{Y})^2}{n-1}} \right]^{1/2} = b_k \left[\frac{\sum_i (X_{ik} - \bar{X}_k)^2}{\sum_i (Y_i - \bar{Y})^2} \right]^{1/2}$$

where s_k and s_Y are the standard deviations of the X_k and Y observations, respectively. The effect of the term in brackets is to make B_k dimensionless. The coefficient B_k reflects the change in the mean response (in units of standard deviations of Y) per unit change in the independent variable X_k (in units of standard deviations of X_k) when all other independent variables are held constant.

For our Zarthan Company example, we obtain the following standardized regression coefficients, using the final expression in (7.69):

$$B_1 = .496 \left[\frac{191,089}{53,902} \right]^{1/2} = .934$$

$$B_2 = .00920 \left[\frac{7,473,616}{53,902} \right]^{1/2} = .108$$

Sometimes, these standardized regression coefficients are interpreted as showing that target population (X_1) has a greater impact than per capita discretionary income (X_2) on sales because B_1 is much larger than B_2. However, as we will see in the next chapter, one must be cautious about interpreting regression coefficients, whether standardized or not. The reason is that when the independent variables are correlated among themselves, the regression coefficients are affected by the other independent variables in the model. We shall discuss this problem in Chapter 8.

Not only does the presence of correlations among the independent variables affect the magnitude of the standardized regression coefficients but the spacings of the observations on the independent variables also affect the standardized regression coefficients. Sometimes the spacings of the observations on the independent variables may be quite arbitrary.

Hence, it is ordinarily not wise to interpret a standardized regression coefficient as reflecting the importance of the independent variable.

7.10 WEIGHTED LEAST SQUARES

Unequal weighting of the observations can be carried out in multiple regression as in simple regression. Let the weights w_i be contained in the diagonal matrix $\mathbf{W}$ of weights:

$$(7.70) \qquad \underset{n \times n}{\mathbf{W}} = \begin{bmatrix} w_1 & & & & \\ & w_2 & & 0 & \\ & & \cdot & & \\ & 0 & & \cdot & \\ & & & & w_n \end{bmatrix}$$

The weighted least squares estimators of the regression coefficients then are:

$$(7.71) \qquad \underset{p \times 1}{\mathbf{b}} = (\mathbf{X'WX})^{-1}\mathbf{X'WY}$$

When the error term variances σ_i^2 are not equal, the weights w_i are chosen to be inversely proportional to σ_i^2, so that $\sigma_i^2 = \sigma^2/w_i$. The estimated variance-covariance matrix of the regression coefficients then is:

$$(7.72) \qquad \underset{p \times p}{s^2(\mathbf{b})} = MSE_w(\mathbf{X'WX})^{-1}$$

where MSE_w is based on the weighted squared deviations:

$$(7.72a) \qquad MSE_w = \frac{\Sigma w_i(Y_i - \hat{Y}_i)^2}{n - p}$$

As for simple regression, the appropriate weights w_i may often be found from basic relationships. For instance, it may be found that the error term variances σ_i^2 are proportional to X_{ik}^2, the level of the kth independent variable squared. Hence, $\sigma_i^2 = \sigma^2 X_{ik}^2$ then and the weights $w_i = 1/X_{ik}^2$ would be used.

PROBLEMS

7.1. Refer to Figure 7.4a. By how much approximately does mean yield increase when rainfall increases from 9 to 11 inches and temperature is held constant? Could you have answered this question if rainfall and temperature interact in their effects on crop yield?

7.2. Consider the response function $E(Y) = 25 + 3X_1 + 4X_2 + 1.5X_1X_2$.
   a. Plot the response function against X_1 when $X_2 = 3$ and when $X_2 = 6$. How is the interaction effect of X_1 and X_2 on Y apparent from this graph?

b. Sketch a set of contour curves for the response surface. How is the interaction effect of X_1 and X_2 on Y apparent from this graph?

7.3. Consider the response function $E(Y) = 14 + 7X_1 - 5X_2$.

a. Plot the response function against X_2 when $X_1 = 1$ and when $X_1 = 4$. How does the graph indicate that the effects of X_1 and X_2 on Y are additive?

b. Sketch a set of contour curves for the response surface. How does the graph indicate that the effects of X_1 and X_2 on Y are additive?

7.4. Set up the **X** matrix and $\boldsymbol{\beta}$ vector for each of the following models (assume $i = 1, \ldots, 4$):

a. $Y_i = \beta_0 + \beta_1 X_{i1} + \beta_2 X_{i1} X_{i2} + \varepsilon_i$

b. $\log Y_i = \beta_0 + \beta_1 X_{i1} + \beta_2 X_{i2} + \varepsilon_i$

7.5. Set up the **X** matrix and $\boldsymbol{\beta}$ vector for each of the following models (assume $i = 1, \ldots, 5$):

a. $Y_i = \beta_1 X_{i1} + \beta_2 X_{i2} + \beta_3 X_{i1}^2 + \varepsilon_i$

b. $\sqrt{Y_i} = \beta_0 + \beta_1 X_{i1} + \beta_2 \log_{10} X_{i2} + \varepsilon_i$

7.6. A student stated: "Adding independent variables to a regression model can never reduce R^2, so we should include all available independent variables in the model." Comment.

7.7. Why is it not meaningful to attach a sign to the coefficient of multiple correlation R, although we do so for the coefficient of simple correlation r?

7.8. **Brand preference.** In a small-scale study of the relation between degree of brand liking (Y) and moisture content (X_1) and sweetness (X_2) of the product, the following results were obtained (data are coded):

i:	1	2	3	4	5	6	7	8	9	10	11	12	13	14	15	16
X_{i1}:	4	4	4	4	6	6	6	6	8	8	8	8	10	10	10	10
X_{i2}:	2	4	2	4	2	4	2	4	2	4	2	4	2	4	2	4
Y_i:	64	73	61	76	72	80	71	83	83	89	86	93	88	95	94	100

Assume that regression model (7.1) with independent normal error terms is appropriate.

a. Find the estimated regression coefficients. State the estimated regression function. How is b_1 interpreted here?

b. Test whether there is a regression relation using a level of significance of .01. State the alternatives, decision rule, and conclusion. What does your test imply about β_1 and β_2?

c. What is the P-value of the test in part (b)?

d. Estimate β_1 and β_2 jointly by the Bonferroni procedure using a 99 percent family confidence coefficient. Interpret your results.

7.9. Refer to **Brand preference** Problem 7.8.

a. Calculate the coefficient of multiple determination R^2. How is it interpreted here?

b. Calculate the coefficient of simple determination r^2 between Y_i and $\hat{Y}_i$. Does it equal R^2?

7.10. Refer to **Brand preference** Problem 7.8.

a. Obtain an interval estimate of $E(Y_h)$ when $X_{h1} = 5$ and $X_{h2} = 4$. Use a 99 percent confidence coefficient. Interpret your interval estimate.

b. Obtain a prediction interval for a new observation $Y_{h(new)}$ when $X_{h1} = 5$ and $X_{h2} = 4$. Use a 99 percent confidence coefficient.

7.11. Refer to **Brand preference** Problem 7.8.
   a. Obtain the residuals.
   b. Plot the residuals against $\hat{Y}$, X_1, and X_2 on separate graphs. Also prepare a normal probability plot. Analyze the plots and summarize your findings.
   c. Conduct a formal test for lack of fit of the first-order regression function; use $\alpha = .01$. State the alternatives, decision rule, and conclusion.

7.12. **Chemical shipment.** The observations to follow, taken on 20 incoming shipments of chemicals in drums arriving at a warehouse, show number of drums in shipment (X_1), total weight of shipment (X_2, in hundred pounds), and number of minutes required to handle shipment (Y).

i:	1	2	3	4	5	6	7	8	9	10
X_{i1}:	7	18	5	14	11	5	23	9	16	5
X_{i2}:	5.11	16.72	3.20	7.03	10.98	4.04	22.07	7.03	10.62	4.76
Y_i:	58	152	41	93	101	38	203	78	117	44

i:	11	12	13	14	15	16	17	18	19	20
X_{i1}:	17	12	6	12	8	15	17	21	6	11
X_{i2}:	11.02	9.51	3.79	6.45	4.60	13.86	13.03	15.21	3.64	9.57
Y_i:	121	112	50	82	48	127	140	155	39	90

Assume that regression model (7.1) with independent normal error terms is appropriate.
   a. Obtain the estimated regression function. How is b_1 here interpreted? How is b_2 here interpreted?
   b. Test whether there is a regression relation, using a level of significance of .05. State the alternatives, decision rule, and conclusion. What does your test result imply about β_1 and β_2? What is the P-value of the test?
   c. Estimate β_1 and β_2 jointly by the Bonferroni procedure using a 95 percent family confidence coefficient. Interpret your results.
   d. Calculate the coefficient of multiple determination R^2. How is this measure interpreted here?

7.13. Refer to **Chemical shipment** Problem 7.12.
   a. Management desires simultaneous interval estimates of the mean handling times for five typical shipments specified to be as follows:

	1	2	3	4	5
X_1:	5	6	10	14	20
X_2:	3.20	4.80	7.00	10.00	18.00

Obtain the family of estimates using a 95 percent family confidence coefficient. Employ the Working-Hotelling type bounds or the Bonferroni procedure, whichever is more efficient.
   b. For the observations in Problem 7.12, would you consider a shipment of 20 drums with a weight of 5 hundred pounds to be within the scope of the model? What about a shipment of 20 drums with a weight of 19 hundred pounds? Support your answer by preparing a relevant plot.

7.14. Refer to **Chemical shipment** Problem 7.12. Four separate shipments with the following characteristics will arrive in the next day or two:

	1	2	3	4
X_1:	9	12	15	18
X_2:	7.20	9.00	12.50	16.50

Management desires predictions of the handling times for these shipments so that the actual handling times can be compared with the predicted times to determine whether any are "out of line." Develop the needed predictions using the most efficient approach and a family confidence coefficient of 95 percent.

7.15. Refer to **Chemical shipment** Problem 7.12.
   a. Obtain the residuals and plot them against $\hat{Y}$, X_1, and X_2 on separate graphs. Also prepare a normal probability plot. Analyze the plots and summarize your findings.
   b. Can you conduct a formal test for lack of fit here?

7.16. Refer to **Chemical shipment** Problem 7.12. Three new shipments are to be received, each with $X_{h1} = 7$ and $X_{h2} = 6$.
   a. Obtain a 95 percent prediction interval for the mean handling time for these shipments.
   b. Convert the interval obtained in part (a) into a 95 percent prediction interval for the total handling time for the three shipments.

7.17. **Patient satisfaction.** A hospital administrator wished to study the relation between patient satisfaction (Y) and patient's age (X_1, in years), severity of illness (X_2, an index), and anxiety level (X_3, an index). She randomly selected 23 patients and collected the data presented below, where larger values of Y, X_2, and X_3 are, respectively, associated with more satisfaction, increased severity of illness, and more anxiety.

i:	1	2	3	4	5	6	7	8	9	10	11	12
X_{i1}:	50	36	40	41	28	49	42	45	52	29	29	43
X_{i2}:	51	46	48	44	43	54	50	48	62	50	48	53
X_{i3}:	2.3	2.3	2.2	1.8	1.8	2.9	2.2	2.4	2.9	2.1	2.4	2.4
Y_i:	48	57	66	70	89	36	46	54	26	77	89	67

i:	13	14	15	16	17	18	19	20	21	22	23
X_{i1}:	38	34	53	36	33	29	33	55	29	44	43
X_{i2}:	55	51	54	49	56	46	49	51	52	58	50
X_{i3}:	2.2	2.3	2.2	2.0	2.5	1.9	2.1	2.4	2.3	2.9	2.3
Y_i:	47	51	57	66	79	88	60	49	77	52	60

Assume that regression model (7.5) for three independent variables with independent normal error terms is appropriate.
   a. Obtain the estimated regression function.
   b. Test whether there is a regression relation; use a .10 level of significance. State the alternatives, decision rule, and conclusion. What does your test imply about β_1, β_2, and β_3? What is the P-value of the test?
   c. Obtain joint interval estimates of β_1, β_2, and β_3 using a 90 percent family confidence coefficient. Interpret your results.
   d. Calculate the coefficient of multiple correlation. What does it indicate here?

7.18. Refer to **Patient satisfaction** Problem 7.17.

a. Obtain an interval estimate of the mean satisfaction when $X_{h1} = 35$, $X_{h2} = 45$, and $X_{h3} = 2.2$. Use a 90 percent confidence coefficient. Interpret your confidence interval.

b. Obtain a prediction interval for a new patient's satisfaction when $X_{h1} = 35$, $X_{h2} = 45$, and $X_{h3} = 2.2$. Use a 90 percent confidence coefficient. Interpret your prediction interval.

7.19. Refer to **Patient satisfaction** Problem 7.17.

a. Obtain the residuals and plot them against $\hat{Y}$ and each of the independent variables on separate graphs. Also prepare a normal probability plot. Analyze your plots and summarize your findings.

b. Can you conduct a formal test for lack of fit here?

7.20. **Mathematicians salaries.** A researcher in a scientific foundation wished to evaluate the relation between intermediate and senior level annual salaries of research mathematicians (Y, in thousand dollars) and an index of publication quality (X_1), number of years of experience (X_2), and an index of success in obtaining grant support (X_3). The data for a sample of 24 intermediate and senior level research mathematicians follow.

i:	1	2	3	4	5	6	7	8	9	10	11	12
X_{i1}:	3.5	5.3	5.1	5.8	4.2	6.0	6.8	5.5	3.1	7.2	4.5	4.9
X_{i2}:	9	20	18	33	31	13	25	30	5	47	25	11
X_{i3}:	6.1	6.4	7.4	6.7	7.5	5.9	6.0	4.0	5.8	8.3	5.0	6.4
Y_i:	33.2	40.3	38.7	46.8	41.4	37.5	39.0	40.7	30.1	52.9	38.2	31.8

i:	13	14	15	16	17	18	19	20	21	22	23	24
X_{i1}:	8.0	6.5	6.6	3.7	6.2	7.0	4.0	4.5	5.9	5.6	4.8	3.9
X_{i2}:	23	35	39	21	7	40	35	23	33	27	34	15
X_{i3}:	7.6	7.0	5.0	4.4	5.5	7.0	6.0	3.5	4.9	4.3	8.0	5.0
Y_i:	43.3	44.1	42.8	33.6	34.2	48.0	38.0	35.9	40.4	36.8	45.2	35.1

Assume that regression model (7.5) for three independent variables with independent normal error terms is appropriate.

a. Obtain the estimated regression function.

b. Test whether there is a regression relation; use $\alpha = .05$. State the alternatives, decision rule, and conclusion. What does your test imply about β_1, β_2, and β_3? What is the P-value of the test?

c. Estimate β_1, β_2, and β_3 jointly by the Bonferroni procedure using a 95 percent family confidence coefficient. Interpret your results.

d. Calculate R^2 and interpret this measure.

7.21. Refer to **Mathematicians salaries** Problem 7.20. The researcher wishes to obtain simultaneous interval estimates of the mean salary levels for four typical research mathematicians specified as follows:

	1	2	3	4
X_1	5.0	6.0	4.0	7.0
X_2	20	30	10	50
X_3	5.0	6.0	4.0	7.0

Obtain the family of estimates using a 95 percent family confidence coefficient. Employ the most efficient procedure.

7.22. Refer to **Mathematicians salaries** Problem 7.20. Three research mathematicians with the following characteristics did not provide any salary information in the study.

	1	2	3
X_1	5.4	6.2	6.4
X_2	17	12	21
X_3	6.0	5.8	6.1

Develop separate prediction intervals for the annual salaries of these mathematicians using a 95 percent statement confidence coefficient in each case. Can the salaries of these three mathematicians be predicted fairly precisely?

7.23. Refer to **Mathematicians salaries** Problem 7.20.
  a. Obtain the residuals and plot them against $\hat{Y}$ and each of the independent variables on separate graphs. Also prepare a normal probability plot. Analyze your plots and summarize your findings.
  b. Can you conduct a formal test for lack of fit here?

7.24. Refer to **Chemical shipment** Problem 7.12.
  a. Obtain the standardized regression coefficients.
  b. Calculate the coefficient of determination between the two independent variables. Is it meaningful here to consider the standardized regression coefficients to reflect the effect of one independent variable when the other is held constant?

7.25. Refer to **Patient satisfaction** Problem 7.17.
  a. Obtain the standardized regression coefficients.
  b. Calculate the coefficients of determination between all pairs of independent variables. Do these indicate that it is meaningful here to consider the standardized regression coefficients as indicating the effect of one independent variable when the others are held constant?

EXERCISES

7.26. For each of the following models, indicate whether it is a general linear regression model. If it is not, state whether it can be expressed in the form of (7.7) by a suitable transformation:
  a. $Y_i = \beta_0 + \beta_1 X_{i1} + \beta_2 \log_{10} X_{i2} + \beta_3 X_{i1}^2 + \varepsilon_i$
  b. $Y_i = \varepsilon_i \exp(\beta_0 + \beta_1 X_{i1} + \beta_2 X_{i2}^2)$
  c. $Y_i = \beta_0 + \log_{10}(\beta_1 X_{i1}) + \beta_2 X_{i2} + \varepsilon_i$
  d. $Y_i = \beta_0 \exp(\beta_1 X_{i1}) + \varepsilon_i$
  e. $Y_i = [1 + \exp(\beta_0 + \beta_1 X_{i1} + \varepsilon_i)]^{-1}$

7.27. (Calculus needed.) Consider the multiple regression model:

$$Y_i = \beta_1 X_{i1} + \beta_2 X_{i2} + \varepsilon_i \qquad i = 1, \ldots, n$$

where the ε_i are uncorrelated, with $E(\varepsilon_i) = 0$ and $\sigma^2(\varepsilon_i) = \sigma^2$.
  a. Derive the least squares estimators of β_1 and β_2.

b. Assuming that the ε_i are independent normal random variables, state the likelihood function and obtain the maximum likelihood estimators of β_1 and β_2. Are these the same as the least squares estimators?

7.28. (Calculus needed.) Consider the multiple regression model:

$$Y_i = \beta_0 + \beta_1 X_{i1} + \beta_2 X_{i1}^2 + \beta_3 X_{i2} + \varepsilon_i \qquad i = 1, \ldots, n$$

where the ε_i are independent $N(0, \sigma^2)$. Derive the least squares normal equations. Will these yield the same estimators of the regression coefficients as the maximum likelihood estimators?

7.29. An analyst wanted to fit the regression model $Y_i = \beta_0 + \beta_1 X_{i1} + \beta_2 X_{i2} + \beta_3 X_{i3} + \varepsilon_i$, $i = 1, \ldots, n$, by the method of least squares when it is known that $\beta_2 = 4$. How can the analyst obtain the desired fit using a multiple regression computer program?

7.30. For regression model (7.1), show that the coefficient of simple determination r^2 between Y_i and $\hat{Y}_i$ equals the coefficient of multiple determination R^2.

7.31. In a small-scale regression study, the following data were obtained:

i:	1	2	3	4	5	6
X_{i1}:	7	4	16	3	21	8
X_{i2}:	33	41	7	49	5	31
Y_i:	42	33	75	28	91	55

Assume that regression model (7.1) with independent normal error terms is appropriate. Using matrix methods, obtain (a) **b**; (b) **e**; (c) SSE; (d) SSR; (e) $s^2(\mathbf{b})$; (f) $\hat{Y}_h$ when $X_{h1} = 10$, $X_{h2} = 30$; (g) $s^2(\hat{Y}_h)$ when $X_{h1} = 10$, $X_{h2} = 30$.

PROJECTS

7.32. Refer to the **SMSA** data set. You have been asked to evaluate two alternative models for predicting the number of active physicians (Y) in an SMSA. Proposed model I includes as independent variables total population (X_1), land area (X_2), and total personal income (X_3). Proposed model II includes as independent variables population density (X_1, total population divided by land area), percent of population in central cities (X_2), and total personal income (X_3).

a. For each of the two proposed models, fit the first-order regression model (7.5) with three independent variables.

b. Calculate R^2 for each model. Is one model clearly preferable in terms of this measure?

c. For each model, obtain the residuals and plot them against $\hat{Y}$ and against each of the three independent variables. Also prepare a normal probability plot for each of the two fitted models. Analyze your plots and state your findings. Is one model clearly preferable in terms of aptness?

7.33. Refer to the **SMSA** data set.

a. For each geographic region, regress the number of serious crimes in an SMSA (Y) against population density (X_1, total population divided by land area), total personal income (X_2), and percent high school graduates (X_3).

Use the first-order regression model (7.5) with three independent variables. State the estimated regression functions.

b. Are the estimated regression functions similar for the four regions? Discuss.

c. Calculate *MSE* and R^2 for each region. Are these measures similar for the four regions? Discuss.

7.34. Refer to the **SENIC** data set. Two models have been proposed for predicting the average length of patient stay in a hospital (Y). Model I utilizes as independent variables age (X_1), infection risk (X_2), and available facilities and services (X_3). Model II uses as independent variables number of beds (X_1), infection risk (X_2), and available facilities and services (X_3).

a. For each of the two proposed models, fit the first-order regression model (7.5) with three independent variables.

b. Calculate R^2 for each model. Is one model clearly preferable in terms of this measure?

c. For each model, obtain the residuals and plot them against $\hat{Y}$ and against each of the three independent variables. Also prepare a normal probability plot for each of the two fitted models. Analyze your plots and state your findings. Is one model clearly preferable in terms of aptness?

7.35. Refer to the **SENIC** data set.

a. For each geographic region, regress infection risk (Y) against the independent variables age (X_1), routine culturing ratio (X_2), average daily census (X_3), and available facilities and services (X_4). Use the first-order regression model (7.5) with four independent variables. State the estimated regression functions.

b. Are the estimated regression functions similar for the four regions? Discuss.

c. Calculate *MSE* and R^2 for each region. Are these measures similar for the four regions? Discuss.

CITED REFERENCE

7.1 *SAS User's Guide*. 1979 ed. Raleigh, N.C.: SAS Institute, 1979.

8

Multiple regression—II

In this chapter, we continue our discussion of multiple regression by first considering multicollinearity and its effects in multiple regression models. Then we take up several other topics in multiple regression, including additional tests of hypotheses concerning the regression coefficients.

8.1 MULTICOLLINEARITY AND ITS EFFECTS

In multiple regression analysis, one is often concerned with the nature and significance of the relations between the independent variables and the dependent variable. Questions that are frequently asked include:

1. What is the relative importance of the effects of the different independent variables?
2. What is the magnitude of the effect of a given independent variable on the dependent variable?
3. Can any independent variable be dropped from the model because it has little or no effect on the dependent variable?
4. Should any independent variables not yet included in the model be considered for possible inclusion?

If the independent variables included in the model are (1) uncorrelated among themselves and (2) uncorrelated with any other independent variables that are related to the dependent variable but omitted from the model, relatively simple answers can be given to these questions. Unfortunately, in many nonexperimental situations in business, economics, and the social and biological sciences, the independent variables tend to be correlated among themselves and with other variables that are related to the dependent variable but are not included in the model.

When the independent variables are correlated among themselves, *intercorrelation* or *multicollinearity* among them is said to exist. (Sometimes the latter term is reserved for those instances when the correlation among independent variables is very high or even perfect.) We shall explore now a variety of interrelated problems that are created by multicollinearity among the independent variables. First, however, we examine the situation when the independent variables are not correlated.

Example of uncorrelated independent variables

Table 8.1 contains data for a small-scale experiment on the effect of work crew size (X_1) and level of bonus pay (X_2) on crew productivity score (Y). It is easy to show that X_1 and X_2 are uncorrelated here, i.e., $r_{12}^2 = 0$, where r_{12}^2 denotes the coefficient of simple determination between X_1 and X_2. Table 8.2a contains the fitted regression function and analysis of variance table when both X_1 and X_2 are included in the model. Table 8.2b contains the same information when only X_1 is included in the model, and Table 8.2c contains this information when only X_2 is in the model. In Table 8.2a, we use the notation $SSR(X_1, X_2)$ and $SSE(X_1, X_2)$ to indicate explicitly the two independent variables in the model. Similarly, in Table 8.2b, we use the notation $SSR(X_1)$ and $SSE(X_1)$ to show that only independent variable X_1 is in the model, so that the regression here is a simple regression. We do likewise in Table 8.2c.

TABLE 8.1 Work crew productivity example with uncorrelated independent variables

Trial i	Crew Size X_{i1}	Bonus Pay X_{i2}	Crew Productivity Score Y_i
1	4	$2	42
2	4	2	39
3	4	3	48
4	4	3	51
5	6	2	49
6	6	2	53
7	6	3	61
8	6	3	60

An important feature to note in Table 8.2 is that the regression coefficients for X_1 and X_2 are the same, whether only the given independent variable is included

TABLE 8.2 ANOVA tables for work crew productivity example with uncorrelated independent variables

(a) Regression of Y on X_1 and X_2

$\hat{Y} = .375 + 5.375X_1 + 9.250X_2$

Source of Variation	SS	df	MS
Regression	$SSR(X_1, X_2) = 402.250$	2	$MSR(X_1, X_2) = 201.125$
Error	$SSE(X_1, X_2) = 17.625$	5	$MSE(X_1, X_2) = 3.525$
Total	$SSTO = 419.875$	7	

(b) Regression of Y on X_1

$\hat{Y} = 23.500 + 5.375X_1$

Source of Variation	SS	df	MS
Regression	$SSR(X_1) = 231.125$	1	$MSR(X_1) = 231.125$
Error	$SSE(X_1) = 188.750$	6	$MSE(X_1) = 31.458$
Total	$SSTO = 419.875$	7	

(c) Regression of Y on X_2

$\hat{Y} = 27.250 + 9.250X_2$

Source of Variation	SS	df	MS
Regression	$SSR(X_2) = 171.125$	1	$MSR(X_2) = 171.125$
Error	$SSE(X_2) = 248.750$	6	$MSE(X_2) = 41.458$
Total	$SSTO = 419.875$	7	

in the model or both independent variables are included. This is a result of the two independent variables being uncorrelated.

Thus, if the independent variables are uncorrelated, the effects ascribed to them by a first-order regression model are the same no matter which other independent variables are included in the model. This is a strong argument for controlled experiments whenever possible, since experimental control permits making the independent variables uncorrelated.

Another important feature of Table 8.2 is related to the error sums of squares. Note from Table 8.2a that the error sum of squares when both X_1 and X_2 are included in the model is $SSE(X_1, X_2) = 17.625$. When only X_1 is included in the model, however, the error sum of squares is $SSE(X_1) = 188.750$ according to Table 8.2b. Since the variation in Y when X_1 alone is considered is 188.750 but is only 17.625 when both X_1 and X_2 are considered, we may ascribe the difference:

$$SSE(X_1) - SSE(X_1, X_2) = 188.750 - 17.625 = 171.125$$

to the effect of X_2. We shall denote this difference by $SSR(X_2|X_1)$:

(8.1) $$SSR(X_2|X_1) = SSE(X_1) - SSE(X_1, X_2)$$

When we fit a regression function containing only X_2, we also obtain a measure of the reduction in the variation of Y associated with X_2, namely, $SSR(X_2)$. Table 8.2c indicates that $SSR(X_2) = 171.125$, which is the same as $SSR(X_2|X_1)$ = 171.125. The reason for this is that X_1 and X_2 are uncorrelated.

The story is the same for independent variable X_1. Let:

(8.2) $$SSR(X_1|X_2) = SSE(X_2) - SSE(X_1, X_2)$$

For our example, we have:

$$SSR(X_1|X_2) = 248.750 - 17.625 = 231.125$$

This sum of squares is the same as $SSR(X_1) = 231.125$ from Table 8.2b, obtained when Y is regressed only on X_1.

In general, when two independent variables are uncorrelated, the marginal contribution of an independent variable in reducing the error sum of squares when the other independent variable is in the model is exactly the same as when this independent variable is in the model alone.

Note

To show that the regression coefficient of X_1 is unchanged when X_2 is added to the regression model in the case where X_1 and X_2 are uncorrelated, consider the algebraic expression for b_1 in the multiple regression model with two independent variables:

(8.3) $$b_1 = \frac{\dfrac{\Sigma(X_{i1} - \bar{X}_1)(Y_i - \bar{Y})}{\Sigma(X_{i1} - \bar{X}_1)^2} - \left[\dfrac{\Sigma(Y_i - \bar{Y})^2}{\Sigma(X_{i1} - \bar{X}_1)^2}\right]^{1/2} r_{Y2}r_{12}}{1 - r_{12}^2}$$

where r_{Y2} denotes the coefficient of simple correlation between Y and X_2, and r_{12}, as before, denotes the coefficient of simple correlation between X_1 and X_2.

If X_1 and X_2 are uncorrelated, $r_{12} = 0$, and (8.3) reduces to:

(8.3a) $$b_1 = \frac{\Sigma(X_{i1} - \bar{X}_1)(Y_i - \bar{Y})}{\Sigma(X_{i1} - \bar{X}_1)^2}$$

But (8.3a) is the estimator of the slope for the simple regression of Y on X_1, per (2.10a).

Hence, if X_1 and X_2 are uncorrelated, adding X_2 to the regression model does not change the regression coefficient for X_1; correspondingly, adding X_1 to the regression model does not change the regression coefficient for X_2.

Example of correlated independent variables

As mentioned earlier, nonexperimental data in many disciplines frequently consist of correlated independent variables. For example, in a regression of family food expenditures on the independent variables family income, family savings, and age of head of the household, the independent variables will be correlated among themselves. Further, the independent variables will also be correlated with other socioeconomic variables not included in the model that do affect family food expenditures, such as family size. We shall now consider the effects of multicollinearity, i.e., correlated independent variables, upon the regression coefficients and upon the regression sums of squares.

Effect of multicollinearity on regression coefficients. Table 8.3 contains data for a study of the relation of body fat (Y) to triceps skinfold thickness (X_1) and thigh circumference (X_2), based on a sample of 20 healthy females 25–34 years old. The triceps skinfold thicknesses and thigh circumferences for these subjects are highly correlated, as the scatter plot in Figure 8.1 suggests. Indeed, the coefficient of simple correlation between these two independent variables for the 20 subjects is $r_{12} = +.92$.

TABLE 8.3 Body fat example with correlated independent variables

Subject i	Triceps Skinfold Thickness X_{i1}	Thigh Circumference X_{i2}	Body Fat Y_i
1	19.5	43.1	11.9
2	24.7	49.8	22.8
3	30.7	51.9	18.7
4	29.8	54.3	20.1
5	19.1	42.2	12.9
6	25.6	53.9	21.7
7	31.4	58.5	27.1
8	27.9	52.1	25.4
9	22.1	49.9	21.3
10	25.5	53.5	19.3
11	31.1	56.6	25.4
12	30.4	56.7	27.2
13	18.7	46.5	11.7
14	19.7	44.2	17.8
15	14.6	42.7	12.8
16	29.5	54.4	23.9
17	27.7	55.3	22.6
18	30.2	58.6	25.4
19	22.7	48.2	14.8
20	25.2	51.0	21.1

Table 8.4a contains the results of a computer run for regressing Y on X_1 and X_2, and presents the fitted regression function and the analysis of variance table. Again, we use the notation $SSR(X_1, X_2)$ and $SSE(X_1, X_2)$ to indicate explicitly that both independent variables are in the fitted model. Table 8.4b contains the fitted regression and the analysis of variance table for the regression of Y on X_1 only, and Table 8.4c contains the results for the regression of Y on X_2 only.

Note first that the regression coefficient for X_1, triceps skinfold thickness, is not the same in Tables 8.4a and b. Thus, the effect ascribed to X_1 by the fitted response function varies here, depending upon whether only X_1 or both X_1 and X_2

FIGURE 8.1 Scatter plot of thigh circumference against triceps skinfold thickness—body fat example

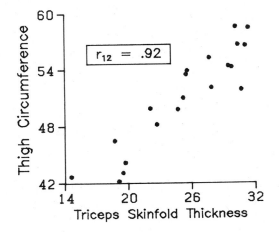

TABLE 8.4 ANOVA table for body fat example with correlated independent variables

(a) Regression of Y on X_1 and X_2

$$\hat{Y} = -19.174 + .2224X_1 + .6594X_2$$

Source of Variation	SS	df	MS
Regression	$SSR(X_1, X_2) = 385.44$	2	$MSR(X_1, X_2) = 192.72$
Error	$SSE(X_1, X_2) = 109.95$	17	$MSE(X_1, X_2) = 6.47$
Total	$SSTO = 495.39$	19	

(b) Regression of Y on X_1

$$\hat{Y} = -1.496 + .8572X_1$$

Source of Variation	SS	df	MS
Regression	$SSR(X_1) = 352.27$	1	$MSR(X_1) = 352.27$
Error	$SSE(X_1) = 143.12$	18	$MSE(X_1) = 7.95$
Total	$SSTO = 495.39$	19	

(c) Regression of Y on X_2

$$\hat{Y} = -23.634 + .8566X_2$$

Source of Variation	SS	df	MS
Regression	$SSR(X_2) = 381.97$	1	$MSR(X_2) = 381.97$
Error	$SSE(X_2) = 113.42$	18	$MSE(X_2) = 6.30$
Total	$SSTO = 495.39$	19	

are being considered in the model. The reason for the different regression coefficients is that the independent variables are correlated here, as we saw earlier from Figure 8.1. If we were to consider as a third independent variable X_3, midarm circumference, the regression coefficient for X_1 in the fitted response function with three independent variables would be different again from the two coefficients in Table 8.4 since midarm circumference is moderately correlated with triceps skinfold thickness.

Next, we turn to the regression coefficient for X_2, thigh circumference. We see again from Table 8.4 that the regression coefficient for X_2 when X_1 is also in the model is different from the regression coefficient when X_2 is the only independent variable in the model.

The important conclusion we must draw is: When independent variables are correlated, the regression coefficient of any independent variable depends on which other independent variables are included in the model and which ones are left out. Thus, a regression coefficient does not reflect any inherent effect of the particular independent variable on the dependent variable but only a marginal or partial effect, given whatever other correlated independent variables are included in the model.

Effect of multicollinearity on regression sums of squares.   Note from Table 8.4a that the error sum of squares when both X_1 and X_2 are included in the model is $SSE(X_1, X_2) = 109.95$. When only X_2 is included in the model, the error sum of squares is $SSE(X_2) = 113.42$ as seen from Table 8.4c. Since the variation in Y when X_2 alone is considered is 113.42 but is 109.95 when both X_1 and X_2 are considered, we ascribe the difference to the effect of X_1. Using (8.2), we obtain:

$$SSR(X_1 | X_2) = SSE(X_2) - SSE(X_1, X_2) = 113.42 - 109.95 = 3.47$$

When we fit a regression function containing only X_1, we also obtain a measure of the reduction in variation of Y associated with X_1, namely, $SSR(X_1)$. For our example, Table 8.4b indicates that $SSR(X_1) = 352.27$ which is not the same as $SSR(X_1 | X_2) = 3.47$. The reason for the large difference is the high positive correlation between X_1 and X_2.

The story is the same for the other independent variable. Using (8.1), we obtain:

$$SSR(X_2 | X_1) = SSE(X_1) - SSE(X_1, X_2) = 143.12 - 109.95 = 33.17$$

This sum of squares is not the same as $SSR(X_2) = 381.97$ from Table 8.4c, obtained when Y is regressed only on X_2.

The important conclusion is: When independent variables are correlated, there is no unique sum of squares which can be ascribed to an independent variable as reflecting its effect in reducing the total variation in Y. The reduction in the total variation ascribed to an independent variable must be viewed in the context of the other independent variables included in the model, whenever the independent variables are correlated.

Let us consider the meaning of $SSR(X_1)$ and $SSR(X_1|X_2)$ further. $SSR(X_1)$ measures the reduction in the variation of Y when X_1 is introduced into the regression model and no other independent variable is present. $SSR(X_1|X_2)$ measures the further reduction in the variation of Y when X_2 is already in the regression model and X_1 is introduced as a second independent variable. Hence, the notation $SSR(X_1|X_2)$ is used to show that the sum of squares measures a reduction in variation of Y associated with X_1, given that X_2 is already included in the model.

The reason why $SSR(X_1|X_2) = 3.47$ is less than $SSR(X_1) = 352.27$ in our example should now be apparent. Since X_1 is highly positively correlated with X_2, much of the power of X_1 to reduce the variation in Y is already accounted for by $SSR(X_2)$ when X_2 alone is included in the model. Hence, the marginal effect of X_1 in reducing the variation in Y, given that X_2 is in the model, is less than the effect if X_1 were introduced into the model without X_2 being present. For the same reason, $SSR(X_2|X_1)$ is less than $SSR(X_2)$.

The terms $SSR(X_2|X_1)$ and $SSR(X_1|X_2)$ are called *extra sums of squares*, since they indicate the additional or extra reduction in the error sum of squares achieved by introducing an additional independent variable.

Simultaneous tests on regression coefficients

Just as multicollinearity among the independent variables leads to regression coefficients that vary depending on which correlated independent variables are included in the model, so does multicollinearity also cause difficulties in statistical tests of the regression coefficients. A not infrequent abuse in the analysis of multiple regression models is to examine the t^* statistic in (7.42b):

$$t^* = \frac{b_k}{s(b_k)}$$

for each regression coefficient in turn to decide whether or not $\beta_k = 0$ for $k = 1,\ldots,p - 1$. Even if a simultaneous inference procedure is used, and often it is not, problems still exist.

Let us consider the first-order regression model with two independent variables:

(8.4) $$Y_i = \beta_0 + \beta_1 X_{i1} + \beta_2 X_{i2} + \varepsilon_i \qquad \text{Full model}$$

If the test on β_1 indicates it is zero, the regression model (8.4) would be:

$$Y_i = \beta_0 + \beta_2 X_{i2} + \varepsilon_i$$

If the test on β_2 indicates it is zero, the regression model (8.4) would be:

$$Y_i = \beta_0 + \beta_1 X_{i1} + \varepsilon_i$$

However, if the separate tests indicate that $\beta_1 = 0$ and $\beta_2 = 0$, that does not necessarily jointly imply that:

$$Y_i = \beta_0 + \varepsilon_i$$

since neither of the tests considers this alternative.

For an example, consider the data in Table 8.5 for 10 ski resorts in New England during a period of normal snow conditions. The computer output for the regression of visitor days (Y) on miles of intermediate trails (X_1) and lift capacity (X_2) is summarized in Table 8.6a. The proper test for the existence of a regression relation:

$$H_0: \beta_1 = 0 \text{ and } \beta_2 = 0$$
$$H_a: \text{not both } \beta_1 \text{ and } \beta_2 \text{ equal } 0$$

is the F test of (7.30). The test statistic for our example is (Table 8.6a):

$$F^* = \frac{MSR}{MSE} = \frac{811,865,088}{2,757,701} = 294$$

Controlling the level of significance at .05, we require $F(.95; 2, 7) = 4.74$. Since $F^* = 294 > 4.74$, we conclude H_a, that there is a regression relation between Y and the independent variables X_1 and X_2. Hence, at least one of the two regression coefficients does not equal zero. The P-value for the test is less than .001 because $F(.999; 2, 7) = 21.7$.

Let us now examine the t^* statistics at a 5 percent family level of significance by the Bonferroni technique. We require $t(.9875; 7) = 2.84$. Since both t^* statistics have absolute values that do not exceed 2.84 (Table 8.6a), we would conclude $\beta_1 = 0$ and $\beta_2 = 0$, contrary to the earlier conclusion that not both coefficients equal zero.

To understand this apparently paradoxical result, let us investigate the test of, say:

$$H_0: \beta_2 = 0$$
$$H_a: \beta_2 \neq 0$$

by the general linear test approach. We first obtain the error sum of squares

TABLE 8.5 Ski resort example with highly correlated independent variables

Ski Resort i	Miles of Intermediate Trails X_{i1}	Lift Capacity (skiers per hour) X_{i2}	Total Visitor Days during Sample Period Y_i
1	10.5	2,200	19,929
2	2.5	1,000	5,839
3	13.1	3,250	23,696
4	4.0	1,475	9,881
5	14.7	3,800	30,011
6	3.6	1,200	7,241
7	7.1	1,900	11,634
8	22.5	5,575	45,684
9	17.0	4,200	36,476
10	6.4	1,850	12,068

TABLE 8.6 Selected computer outputs for ski resort example with highly correlated independent variables

(a)   Regression of Y on X_1 and X_2

$$\hat{Y} = -1,806.82 + 1,131.03 X_1 + 4.00 X_2$$

Source of Variation	SS	df	MS
Regression	$SSR(X_1, X_2) = 1,623,730,176$	2	$MSR(X_1, X_2) = 811,865,088$
Error	$SSE(X_1, X_2) = \quad 19,303,908$	7	$MSE(X_1, X_2) = \quad 2,757,701$
Total	$SSTO = 1,643,034,084$	9	

Variable	Estimated Regression Coefficient	Estimated Standard Deviation	t^*
X_1	$b_1 = 1,131.03$	$s(b_1) = 615.76$	1.837
X_2	$b_2 = \quad 4.002$	$s(b_2) = \quad 2.71$	1.477

(b)   Regression of Y on X_1

$$\hat{Y} = -363.98 + 2,032.53 X_1$$

Source of Variation	SS	df	MS
Regression	$SSR(X_1) = 1,617,707,400$	1	$MSR(X_1) = 1,617,707,400$
Error	$SSE(X_1) = \quad 25,326,684$	8	$MSE(X_1) = \quad 3,165,836$
Total	$SSTO = 1,643,034,084$	9	

$SSE(F)$ for the full model in (8.4), associated with $n - 3$ degrees of freedom. Next, we formulate the reduced model under H_0:

$$(8.5) \qquad\qquad Y_i = \beta_0 + \beta_1 X_{i1} + \varepsilon_i \qquad\qquad \text{Reduced model}$$

and obtain the error sum of squares $SSE(R)$ for the reduced model. Associated with $SSE(R)$ are $n - 2$ degrees of freedom. The test statistic (3.68) then leads to:

$$F^* = \frac{SSE(R) - SSE(F)}{(n - 2) - (n - 3)} \div \frac{SSE(F)}{n - 3}$$

In our notation recognizing explicitly the independent variables in the model, we have from the two parts of Table 8.6:

$$SSE(F) = SSE(X_1, X_2) = 19,303,908$$
$$SSE(R) = SSE(X_1) = 25,326,684$$

so that by (8.1):

$$SSE(R) - SSE(F) = SSE(X_1) - SSE(X_1, X_2) = SSR(X_2 | X_1)$$

Hence:

(8.6)
$$F^* = \frac{SSR(X_2|X_1)}{1} \div \frac{SSE(X_1, X_2)}{n-3}$$

For our example in Table 8.6, we have:

$$SSR(X_2|X_1) = 25{,}326{,}684 - 19{,}303{,}908 = 6{,}022{,}776$$

and hence:

$$F^* = \frac{6{,}022{,}776}{1} \div \frac{19{,}303{,}908}{7} = 2.18$$

Recall from Table 8.6a that the t^* statistic for testing $\beta_2 = 0$ is $t^* = 1.477$. Since:

$$(t^*)^2 = (1.477)^2 = 2.18 = F^*$$

always t test, true also for ...
partial F ?

we see that the two test statistics are equivalent. We already knew this holds for simple regression, and now we can see it also holds for multiple regression.

The F^* test statistic (8.6) to test whether or not $\beta_2 = 0$ is called a *partial F test* statistic to distinguish it from the F^* statistic in (7.30b) for testing whether all $\beta_k = 0$, i.e., whether or not there is a regression relation between Y and the set of independent variables. The latter test is called the *overall F test*.

The test statistic (8.6) for the partial F test indicates clearly that a test on whether or not $\beta_2 = 0$ is a *marginal* test, given that X_1 is in the model. Similarly, a test on whether or not $\beta_1 = 0$ is a *marginal* test, given that X_2 is in the model. It is apparent now why the simultaneous tests with the t^* statistics for the two different regression coefficients both led to the conclusion that the regression coefficient equals zero. The independent variables X_1 and X_2 in Table 8.5 are highly positively correlated; indeed, the coefficient of simple correlation is $r_{12} = +.99$. Hence, if X_1 is already in the model, adding X_2 achieves little more reduction in the variation of Y, and $SSR(X_2|X_1)$ is small. When $SSR(X_2|X_1)$ is small, the test for β_2 leads to a small partial F test statistic F^* and therefore to the conclusion that $\beta_2 = 0$.

Similarly, the explanation why the t test led to the conclusion that $\beta_1 = 0$ is that X_1 is highly positively correlated with X_2, and X_2 is assumed to be in the model when the t test for β_1 is employed.

Thus, despite the fact that there is a clear relation between the dependent variable Y and the set of independent variables X_1 and X_2, the separate t tests indicated the respective regression coefficients equal zero because each test considers only the marginal contribution of the independent variable, given that the other is included in the model.

Note

We have just seen that it is possible that a set of independent variables is related to the dependent variable, yet all of the individual tests on the regression coefficients will lead to the conclusion that they equal zero because of the multicollinearity among the independent variables. This apparently paradoxical result is also possible under special circumstances when there is no multicollinearity among the independent variables. The special circumstances are not likely to be found in practice, however.

A final comment

The discussion in this section has indicated that the choice of the particular set of independent variables which are to be included in the model is highly important and that in the presence of multicollinearity, the interpretation of regression coefficients and regression sums of squares must be undertaken with caution. The regression coefficients are affected not only by the other intercorrelated variables in the model but also by intercorrelated variables omitted from the model. For instance, an analyst was perplexed to find in a regression of territory company sales on territory population size, per capita income, and some other independent variables that the confidence interval for the regression coefficient for population size indicated it is negative. The analyst should have considered some of the omitted independent variables in search of an explanation. A consultant noted that the analyst did not include the major competitor's market penetration in the model. Since the competitor was most active and effective in territories with large populations, and thereby kept company sales down in these territories, the result of the omission of this independent variable from the model was a negative coefficient for the population size variable.

In view of the importance of the problems caused by multicollinearity among the independent variables, we take up this subject in more detail in Chapter 11. We consider there how to identify the presence of multicollinearity and take up several measures which may help to overcome some of the problems caused by multicollinearity.

8.2 DECOMPOSITION OF *SSR* INTO EXTRA SUMS OF SQUARES

Extra sums of squares

We defined the extra sum of squares $SSR(X_1|X_2)$ in (8.2):

(8.7a) $$SSR(X_1|X_2) = SSE(X_2) - SSE(X_1, X_2)$$

Likewise, we defined in (8.1):

(8.7b) $$SSR(X_2|X_1) = SSE(X_1) - SSE(X_1, X_2)$$

These extra sums of squares reflect the reduction in the error sum of squares by adding an independent variable to the model, given that another independent variable is already in the model.

Any reduction in the error sum of squares, of course, is equal to the same increase in the regression sum of squares since always:

$$SSTO = SSR + SSE$$

Hence, an extra sum of squares can also be thought of as the increase in the regression sum of squares achieved by introducing the new variable. We can therefore state, equivalently:

(8.8a) $$SSR(X_1|X_2) = SSR(X_1, X_2) - SSR(X_2)$$
(8.8b) $$SSR(X_2|X_1) = SSR(X_1, X_2) - SSR(X_1)$$

We show in Figure 8.2 a schematic representation of the extra sums of squares for our body fat example. The total bar on the left represents $SSTO$. The un-shaded component of this bar is $SSR(X_2)$, and the combined shaded area represents $SSE(X_2)$. The latter area in turn is the combination of the extra sum of squares $SSR(X_1|X_2)$ and the error sum of squares $SSE(X_1, X_2)$ when both X_1 and X_2 are included in the model. Similarly, the bar on the right in Figure 8.2 shows the decomposition containing $SSR(X_2|X_1)$. Note in both cases how the extra sum of squares can be viewed either as a reduction in the error sum of squares or as an increase in the regression sum of squares when the second variable is added to the regression model.

FIGURE 8.2   Schematic representation of extra sums of squares—body fat example

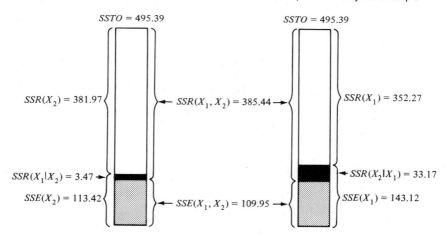

Extensions for three or more X variables are straightforward. For instance, we define:

(8.9) $$SSR(X_3|X_1, X_2) = SSE(X_1, X_2) - SSE(X_1, X_2, X_3)$$

$SSR(X_3|X_1, X_2)$ measures the reduction in the remaining variation of Y which is achieved by introducing X_3 into the regression model when X_1 and X_2 are already in the model.

Each extra sum of squares involving the addition of one independent variable into the regression model has associated with it one degree of freedom.

Decomposition of *SSR*

We can now obtain a variety of decompositions for the regression sum of squares *SSR*. Let us consider the case of three *X* variables. We begin with the identity (3.48a) for variable X_1:

$$(8.10) \qquad SSTO = SSR(X_1) + SSE(X_1)$$

where the notation now shows explicitly that X_1 is the *X* variable in the model. Replacing $SSE(X_1)$ by its equivalent in (8.7b), we obtain:

$$(8.10a) \qquad SSTO = SSR(X_1) + SSR(X_2|X_1) + SSE(X_1, X_2)$$

Replacing $SSE(X_1, X_2)$ by its equivalent in (8.9), we obtain:

$$(8.10b) \qquad SSTO = SSR(X_1) + SSR(X_2|X_1) + SSR(X_3|X_1, X_2) \\ + SSE(X_1, X_2, X_3)$$

Since we have the same identity for multiple regression with three independent variables as in (8.10) for a single independent variable, namely:

$$(8.11) \qquad SSTO = SSR(X_1, X_2, X_3) + SSE(X_1, X_2, X_3)$$

equation (8.10b) therefore reduces to:

$$(8.12) \quad SSR(X_1, X_2, X_3) = SSR(X_1) + SSR(X_2|X_1) + SSR(X_3|X_1, X_2)$$

Thus, the regression sum of squares has been decomposed into marginal components, each associated with one degree of freedom. Of course, the order of the independent variables is arbitrary, and other orders can be developed. For instance:

$$(8.13) \quad SSR(X_1, X_2, X_3) = SSR(X_3) + SSR(X_1|X_3) + SSR(X_2|X_1, X_3)$$

Indeed, we can define extra sums of squares for two or more independent variables at a time and obtain still other decompositions. For instance, we define:

$$(8.14) \qquad SSR(X_2, X_3|X_1) = SSE(X_1) - SSE(X_1, X_2, X_3)$$

Thus, $SSR(X_2, X_3|X_1)$ represents the reduction in the variation of *Y* gained when X_2 and X_3 are added to the model already containing X_1. There are two degrees of freedom associated with $SSR(X_2, X_3|X_1)$, as may be seen from the following relation which is based directly on the definitions of the extra sums of squares:

$$(8.14a) \qquad SSR(X_2, X_3|X_1) = SSR(X_2|X_1) + SSR(X_3|X_1, X_2)$$

Thus, extra sums of squares for two or more *X* variables can be obtained from extra sums of squares where one *X* variable is added at a time.

With $SSR(X_2, X_3|X_1)$ we can then make use of the decomposition:

$$(8.15) \qquad SSR(X_1, X_2, X_3) = SSR(X_1) + SSR(X_2, X_3|X_1)$$

It is obvious that the number of possible decompositions becomes vast as the number of independent variables increases. Table 8.7 contains the ANOVA table for one possible decomposition for the case of three independent variables, and Table 8.8 contains two possible decompositions for our earlier body fat example.

Uses of extra sums of squares for tests concerning regression coefficients

One of the major uses of extra sums of squares is for conducting tests concerning regression coefficients without having to fit both the full and reduced models

TABLE 8.7   Example of ANOVA table with decomposition of *SSR* for three independent variables

Source of Variation	SS	df	MS
Regression	$SSR(X_1, X_2, X_3)$	3	$MSR(X_1, X_2, X_3)$
X_1	$SSR(X_1)$	1	$MSR(X_1)$
$X_2 \mid X_1$	$SSR(X_2 \mid X_1)$	1	$MSR(X_2 \mid X_1)$
$X_3 \mid X_1, X_2$	$SSR(X_3 \mid X_1, X_2)$	1	$MSR(X_3 \mid X_1, X_2)$
Error	$SSE(X_1, X_2, X_3)$	$n-4$	$MSE(X_1, X_2, X_3)$
Total	$SSTO$	$n-1$	

TABLE 8.8   ANOVA tables with different decompositions of *SSR*— body fat example

(a)

Source of Variation	SS	df	MS
Regression	$SSR(X_1, X_2) = 385.44$	2	192.72
X_1	$SSR(X_1) = 352.27$	1	352.27
$X_2 \mid X_1$	$SSR(X_2 \mid X_1) = 33.17$	1	33.17
Error	$SSE(X_1, X_2) = 109.95$	17	6.47
Total	$SSTO = 495.39$	19	

(b)

Source of Variation	SS	df	MS
Regression	$SSR(X_1, X_2) = 385.44$	2	192.72
X_2	$SSR(X_2) = 381.97$	1	381.97
$X_1 \mid X_2$	$SSR(X_1 \mid X_2) = 3.47$	1	3.47
Error	$SSE(X_1, X_2) = 109.95$	17	6.47
Total	$SSTO = 495.39$	19	

parately. Many computer packages provide extra sums of squares in any de-
red order when fitting a regression model. Often, the decomposition provided
that corresponding to the order in which the X variables are entered in the
regression fit.

In our ski resort example, for instance, the independent variables were entered
in the order X_1 and X_2 and the computer output yielded the following results:

Source of Variation	SS	df	MS
Regression	1,623,730,176	2	811,865,088
X_1	1,617,707,400	1	1,617,707,400
$X_2 \mid X_1$	6,022,776	1	6,022,776
Error	19,303,908	7	2,757,701

Hence, to test whether or not $\beta_2 = 0$, we need not actually fit the reduced model
since the partial F test statistic (8.6) can be calculated immediately from the
above results:

$$F^* = \frac{SSR(X_2 \mid X_1)}{1} \div \frac{SSE(X_1, X_2)}{n - 3}$$

$$= \frac{6,022,776}{1} \div \frac{19,303,908}{7} = 2.18$$

As we shall see later, judicious choices in obtaining extra sums of squares will
permit a variety of tests on the regression coefficients without requiring extra
computer runs for fitting reduced models.

8.3 COEFFICIENTS OF PARTIAL DETERMINATION

A coefficient of multiple determination R^2, it will be recalled, measures the
proportionate reduction in the variation of Y achieved by the introduction of the
entire set of X variables considered in the model. A *coefficient of partial determination*, in contrast, measures the marginal contribution of one X variable, when
all others are already included in the model.

Two independent variables

Let us consider a first-order multiple regression model with two independent
variables, as given in (7.1):

$$Y_i = \beta_0 + \beta_1 X_{i1} + \beta_2 X_{i2} + \varepsilon_i$$

$SSE(X_2)$ measures the variation in Y when X_2 is included in the model.
$SSE(X_1, X_2)$ measures the variation in Y when both X_1 and X_2 are included in the
model. Hence, the relative marginal reduction in the variation in Y associated
with X_1 when X_2 is already in the model is:

$$\frac{SSE(X_2) - SSE(X_1, X_2)}{SSE(X_2)}$$

This measure is the coefficient of partial determination between Y and X_1, given that X_2 is in the model. We denote this measure by $r_{Y1.2}^2$:

(8.16) $\qquad r_{Y1.2}^2 = \dfrac{SSE(X_2) - SSE(X_1, X_2)}{SSE(X_2)} = 1 - \dfrac{SSE(X_1, X_2)}{SSE(X_2)}$

Using (8.7a), we can express the coefficient of partial determination in terms of an extra sum of squares:

(8.16a) $\qquad\qquad\qquad r_{Y1.2}^2 = \dfrac{SSR(X_1 | X_2)}{SSE(X_2)}$

$r_{Y1.2}^2$ thus measures the proportionate reduction in the variation of Y remaining after X_2 is included in the model which is gained by also including X_1 in the model.

The coefficient of partial determination between Y and X_2, given that X_1 is in the model, is defined:

(8.17) $\qquad\qquad\qquad r_{Y2.1}^2 = \dfrac{SSR(X_2 | X_1)}{SSE(X_1)}$

For our earlier body fat example, these two coefficients of partial determination are (Tables 8.4 and 8.8):

$$r_{Y1.2}^2 = \frac{3.47}{113.42} = .031$$

$$r_{Y2.1}^2 = \frac{33.17}{143.12} = .232$$

Thus, when X_1 here is added to the model containing X_2, the error sum of squares $SSE(X_2)$ is reduced by 3.1 percent. Correspondingly, when X_2 is added to the model containing X_1, the error sum of squares $SSE(X_1)$ is reduced by 23.2 percent.

General case

The generalization of coefficients of partial determination to three or more independent variables in the model is immediate. For instance:

(8.18a) $\qquad\qquad\qquad r_{Y1.23}^2 = \dfrac{SSR(X_1 | X_2, X_3)}{SSE(X_2, X_3)}$

(8.18b) $\qquad\qquad\qquad r_{Y2.13}^2 = \dfrac{SSR(X_2 | X_1, X_3)}{SSE(X_1, X_3)}$

(8.18c) $\qquad\qquad\qquad r_{Y3.12}^2 = \dfrac{SSR(X_3 | X_1, X_2)}{SSE(X_1, X_2)}$

Note that in the subscripts to r^2, the entries to the left of the dot show in turn the variable taken as the response and the variable being added. The entries to the right of the dot show the X variables already in the model.

Comments

1. The coefficients of partial determination can take on values between 0 and 1, as the definitions readily indicate.

2. A coefficient of partial determination can be interpreted as a coefficient of simple determination. Consider a multiple regression model with two independent variables. Suppose we regress Y on X_2 and obtain the residuals:

$$Y_i - \hat{Y}_i(X_2)$$

where $\hat{Y}_i(X_2)$ denotes the fitted values of Y when X_2 is in the model.

Suppose we further regress X_1 on X_2 and obtain the residuals:

$$X_{i1} - \hat{X}_{i1}(X_2)$$

where $\hat{X}_{i1}(X_2)$ denotes the fitted values of X_1 in the regression of X_1 on X_2. The coefficient of simple determination r^2 between these two sets of residuals equals the coefficient of partial determination $r^2_{Y1.2}$. Thus, this coefficient measures the relation between Y and X_1 when both of these have been adjusted for their linear relationships to X_2.

Coefficients of partial correlation

The square root of the coefficient of partial determination is called the *coefficient of partial correlation*. This coefficient is frequently used in practice, although it does not have as clear a meaning as the coefficient of partial determination.

For our body fat example, we have:

$$r_{Y1.2} = \sqrt{.031} = .176$$
$$r_{Y2.1} = \sqrt{.232} = .482$$

Usually the partial correlation coefficient is given the same sign as that of the corresponding regression coefficient in the fitted regression function. For our body fat example, we had (Table 8.4) $b_1 = +.2224$ and $b_2 = +.6594$. Since these are both positive, each of the partial correlation coefficients above is taken as positive.

Partial correlation coefficients are frequently used in computer routines for finding the best independent variable to be selected next for inclusion in the regression model. We shall discuss this use in Chapter 12.

Note

The coefficients of partial determination can be expressed in terms of simple or other partial correlation coefficients. For example:

$$(8.19) \qquad r_{Y2.1}^2 = \frac{(r_{Y2} - r_{12}r_{Y1})^2}{(1 - r_{12}^2)(1 - r_{Y1}^2)}$$

$$(8.20) \qquad r_{Y2.13}^2 = \frac{(r_{Y2.3} - r_{12.3}r_{Y1.3})^2}{(1 - r_{12.3}^2)(1 - r_{Y1.3}^2)}$$

Extensions are straightforward.

8.4 TESTING HYPOTHESES CONCERNING REGRESSION COEFFICIENTS IN MULTIPLE REGRESSION

We have already discussed how to conduct two types of tests concerning the regression coefficients in a multiple regression model. For completeness, we summarize these tests here and then take up some additional types of tests.

Test whether all $\beta_k = 0$

This is the *overall F test* (7.30) of whether or not there is a regression relation between the dependent variable Y and the set of independent variables. The alternatives are:

$$(8.21) \qquad \begin{aligned} H_0&: \beta_1 = \beta_2 = \cdots = \beta_k = 0 \\ H_a&: \text{not all } \beta_k \ (k = 1, \ldots, p-1) \text{ equal } 0 \end{aligned}$$

and the test statistic is:

$$(8.22) \qquad \begin{aligned} F^* &= \frac{SSR(X_1, \ldots, X_{p-1})}{p-1} \div \frac{SSE(X_1, \ldots, X_{p-1})}{n-p} \\ &= \frac{MSR}{MSE} \end{aligned}$$

If H_0 holds, $F^* \sim F(p-1, n-p)$. Large values of F^* lead to conclusion H_a.

Test whether a single $\beta_k = 0$

This is the *partial F test* of whether a particular regression coefficient β_k equals zero. The alternatives are:

$$(8.23) \qquad \begin{aligned} H_0&: \beta_k = 0 \\ H_a&: \beta_k \neq 0 \end{aligned}$$

and the test statistic is:

$$(8.24) \qquad \begin{aligned} F^* &= \frac{SSR(X_k | X_1, \ldots, X_{k-1}, X_{k+1}, \ldots, X_{p-1})}{1} \div \frac{SSE(X_1, \ldots, X_{p-1})}{n-p} \\ &= \frac{MSR(X_k | X_1, \ldots, X_{k-1}, X_{k+1}, \ldots, X_{p-1})}{MSE} \end{aligned}$$

If H_0 holds, $F^* \sim F(1, n-p)$. Large values of F^* lead to conclusion H_a. Computer packages which provide extra sums of squares permit use of this test without having to fit the reduced model.

An equivalent test statistic is, as we have seen, (7.42b):

$$(8.25) \qquad t^* = \frac{b_k}{s(b_k)}$$

If H_0 holds, $t^* \sim t(n - p)$. Large values of $|t^*|$ lead to conclusion H_a.

Since the two tests are equivalent, the choice is usually made in terms of available information provided by the computer package output.

Test whether some regression coefficients equal zero

Sometimes we wish to determine whether or not some regression coefficients equal zero. The approach is that of a general linear test, and no new problems arise. We first illustrate the approach by an example and then state the general test statistic.

Example. We consider the multiple regression application in Table 8.9. The data pertain to 14 different computer simulations conducted in designing the layout for a chemical warehouse. The dependent variable is CPU time (the operating time required by the computer's central processing unit to run the simulation), the first independent variable is number of trials in the simulation, and the second independent variable is number of statements in the computer program. Since it is hypothesized that the effect of number of statements (X_2) on CPU time (Y) may be curvilinear, the model being considered is:

$$(8.26) \qquad Y_i = \beta_0 + \beta_1 X_{i1} + \beta_2 X_{i2} + \beta_3 X_{i2}^2 + \varepsilon_i \qquad \text{Full model}$$

TABLE 8.9 Data for computer simulation example

Simulation i	Number of Trials X_{i1}	Number of Statements in Program X_{i2}	X_{i2}^2	CPU Time (seconds) Y_i
1	550	458	209,764	445.37
2	600	152	23,104	408.88
3	200	635	403,225	264.61
4	50	128	16,384	73.90
5	350	100	10,000	246.07
6	300	550	302,500	312.00
7	200	577	332,929	250.36
8	175	234	54,756	179.95
9	200	500	250,000	243.76
10	200	491	241,081	243.63
11	175	580	336,400	240.72
12	700	135	18,225	462.60
13	800	162	26,244	531.45
14	650	176	30,976	445.48

It is desired that we test whether the number of statements variable can be dropped from the model. Hence, we wish to choose between the alternatives:

$$H_0: \beta_2 = \beta_3 = 0$$
$$H_a: \text{not both } \beta_2 \text{ and } \beta_3 \text{ equal zero}$$

We first fit the full model and obtain (results are shown in Table 8.10a):

$$SSE(F) = SSE(X_1, X_2, X_2^2) = 44.1$$

The model under H_0 is:

(8.27) $$Y_i = \beta_0 + \beta_1 X_{i1} + \varepsilon_i \qquad \text{Reduced model}$$

When we fit the reduced model, we obtain (Table 8.10b):

$$SSE(R) = SSE(X_1) = 17,240.3$$

The general test statistic (3.68):

$$F^* = \frac{SSE(R) - SSE(F)}{(n-2) - (n-4)} \div \frac{SSE(F)}{n-4}$$

can be simplified here because:

$$SSE(R) - SSE(F) = SSE(X_1) - SSE(X_1, X_2, X_2^2) = SSR(X_2, X_2^2 | X_1)$$

Hence, we can write:

(8.28) $$F^* = \frac{SSR(X_2, X_2^2 | X_1)}{2} \div \frac{SSE(F)}{n-4}$$

TABLE 8.10 Computer results for computer simulation example

(a) Regression of Y on X_1, X_2, and X_2^2
$$\hat{Y} = 7.028 + .595 X_1 + .323 X_2 - .000173 X_2^2$$

Source of Variation	SS	df	MS
Regression	$SSR(X_1, X_2, X_2^2) = 214{,}691.2$	3	71,563.7
Error	$SSE(X_1, X_2, X_2^2) = \quad\ \ 44.1$	10	4.41
Total	$SSTO = 214{,}735.3$	13	

(b) Regression of Y on X_1
$$\hat{Y} = 122.71 + .511 X_1$$

Source of Variation	SS	df	MS
Regression	$SSR(X_1) = 197{,}495.1$	1	197,495.1
Error	$SSE(X_1) = \quad 17{,}240.3$	12	1,436.7
Total	$SSTO = 214{,}735.4$	13	

For our example, we have (Table 8.10):

$$SSR(X_2, X_2^2|X_1) = 17,240.3 - 44.1 = 17,196.2$$

$$F^* = \frac{17,196.2}{2} \div \frac{44.1}{10} = 1,950$$

Suppose the α risk is to be .01. We require $F(.99; 2, 10) = 7.56$. Since $F^* = 1,950 > 7.56$, we conclude H_a, that the number of statements variable should not be dropped from the model.

Actually, we did not need to fit the reduced model separately to conduct the test; we used this approach only to explain the logic of the test. The computer run for fitting the full model provided the following decomposition of SSR:

Component	SS
X_1	197,495
$X_2|X_1$	17,053
$X_2^2|X_1, X_2$	144

Since, by (8.14a):

$$SSR(X_2, X_2^2|X_1) = SSR(X_2|X_1) + SSR(X_2^2|X_1, X_2)$$

we can obtain directly:

$$SSR(X_2, X_2^2|X_1) = 17,053 + 144 = 17,197$$

The difference in the final digit is due to rounding effects.

In our example, the regression model contained both X_2 and X_2^2 terms. Computational difficulties can at times arise in such cases when obtaining the least squares regression coefficients. Chapter 9 discusses polynomial regression models and how to avoid these computational difficulties.

Test statistic. For the general multiple regression model:

$$(8\ 29) \qquad Y_i = \beta_0 + \beta_1 X_{i1} + \cdots + \beta_{p-1} X_{i,p-1} + \varepsilon_i \qquad \text{Full model}$$

we wish to test:

$$(8.30) \qquad \begin{aligned} &H_0: \beta_q = \beta_{q+1} = \cdots = \beta_{p-1} = 0 \\ &H_a: \text{not all of the } \beta\text{'s in } H_0 \text{ equal zero} \end{aligned}$$

where for convenience, we arrange the model so that the last $p - q$ coefficients are the ones to be tested. We first fit the full model and thereby obtain $SSE(X_1, \ldots, X_{p-1})$. Then we fit the reduced model:

$$(8.31) \qquad Y_i = \beta_0 + \beta_1 X_{i1} + \cdots + \beta_{q-1} X_{i,q-1} + \varepsilon_i \qquad \text{Reduced model}$$

and obtain $SSE(X_1, \ldots, X_{q-1})$. Finally, we set up the general linear test statistic (3.68), which here is:

$$(8.32)\ F^* = \frac{SSE(X_1, \ldots, X_{q-1}) - SSE(X_1, \ldots, X_{p-1})}{(n - q) - (n - p)} \div \frac{SSE(X_1, \ldots, X_{p-1})}{n - p}$$

or equivalently:

$$(8.32a) \qquad F* = \frac{SSR(X_q, \ldots, X_{p-1} | X_1, \ldots, X_{q-1})}{p - q} \div MSE(F)$$

Note that test statistic (8.32a) actually encompasses the two earlier cases. If $q = 1$, the test is whether all regression coefficients equal zero. If $q = p - 1$, the test is whether a single regression coefficient equals zero. Also note that test statistic (8.32a) can be calculated without having to fit the reduced model if the computer program provides the needed extra sums of squares:

$$(8.33) \qquad SSR(X_q, \ldots, X_{p-1} | X_1, \ldots, X_{q-1}) = SSR(X_q | X_1, \ldots, X_{q-1})$$
$$+ \cdots + SSR(X_{p-1} | X_1, \ldots, X_{p-2})$$

Other tests

Other types of tests are occasionally required. For instance, for the full model containing three X variables:

$$(8.34) \qquad Y_i = \beta_0 + \beta_1 X_{i1} + \beta_2 X_{i2} + \beta_3 X_{i3} + \varepsilon_i \qquad \text{Full model}$$

we might wish to test:

$$(8.35) \qquad\qquad \begin{aligned} H_0 &: \beta_1 = \beta_2 \\ H_a &: \beta_1 \neq \beta_2 \end{aligned}$$

The procedure would be to fit the full model (8.34), and then the reduced model:

$$(8.36) \qquad Y_i = \beta_0 + \beta_c(X_{i1} + X_{i2}) + \beta_3 X_{i3} + \varepsilon_i \qquad \text{Reduced model}$$

where β_c denotes the common coefficient for β_1 and β_2 and $X_{i1} + X_{i2}$ is the corresponding new independent variable. We then use the general $F*$ test statistic (3.68) with 1 and $n - 4$ degrees of freedom. Since this test does not involve alternatives where one or more regression coefficients equal zero, extra sums of squares are not applicable and the reduced model must be fitted for the test.

8.5 MATRIX FORMULATION OF GENERAL LINEAR TEST

The general linear test is based on the test statistic (3.68):

$$(8.37) \qquad\qquad F* = \frac{SSE(R) - SSE(F)}{df_R - df_F} \div \frac{SSE(F)}{df_F}$$

which, when H_0 holds, follows the F distribution with $df_R - df_F$ degrees of freedom for the numerator and df_F degrees of freedom for the denominator. We now explain how to represent this test statistic in matrix form.

Full model

The full regression model with $p - 1$ predictor variables is given by (7.18):

$$(8.38) \qquad \mathbf{Y} = \mathbf{X\beta} + \mathbf{\epsilon}$$

The least squares estimators for the full model will now be denoted by $\mathbf{b}_F$ and are, as before, given by (7.21):

$$(8.39) \qquad \mathbf{b}_F = (\mathbf{X'X})^{-1}\mathbf{X'Y}$$

Also, the error sum of squares is given by (7.27):

$$(8.40) \qquad SSE(F) = (\mathbf{Y} - \mathbf{Xb}_F)'(\mathbf{Y} - \mathbf{Xb}_F) = \mathbf{Y'Y} - \mathbf{b}'_F\mathbf{X'Y}$$

which has associated with it $df_F = n - p$ degrees of freedom.

Statement of hypothesis H_0

A linear test hypothesis H_0 is represented in matrix form as follows:

$$(8.41) \qquad H_0: \underset{s \times p}{\mathbf{C}} \ \underset{p \times 1}{\mathbf{\beta}} = \underset{s \times 1}{\mathbf{h}}$$

where $\mathbf{C}$ is a specified $s \times p$ matrix of rank s and $\mathbf{h}$ is a specified $s \times 1$ vector.

Example 1. The regression model contains two X variables, and we wish to test $H_0: \beta_1 = 2$. Then:

$$\underset{1 \times 3}{\mathbf{C}} = [0 \quad 1 \quad 0] \qquad \underset{1 \times 1}{\mathbf{h}} = [2]$$

and we have:

$$H_0: \mathbf{C\beta} = [0 \quad 1 \quad 0] \begin{bmatrix} \beta_0 \\ \beta_1 \\ \beta_2 \end{bmatrix} = [2]$$

or $H_0: \beta_1 = 2$.

Example 2. The regression model contains two X variables, and we wish to test $H_0: \beta_1 = \beta_2 = 0$. Then:

$$\underset{2 \times 3}{\mathbf{C}} = \begin{bmatrix} 0 & 1 & 0 \\ 0 & 0 & 1 \end{bmatrix} \qquad \underset{2 \times 1}{\mathbf{h}} = \begin{bmatrix} 0 \\ 0 \end{bmatrix}$$

and we have:

$$H_0: \mathbf{C\beta} = \begin{bmatrix} 0 & 1 & 0 \\ 0 & 0 & 1 \end{bmatrix} \begin{bmatrix} \beta_0 \\ \beta_1 \\ \beta_2 \end{bmatrix} = \begin{bmatrix} 0 \\ 0 \end{bmatrix}$$

or $H_0: \beta_1 = 0, \beta_2 = 0$.

Example 3.   The regression model contains three X variables, and we wish to test $H_0: \beta_1 = \beta_2$. Then:

$$\mathbf{C}_{\;1 \times 4} = [0 \quad 1 \quad -1 \quad 0] \qquad \mathbf{h}_{\;1 \times 1} = [0]$$

and we have:

$$H_0: \mathbf{C\beta} = [0 \quad 1 \quad -1 \quad 0] \begin{bmatrix} \beta_0 \\ \beta_1 \\ \beta_2 \\ \beta_3 \end{bmatrix} = [0]$$

or $H_0: \beta_1 - \beta_2 = 0$.

Reduced model

The reduced model is:

$$(8.42) \qquad\qquad \mathbf{Y = X\beta + \varepsilon} \qquad \text{where } \mathbf{C\beta = h}$$

It can be shown that the least squares estimators under the reduced model, to be denoted by $\mathbf{b}_R$, are:

$$(8.43) \qquad \mathbf{b}_R = \mathbf{b}_F - (\mathbf{X'X})^{-1}\mathbf{C'}(\mathbf{C(X'X)}^{-1}\mathbf{C'})^{-1}(\mathbf{Cb}_F - \mathbf{h})$$

and the error sum of squares is:

$$(8.44) \qquad\qquad SSE(R) = (\mathbf{Y - Xb}_R)'(\mathbf{Y - Xb}_R)$$

which has associated with it $df_R = n - (p - s)$ degrees of freedom.

Test statistic

It can be shown that the difference $SSE(R) - SSE(F)$ can be expressed as follows:

$$(8.45) \quad SSE(R) - SSE(F) = (\mathbf{Cb}_F - \mathbf{h})'(\mathbf{C(X'X)}^{-1}\mathbf{C'})^{-1}(\mathbf{Cb}_F - \mathbf{h})$$

which has associated with it $df_R - df_F = (n - p + s) - (n - p) = s$ degrees of freedom.

Hence, the test statistic is:

$$(8.46) \qquad\qquad F^* = \frac{SSE(R) - SSE(F)}{s} \div \frac{SSE(F)}{n - p}$$

where $SSE(R) - SSE(F)$ is given by (8.45) and $SSE(F)$ is given by (8.40).

To confirm that the degrees of freedom associated with $SSE(R) - SSE(F)$ are s, consider the earlier three examples.

1. In Example 1, $s = 1$. This is consistent with the numerator degrees of freedom in test statistic (8.24).

2. In Example 2, $s = 2$. This is consistent with the numerator degrees of freedom in test statistic (8.22).
3. In Example 3, $s = 1$. This is consistent with the numerator degrees of freedom in the test statistic for the example on page 293.

Note

The least squares estimators $\mathbf{b}_R$ under the reduced model, given in (8.43), can be derived by minimizing the least squares criterion $Q = (\mathbf{Y} - \mathbf{X}\boldsymbol{\beta})'(\mathbf{Y} - \mathbf{X}\boldsymbol{\beta})$ subject to the constraint $\mathbf{C}\boldsymbol{\beta} - \mathbf{h} = \mathbf{0}$, using Lagrangian multipliers.

PROBLEMS

8.1. A speaker stated in a workshop on applied regression analysis: "In business and the social sciences, some degree of multicollinearity in survey data is practically inevitable." Does this statement apply equally to experimental data?

8.2. Refer to the Zarthan Company example on page 247. The company's sales manager has suggested that the predictive ability of the model could be greatly improved if promotional expenditures were added to the model, since these expenditures are known to have a substantial impact on sales. The company allocates its total promotional budget proportionately to the target population in the districts. Thus, a district containing 4.7 percent of the total target population receives 4.7 percent of the total promotional budget. Evaluate the sales manager's suggestion.

8.3. Refer to **Brand preference** Problem 7.8.
  a. Fit the first-order simple linear regression model (3.1) for relating brand liking (Y) to moisture content (X_1). State the fitted regression function.
  b. Compare the estimated regression coefficient for moisture content obtained in part (a) with the corresponding coefficient obtained in Problem 7.8a. What do you find?
  c. Does $SSR(X_1)$ equal $SSR(X_1 \mid X_2)$ here? If not, is the difference substantial?
  d. Calculate the coefficient of simple correlation between X_1 and X_2. What bearing does this have on your findings in parts (b) and (c)?

8.4. Refer to **Chemical shipment** Problem 7.12.
  a. Fit the first-order simple linear regression model (3.1) for relating number of minutes required to handle shipment (Y) to total weight of shipment (X_2). State the fitted regression function.
  b. Compare the estimated regression coefficient for total weight of shipment obtained in part (a) with the corresponding coefficient obtained in Problem 7.12a. What do you find?
  c. Does $SSR(X_2)$ equal $SSR(X_2 \mid X_1)$ here? If not, is the difference substantial?
  d. Calculate the coefficient of simple correlation between X_1 and X_2. What bearing does this have on your findings in parts (b) and (c)?

8.5. Refer to **Patient satisfaction** Problem 7.17.

    a. Fit the first-order linear regression model (7.1) for relating patient satisfaction (Y) to patient's age (X_1) and severity of illness (X_2). State the fitted regression function.

    b. Compare the estimated regression coefficients for patient's age and severity of illness obtained in part (a) with the corresponding coefficients obtained in Problem 7.17a. What do you find?

    c. Does $SSR(X_1)$ equal $SSR(X_1 | X_3)$ here? Does $SSR(X_2)$ equal $SSR(X_2 | X_3)$?

    d. Calculate the coefficients of simple correlation between pairs of X_1, X_2, and X_3. What bearing do these have on your findings in parts (b) and (c)?

8.6. Refer to **Chemical shipment** Problem 7.12. Does $SSR(X_1) + SSR(X_2 | X_1)$ equal $SSR(X_2) + SSR(X_1 | X_2)$ here? Must this always be the case?

8.7. Refer to **Brand preference** Problem 7.8. Test whether X_2 can be dropped from the regression model given that X_1 is retained. Use the F^* test statistic and level of significance .01. State the alternatives, decision rule, and conclusion.

8.8. Refer to **Chemical shipment** Problem 7.12. Test whether X_1 can be dropped from the regression model given that X_2 is retained. Use the F^* test statistic and $\alpha = .05$. State the alternatives, decision rule, and conclusion.

8.9. Refer to **Patient satisfaction** Problem 7.17. Test whether X_3 can be dropped from the regression model given that X_1 and X_2 are retained. Use the F^* test statistic and level of significance .025. State the alternatives, decision rule, and conclusion.

8.10. Refer to the work crew productivity example on page 272.

    a. Calculate r_{Y1}^2, r_{Y2}^2, r_{12}^2, $r_{Y1.2}^2$, $r_{Y2.1}^2$, and R^2. Explain what each coefficient measures and interpret your results.

    b. Are any of the results obtained in part (a) special because the two independent variables are uncorrelated?

    c. Obtain the standardized regression coefficients. How do you interpret these coefficients? Do they have a special meaning here because the independent variables are uncorrelated? (*Hint:* Obtain r_{Y1}, r_{Y2}.)

8.11. Refer to **Brand preference** Problem 7.8. Calculate r_{Y1}^2, r_{Y2}^2, r_{12}^2, $r_{Y1.2}^2$, $r_{Y2.1}^2$, and R^2. Explain what each coefficient measures and interpret your results.

8.12. Refer to **Chemical shipment** Problem 7.12. Calculate r_{Y1}^2, r_{Y2}^2, r_{12}^2, $r_{Y1.2}^2$, $r_{Y2.1}^2$, and R^2. Explain what each coefficient measures and interpret your results.

8.13. Refer to **Patient satisfaction** Problem 7.17.

    a. Calculate r_{Y1}^2, $r_{Y1.2}^2$, and $r_{Y1.23}^2$. How is the degree of linear association between Y and X_1 affected as other independent variables have already been included in the model?

    b. Make a similar analysis as in part (a) for the degree of linear association between Y and X_2. Are your findings similar to those in part (a) for Y and X_1?

8.14. Refer to the computer simulation example on page 290. An observer states that both the number of trials variable (X_1) and the second-order term for the number of statements variable (X_2^2) can be dropped from model (8.26). Conduct the appropriate test using a level of significance of .01. State the alternatives, decision rule, and conclusion.

8.15. Refer to **Patient satisfaction** Problem 7.17. Test whether both X_2 and X_3 can be dropped from the regression model given that X_1 is retained. Use $\alpha = .025$. State the alternatives, decision rule, and conclusion.

8.16. Refer to **Mathematicians salaries** Problem 7.20. Test whether both X_1 and X_3 can be dropped from the regression model given that X_2 is retained; use $\alpha = .01$. State the alternatives, decision rule, and conclusion.

EXERCISES

8.17. a. Define each of the following extra sums of squares: (1) $SSR(X_5 \mid X_1)$; (2) $SSR(X_3, X_4 \mid X_1)$; (3) $SSR(X_4 \mid X_1, X_2, X_3)$.

   b. For a multiple regression model with five X variables, what is the relevant extra sum of squares for testing whether or not $\beta_5 = 0$? Whether or not $\beta_2 = \beta_4 = 0$?

8.18. Show that:

   a. $SSR(X_1, X_2, X_3, X_4) = SSR(X_1) + SSR(X_2, X_3 \mid X_1) + SSR(X_4 \mid X_1, X_2, X_3)$

   b. $SSR(X_1, X_2, X_3, X_4) = SSR(X_2, X_3) + SSR(X_1 \mid X_2, X_3)$
   $+ SSR(X_4 \mid X_1, X_2, X_3)$

8.19. Refer to **Brand preference** Problem 7.8.

   a. Regress Y on X_2 using the simple linear regression model (3.1) and obtain the residuals.

   b. Regress X_1 on X_2 using the simple linear regression model (3.1) and obtain the residuals.

   c. Calculate the coefficient of simple correlation between the two sets of residuals and show that it equals $r_{Y1.2}$.

8.20. An undergraduate working for a campus apparel shop serving student customers studied the relation between monthly allowance received by customer (X_1), number of years customer is in college (X_2), and dollar sales to customer to date (Y). The predictive model considered was:

$$Y_i = \beta_0 + \beta_1 X_{i1} + \beta_2 X_{i2} + \beta_3 X_{i1}^2 + \varepsilon_i$$

State the reduced models for testing whether or not: (1) $\beta_1 = \beta_3 = 0$, (2) $\beta_0 = 0$, (3) $\beta_3 = 5$, (4) $\beta_0 = 10$, (5) $\beta_1 = \beta_2$.

8.21. Refer to Exercise 8.20. For each of the cases, state the hypothesis H_0 using the matrix formulation (8.41).

8.22. The following regression model is being considered in a water resources study:

$$Y_i = \beta_0 + \beta_1 X_{i1} + \beta_2 X_{i2} + \beta_3 X_{i1} X_{i2} + \beta_4 \sqrt{X_{i3}} + \varepsilon_i$$

State the reduced models for testing whether or not: (1) $\beta_3 = \beta_4 = 0$, (2) $\beta_3 = 0$, (3) $\beta_1 = \beta_2 = 5$, (4) $\beta_4 = 7$.

8.23. Refer to Exercise 8.22. For each of the cases, state the hypothesis H_0 using the matrix formulation (8.41).

8.24. (Calculus needed.) Derive the least squares estimator under the reduced model (8.43), where $\mathbf{C\beta} = \mathbf{h}$. [*Hint:* The Lagrangian function is:

$$L = (\mathbf{Y} - \mathbf{X\beta})'(\mathbf{Y} - \mathbf{X\beta}) + \boldsymbol{\lambda}'(\mathbf{C\beta} - \mathbf{h}), \text{ where } \boldsymbol{\lambda}' = (\lambda_1, \ldots, \lambda_s).]$$

8.25. Derive (8.45). [*Hint:* Show that $SSE(R) - SSE(F) = (\mathbf{b}_F - \mathbf{b}_R)'\mathbf{X}'\mathbf{X}(\mathbf{b}_F - \mathbf{b}_R)$ and obtain an expression for $\mathbf{b}_F - \mathbf{b}_R$ from (8.43).]

PROJECTS

8.26. Refer to the **SMSA** data set. For predicting the number of active physicians (Y) in an SMSA, it has been decided to include total population (X_1) and total personal income (X_2) as independent variables. The question now is whether an additional independent variable would be helpful in the model, and if so, which variable would be most helpful. Assume that a first-order multiple regression model is appropriate.

    a. For each of the following variables, calculate the coefficient of partial determination given that X_1 and X_2 are included in the model: land area (X_3), percent of population 65 or older (X_4), number of hospital beds (X_5), and total serious crimes (X_6).

    b. On the basis of the results in part (a), which of the four additional independent variables is best? Is the extra sum of squares associated with this variable larger than those for the other three variables?

    c. Using the F^* test statistic, test whether or not the variable determined to be best in part (b) is helpful in the regression model when X_1 and X_2 are included in the model; use $\alpha = .01$. State the alternatives, decision rule, and conclusion. Would the F^* test statistics for the other three potential independent variables be as large as the one here? Discuss.

8.27. Refer to the **SENIC** data set. For predicting the average length of stay of patients in a hospital (Y), it has been decided to include age (X_1) and infection risk (X_2) as independent variables. The question now is whether an additional independent variable would be helpful in the model, and if so, which variable would be most helpful. Assume that a first-order multiple regression model is appropriate.

    a. For each of the following variables, calculate the coefficient of partial determination given that X_1 and X_2 are included in the model: routine culturing ratio (X_3), average daily census (X_4), number of nurses (X_5), and available facilities and services (X_6).

    b. On the basis of the results in part (a), which of the four additional independent variables is best? Is the extra sum of squares associated with this variable larger than those for the other three variables?

    c. Using the F^* test statistic, test whether or not the variable determined to be best in part (b) is helpful in the regression model when X_1 and X_2 are included in the model; use $\alpha = .05$. State the alternatives, decision rule, and conclusion. Would the F^* test statistics for the other three potential independent variables be as large as the one here? Discuss.

8.28. Refer to Exercise 7.31. It is desired to test whether or not $\beta_1 = \beta_2$. Using matrix methods, obtain $SSE(R) - SSE(F)$ according to (8.45).

9

Polynomial regression

In this chapter, we consider one important type of curvilinear response model, namely, the polynomial regression model. This is the most frequently used curvilinear response model in practice, because of its ease in handling as a special case of the general linear regression model (7.18). First, we discuss some commonly used polynomial regression models. Then we present two cases to illustrate some of the major problems encountered with polynomial regression.

9.1 POLYNOMIAL REGRESSION MODELS

Polynomial regression models can contain one, two, or more than two independent variables. Further, the independent variable can be present in various powers. We illustrate now some major possibilities.

One independent variable—second order

The model:

(9.1)
$$Y_i = \beta_0 + \beta_1 x_i + \beta_2 x_i^2 + \varepsilon_i$$

where:

$$x_i = X_i - \overline{X}$$

is called a *second-order model with one independent variable* because the single independent variable appears to the first and second powers. Note that the independent variable is expressed as a deviation around its mean $\bar{X}$, and that the ith observation deviation is denoted by x_i. The reason for using deviations around the mean in polynomial regression models is that X, X^2, and higher-power terms often will be highly correlated. This can cause serious computational difficulties when the $\mathbf{X'X}$ matrix is inverted for estimating the regression coefficients. Expressing the independent variable as a deviation from its mean reduces the multicollinearity substantially and tends to avoid computational difficulties.

The regression coefficients in polynomial regression are frequently written in a slightly different fashion, to reflect the pattern of the exponents:

$$(9.1a) \qquad Y_i = \beta_0 + \beta_1 x_i + \beta_{11} x_i^2 + \varepsilon_i$$

We shall employ this latter notation in this chapter.

The response function for model (9.1a) is:

$$(9.2) \qquad E(Y) = \beta_0 + \beta_1 x + \beta_{11} x^2$$

which is a parabola and is frequently called a *quadratic* response function. Figure 9.1 contains two examples of second-order polynomial response functions.

The regression coefficient β_0 represents the mean response of Y when $x = 0$, i.e., when $X = \bar{X}$. The regression coefficient β_1 is often called the *linear effect coefficient* while β_{11} is called the *quadratic effect coefficient*.

FIGURE 9.1 Examples of second-order polynomial response functions

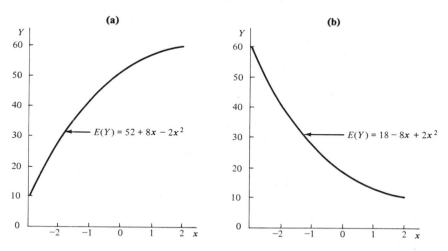

Uses of second-order model. The second-order polynomial response function (9.2) has two basic types of uses:

1. When the true response function is indeed a second-degree polynomial, containing additive linear and quadratic effect components.

2. When the true response function is unknown (or complex) but a second-order polynomial is a good approximation to the true function.

The second type of use is the more common one, but it entails a special danger, that of extrapolation. Consider again Figure 9.1a. This response function may fit the data at hand very well. If, however, information about $E(Y)$ is sought for a larger value of x, extrapolation of this response function leads to the result shown in Figure 9.2, namely, a turning down of the response function, which may not be in accord with reality. Polynomial regressions of all types, especially those of higher order, share this danger of extrapolation. They may provide good fits for the data at hand, but may turn in unexpected directions when extrapolated beyond the range of the data.

FIGURE 9.2 Extrapolation of second-order polynomial response function in Figure 9.1a

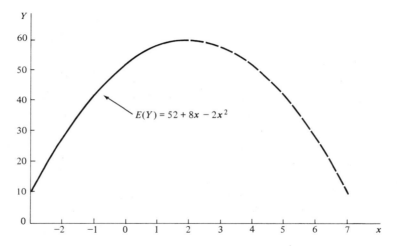

One independent variable—third order

The model:

(9.3) $$Y_i = \beta_0 + \beta_1 x_i + \beta_{11} x_i^2 + \beta_{111} x_i^3 + \varepsilon_i$$

where:

$$x_i = X_i - \bar{X}$$

is a *third-order model with one independent variable*. The response function for model (9.3) is:

(9.4) $$E(Y) = \beta_0 + \beta_1 x + \beta_{11} x^2 + \beta_{111} x^3$$

Figure 9.3 contains two examples of third-order polynomial response functions.

FIGURE 9.3 Examples of third-order polynomial response functions

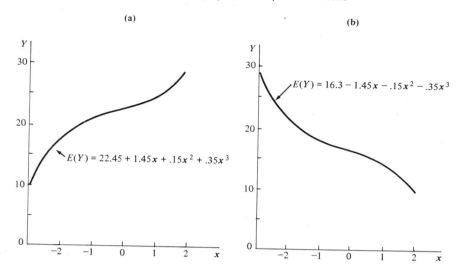

(a)

(b)

$E(Y) = 22.45 + 1.45x + .15x^2 + .35x^3$

$E(Y) = 16.3 - 1.45x - .15x^2 - .35x^3$

One independent variable—higher orders

Polynomial models with the independent variable present in higher powers than the third are not often employed. The interpretation of the coefficients becomes difficult for such models, and they may be highly erratic for even small extrapolations. It must be recognized in this connection that a polynomial model of sufficiently high order can always be found to fit the data perfectly. For instance, the fitted polynomial regression function for one independent variable of order $n - 1$ will pass through all n observed Y values. One needs to be wary therefore of using high order polynomials for the sole purpose of obtaining a good fit. Such regression functions may not show clearly the basic elements of the regression relation between X and Y and may lead to erratic extrapolations.

Two independent variables—second order

The model:

$$(9.5) \qquad Y_i = \beta_0 + \beta_1 x_{i1} + \beta_2 x_{i2} + \beta_{11} x_{i1}^2 + \beta_{22} x_{i2}^2 + \beta_{12} x_{i1} x_{i2} + \varepsilon_i$$

where:

$$x_{i1} = X_{i1} - \bar{X}_1$$
$$x_{i2} = X_{i2} - \bar{X}_2$$

is a *second-order model with two independent variables*. The response surface is:

$$(9.6) \qquad E(Y) = \beta_0 + \beta_1 x_1 + \beta_2 x_2 + \beta_{11} x_1^2 + \beta_{22} x_2^2 + \beta_{12} x_1 x_2$$

which is the equation of a conic section. Note that model (9.5) contains separate linear and quadratic components for each of the two independent variables and a cross-product term. The latter represents the interaction effects between x_1 and x_2, as we noted in Chapter 7. The coefficient β_{12} is often called the *interaction effect coefficient.*

The second-order response surface for two independent variables in (9.6) represents the two basic types of surfaces illustrated in Figure 7.3. Stationary and rising ridges constitute limiting cases of these two basic types of response surfaces.

Usually, it is easiest to portray the second-order response surface (9.6) in terms of contour lines. Figure 9.4 contains a representation of the response function in terms of contour curves:

$$(9.7) \qquad\qquad E(Y) = 1,740 - 4x_1^2 - 3x_2^2 - 3x_1x_2$$

Note that this response surface has a maximum at $x_1 = 0$ and $x_2 = 0$.

FIGURE 9.4 Example of a quadratic response surface:
$E(Y) = 1,740 - 4x_1^2 - 3x_2^2 - 3x_1x_2$

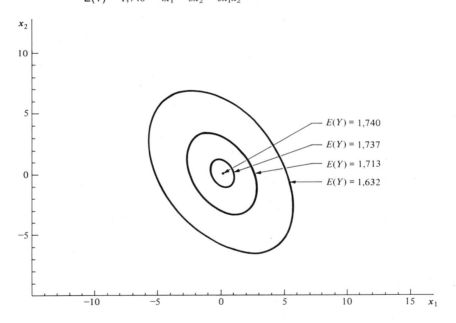

Polynomial models in two (or more) independent variables are well adapted to situations where the response function is unknown and a suitable model is to be developed empirically.

Note

The cross-product term $\beta_{12}x_1x_2$ in (9.6) is considered to be a second-order term, the same as $\beta_{11}x_1^2$ or $\beta_{22}x_2^2$. The reason can be seen readily by writing the latter terms as $\beta_{11}x_1x_1$ and $\beta_{22}x_2x_2$, respectively.

Three independent variables—second order

The *second-order model with three independent variables* is:

$$(9.8) \quad Y_i = \beta_0 + \beta_1 x_{i1} + \beta_2 x_{i2} + \beta_3 x_{i3} + \beta_{11} x_{i1}^2 + \beta_{22} x_{i2}^2 + \beta_{33} x_{i3}^2$$
$$+ \beta_{12} x_{i1} x_{i2} + \beta_{13} x_{i1} x_{i3} + \beta_{23} x_{i2} x_{i3} + \varepsilon_i$$

where:

$$x_{i1} = X_{i1} - \bar{X}_1$$
$$x_{i2} = X_{i2} - \bar{X}_2$$
$$x_{i3} = X_{i3} - \bar{X}_3$$

The response surface for this model is:

$$(9.9) \quad E(Y) = \beta_0 + \beta_1 x_1 + \beta_2 x_2 + \beta_3 x_3 + \beta_{11} x_1^2 + \beta_{22} x_2^2 + \beta_{33} x_3^2$$
$$+ \beta_{12} x_1 x_2 + \beta_{13} x_1 x_3 + \beta_{23} x_2 x_3$$

The coefficients β_{12}, β_{13}, and β_{23} are interaction effects coefficients for interactions between pairs of independent variables.

Use of polynomial regression models

Fitting of polynomial regression models presents no new problems since, as we have seen in Chapter 7, they are special cases of the general linear regression model (7.18). Hence, all earlier results on fitting apply, as do the earlier results on making inferences.

When using a polynomial regression model as an approximation to the true regression function, one will often fit a second-order or third-order model and then explore whether a lower-order model is adequate. For instance, with one independent variable, the model:

$$Y_i = \beta_0 + \beta_1 x_i + \beta_{11} x_i^2 + \beta_{111} x_i^3 + \varepsilon_i$$

may be fitted with the hope that the cubic term and perhaps even the quadratic term can be dropped. Thus, one would wish to test whether or not $\beta_{111} = 0$, or whether or not both $\beta_{11} = 0$ and $\beta_{111} = 0$. Similar tests would often be conducted with polynomial regression models for two or more independent variables.

9.2   EXAMPLE 1—ONE INDEPENDENT VARIABLE

We illustrate now some of the major types of analyses usually conducted with polynomial regression models with one independent variable.

Setting

A staff analyst for a cafeteria chain wishes to investigate the relation between the number of self-service coffee dispensers in a cafeteria line and sales of coffee. Fourteen cafeterias that are similar in such respects as volume of business, type of clientele, and location are chosen for the experiment. The number of self-service dispensers that are placed in the test cafeterias varies from zero (coffee is dispensed here by a line attendant) to six and is assigned randomly to each cafeteria.

Table 9.1 contains the results of the experimental study. Sales are measured in hundreds of gallons of coffee sold.

TABLE 9.1 Data for cafeteria coffee sales example

Cafeteria *i*	Number of Dispensers X_i	Coffee Sales (hundred gallons) Y_i
1	0	508.1
2	0	498.4
3	1	568.2
4	1	577.3
5	2	651.7
6	2	657.0
7	3	713.4
8	3	697.5
9	4	755.3
10	4	758.9
11	5	787.6
12	5	792.1
13	6	841.4
14	6	831.8

Fitting of model

The analyst believes that the relation between sales and number of self-service dispensers is quadratic in the range of observations; sales should increase as the number of dispensers is greater, but if the space is cluttered with dispensers, this increase becomes retarded. Hence, she would like to fit the quadratic model:

$$(9.10) \qquad Y_i = \beta_0 + \beta_1 x_i + \beta_{11} x_i^2 + \varepsilon_i$$

where:

$$x_i = X_i - \bar{X}$$

She further anticipates that the error terms ε_i will be fairly normally distributed with constant variance.

The **Y** and **X** matrices for this application are given in Table 9.2. Note that the **X** matrix contains a column of 1's, a column of the independent variable observations x (expressed as deviations around their mean $\bar{X} = 3$), and a column of the x^2 values. From this point on, the calculations are routine. We could do the matrix calculations manually, as illustrated in Chapter 7, or use a computer multiple regression program. Since no new problems are encountered, we simply present the basic computer output in Table 9.3, including needed extra sums of squares and the $s^2(\mathbf{b})$ matrix.

TABLE 9.2 Data matrices for cafeteria coffee sales example

$$
Y = \begin{bmatrix} 508.1 \\ 498.4 \\ 568.2 \\ 577.3 \\ 651.7 \\ 657.0 \\ 713.4 \\ 697.5 \\ 755.3 \\ 758.9 \\ 787.6 \\ 792.1 \\ 841.4 \\ 831.8 \end{bmatrix}
\qquad
X = \begin{bmatrix}
 & x & x^2 \\
1 & -3 & 9 \\
1 & -3 & 9 \\
1 & -2 & 4 \\
1 & -2 & 4 \\
1 & -1 & 1 \\
1 & -1 & 1 \\
1 & 0 & 0 \\
1 & 0 & 0 \\
1 & 1 & 1 \\
1 & 1 & 1 \\
1 & 2 & 4 \\
1 & 2 & 4 \\
1 & 3 & 9 \\
1 & 3 & 9
\end{bmatrix}
$$

TABLE 9.3 Regression results for cafeteria coffee sales example

(a) Regression Coefficients

Regression Coefficient	Estimated Regression Coefficient	Estimated Standard Deviation	t^*
β_0	705.474	3.208	219.91
β_1	54.893	1.050	52.28
β_{11}	-4.249	.606	-7.01

(b) Analysis of Variance

Source of Variation	SS	df	MS
Regression	171,773	2	85,887
x	168,741	1	168,741
$x^2 \mid x$	3,033	1	3,033
Error	679	11	61.7
Total	172,453	13	

(c) $s^2(b)$ Matrix

$$
\begin{bmatrix}
10.2912 & 0 & -1.4702 \\
0 & 1.1026 & 0 \\
-1.4702 & 0 & .3675
\end{bmatrix}
$$

The fitted regression function is:

(9.11) $$\hat{Y} = 705.47 + 54.89x - 4.25x^2$$

This response function is plotted in Figure 9.5, together with the original data. We show the horizontal scale expressed at the bottom in the deviation units x and at the top in the original units X.

FIGURE 9.5 Fitted second-order polynomial regression—cafeteria coffee sales example

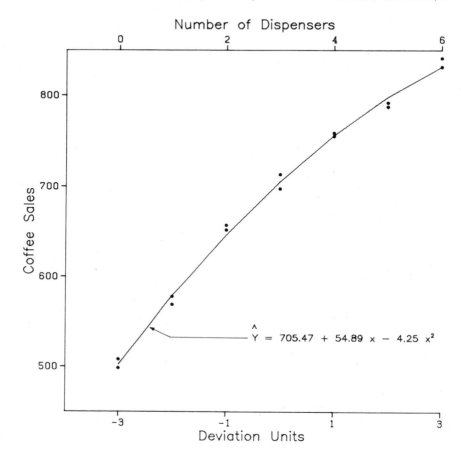

Algebraic version of normal equations. The algebraic version of the least squares normal equations:

$$\mathbf{X'Xb} = \mathbf{X'Y}$$

for the second-order polynomial model (9.10) can be readily obtained from (7.65) by replacing X_{i1} by x_i and X_{i2} by x_i^2. Since $\Sigma x_i = 0$, this yields the normal equations:

(9.12)
$$\Sigma Y_i = nb_0 + b_{11}\Sigma x_i^2$$
$$\Sigma x_i Y_i = b_1 \Sigma x_i^2 + b_{11}\Sigma x_i^3$$
$$\Sigma x_i^2 Y_i = b_0 \Sigma x_i^2 + b_1 \Sigma x_i^3 + b_{11}\Sigma x_i^4$$

Analysis of aptness of model

Residual analysis. To study the aptness of model (9.10) for her data, the analyst plotted the residuals e_i against the fitted values, as shown in Figure 9.6a, and also against the independent variable x_i expressed in deviation units, as shown in Figure 9.6b. We do not present the calculations of the e_i, as these are routine.

FIGURE 9.6 Residual plots for cafeteria coffee sales example

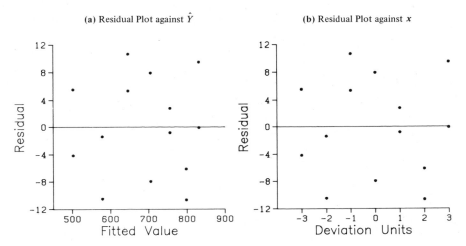

There are no systematic departures from 0 evident in the residuals as either $\hat{Y}$ or x increases, suggesting that the quadratic response function is a good fit. Figure 9.5 makes this point also. Further, there is no tendency in Figures 9.6a and 9.6b for the spread in the residuals to vary systematically, so it appears that the constant error variance assumption is reasonable. A normal probability plot, not shown here, did not provide any strong evidence that the distribution of the error terms is far from normal.

Based on this study of the aptness of the model, the analyst was willing to conclude that the normal error model (9.10) with constant error variance is appropriate here.

Test for quadratic response function. Since there are two repeat observations for each level of x, the analyst could have made a formal test of the aptness of the model, the alternatives being:

(9.13)
$$H_0: E(Y) = \beta_0 + \beta_1 x + \beta_{11} x^2$$
$$H_a: E(Y) \neq \beta_0 + \beta_1 x + \beta_{11} x^2$$

The basic ANOVA results were presented earlier in Table 9.3b. The pure error sum of squares is obtained as follows from the data in Table 9.2:

$$SSPE = (508.1 - 503.25)^2 + (498.4 - 503.25)^2 + (568.2 - 572.75)^2$$
$$+ \cdots + (831.8 - 836.6)^2 = 292$$

Note that $\bar{Y}_1 = 503.25$ for $x = -3$, $\bar{Y}_2 = 572.75$ for $x = -2$, and so on. There are $14 - 7 = 7$ degrees of freedom associated with $SSPE$. Hence, we have:

$$MSPE = \frac{SSPE}{7} = \frac{292}{7} = 41.7$$

Now we are in a position to obtain the lack of fit sum of squares by (4.12):

$$SSLF = SSE - SSPE = 679 - 292 = 387$$

There are $7 - 3 = 4$ degrees of freedom associated with $SSLF$. (Remember that three parameters had to be estimated for the fitted regression equation.) Hence, we have:

$$MSLF = \frac{SSLF}{4} = \frac{387}{4} = 96.8$$

Thus, test statistic (4.15) here is:

$$F^* = \frac{MSLF}{MSPE} = \frac{96.8}{41.7} = 2.32$$

Assuming the level of significance is to be .05, we require $F(.95; 4, 7) = 4.12$. Since $F^* = 2.32 \le 4.12$, we conclude H_0, that the quadratic response function is appropriate.

Test whether β_{11} equals zero

t test. The analyst next studied whether the quadratic term could be dropped from the model. She therefore wished to test:

(9.14)
$$H_0: \beta_{11} = 0$$
$$H_a: \beta_{11} \ne 0$$

H_0 implies that there is no quadratic effect in the response function.
Table 9.3a indicates that:

$$t^* = \frac{b_{11}}{s(b_{11})} = \frac{-4.249}{.606} = -7.012$$

For a level of significance of .05, we require $t(.975; 11) = 2.201$. The decision rule is:

$$\text{If } |t^*| \le 2.201, \text{ conclude } H_0$$
$$\text{If } |t^*| > 2.201, \text{ conclude } H_a$$

Since $|t*| = 7.012 > 2.201$, we conclude H_a, that a quadratic effect does exist, so that the quadratic term should be retained in the model.

Partial F test. The analyst could also have used the partial F test to choose the appropriate conclusion in (9.14). Indeed, she had specified the order of entering the variables x and x^2 into the computer regression fit so that she would obtain the extra sums of squares $SSR(x)$ and $SSR(x^2|x)$ in the output. Utilizing the partial F test statistic (8.24) and the results in Table 9.3b, we obtain:

$$F* = \frac{MSR(x^2|x)}{MSE} = \frac{3,033}{61.7} = 49.2$$

For a 5 percent level of significance, we need $F(.95; 1, 11) = 4.84$. Since $F* = 49.2 > 4.84$, we are led to conclude H_a, as by the t test.

Note

One may observe here the relation discussed in the previous chapter between the t and partial F tests as to whether a regression coefficient equals zero. We have for the two test statistics:

$$(t*)^2 = (-7.012)^2 = 49.2 = F*$$

Estimation of regression coefficients

The analyst next wished to obtain confidence bounds on the two regression coefficients β_1 and β_{11} with family confidence coefficient .90. The Bonferroni method is to be used, in view of its simplicity and the ease of interpreting the results.

Here $g = 2$ statements are desired; hence, by (7.44a), we have:

$$B = t(1 - .10/2(2); 11) = t(.975; 11) = 2.201$$

From Table 9.3a, we find:

$$b_1 = 54.893 \qquad s(b_1) = 1.050$$
$$b_{11} = -4.249 \qquad s(b_{11}) = .606$$

The Bonferroni confidence intervals therefore are by (7.44):

$$54.893 - 2.201(1.050) \le \beta_1 \le 54.893 + 2.201(1.050)$$

or:

$$52.58 \le \beta_1 \le 57.20$$

$$-4.249 - 2.201(.606) \le \beta_{11} \le -4.249 + 2.201(.606)$$

or:

$$-5.58 \le \beta_{11} \le -2.92$$

The analyst was satisfied with the precision of these two statements, feeling that the intervals are narrow enough to give her reliable simultaneous information about the comparative magnitudes of the linear and quadratic effects.

Coefficient of multiple determination

For a descriptive measure of the degree of relation between coffee sales and number of dispensing machines, the analyst calculated the coefficient of multiple determination using the data in Table 9.3b:

$$R^2 = \frac{SSR}{SSTO} = \frac{171,773}{172,453} = .996$$

This measure shows that the variation in coffee sales is reduced by 99.6 percent when the quadratic relation to the number of dispensing machines is utilized.

Note that the coefficient of multiple determination R^2 is the relevant measure here, not the coefficient of simple determination r^2, since model (9.10) is a multiple regression model even though it contains only one independent variable. Sometimes in curvilinear regression, the coefficient of multiple correlation R is called the *correlation index*.

Estimation of mean response

The analyst was particularly interested in the mean response for $X_h = 3$ dispensing machines. She wished to estimate this mean response with a 98 percent confidence coefficient. The proper interval estimate is given by (7.51). For our example, where $x_h = X_h - \bar{X} = 3 - 3 = 0$, we have:

$$\mathbf{X}_h = \begin{bmatrix} 1 \\ x_h \\ x_h^2 \end{bmatrix} = \begin{bmatrix} 1 \\ 0 \\ 0 \end{bmatrix}$$

The estimated mean response $\hat{Y}_h$ corresponding to $\mathbf{X}_h$ is by (7.47):

$$\hat{Y}_h = \mathbf{X}_h'\mathbf{b} = \begin{bmatrix} 1 & 0 & 0 \end{bmatrix} \begin{bmatrix} 705.474 \\ 54.893 \\ -4.249 \end{bmatrix} = 705.474$$

Next, using the results in Table 9.3c for $s^2(\mathbf{b})$, we obtain when substituting into (7.50):

$$s^2(\hat{Y}_h) = \mathbf{X}_h's^2(\mathbf{b})\mathbf{X}_h$$

$$= \begin{bmatrix} 1 & 0 & 0 \end{bmatrix} \begin{bmatrix} 10.2912 & 0 & -1.4702 \\ 0 & 1.1026 & 0 \\ -1.4702 & 0 & .3675 \end{bmatrix} \begin{bmatrix} 1 \\ 0 \\ 0 \end{bmatrix}$$

$$= 10.2912$$

or:

$$s(\hat{Y}_h) = \sqrt{10.2912} = 3.208$$

We require $t(.99; 11) = 2.718$. Hence, we obtain:

$$705.474 - 2.718(3.208) \leq E(Y_h) \leq 705.474 + 2.718(3.208)$$

or:

$$696.8 \le E(Y_h) \le 714.2$$

With confidence coefficient .98, the analyst can conclude that the mean of the distribution of coffee sales when three dispensing machines are used is somewhere between 696.8 and 714.2 hundred gallons.

Regression function in terms of X

The analyst wishes for reporting purposes to express the fitted regression function in terms of X rather than in terms of deviations $x = X - \bar{X}$. The following formulas provide the appropriate coefficients for model (9.10); the primes on the coefficients denote the new coefficients in terms of X:

(9.15a) $$b_0' = b_0 - b_1\bar{X} + b_{11}\bar{X}^2$$

(9.15b) $$b_1' = b_1 - 2b_{11}\bar{X}$$

(9.15c) $$b_{11}' = b_{11}$$

For our example, where $\bar{X} = 3$, we obtain:

$$b_0' = 705.474 - 54.893(3) + (-4.249)(3)^2 = 502.554$$
$$b_1' = 54.893 - 2(-4.249)(3) = 80.387$$
$$b_{11}' = -4.249$$

so that the fitted regression function in terms of X is:

$$\hat{Y} = 502.554 + 80.387X - 4.249X^2$$

The fitted values and residuals for the regression function in terms of X are exactly the same as for the regression function in terms of the deviations x. The reason for utilizing model (9.10), which is expressed in terms of deviations x, is to avoid potential calculational difficulties due to multicollinearity between X and X^2, inherent in polynomial regression.

Note

The estimated standard deviations of the regression coefficients in Table 9.3a do not apply to the regression coefficients in terms of X obtained by (9.15). If the estimated standard deviations for the regression coefficients in terms of X are desired, they may be obtained from $s^2(\mathbf{b})$ in Table 9.3c by using theorem (6.49), where the transformation matrix $\mathbf{A}$ is developed from (9.15).

9.3 EXAMPLE 2—TWO INDEPENDENT VARIABLES

We shall discuss now another example of polynomial regression, this one involving two independent variables. Rather than carrying this example through all of the various analytical stages as we did the first example, we shall focus here on the analysis of interaction effects and quadratic effects.

Setting

For a sample of 18 managers in the 35–44 age group, Table 9.4 shows the average annual income during the past two years (X_1), risk aversion score (X_2), and amount of life insurance carried (Y). Risk aversion was measured by a standard questionnaire administered to each manager; the higher the score the greater the degree of risk aversion.

TABLE 9.4 Data for life insurance example

Manager i	Average Annual Income (thousand dollars) X_{i1}	Risk Aversion Score X_{i2}	Amount of Life Insurance Carried (thousand dollars) Y_i
1	66.290	7	196
2	40.964	5	63
3	72.996	10	252
4	45.010	6	84
5	57.204	4	126
6	26.852	5	14
7	38.122	4	49
8	35.840	6	49
9	75.796	9	266
10	37.408	5	49
11	54.376	2	105
12	46.186	7	98
13	46.130	4	77
14	30.366	3	14
15	39.060	5	56
16	79.380	1	245
17	52.766	8	133
18	55.916	6	133
	$\bar{X}_1 = 50.037$	$\bar{X}_2 = 5.389$	

A researcher was studying the relation of average annual income and risk aversion to amount of life insurance carried by managers in the given age group. He expected that a quadratic relation would hold between income and amount of life insurance carried. However, he would not have been surprised if aversion to risk showed only linear effects and no quadratic effects on amount of life insurance carried, and he was quite uncertain whether or not the two variables interact in their effects on amount of life insurance carried. Hence, he fitted the second-order polynomial regression model:

$$(9.16) \quad Y_i = \beta_0 + \beta_1 x_{i1} + \beta_2 x_{i2} + \beta_{11} x_{i1}^2 + \beta_{22} x_{i2}^2 + \beta_{12} x_{i1} x_{i2} + \varepsilon_i$$

where:

$$x_{i1} = X_{i1} - \bar{X}_1$$
$$x_{i2} = X_{i2} - \bar{X}_2$$

with the intention of first testing for the presence of interaction effects and then for quadratic effects of aversion to risk.

Development of model

Table 9.5a contains the basic results for the fit of model (9.16). Since the researcher wished to test first for the interaction effects ($\beta_{12}x_1x_2$), he entered the variables for the regression fit so as to obtain the extra sum of squares $SSR(x_1x_2|x_1, x_2, x_1^2, x_2^2)$ for the partial F test. The ANOVA table and decomposition of SSR into extra sums of squares is shown in Table 9.5b.

TABLE 9.5 Regression results for model (9.16)—life insurance example

(a) Regression Coefficients

Regression Coefficient	Estimated Regression Coefficient	Estimated Standard Deviation	$t*$
β_0	102.768	.662	155.15
β_1	4.4930	.0475	94.54
β_2	6.028	.301	20.00
β_{11}	.03579	.00219	16.34
β_{22}	.166	.120	1.38
β_{12}	−.0196	.0140	−1.40

(b) Analysis of Variance

Source of Variation	SS	df	MS
Regression	108,006	5	21,601
x_1	104,474	1	104,474
$x_2\|x_1$	2,284	1	2,284
$x_1^2\|x_1, x_2$	1,238	1	1,238
$x_2^2\|x_1, x_2, x_1^2$	3	1	3
$x_1x_2\|x_1, x_2, x_1^2, x_2^2$	6	1	6
Error	36	12	3.00
Total	108,042	17	

The test for the presence of interaction effects involves the alternatives:

$$(9.17) \qquad \begin{aligned} H_0&: \beta_{12} = 0 \\ H_a&: \beta_{12} \neq 0 \end{aligned}$$

Using the partial F test statistic (8.24), the researcher obtained:

$$F* = \frac{MSR(x_1x_2|x_1, x_2, x_1^2, x_2^2)}{MSE} = \frac{6}{3} = 2.00$$

For level of significance $\alpha = .05$, we require $F(.95; 1, 12) = 4.75$. Since $F* = 2.00 \leq 4.75$, we conclude H_0, that no interaction effects exist. This result was welcome to the researcher, as it simplifies the interpretation of the effects of the two independent variables.

Note that the researcher could also have tested whether or not $\beta_{12} = 0$ by using $|t*| = |-1.40| = 1.40$ (Table 9.5a).

At this point, the researcher tentatively decided to adopt the no-interaction model:

(9.18) $$Y_i = \beta_0 + \beta_1 x_{i1} + \beta_2 x_{i2} + \beta_{11} x_{i1}^2 + \beta_{22} x_{i2}^2 + \varepsilon_i$$

but he still wished to examine whether a quadratic effect for risk aversion exists. This test can be conducted by a partial F test without fitting a new model. We utilize the definition of $SSR(x_1 x_2 | x_1, x_2, x_1^2, x_2^2)$:

$$SSR(x_1 x_2 | x_1, x_2, x_1^2, x_2^2) = SSE(x_1, x_2, x_1^2, x_2^2) - SSE(x_1, x_2, x_1^2, x_2^2, x_1 x_2)$$

Hence:

$$SSE(x_1, x_2, x_1^2, x_2^2) = SSR(x_1 x_2 | x_1, x_2, x_1^2, x_2^2) + SSE(x_1, x_2, x_1^2, x_2^2, x_1 x_2)$$
$$= 6 + 36 = 42$$

When testing model (9.18) for:

(9.19) $$\begin{aligned} H_0: \beta_{22} = 0 \\ H_a: \beta_{22} \neq 0 \end{aligned}$$

the partial F test statistic (8.24) is:

$$F^* = \frac{SSR(x_2^2 | x_1, x_2, x_1^2)}{1} \div \frac{SSE(x_1, x_2, x_1^2, x_2^2)}{18 - 5}$$

$$= \frac{3}{1} \div \frac{42}{13} = .93$$

For a 5 percent level of significance, we require $F(.95; 1, 13) = 4.67$. Since $F^* = .93 \leq 4.67$, we conclude H_0, that there is no quadratic effect for aversion to risk.

Hence, the researcher decided to adopt the revised model:

(9.20) $$Y_i = \beta_0 + \beta_1 x_{i1} + \beta_2 x_{i2} + \beta_{11} x_{i1}^2 + \varepsilon_i$$

where:

$$\begin{aligned} x_{i1} = X_{i1} - \bar{X}_1 \\ x_{i2} = X_{i2} - \bar{X}_2 \end{aligned}$$

and fitted this model to the data. He obtained the estimated response function:

$$\hat{Y} = 103.136 + 4.551 x_1 + 5.685 x_2 + .0371 x_1^2$$

which in the original units is:

$$\hat{Y} = -74.583 + .8383 X_1 + 5.685 X_2 + .0371 X_1^2$$

Figure 9.7 contains a three-dimensional computer-generated plot of this fitted response surface. The researcher then used this fitted response function for further investigation of the effects of average annual income and aversion to risk on amount of life insurance carried in the population under study.

FIGURE 9.7 Three-dimensional computer-generated plot of response function
$\hat{Y} = -74.583 + .8383X_1 + 5.685X_2 + .0371X_1^2$—life insurance example

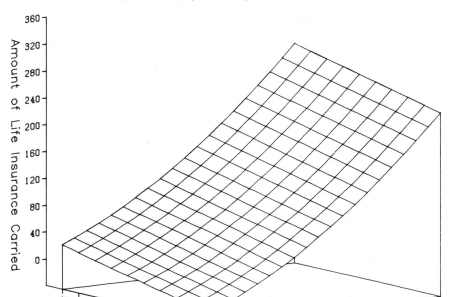

Comments

1. Note the advantage of computer packages which provide extra sums of squares in appropriate order. The researcher was able to conduct a F test about the revised model (9.18) using the regression run for model (9.16). The equivalent t test would require a new run for fitting model (9.18).

2. When multiple tests on the same data are conducted, there exist, as noted earlier, problems with respect to the level of significance for the family of inferences. In the example just cited, the researcher was willing to conduct two tests on the same data, one for interaction effects and one for quadratic effects of aversion to risk. The reason was that he knew by the Bonferroni inequality (5.6a) that the family level of significance for the two tests (each conducted at the 5 percent level of significance) could not exceed 10 percent.

In Chapter 12, we shall discuss more fully the empirical determination of an appropriate regression model.

9.4 ESTIMATING THE MAXIMUM OR MINIMUM OF A QUADRATIC REGRESSION FUNCTION

Sometimes in quadratic regression, we wish to estimate the maximum (or minimum) mean response of the regression function, and/or the level of X at which the maximum (or minimum) occurs. Figure 9.2 illustrates a quadratic response function with a maximum mean response.

Given the estimated quadratic response function:

(9.21) $$\hat{Y} = b_0 + b_1 x + b_{11} x^2$$

the maximum (minimum) occurs at the level x_m:

(9.22) $$x_m = -\frac{b_1}{2b_{11}}$$

In terms of the original variable X, the maximum (minimum) occurs at the level X_m:

(9.22a) $$X_m = \bar{X} - \frac{b_1}{2b_{11}}$$

The estimated mean response at X_m is:

(9.23) $$\hat{Y}_m = b_0 - \frac{b_1^2}{4b_{11}}$$

$\hat{Y}_m$ is a maximum if b_{11} is negative, and a minimum if b_{11} is positive.

Example

For our earlier cafeteria coffee sales example, the fitted regression curve was:

$$\hat{Y} = 705.47 + 54.89x - 4.25x^2$$

If the quadratic regression function were appropriate for larger x values than those in the study, we could estimate that maximum mean coffee sales occur at:

$$X_m = 3 - \frac{54.89}{2(-4.25)} = 9$$

and the estimated mean response there is:

$$\hat{Y}_m = 705.47 + 54.89(9) - 4.25(9)^2 = 855$$

Comments

1. To derive (9.22), we differentiate $\hat{Y}$ in (9.21) with respect to x, and set this derivative equal to 0:

$$\frac{d\hat{Y}}{dx} = \frac{d}{dx}(b_0 + b_1 x + b_{11} x^2) = b_1 + 2b_{11}x = 0$$

and obtain:

$$x_m = -\frac{b_1}{2b_{11}}$$

Substituting this value into the fitted response function (9.21), we find:

$$\hat{Y}_m = b_0 + b_1\left(\frac{-b_1}{2b_{11}}\right) + b_{11}\left(\frac{-b_1}{2b_{11}}\right)^2$$

$$= b_0 - \frac{b_1^2}{4b_{11}}$$

2. For large samples, the approximate estimated variance of X_m is:

$$(9.24) \qquad s^2(X_m) = \frac{b_1^2}{4b_{11}^2}\left[\frac{s^2(b_1)}{b_1^2} + \frac{s^2(b_{11})}{b_{11}^2} - \frac{2s(b_1, b_{11})}{b_1 b_{11}}\right]$$

This approximate estimated variance can be used to construct a confidence interval for the true X level at which the maximum (minimum) occurs. Approximate confidence intervals for $E(Y_m)$ can also be obtained. These are discussed in Reference 9.1.

9.5 SOME FURTHER COMMENTS ON POLYNOMIAL REGRESSION

1. The use of polynomial models in X is not without drawbacks. Such models can be more expensive in degrees of freedom than alternative nonlinear models or linear models with transformed variables. Another potential drawback is that multicollinearity is unavoidable. Indeed, if the levels of X are restricted to a narrow range, the degree of multicollinearity in the columns of the $\mathbf{X}$ matrix can be quite high, especially for higher-degree polynomials. It is for this reason that all polynomial regression models in this chapter are formulated in terms of deviations $x_i = X_i - \bar{X}$. To illustrate how helpful the use of deviation variables can be, in our life insurance example the coefficient of correlation between X_1 and X_1^2 is .991, but it is only .445 between x_1 and x_1^2. As noted earlier, when multicollinearity is high, serious calculational difficulties in inverting the $\mathbf{X'X}$ matrix can arise.

2. An alternative to using variables expressed in deviations from the mean in polynomial regression is to use *orthogonal polynomials*. Orthogonal polynomials are uncorrelated. Some computer packages use orthogonal polynomials in their polynomial regression routines and present the final fitted results in terms of both the orthogonal polynomials and the original polynomials. Orthogonal polynomials are discussed in specialized texts such as Reference 9.2.

3. Sometimes a quadratic response function is fitted for the purpose of establishing the linearity of the response function when repeat observations are not available for directly testing the linearity of the response function. Fitting the quadratic model:

$$(9.25) \qquad Y_i = \beta_0 + \beta_1 x_i + \beta_{11} x_i^2 + \varepsilon_i$$

and testing whether $\beta_{11} = 0$ does not, however, necessarily establish that a linear response function is appropriate. Figure 9.8 provides an example. If sample data were obtained for the response function in Figure 9.8, model (9.25) fitted, and a test on β_{11} made, it likely would lead to the conclusion that $\beta_{11} = 0$. Yet a linear

FIGURE 9.8 Example of curvilinear response function

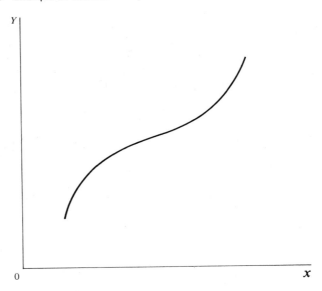

response function clearly might not be appropriate. Examination of residuals would disclose this lack of fit, and should always accompany formal testing of polynomial regression coefficients.

4. When a polynomial regression model with one independent variable is employed, one ordinarily fits a polynomial to the highest power expected to be appropriate, and the decomposition of SSR into extra sum of squares components proceeds as follows:

$$SSR(x)$$
$$SSR(x^2|x)$$
$$SSR(x^3|x, x^2)$$

etc.

The reason for this approach is that generally one is most interested in whether higher-order terms can be dropped from the model. Thus, when a cubic model is fitted because it is expected that a third-order model will be sufficient and one wishes to test whether $\beta_{111} = 0$, the appropriate extra sum of squares is $SSR(x^3|x, x^2)$. If, instead, one wishes to test whether a linear term is adequate so that $\beta_{11} = \beta_{111} = 0$, the appropriate extra sum of squares is $SSR(x^2, x^3|x) = SSR(x^2|x) + SSR(x^3|x, x^2)$.

Ordinarily, one would not fit a third-order model and test first whether a lower-order coefficient is zero, say, whether $\beta_{11} = 0$. The reason is that it is usually desired to employ as simple a regression model as possible, which in the case of polynomial regression means a lower-order model.

PROBLEMS

9.1. A speaker stated: "In developing third-order or higher-order polynomial regression models in social science and managerial applications, inferences on the β's usually take the form of direct tests. There is relatively little interest in estimating the β's to assess effects of the individual polynomial terms." Why might this be so?

9.2. Plot several contour curves for the quadratic response surface $E(Y) = 140 + 4x_1^2 - 2x_2^2 + 5x_1x_2$.

9.3. A junior investment analyst used a polynomial regression model of relatively high order in a research seminar on municipal bonds. She obtained an R^2 of .991 in the regression of net interest yield of bond (Y) on industrial diversity index of municipality (X) for seven bond issues. A classmate, unimpressed, said: "You overfitted. Your curve follows the random effects in the data."

  a. Comment on the criticism.
  b. Might R_a^2 defined in (7.33) be more appropriate than R^2 as a descriptive measure here?

9.4. A student in a class demonstration of how to fit a second-order polynomial model in one independent variable entered the X variables in the form X, X^2. He was disturbed when the computer program would not include X^2 in the regression model and regressed Y on X only. The output contained the message: X-SQUARE IS REDUNDANT VARIABLE. X'X IS NEAR-SINGULAR WHEN X-SQUARE IS INCLUDED. Explain the situation. What should the student have done?

9.5. Refer to the life insurance example on page 314. A student observed that the interaction term $\beta_{12}x_1x_2$ and the quadratic effect term $\beta_{22}x_2^2$ were each dropped from model (9.16) at an .05 level of significance and suggested that this same result could have been obtained from a glance at Table 9.5a, since each relevant $|t^*|$ statistic does not exceed $t(.975; 12) = 2.179$. Do you agree with the student's suggestion? Explain.

9.6. **Mileage study.** The effectiveness of a new experimental overdrive gear in reducing gasoline consumption was studied in 12 trials with a light truck equipped with this gear. In the data that follow, X_i denotes the constant speed (in miles per hour) on the test track in the ith trial and Y_i denotes miles per gallon obtained.

i:	1	2	3	4	5	6	7	8	9	10	11	12
X_i:	35	35	40	40	45	45	50	50	55	55	60	60
Y_i:	22	20	28	31	37	38	41	39	34	37	27	30

The second-order regression model (9.1a) with independent normal error terms is expected to be appropriate.

  a. Fit regression model (9.1a). Plot the fitted regression function and the data. Does the quadratic regression function appear to be a good fit here? Find R^2.
  b. Test whether or not there is a regression relation. Control the risk of a Type I error at .05. State the alternatives, decision rule, and conclusion.
  c. Estimate the mean miles per gallon for test runs at a speed of 48 miles per hour; use a 95 percent confidence interval.
  d. Predict the miles per gallon in the next test run at 48 miles per hour; use a 95 percent prediction interval.

e. Test whether the quadratic term can be dropped from the regression model; use $\alpha = .05$. State the alternatives, decision rule, and conclusion.

f. Express the fitted regression function obtained in part (a) in terms of the original variable X.

9.7. Refer to **Mileage study** Problem 9.6.

a. Obtain the residuals and plot them against $\hat{Y}$. Also prepare a normal probability plot. Interpret your plots.

b. Test formally for lack of fit of the quadratic regression function; use $\alpha = .05$. State the alternatives, decision rule, and conclusion. What assumptions did you implicitly make in this test?

c. Fit the third-order model (9.3) and test whether or not $\beta_{111} = 0$; use $\alpha = .05$. State the alternatives, decision rule, and conclusion. Is your conclusion consistent with your finding in part (b)?

9.8. **Piecework operation.** An operations analyst in a multinational electronics firm studied factors affecting production in a piecework operation where earnings are based on the number of pieces produced. Two employees each were selected from various age groups and data on their productivity last year were obtained (X is age of employee, in years; Y is employee's productivity, coded):

i:	1	2	3	4	5	6	7	8	9
X_i:	20	20	25	25	30	30	35	35	40
Y_i:	97	93	99	105	109	106	109	111	100

i:	10	11	12	13	14	15	16	17	18
X_i:	40	45	45	50	50	55	55	60	60
Y_i:	105	97	101	105	103	105	109	112	110

The analyst recognized that the relation between age and productivity is complex, in part because earnings targets (which he could not measure) shift in complex ways with age. However, he believed that for purposes of estimating mean responses, the response function can be approximated suitably by a polynomial of third order and that the error terms are independent and approximately normally distributed.

a. Fit regression model (9.3). Plot the fitted regression function and the data. Does the cubic regression function appear to be a good fit here? Find R^2.

b. Test whether or not there is a regression relation; use a level of significance of .01. State the alternatives, decision rule, and conclusion. What is the P-value of the test?

c. Obtain joint interval estimates for the mean productivity of employees aged 53, 58, and 62, respectively. Use the most efficient simultaneous estimation procedure and a 99 percent family confidence coefficient.

d. Predict the productivity of an employee aged 53 using a 99 percent prediction interval.

e. Express the fitted regression function obtained in part (a) in terms of the original variable X.

9.9. Refer to **Piecework operation** Problem 9.8.

a. Test whether both the quadratic and cubic terms can be dropped from the regression model; use $\alpha = .01$. State the alternatives, decision rule, and conclusion.

b. Test whether the cubic term alone can be dropped from the regression model; use $\alpha = .01$. State the alternatives, decision rule, and conclusion.

9.10. Refer to **Piecework operation** Problem 9.8.
  a. Obtain the residuals and plot them against the fitted values. Also prepare a normal probability plot. What do your plots show?
  b. Test formally for lack of fit. Control the risk of a Type I error at .01. State the alternatives, decision rule, and conclusion. What assumptions did you implicitly make in this test?

9.11. Sales forecasting. The Wheaton Company introduced a new product in 1975. Annual sales of this product (Y, in thousand units) follow; the time period (X) is coded, with $X = 1$ for 1975.

i:	1	2	3	4	5	6	7	8	9
X_i:	1	2	3	4	5	6	7	8	9
Y_i:	3.49	3.78	4.05	4.41	4.73	5.12	5.56	5.99	6.44

Assume that the second-order polynomial regression model (9.1a) with independent normal error terms is appropriate.
  a. Fit regression model (9.1a). Plot the fitted regression function and the data. Does the quadratic regression function appear to be a good fit here? What is R^2? Do you believe that the quadratic regression function is appropriate for making projections to 1995? Discuss.
  b. Obtain simultaneous Bonferroni confidence intervals for β_1 and β_{11} with a 90 percent family confidence coefficient.
  c. Predict sales of the product for 1985 using a 90 percent prediction interval.
  d. Express the fitted regression function obtained in part (a) in the original X units.

9.12. Refer to **Sales forecasting** Problem 9.11.
  a. Test whether the quadratic term can be dropped from the regression model. Control the risk of a Type I error at .10. State the alternatives, decision rule, and conclusion. What is the P-value of the test?
  b. Obtain the residuals. Plot the residuals against the fitted values. Also plot them against time. What do these plots show?

9.13. Crop yield. An agronomist studied the effects of moisture (X_1, in inches) and temperature (X_2, in °C) on the yield of a new hybrid tomato (Y). The experimental data follow.

i:	1	2	3	4	5	6	7	8	9	10	11	12	13
X_{i1}:	6	6	6	6	6	8	8	8	8	8	10	10	10
X_{i2}:	20	21	22	23	24	20	21	22	23	24	20	21	22
Y_i:	49.2	48.1	48.0	49.6	47.0	51.5	51.7	50.4	51.2	48.4	51.1	51.5	50.3

i:	14	15	16	17	18	19	20	21	22	23	24	25
X_{i1}:	10	10	12	12	12	12	12	14	14	14	14	14
X_{i2}:	23	24	20	21	22	23	24	20	21	22	23	24
Y_i:	48.9	48.7	48.6	47.0	48.0	46.4	46.2	43.2	42.6	42.1	43.9	40.5

The agronomist expects that the second-order polynomial regression model (9.5) with independent normal error terms is appropriate here.
  a. Fit regression model (9.5). Plot the Y observations against the fitted values. Does the response function provide a good fit?

    b.   Calculate R^2. What information does this measure provide?

    c.   Test whether or not there is a regression relation; use $\alpha = .05$. State the alternatives, decision rule, and conclusion. What is the P-value of the test?

    d.   Estimate the mean yield when $X_1 = 7$ and $X_2 = 22$; use a 95 percent confidence interval.

    e.   Express the fitted response function obtained in part (a) in the original X variables.

9.14. Refer to **Crop yield** Problem 9.13.

    a.   Test whether or not the interaction term can be dropped from the model. Control the α risk at .005. State the alternatives, decision rule, and conclusion.

    b.   Assuming that the interaction term has been dropped from the model, test whether or not the quadratic effect term for temperature can be dropped from the model; control the α risk at .005. State the alternatives, decision rule, and conclusion. What is the combined α risk for both the test here and the one in part (a)?

    c.   Fit the second-order polynomial model omitting the interaction term and the quadratic effect term for temperature. Obtain the residuals and plot them against the fitted values. What does your plot show?

9.15. **Computerized game.** Students comprising firm A in a computerized marketing game have approached you for assistance in analyzing the relation between promotional expenditures (X) and demand for their firm's product (Y) in the firm's home territory. They believe that the following characteristics hold in this relation: (1) demand in the home territory is affected primarily by promotional expenditures, (2) the relation is either quadratic or linear within the range of X levels of interest to the firm. The team has provided the observations shown below for the 14 periods covered in the game to date (X in thousand dollars, Y in thousand units), and has stated that these observations span the X levels of interest.

i:	1	2	3	4	5	6	7
X_i:	17	15	25	10	18	15	20
Y_i:	56.15	54.50	55.27	52.54	56.23	55.97	55.55

i:	8	9	10	11	12	13	14
X_i:	25	17	13	20	23	25	16
Y_i:	54.32	55.14	54.28	55.78	55.65	54.96	55.06

Assume that the second-order model (9.1a) with independent normal error terms applies.

    a.   Fit this model and test whether a regression relation exists. Use a level of significance of .01. State the alternatives, decision rule, and conclusion.

    b.   Test whether the quadratic term can be dropped from the model. Use a level of significance of .01. State the alternatives, decision rule, and conclusion.

    c.   Obtain the residuals and plot them against $\hat{Y}$. Also obtain a normal probability plot. What do your plots show?

    d.   Conduct a formal test for lack of fit using a level of significance of .01. State the alternatives, decision rule, and conclusion. Does your conclusion imply that the model cannot be improved further? Discuss.

9.16. Refer to **Computerized game** Problem 9.15. Someone who is familiar with this computerized marketing game enters the discussion. She states that in the system of equations on which the game is based, a quadratic relation does hold between promotional expenditures and mean demand in the firm's home territory. She believes that another significant variable related to expected demand in the home territory is the ratio of the firm's selling price to the average competitive selling price; however she does not recall whether this price ratio has both linear and quadratic effects. She also does not recall whether price ratio and promotional expenditures interact in affecting demand. The firm's price ratios for the 14 periods are as follows:

i:	1	2	3	4	5	6	7
Ratio:	.931	.976	1.045	.939	1.010	1.059	1.000

i:	8	9	10	11	12	13	14
Ratio:	.950	.995	1.011	1.008	.947	1.000	1.017

a. Fit the second-order polynomial regression model (9.5) with promotional expenditures (X_1) and price ratio (X_2) as independent variables. How much has R^2 increased by adding the price ratio as an independent variable?

b. Test whether the price ratio variable should be retained in the model. Control the risk of Type I error at .05. State the alternatives, decision rule, and conclusion.

c. Assuming that the price ratio variable is to be retained in the model, test whether the interaction term is needed in the model; use $\alpha = .01$. State the alternatives, decision rule, and conclusion. What is the P-value of the test?

d. The team has decided to adopt regression model (9.5) without interaction effects. Fit this model and obtain the residuals. Plot the residuals against $\hat{Y}$ and against the time order of the observations. Also, prepare a normal probability plot. Analyze these plots and state your findings.

9.17. Refer to **Mileage study** Problem 9.6.
a. At what speed is the estimated quadratic response function a maximum? What is the estimated mean mileage at this speed?
b. Does the maximum of the response function occur within the scope of the model?

9.18. Refer to **Sales forecasting** Problem 9.11.
a. In what year does the minimum of the estimated quadratic response function occur? What is the estimated mean sales for this year?
b. Does the minimum of the response function occur within the scope of the model?

EXERCISES

9.19. Consider the second-order regression model with one independent variable (9.1a) and the following two sets of X values:

Set 1: 1.0 1.5 1.1 1.3 1.9 .8 1.2 1.4
Set 2: 12 1 123 17 415 71 283 38

For each set, calculate the coefficient of correlation between X and X^2, then between x and x^2. Also calculate the coefficients of correlation between X and X^3 and between x and x^3. What generalizations are suggested by your results?

9.20. (Calculus needed.) Refer to the second-order response function (9.2). Explain precisely the meaning of the linear effect coefficient β_1 and the quadratic effect coefficient β_{11}.

9.21. a. Derive the expressions for b_0', b_1', and b_{11}' in (9.15).
      b. Using theorem (6.49), obtain the variance-covariance matrix for the regression coefficients pertaining to the original X variable in terms of the variance-covariance matrix for the regression coefficients pertaining to the transformed x variable.

9.22. How are the normal equations (9.12) simplified if the X values are equally spaced, such as the time series representation $X_1 = 1$, $X_2 = 2, \ldots, X_n = n$?

PROJECTS

9.23. Refer to the **SMSA** data set. It is desired to fit the second-order regression model (9.1a) for relating number of active physicians (Y) against total population (X).
    a. Fit the second-order regression model. Plot the residuals against the fitted values. How well does the second-order model appear to fit the data?
    b. Obtain R^2 for the second-order regression model. Also obtain the coefficient of simple determination r^2 for the first-order regression model. Has the addition of the quadratic term in the regression model substantially increased the coefficient of determination?
    c. Test whether the quadratic term can be dropped from the regression model; use $\alpha = .05$. State the alternatives, decision rule, and conclusion.
    d. Omit observation 1 (New York City) from the data set. Fit the second-order regression model (9.1a) based on the remaining 140 SMSA's. Repeat the test in part (c). Has the omission of the outlying observation affected your conclusion about whether the quadratic term can be dropped from the model?

9.24. Refer to the **SMSA** data set. A regression model relating serious crime rate (Y, total serious crimes divided by total population) to population density (X_1, total population divided by land area) and percent of population in central cities (X_3) is to be constructed.
    a. Fit the second-order regression model (9.5). Plot the residuals against the fitted values. How well does the second-order model appear to fit the data? What is R^2?
    b. Test whether or not all quadratic and interaction terms can be dropped from the model; use $\alpha = .01$. State the alternatives, decision rule, and conclusion.
    c. Instead of using the independent variable population density, total population (X_1) and land area (X_2) are to be employed as separate independent variables, in addition to percent of population in central cities (X_3). The regression model should contain linear and quadratic terms for total population, and linear terms only for land area and percent of population in central cities. (No interaction terms are to be included in this model.) Fit this regression model and obtain R^2. Is this coefficient of multiple determination substantially different from the one for the model in part (a)?

9.25. Refer to the **SENIC** data set. The second-order regression model (9.1a) is to be fitted for relating number of nurses (Y) to available facilities and services (X).

a. Fit the second-order regression model. Plot the residuals against the fitted values. How well does the second-order model appear to fit the data?

b. Obtain R^2 for the second-order regression model. Also obtain the coefficient of simple determination r^2 for the first-order regression model. Has the addition of the quadratic term in the regression model substantially increased the coefficient of determination?

c. Test whether the quadratic term can be dropped from the regression model; use $\alpha = .10$. State the alternatives, decision rule, and conclusion.

9.26. Refer to **Sales forecasting** Problem 9.11. Instead of using a polynomial regression model here, it has been suggested that a transformation of variables might yield an equally good fit and would be more desirable since forecasting requires extrapolation.

a. Fit a linear regression model for relating $Y' = \sqrt{Y}$ against X. Plot the fitted regression function and the transformed data. How effective does the use of the transformed variable appear to be here?

b. Obtain the fitted values and transform them to the original variable Y. Then calculate the residuals in the original variable. Plot these residuals against X and analyze your plot.

c. Square the residuals in the original variable obtained in part (b), sum, and obtain MSE. Compare this with MSE for the quadratic model in Problem 9.11a. How does the variability around the fitted regression function, as measured by MSE, compare for the two approaches?

d. Repeat parts (a) through (c) using the transformation $Y' = \log_{10} Y$. Is either the square-root or the logarithmic transformation clearly preferable here?

CITED REFERENCES

9.1 Williams, E. J. *Regression Analysis*. New York: John Wiley & Sons, 1959.

9.2 Draper, N. R., and H. Smith. *Applied Regression Analysis*. 2d ed. New York: John Wiley & Sons, 1981.

10

Indicator variables

Throughout the previous chapters on regression analysis, we have utilized quantitative variables in the regression models considered. Quantitative variables take on values on a well-defined scale; examples are income, age, temperature, and amounts of liquid assets.

Many variables of interest in business, economics, and the social and biological sciences, however, are not quantitative but are qualitative. Examples of qualitative variables are sex (male, female), purchase status (purchase, no purchase), and disability status (not disabled, partly disabled, fully disabled).

Qualitative variables can be used in a multiple regression model just as quantitative variables can, as we shall explain in this chapter. First, we take up the case where some or all of the independent variables are qualitative. Then we turn to the case where the dependent variable is qualitative.

10.1 ONE INDEPENDENT QUALITATIVE VARIABLE

An economist wished to relate the speed with which a particular insurance innovation is adopted (Y) to the size of the insurance firm (X_1) and the type of firm. The dependent variable is measured by the number of months elapsed

328

between the time the first firm adopted the innovation and the time the given firm adopted the innovation. The first independent variable, size of firm, is quantitative, and is measured by the amount of total assets of the firm. The second independent variable, type of firm, is qualitative and is composed of two classes—stock companies and mutual companies. In order that such a qualitative variable can be used in a regression model, quantitative indicators for the classes of the qualitative variable must be found.

Indicator variables

There are many ways of quantitatively identifying the classes of a qualitative variable. We shall use indicator variables that take on the values 0 and 1. These indicator variables are easy to use and are widely employed, but they are by no means the only way to quantify a qualitative variable.

For our example, where the qualitative variable has two classes, we might define two indicator variables X_2 and X_3 as follows:

(10.1)
$$X_2 = \begin{matrix} 1 \text{ if stock company} \\ 0 \text{ otherwise} \end{matrix}$$

$$X_3 = \begin{matrix} 1 \text{ if mutual company} \\ 0 \text{ otherwise} \end{matrix}$$

Assuming that a first-order model is to be employed, it would be:

(10.2) $\quad Y_i = \beta_0 X_{i0} + \beta_1 X_{i1} + \beta_2 X_{i2} + \beta_3 X_{i3} + \varepsilon_i \qquad$ where $X_{i0} \equiv 1$

This intuitive approach of setting up an indicator variable for each class of the qualitative variable unfortunately leads to computational difficulties. To see why, suppose we have $n = 4$ observations, the first two being stock firms for which $X_2 = 1$ and $X_3 = 0$, and the second two being mutual firms for which $X_2 = 0$ and $X_3 = 1$. The $\mathbf{X}$ matrix would then be:

$$\mathbf{X} = \begin{matrix} X_0 & X_1 & X_2 & X_3 \\ \begin{bmatrix} 1 & X_{11} & 1 & 0 \\ 1 & X_{21} & 1 & 0 \\ 1 & X_{31} & 0 & 1 \\ 1 & X_{41} & 0 & 1 \end{bmatrix} \end{matrix}$$

Note that the X_0 column is equal to the sum of the X_2 and X_3 columns, so that the columns are linearly dependent according to definition (6.22). This has a serious effect on the $\mathbf{X'X}$ matrix:

$$\mathbf{X'X} = \begin{bmatrix} 1 & 1 & 1 & 1 \\ X_{11} & X_{21} & X_{31} & X_{41} \\ 1 & 1 & 0 & 0 \\ 0 & 0 & 1 & 1 \end{bmatrix} \begin{bmatrix} 1 & X_{11} & 1 & 0 \\ 1 & X_{21} & 1 & 0 \\ 1 & X_{31} & 0 & 1 \\ 1 & X_{41} & 0 & 1 \end{bmatrix}$$

$$
= \begin{bmatrix}
4 & \sum_{i=1}^{4} X_{i1} & 2 & 2 \\[2ex]
\sum_{i=1}^{4} X_{i1} & \sum_{i=1}^{4} X_{i1}^2 & \sum_{i=1}^{2} X_{i1} & \sum_{i=3}^{4} X_{i1} \\[2ex]
2 & \sum_{i=1}^{2} X_{i1} & 2 & 0 \\[2ex]
2 & \sum_{i=3}^{4} X_{i1} & 0 & 2
\end{bmatrix}
$$

It is quickly apparent that the first column of the $\mathbf{X'X}$ matrix equals the sum of the last two columns, so that the columns are linearly dependent. Hence, the $\mathbf{X'X}$ matrix does not have an inverse, and no unique estimators of the regression coefficients can be found.

A simple way out of this difficulty is to drop one of the indicator variables. In our example, for instance, we might drop X_3. While dropping one indicator variable is not the only way out of the difficulty, it leads to simple interpretations of the parameters. In general, therefore, we shall follow the principle:

(10.3)   A qualitative variable with c classes will be represented by $c - 1$ indicator variables, each taking on the values 0 and 1.

Note

Indicator variables are frequently also called *dummy variables* or *binary variables*. The latter term has reference to the binary number system containing only 0 and 1.

Interpretation of regression parameters

Returning to our example, suppose that we drop the indicator variable X_3 from model (10.2) so that the model becomes:

(10.4) $Y_i = \beta_0 + \beta_1 X_{i1} + \beta_2 X_{i2} + \varepsilon_i$

where:

$$X_{i1} = \text{Size of firm}$$
$$X_{i2} = \begin{array}{l} 1 \text{ if stock company} \\ 0 \text{ otherwise} \end{array}$$

The response function for this model is:

(10.5) $E(Y) = \beta_0 + \beta_1 X_1 + \beta_2 X_2$

To understand the meaning of the parameters of this model, consider first the case of a mutual firm. For such a firm, $X_2 = 0$ and we have:

(10.5a) $E(Y) = \beta_0 + \beta_1 X_1 + \beta_2(0) = \beta_0 + \beta_1 X_1$ Mutual firms

Thus, the response function for mutual firms is a straight line, with Y intercept β_0 and slope β_1. This response function is shown in Figure 10.1.

For a stock firm, $X_2 = 1$ and the response function (10.5) is:

(10.5b) $E(Y) = \beta_0 + \beta_1 X_1 + \beta_2(1) = (\beta_0 + \beta_2) + \beta_1 X_1$ Stock firms

This also is a straight line, with the same slope β_1 but with Y intercept $\beta_0 + \beta_2$. This response function is also shown in Figure 10.1.

The meaning of the parameters in the response function (10.5) is now clear. With reference to our earlier example, the mean time elapsed before the innovation is adopted, $E(Y)$, is a linear function of size of firm (X_1), with the same slope β_1 for both types of firms. β_2 indicates how much higher (lower) the response function for stock firms is than the one for mutual firms. Thus, β_2 measures the differential effect of type of firm. In general, β_2 shows how much higher (lower) the mean response line is for the class coded 1 than the line for the class coded 0.

FIGURE 10.1   Illustration of meaning of regression parameters for model (10.4) with indicator variable X_2—insurance innovation example

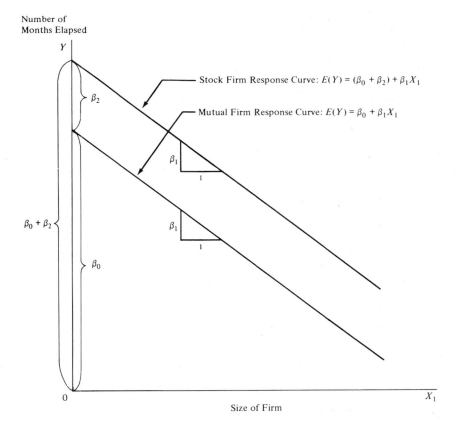

Number of Months Elapsed

Example

With reference to our earlier illustration, the economist studied 10 mutual firms and 10 stock firms. The data are shown in Table 10.1. The $\mathbf{Y}$ and $\mathbf{X}$ data matrices are shown in Table 10.2. Note that $X_2 = 1$ for each stock firm and $X_2 = 0$ for each mutual firm.

Given the $\mathbf{Y}$ and $\mathbf{X}$ matrices in Table 10.2, fitting the regression model (10.4) is straightforward. Table 10.3 presents the key results from a computer run. The fitted response function is:

$$\hat{Y} = 33.87407 - .10174X_1 + 8.05547X_2$$

Figure 10.2 (p. 334) contains the fitted response function for each type of firm, together with the actual observations.

The economist was most interested in the effect of type of firm (X_2) on the elapsed time for the innovation to be adopted. He therefore desired to obtain a 95 percent confidence interval for β_2. We require $t(.975; 17) = 2.110$ and obtain from the data in Table 10.3:

$$4.97675 = 8.05547 - 2.110(1.45911) \le \beta_2$$
$$\le 8.05547 + 2.110(1.45911) = 11.13419$$

Thus, with 95 percent confidence, we conclude that stock companies tend to adopt the particular innovation studied somewhere between 5 and 11 months later, on the average, than mutual companies.

TABLE 10.1 Data for insurance innovation study

Firm i	Number of Months Elapsed Y_i	Size of Firm (million dollars) X_{i1}	Type of Firm
1	17	151	Mutual
2	26	92	Mutual
3	21	175	Mutual
4	30	31	Mutual
5	22	104	Mutual
6	0	277	Mutual
7	12	210	Mutual
8	19	120	Mutual
9	4	290	Mutual
10	16	238	Mutual
11	28	164	Stock
12	15	272	Stock
13	11	295	Stock
14	38	68	Stock
15	31	85	Stock
16	21	224	Stock
17	20	166	Stock
18	13	305	Stock
19	30	124	Stock
20	14	246	Stock

TABLE 10.2 Data matrices for insurance innovation study

$$
Y = \begin{bmatrix} 17 \\ 26 \\ 21 \\ 30 \\ 22 \\ 0 \\ 12 \\ 19 \\ 4 \\ 16 \\ 28 \\ 15 \\ 11 \\ 38 \\ 31 \\ 21 \\ 20 \\ 13 \\ 30 \\ 14 \end{bmatrix}
\qquad
X = \begin{bmatrix}
1 & 151 & 0 \\
1 & 92 & 0 \\
1 & 175 & 0 \\
1 & 31 & 0 \\
1 & 104 & 0 \\
1 & 277 & 0 \\
1 & 210 & 0 \\
1 & 120 & 0 \\
1 & 290 & 0 \\
1 & 238 & 0 \\
1 & 164 & 1 \\
1 & 272 & 1 \\
1 & 295 & 1 \\
1 & 68 & 1 \\
1 & 85 & 1 \\
1 & 224 & 1 \\
1 & 166 & 1 \\
1 & 305 & 1 \\
1 & 124 & 1 \\
1 & 246 & 1
\end{bmatrix}
$$

with column headers X_0 X_1 X_2

TABLE 10.3 Regression results for model (10.4) fit—insurance innovation example

(a)   Regression Coefficients

Regression Coefficient	Estimated Regression Coefficient	Estimated Standard Deviation	t^*
β_0	33.87407	1.81386	18.68
β_1	$-.10174$	.00889	-11.44
β_2	8.05547	1.45911	5.52

(b)   Analysis of Variance

Source of Variation	SS	df	MS
Regression	1,504.41	2	752.20
Error	176.39	17	10.38
Total	1,680.80	19	

A formal test of:

$$H_0: \beta_2 = 0$$
$$H_a: \beta_2 \neq 0$$

with level of significance .05 would lead to H_a, that type of firm has an effect, since the confidence interval for β_2 does not include zero.

FIGURE 10.2 Fitted regression functions for model (10.4)—insurance innovation example

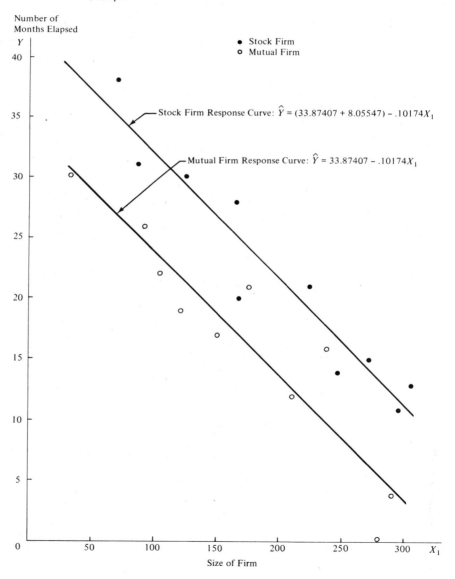

Number of Months Elapsed

Stock Firm Response Curve: $\hat{Y} = (33.87407 + 8.05547) - .10174X_1$

Mutual Firm Response Curve: $\hat{Y} = 33.87407 - .10174X_1$

● Stock Firm
○ Mutual Firm

Size of Firm

The economist also carried out other analyses, some of which will be described shortly.

Note

The reader may wonder why we did not simply fit separate regressions for stock firms and mutual firms in our example, and instead adopted the approach of fitting one regression with an indicator variable. There are two reasons for this. Since the model assumed equal slopes and the same constant error term variance for each type of firm, the common slope β_1 can best be estimated by pooling the two types of firms. Also, other inferences, such as ones pertaining to β_0 and β_2, can be made more precisely by working with one regression model containing an indicator variable since more degrees of freedom will then be associated with MSE.

10.2   MODEL CONTAINING INTERACTION EFFECTS

In our earlier illustration, the economist actually did not begin his analysis with model (10.4) since he expected interaction effects between size and type of firm. Even though one of the independent variables in the regression model is qualitative, interaction effects are introduced into the model in the usual manner, by including cross-product terms. A first-order model with an interaction term for our example is:

$$(10.6) \qquad Y_i = \beta_0 + \beta_1 X_{i1} + \beta_2 X_{i2} + \beta_3 X_{i1} X_{i2} + \varepsilon_i$$

where:

$$X_{i1} = \text{Size of firm}$$

$$X_{i2} = \begin{array}{l} 1 \text{ if stock company} \\ 0 \text{ otherwise} \end{array}$$

The response function for this model is:

$$(10.7) \qquad E(Y) = \beta_0 + \beta_1 X_1 + \beta_2 X_2 + \beta_3 X_1 X_2$$

Meaning of regression parameters

The meaning of the regression parameters in the response function (10.7) can best be understood by examining the nature of this function for each type of firm. For a mutual firm, $X_2 = 0$ and hence $X_1 X_2 = 0$. The response function for mutual firms therefore is:

$$(10.7a) \quad E(Y) = \beta_0 + \beta_1 X_1 + \beta_2(0) + \beta_3(0) = \beta_0 + \beta_1 X_1 \quad \text{Mutual firms}$$

This response function is shown in Figure 10.3. Note that the Y intercept is β_0 and the slope is β_1 for the response function for mutual firms.

For stock firms, $X_2 = 1$ and hence $X_1 X_2 = X_1$. The response function for stock firms therefore is:

$$E(Y) = \beta_0 + \beta_1 X_1 + \beta_2(1) + \beta_3 X_1$$

or:

$$(10.7b) \qquad E(Y) = (\beta_0 + \beta_2) + (\beta_1 + \beta_3)X_1 \qquad \text{Stock firms}$$

FIGURE 10.3 Illustration of meaning of regression parameters for model (10.6) with indicator variable X_2 and interaction term—insurance innovation example

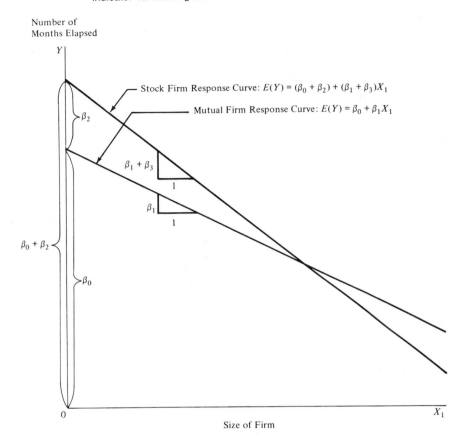

Number of
Months Elapsed

Stock Firm Response Curve: $E(Y) = (\beta_0 + \beta_2) + (\beta_1 + \beta_3)X_1$

Mutual Firm Response Curve: $E(Y) = \beta_0 + \beta_1 X_1$

β_2

$\beta_1 + \beta_3$

β_1

$\beta_0 + \beta_2$

β_0

Size of Firm

This response function is also shown in Figure 10.3. Note that the response function for stock firms has Y intercept $\beta_0 + \beta_2$ and slope $\beta_1 + \beta_3$.

Thus, β_2 indicates how much greater (smaller) is the Y intercept for the class coded 1 than that for the class coded 0, and similarly β_3 indicates how much greater (smaller) is the slope for the class coded 1 than that for the class coded 0. Because both the intercept and the slope differ for the two classes in model (10.6), it is no longer true that β_2 indicates how much higher (lower) one response line is than the other. Figure 10.3 makes it clear that the effect of type of firm with model (10.6) depends on the size of the firm. For smaller firms, according to Figure 10.3, mutual firms tend to innovate more quickly, but for larger firms stock firms tend to innovate more quickly. Thus, when interaction effects are present, the effect of the qualitative variable can only be studied by comparing the regression functions for each class of the qualitative variable.

FIGURE 10.4   Another illustration of model (10.6) with indicator variable X_2 and interaction term—insurance innovation example

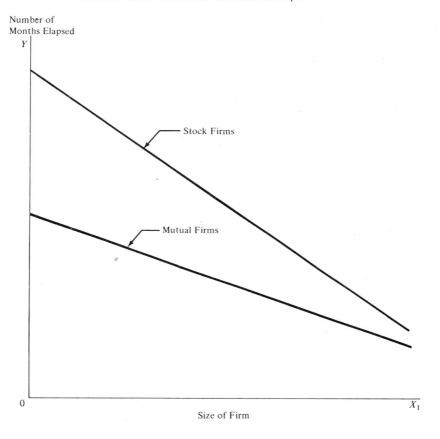

Figure 10.4 illustrates another possible interaction situation. Here, mutual firms tend to introduce the innovation more quickly than stock firms for all sizes of firms in the scope of the model, but the differential effect is much smaller for large firms than for small ones.

Example

Since the economist anticipated that interaction effects between size and type of firm may be present, he actually first wished to fit model (10.6):

$$Y_i = \beta_0 + \beta_1 X_{i1} + \beta_2 X_{i2} + \beta_3 X_{i1}X_{i2} + \varepsilon_i$$

Table 10.4 shows the **X** matrix for this model. The **Y** matrix is the same as in Table 10.2. Note that the X_1X_2 column in the **X** matrix in Table 10.4 contains 0 for mutual companies and X_{i1} for stock companies.

TABLE 10.4 **X** matrix for fitting model (10.6) with interaction term—insurance innovation example

$$\mathbf{X} = \begin{bmatrix} X_0 & X_1 & X_2 & X_1X_2 \\ 1 & 151 & 0 & 0 \\ 1 & 92 & 0 & 0 \\ 1 & 175 & 0 & 0 \\ 1 & 31 & 0 & 0 \\ 1 & 104 & 0 & 0 \\ 1 & 277 & 0 & 0 \\ 1 & 210 & 0 & 0 \\ 1 & 120 & 0 & 0 \\ 1 & 290 & 0 & 0 \\ 1 & 238 & 0 & 0 \\ 1 & 164 & 1 & 164 \\ 1 & 272 & 1 & 272 \\ 1 & 295 & 1 & 295 \\ 1 & 68 & 1 & 68 \\ 1 & 85 & 1 & 85 \\ 1 & 224 & 1 & 224 \\ 1 & 166 & 1 & 166 \\ 1 & 305 & 1 & 305 \\ 1 & 124 & 1 & 124 \\ 1 & 246 & 1 & 246 \end{bmatrix}$$

Given the **Y** and **X** matrices, the regression fit is routine. Basic results from a computer run are shown in Table 10.5. To test for the presence of interaction effects:

$$H_0: \beta_3 = 0$$
$$H_a: \beta_3 \neq 0$$

the economist used the t^* statistic from Table 10.5a:

$$t^* = \frac{b_3}{s(b_3)} = \frac{-.0004171}{.01833} = -.02$$

For level of significance .05, we require $t(.975; 16) = 2.120$. Since $|t^*| = .02 \leq 2.120$, we conclude H_0, i.e., $\beta_3 = 0$, so that no interaction effects are present. The two-sided P-value for the test is very high, namely, .98. It was because of this result that the economist adopted model (10.4) with no interaction term, which we discussed earlier.

Note

Fitting model (10.6) yields the same response functions as fitting separate regressions for stock firms and mutual firms. An advantage of using model (10.6) with an indicator

TABLE 10.5 Regression results for fit of model (10.6) with interaction term—insurance innovation example

(a) Regression Coefficients

Regression Coefficient	Estimated Regression Coefficient	Estimated Standard Deviation	t^*
β_0	33.83837	2.44065	13.86
β_1	$-.10153$	.01305	-7.78
β_2	8.13125	3.65405	2.23
β_3	$-.0004171$	.01833	$-.02$

(b) Analysis of Variance

Source of Variation	SS	df	MS
Regression	1,504.42	3	501.47
Error	176.38	16	11.02
Total	1,680.80	19	

variable is that one regression run on the computer will yield both fitted regressions.

Another advantage is that tests for comparing the regression functions for the different classes of the qualitative variable can be clearly seen to be tests of regression coefficients in a general linear model. For instance, Figure 10.3 makes it clear for our example that the test whether the two regression functions have the same slope involves:

$$H_0: \beta_3 = 0$$
$$H_a: \beta_3 \neq 0$$

Similarly, the test whether the two regression functions in our example are identical would involve:

$$H_0: \beta_2 = \beta_3 = 0$$
$$H_a: \text{not both } \beta_2 = 0 \text{ and } \beta_3 = 0$$

10.3 MORE COMPLEX MODELS

We now briefly consider more complex models involving qualitative independent variables.

Qualitative variable with more than two classes

If a qualitative independent variable has more than two classes, we require additional indicator variables in the regression model. Consider the regression of tool wear (Y) on tool speed (X_1), where we wish to include also tool model (M1, M2, M3, M4) as an independent variable. Since the qualitative variable (tool model) has four classes, we require three indicator variables. Let us define them as follows:

$$X_2 = \begin{array}{l} 1 \text{ if tool model M1} \\ 0 \text{ otherwise} \end{array}$$

(10.8)
$$X_3 = \begin{array}{l} 1 \text{ if tool model M2} \\ 0 \text{ otherwise} \end{array}$$

$$X_4 = \begin{array}{l} 1 \text{ if tool model M3} \\ 0 \text{ otherwise} \end{array}$$

First-order model. A first-order model is:

(10.9) $\quad Y_i = \beta_0 + \beta_1 X_{i1} + \beta_2 X_{i2} + \beta_3 X_{i3} + \beta_4 X_{i4} + \varepsilon_i$

For this model, the data input for the **X** matrix would be as follows:

Tool Model	X_0	X_1	X_2	X_3	X_4
M1	1	X_{i1}	1	0	0
M2	1	X_{i1}	0	1	0
M3	1	X_{i1}	0	0	1
M4	1	X_{i1}	0	0	0

The response function for model (10.9) is:

(10.10) $\quad E(Y) = \beta_0 + \beta_1 X_1 + \beta_2 X_2 + \beta_3 X_3 + \beta_4 X_4$

To see the meaning of the regression parameters, consider first the response function for tool models M4 for which $X_2 = 0$, $X_3 = 0$, and $X_4 = 0$:

(10.10a) $\qquad\qquad E(Y) = \beta_0 + \beta_1 X_1 \qquad\qquad$ Tool models M4

For tool models M1, $X_2 = 1$, $X_3 = 0$, and $X_4 = 0$, and the response function is:

(10.10b) $\quad E(Y) = \beta_0 + \beta_1 X_1 + \beta_2 = (\beta_0 + \beta_2) + \beta_1 X_1 \quad$ Tool models M1

Similarly, the response functions for tool models M2 and M3 are:

(10.10c) $\qquad\qquad E(Y) = (\beta_0 + \beta_3) + \beta_1 X_1 \qquad$ Tool models M2

(10.10d) $\qquad\qquad E(Y) = (\beta_0 + \beta_4) + \beta_1 X_1 \qquad$ Tool models M3

Thus, the response function (10.10) implies that the regression of tool wear on tool speed is linear, with the same slope for all types of tool models. The coefficients β_2, β_3, and β_4 indicate, respectively, how much higher (lower) the response functions for tool models M1, M2, and M3 are than the one for tool models M4. Thus, β_2, β_3, and β_4 measure the differential effects of the qualitative variable classes on the height of the response function, always compared with the class for which $X_2 = X_3 = X_4 = 0$. Figure 10.5 illustrates a possible arrangement of the response functions.

When using model (10.9), one may wish to estimate differential effects other than against tool models M4. For instance, $\beta_4 - \beta_3$ measures how much higher (lower) the response function for tool models M3 is than the response function for tool models M2, as may be seen by comparing (10.10c) and (10.10d). The point estimator of this quantity is, of course, $b_4 - b_3$, and the estimated variance

FIGURE 10.5 Illustration of model (10.9)—tool wear example

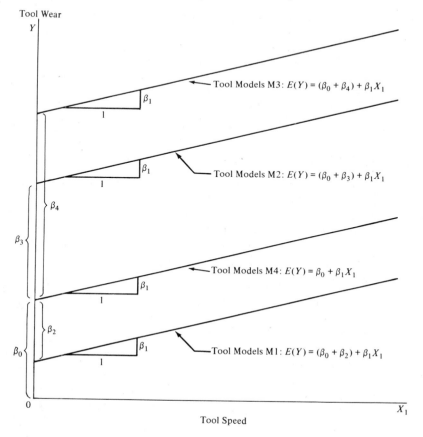

of this estimator is:

(10.11) $$s^2(b_4 - b_3) = s^2(b_4) + s^2(b_3) - 2s(b_4, b_3)$$

The needed variances and covariance can be readily obtained from the estimated variance-covariance matrix of the regression coefficients.

First-order model with interactions added. If interaction effects between tool speed and tool model are present in our previous illustration, model (10.9) would be modified as follows:

(10.12) $$Y_i = \beta_0 + \beta_1 X_{i1} + \beta_2 X_{i2} + \beta_3 X_{i3} + \beta_4 X_{i4} + \beta_5 X_{i1}X_{i2}$$
$$+ \beta_6 X_{i1}X_{i3} + \beta_7 X_{i1}X_{i4} + \varepsilon_i$$

The response function for tool models M4, for which $X_2 = 0$, $X_3 = 0$, and $X_4 = 0$, is as follows:

(10.13a) $$E(Y) = \beta_0 + \beta_1 X_1 \qquad \text{Tool models M4}$$

Similarly, we find for the other tool models:

(10.13b) $$E(Y) = (\beta_0 + \beta_2) + (\beta_1 + \beta_5)X_1 \qquad \text{Tool models M1}$$

(10.13c) $$E(Y) = (\beta_0 + \beta_3) + (\beta_1 + \beta_6)X_1 \qquad \text{Tool models M2}$$

(10.13d) $$E(Y) = (\beta_0 + \beta_4) + (\beta_1 + \beta_7)X_1 \qquad \text{Tool models M3}$$

Thus, the interaction model (10.12) implies that each tool model has its own regression line, with different intercepts and slopes for the different tool models.

More than one qualitative independent variable

Models can readily be constructed for cases where two or more of the independent variables are qualitative. Consider the regression of advertising expenditures (Y) on sales (X_1), type of firm (incorporated, not incorporated), and quality of sales management (high, low). We may define:

(10.14)
$$X_2 = \begin{array}{l} 1 \text{ if firm incorporated} \\ 0 \text{ otherwise} \end{array}$$

$$X_3 = \begin{array}{l} 1 \text{ if quality of sales management high} \\ 0 \text{ otherwise} \end{array}$$

First-order model. A first-order model for the above example is:

(10.15) $$Y_i = \beta_0 + \beta_1 X_{i1} + \beta_2 X_{i2} + \beta_3 X_{i3} + \varepsilon_i$$

This model implies that the response function of advertising expenditures on sales is linear, with the same slope for all "type of firm–quality of sales management" combinations, and β_2 and β_3 indicate the additive differential effects of type of firm and quality of sales management on the height of the regression line.

First-order model with certain interactions added. A first-order model with interaction effects between pairs of the independent variables added is:

(10.16) $$Y_i = \beta_0 + \beta_1 X_{i1} + \beta_2 X_{i2} + \beta_3 X_{i3} + \beta_4 X_{i1}X_{i2}$$
$$+ \beta_5 X_{i1}X_{i3} + \beta_6 X_{i2}X_{i3} + \varepsilon_i$$

Note the implications of this model:

Type of Firm	Quality of Sales Management	Response Function
Incorp.	High	$E(Y) = (\beta_0 + \beta_2 + \beta_3 + \beta_6) + (\beta_1 + \beta_4 + \beta_5)X_1$
Not incorp.	High	$E(Y) = (\beta_0 + \beta_3) + (\beta_1 + \beta_5)X_1$
Incorp.	Low	$E(Y) = (\beta_0 + \beta_2) + (\beta_1 + \beta_4)X_1$
Not incorp.	Low	$E(Y) = \beta_0 + \beta_1 X_1$

Not only are all response functions different for the various "type of firm–quality of sales management" combinations, but the differential effects of one qualitative variable on the intercept depend on the particular class of the other qualitative variable. For instance, when we move from "not incorporated–low quality" to "incorporated–low quality," the intercept changes by β_2. But if we move from "not incorporated–high quality" to "incorporated–high quality," the intercept changes by $\beta_2 + \beta_6$.

Qualitative independent variables only

Regression models containing only qualitative independent variables can also be constructed. With reference to our previous example, we could regress advertising expenditures only on type of firm and quality of sales management. The first-order model then would be:

$$(10.17) \qquad Y_i = \beta_0 + \beta_2 X_{i2} + \beta_3 X_{i3} + \varepsilon_i$$

where X_{i2} and X_{i3} are defined in (10.14).

Comments

1. Models in which all independent variables are qualitative are called *analysis of variance models*.

2. Models containing some quantitative and some qualitative independent variables, where the chief independent variables of interest are qualitative and the quantitative independent variables are introduced primarily to reduce the variance of the error terms, are called *covariance models*.

10.4   OTHER USES OF INDEPENDENT INDICATOR VARIABLES

Comparison of two or more regression functions

Frequently we encounter regressions for two or more populations and wish to examine their similarities and differences. We present two examples.

1. An economist is studying the relation between amount of savings and level of income for middle-income families from urban and rural areas, based on independent samples from the two populations. Each of the two relations can be modeled by linear regression. She wishes to compare whether, at given income levels, urban and rural families tend to save the same amount—i.e., whether the two regression lines are the same. If they are not, she wishes to explore whether at least the amounts of savings out of an additional dollar of income are the same for the two groups—i.e., whether the slopes of the two regression lines are the same.

2. A company has two instruments constructed to identical specifications to measure pressure in an industrial process. A study has been made for each instrument of the relation between its gauge readings and actual pressures as determined by an almost exact but slow and costly method. If the two regression lines are the same, a single calibration schedule can be developed for the two instruments; otherwise, two different calibration schedules would be required.

When it is reasonable to assume that the error term variances in the regression models for the different populations are equal, use of indicator variables permits us to test the equality of the different regression functions. If the error variances are not equal, transformations may equalize them at least approximately.

We have already seen that regression models with indicator variables that contain interaction terms permit testing of the equality of regression functions for the different classes of a qualitative variable. This methodology can be used directly for testing the equality of regression functions for different populations. One simply considers the different populations studied as classes of an independent variable, defines indicator variables for the different populations, and develops a regression model containing appropriate interaction terms. Since no new principles arise in the testing of the equality of regression functions for different populations, we shall utilize the two earlier examples to illustrate the approach.

Example 1—Savings study. In the savings study example, the economist was willing to assume that the regression relation between savings (Y) and income (X_1) for middle-income families is linear for both rural and urban families, and that the error term variances for the two populations are the same. Since she wished to test whether the two regression lines are the same, she fitted model (10.6) which permits both the slopes and the intercepts to be different in the two regressions:

$$(10.18) \qquad Y_i = \beta_0 + \beta_1 X_{i1} + \beta_2 X_{i2} + \beta_3 X_{i1} X_{i2} + \varepsilon_i$$

where:

$$X_{i1} = \text{Family income}$$

$$X_{i2} = \begin{array}{l} 1 \text{ if urban family} \\ 0 \text{ otherwise} \end{array}$$

Identity of the regression functions is tested by considering the alternatives:

$$(10.19) \qquad \begin{array}{l} H_0\colon \beta_2 = \beta_3 = 0 \\ H_a\colon \text{Not both } \beta_2 = 0 \text{ and } \beta_3 = 0 \end{array}$$

The appropriate test statistic is given by (8.32a):

$$(10.19a) \qquad F^* = \frac{SSR(X_2, X_1 X_2 \mid X_1)}{2} \div \frac{SSE(X_1, X_2, X_1 X_2)}{n - 4}$$

where n represents the combined sample size for both populations.

If the economist only wished to test whether the slopes of the regression lines are the same, the alternatives would be:

$$(10.20) \qquad \begin{array}{l} H_0\colon \beta_3 = 0 \\ H_a\colon \beta_3 \neq 0 \end{array}$$

and the appropriate test statistic is either the t^* statistic (8.25) or the partial F test statistic (8.24):

(10.20a) $F^* = \dfrac{SSR(X_1X_2|X_1, X_2)}{1} \div \dfrac{SSE(X_1, X_2, X_1X_2)}{n - 4}$

If the economist wishes to estimate the difference in the slopes of the regression lines for urban and rural families, she would construct a confidence interval for β_3 in the usual fashion.

Example 2—Instrument calibration study.   The engineer making the calibration study believed that the regression functions relating gauge reading (Y) and actual pressure (X_1) for both instruments are second-order polynomials:

$$E(Y) = \beta_0 + \beta_1 X_1 + \beta_2 X_1^2$$

but that they might differ for the two instruments. Hence, he employed the model (using a deviation variable for X_1 to reduce multicollinearity problems—see Chapter 9):

(10.21) $Y_i = \beta_0 + \beta_1 x_{i1} + \beta_2 x_{i1}^2 + \beta_3 X_{i2} + \beta_4 x_{i1} X_{i2} + \beta_5 x_{i1}^2 X_{i2} + \varepsilon_i$

where:

$x_{i1} = X_{i1} - \bar{X}_1 = $ deviation of actual pressure

$X_{i2} = \begin{matrix} 1 \text{ if instrument B} \\ 0 \text{ otherwise} \end{matrix}$

Note that for instrument A, where $X_2 = 0$, the response function is:

(10.22a) $$E(Y) = \beta_0 + \beta_1 x_1 + \beta_2 x_1^2$$

and for instrument B, where $X_2 = 1$, the response function is:

(10.22b) $E(Y) = (\beta_0 + \beta_3) + (\beta_1 + \beta_4)x_1 + (\beta_2 + \beta_5)x_1^2$

Hence, the test for equality of the two response functions involves the alternatives:

(10.23) $H_0: \beta_3 = \beta_4 = \beta_5 = 0$
$H_a:$ Not all three $\beta_k = 0 \ (k = 3, 4, 5)$

and the appropriate test statistic is (8.32a):

(10.23a) $F^* = \dfrac{SSR(X_2, x_1X_2, x_1^2X_2|x_1, x_1^2)}{3} \div \dfrac{SSE(x_1, x_1^2, X_2, x_1X_2, x_1^2X_2)}{n - 6}$

where n represents the combined sample size for both populations.

Comments

1. The approach just described is completely general. If three or more populations are involved, additional indicator variables would simply be added to the model.
2. The use of indicator variables for testing whether two or more regression functions are the same is equivalent to the general linear test approach where fitting the full model involves fitting separate regressions to the data from each population and fitting the reduced model involves fitting one regression to the combined data.

Piecewise linear regression

Sometimes the regression of Y on X follows a particular linear relation in some range of X, but follows a different linear relation elsewhere. For instance, unit cost (Y) regressed on lot size may follow a certain linear regression up to $X_p = 500$, at which point the slope changes because of some operating efficiencies only possible with lot sizes of more than 500. Figure 10.6 illustrates this situation.

FIGURE 10.6   Illustration of piecewise linear regression

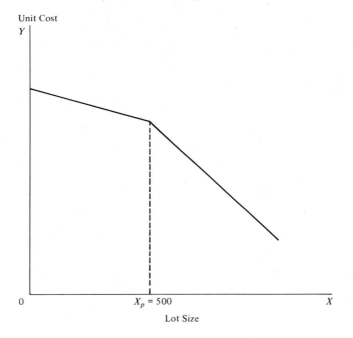

We consider now how indicator variables may be used to fit piecewise linear regressions consisting of two pieces. We take up the case where X_p, the point where the slope changes, is known.

We return to our lot size illustration, for which it is known that the slope changes at $X_p = 500$. The model for our illustration may be expressed as follows:

$$(10.24) \qquad Y_i = \beta_0 + \beta_1 X_{i1} + \beta_2(X_{i1} - 500)X_{i2} + \varepsilon_i$$

where:

$$X_{i1} = \text{Lot size}$$

$$X_{i2} = \begin{array}{l} 1 \text{ if } X_{i1} > 500 \\ 0 \text{ otherwise} \end{array}$$

To check that model (10.24) does provide a two-piecewise linear regression, consider the response function:

$$(10.25) \qquad E(Y) = \beta_0 + \beta_1 X_1 + \beta_2(X_1 - 500)X_2$$

When $X_1 \leq 500$, $X_2 = 0$ so that (10.25) becomes:

$$(10.25a) \qquad E(Y) = \beta_0 + \beta_1 X_1 \qquad X_1 \leq 500$$

On the other hand, when $X_1 > 500$, $X_2 = 1$ and we obtain:

$$(10.25b) \qquad E(Y) = (\beta_0 - 500\beta_2) + (\beta_1 + \beta_2)X_1 \qquad X_1 > 500$$

Thus, β_1 and $\beta_1 + \beta_2$ are the slopes of the two regression lines, and β_0 and $(\beta_0 - 500\beta_2)$ are the two Y intercepts. These parameters are shown in Figure 10.7.

Example. Table 10.6a contains eight observations on unit costs for given lot sizes. It is known that the response function slope changes at $X_p = 500$ so that

FIGURE 10.7 Illustration of parameters of piecewise linear regression model (10.24)

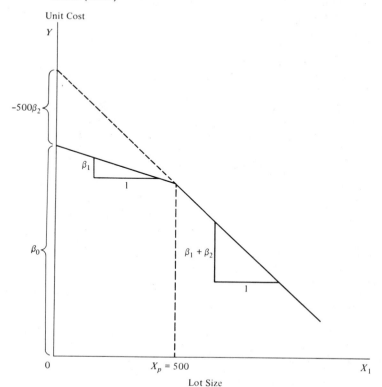

TABLE 10.6 Data and **X** matrix for piecewise linear regression—lot size example

	(a)				(b)	
Lot i	Unit Cost (dollars) Y_i	Lot Size X_i		X_0	X_1	$(X_1 - 500)X_2$
1	2.57	650		1	650	150
2	4.40	340		1	340	0
3	4.52	400		1	400	0
4	1.39	800	$X =$	1	800	300
5	4.75	300		1	300	0
6	3.55	570		1	570	70
7	2.49	720		1	720	220
8	3.77	480		1	480	0

model (10.24) is to be employed. Table 10.6b contains the **X** matrix for our example. The left column of **X** is a column of 1's as usual. The next column contains X_{i1}. The final column on the right contains $(X_{i1} - 500)X_{i2}$, which consists of 0's for all lot sizes up to 500, and of $X_{i1} - 500$ for all lot sizes above 500. The fitting of regression model (10.24) at this point becomes routine. The fitted response function is:

$$\hat{Y} = 5.89545 - .00395X_1 - .00389(X_1 - 500)X_2$$

From this fitted model, expected unit cost is estimated to decline by .00395 for a lot size increase of one when the lot size is less than 500 and by .00395 + .00389 = .00784 when the lot size is 500 or more.

Note

The extension of model (10.24) to more than two-piecewise regression lines is straightforward. For instance, if the slope of the regression line in our earlier lot size illustration actually changes at both $X = 500$ and $X = 800$, the model would be:

(10.26) $\quad Y_i = \beta_0 + \beta_1 X_{i1} + \beta_2(X_{i1} - 500)X_{i2} + \beta_3(X_{i1} - 800)X_{i3} + \varepsilon_i$

where:

X_{i1} = Lot size

$$X_{i2} = \begin{cases} 1 \text{ if } X_{i1} > 500 \\ 0 \text{ otherwise} \end{cases}$$

$$X_{i3} = \begin{cases} 1 \text{ if } X_{i1} > 800 \\ 0 \text{ otherwise} \end{cases}$$

Discontinuity in regression function

Sometimes the linear regression function may not only change its slope at some value X_p but may also have a jump point there. Figure 10.8 illustrates this case. Another indicator variable must now be introduced to take care of the jump. Suppose time required to solve a task successfully (Y) is to be regressed on complexity of task (X), when complexity of task is measured on a quantitative scale from 0 to 100. It is known that the slope of the response line changes at $X_p = 40$, and it is believed that the regression relation may be discontinuous there. We therefore set up the model:

(10.27) $Y_i = \beta_0 + \beta_1 X_{i1} + \beta_2(X_{i1} - 40)X_{i2} + \beta_3 X_{i3} + \varepsilon_i$

where:

$$X_{i1} = \text{Complexity of task}$$

$$X_{i2} = \begin{array}{l} 1 \text{ if } X_{i1} > 40 \\ 0 \text{ otherwise} \end{array}$$

$$X_{i3} = \begin{array}{l} 1 \text{ if } X_{i1} > 40 \\ 0 \text{ otherwise} \end{array}$$

The response function for model (10.27) is:

(10.28) $E(Y) = \beta_0 + \beta_1 X_1 + \beta_2(X_1 - 40)X_2 + \beta_3 X_3$

When $X_1 \leq 40$, then $X_2 = 0$ and $X_3 = 0$, so (10.28) becomes:

(10.28a) $E(Y) = \beta_0 + \beta_1 X_1 \qquad X_1 \leq 40$

FIGURE 10.8   Illustration of model (10.27) for discontinuous piecewise linear regression

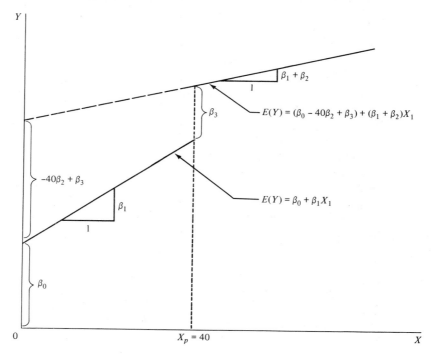

Similarly, when $X_1 > 40$, then $X_2 = 1$ and $X_3 = 1$, so (10.28) becomes:

(10.28b) $E(Y) = (\beta_0 - 40\beta_2 + \beta_3) + (\beta_1 + \beta_2)X_1$ $X_1 > 40$

These two response functions are shown in Figure 10.8, together with the parameters involved. Note that β_3 represents the difference in the mean responses for the two regression lines at $X_p = 40$ and β_2 represents the difference in the two slopes.

The estimation of the regression coefficients for model (10.27) presents no new problems. One may test whether or not $\beta_3 = 0$ in the usual manner. If it is concluded that $\beta_3 = 0$, the regression function is continuous at X_p so that the earlier piecewise linear regression model applies.

Time series applications

Economists and business analysts frequently use time series data in regression analysis. For instance, savings (Y) may be regressed on income (X), where both the savings and income data pertain to a number of years. The model employed might be:

(10.29) $Y_t = \beta_0 + \beta_1 X_t + \varepsilon_t$ $t = 1, \ldots, n$

where Y_t and X_t are savings and income, respectively, for time period t. Suppose that the period covered includes both peacetime and wartime years, and that this factor should be recognized since it is anticipated that savings in wartime years tend to be higher. The following model might then be appropriate:

(10.30) $Y_t = \beta_0 + \beta_1 X_{t1} + \beta_2 X_{t2} + \varepsilon_t$

where:

$X_{t1} = $ Income

$X_{t2} = \begin{matrix} 1 \text{ if period } t \text{ peacetime} \\ 0 \text{ otherwise} \end{matrix}$

Note that model (10.30) assumes that the marginal propensity to save (β_1) is constant in both peacetime and wartime years, and that only the height of the response curve is affected by this qualitative variable.

Another use of indicator variables in time series applications occurs when monthly or quarterly data are used. Suppose that quarterly sales (Y) are regressed on quarterly advertising expenditures (X_1) and quarterly disposable personal income (X_2). If seasonal effects also have an influence on quarterly sales, a first-order model incorporating seasonal effects would be:

(10.31) $Y_t = \beta_0 + \beta_1 X_{t1} + \beta_2 X_{t2} + \beta_3 X_{t3} + \beta_4 X_{t4} + \beta_5 X_{t5} + \varepsilon_t$

where:

$X_{t1} = $ Quarterly advertising expenditures
$X_{t2} = $ Quarterly disposable personal income

$$X_{t3} = \begin{array}{l} 1 \text{ if first quarter} \\ 0 \text{ otherwise} \end{array}$$

$$X_{t4} = \begin{array}{l} 1 \text{ if second quarter} \\ 0 \text{ otherwise} \end{array}$$

$$X_{t5} = \begin{array}{l} 1 \text{ if third quarter} \\ 0 \text{ otherwise} \end{array}$$

10.5 SOME CONSIDERATIONS IN USING INDEPENDENT INDICATOR VARIABLES

Indicator variables versus allocated codes

An alternative to the use of indicator variables for a qualitative independent variable is to employ *allocated codes*. Consider, for instance, the independent variable "frequency of product use" which has three classes: frequent user, occasional user, nonuser. With the allocated codes approach, a single independent variable is employed and values are assigned to the classes; for instance:

Class	X_1
Frequent user	3
Occasional user	2
Nonuser	1

The allocated codes are, of course, arbitrary and could be other sets of numbers. The model with allocated codes for our example, assuming no other independent variables, would be:

(10.32)
$$Y_i = \beta_0 + \beta_1 X_{i1} + \varepsilon_i$$

The basic difficulty with allocated codes is that they define a metric for the classes of the qualitative variable which may not be reasonable. To see this concretely, consider the mean responses with model (10.32) for the three classes of the qualitative variable:

Class	$E(Y)$
Frequent user	$E(Y) = \beta_0 + \beta_1(3) = \beta_0 + 3\beta_1$
Occasional user	$E(Y) = \beta_0 + \beta_1(2) = \beta_0 + 2\beta_1$
Nonuser	$E(Y) = \beta_0 + \beta_1(1) = \beta_0 + \beta_1$

Note the key implication:

$$E(Y|\text{frequent user}) - E(Y|\text{occasional user})$$
$$= E(Y|\text{occasional user}) - E(Y|\text{nonuser}) = \beta_1$$

Thus, the coding 1, 2, 3 implies equal distances between the three user classes, which may not be in accord with reality. Other allocated codes may, of course, imply different spacings of the classes of the qualitative variable, but these would ordinarily still be arbitrary.

Indicator variables, in contrast, make no assumptions about the spacing of the classes and rely on the data to show the differential effects that occur. If, for the same example, two indicator variables, say, X_1 and X_2, are employed to represent the qualitative variable, as follows:

Class	X_1	X_2
Frequent user	1	0
Occasional user	0	1
Nonuser	0	0

the model would be:

$$(10.33) \qquad Y_i = \beta_0 + \beta_1 X_{i1} + \beta_2 X_{i2} + \varepsilon_i$$

Here β_1 measures the differential effect:

$$E(Y|\text{frequent user}) - E(Y|\text{nonuser})$$

and β_2 measures:

$$E(Y|\text{occasional user}) - E(Y|\text{nonuser})$$

Thus, β_1 and β_2 measure the differential effects of user status relative to the class of nonusers without any arbitrary restrictions to be satisfied by these differential effects.

Indicator variables versus quantitative variables

If an independent variable is quantitative, such as age, one can nevertheless use indicator variables instead. For instance, the quantitative variable age may be transformed by grouping ages into classes such as under 21, 21–34, 35–49, etc. Indicator variables may then be used for the classes of this new independent variable. At first sight, this may seem to be a questionable approach because information about the actual ages is thrown away. Furthermore, additional parameters are placed into the model, which leads to a reduction of the degrees of freedom associated with MSE.

Nevertheless, there are occasions when replacement of a quantitative variable by indicator variables may be appropriate. Consider, for example, a large-scale survey in which the relation between liquid assets (Y) and age (X) of head of household is to be studied. Two thousand households will be included in the study, so that the loss of 10 or 20 degrees of freedom is immaterial. The analyst is very much in doubt about the shape of the regression function, which could be highly complex, and hence may prefer the indicator variable approach in order to obtain information about the shape without making any assumptions about the functional form of the regression function.

Another alternative, also utilizing indicator variables, is available to the analyst in doubt about the functional form of a possibly complex regression function. He or she could use the quantitative variable age, but employ piecewise linear regression with a number of pieces. Again, this approach loses degrees of freedom for estimating *MSE*, but this is of no concern in large-scale studies. The benefit would be that information about the shape of the regression function is obtained without making strong assumptions about its functional form.

Other codings for indicator variables

As stated earlier, many different codings of indicator variables are possible. We mention here two alternative codings to our 0, 1 coding with $c - 1$ indicator variables for a qualitative variable with c classes.

For the insurance innovation example, where Y is time to adopt an innovation, X_1 is size of insurance firm, and the second independent variable is type of company (stock, mutual), we could use the coding:

$$(10.34) \qquad X_2 = \begin{array}{l} 1 \text{ if stock company} \\ -1 \text{ if mutual company} \end{array}$$

In that case, the first-order linear regression model would be:

$$(10.35) \qquad Y_i = \beta_0 + \beta_1 X_{i1} + \beta_2 X_{i2} + \varepsilon_i$$

which has response function:

$$(10.36) \qquad E(Y) = \beta_0 + \beta_1 X_1 + \beta_2 X_2$$

This response function is as follows for the two types of companies:

$$(10.36a) \qquad E(Y) = (\beta_0 + \beta_2) + \beta_1 X_1 \qquad \text{Stock firms}$$

$$(10.36b) \qquad E(Y) = (\beta_0 - \beta_2) + \beta_1 X_1 \qquad \text{Mutual firms}$$

Thus, β_0 here may be viewed as an "average" intercept of the regression line, from which the stock company and mutual company intercepts differ by β_2 in opposite directions. A test whether the regression lines are the same for both types of companies involves $H_0: \beta_2 = 0$, $H_a: \beta_2 \neq 0$.

A second alternative coding scheme is to use a 0, 1 indicator variable for each of the c classes of the qualitative variable and drop the intercept term in the regression model. For our insurance innovation example, we would have:

$$(10.37) \qquad Y_i = \beta_1 X_{i1} + \beta_2 X_{i2} + \beta_3 X_{i3} + \varepsilon_i$$

where:

$$X_{i1} = \text{Size of firm}$$

$$X_{i2} = \begin{array}{l} 1 \text{ if stock company} \\ 0 \text{ otherwise} \end{array}$$

$$X_{i3} = \begin{array}{l} 1 \text{ if mutual company} \\ 0 \text{ otherwise} \end{array}$$

Here, the two response functions would be:

$$(10.38a) \qquad E(Y) = \beta_2 + \beta_1 X_1 \qquad \text{Stock firms}$$

$$(10.38b) \qquad E(Y) = \beta_3 + \beta_1 X_1 \qquad \text{Mutual firms}$$

A test of whether or not the two regression lines are the same would involve the alternatives H_0: $\beta_2 = \beta_3$, H_a: $\beta_2 \neq \beta_3$. This type of test is discussed in Section 8.4.

10.6 DEPENDENT INDICATOR VARIABLE

In a variety of applications, the dependent variable of interest has only two possible outcomes, and therefore can be represented by an indicator variable taking on values 0 and 1.

1. In an analysis of whether or not business firms have an industrial relations department, according to size of firm, the dependent variable was defined to have the two possible outcomes: firm has industrial relations department, firm does not have industrial relations department. These outcomes may be coded 1 and 0, respectively (or vice versa).

2. In a study of labor force participation of wives, as a function of age of wife, number of children, and husband's income, the dependent variable Y was defined to have the two possible outcomes: wife in labor force, wife not in labor force. Again, these outcomes may be coded 1 and 0, respectively.

3. In a study of liability insurance possession, according to age of head of household, amount of liquid assets, and type of occupation of head of household, the dependent variable Y was defined to have the two possible outcomes: household has liability insurance policy, household does not have liability insurance policy. These outcomes again can be coded 1 and 0, respectively.

These examples show the wide range of applications in which the dependent variable is dichotomous, and hence may be represented by an indicator variable. A dependent dichotomous variable, taking on the values 0 and 1, is sometimes said to involve *quantal responses* or *binary responses*.

Meaning of response function when dependent variable is binary

Consider the simple linear regression model:

$$(10.39) \qquad Y_i = \beta_0 + \beta_1 X_i + \varepsilon_i \qquad Y_i = 0, 1$$

where the responses Y_i are binary 0, 1 observations. The expected response $E(Y_i)$ has a special meaning in this case. Since $E(\varepsilon_i) = 0$ as usual, we have:

$$(10.40) \qquad E(Y_i) = \beta_0 + \beta_1 X_i$$

Consider now Y_i as an ordinary Bernoulli random variable for which we can state the probability distribution as follows:

Y_i	Probability
1	$P(Y_i = 1) = \pi_i$
0	$P(Y_i = 0) = 1 - \pi_i$

Thus, π_i is the probability that $Y_i = 1$ and $1 - \pi_i$ is the probability that $Y_i = 0$. By the ordinary definition of expected value of a random variable in (1.12), we obtain:

(10.41) $$E(Y_i) = 1(\pi_i) + 0(1 - \pi_i) = \pi_i$$

Equating (10.40) and (10.41), we thus find:

(10.42) $$E(Y_i) = \beta_0 + \beta_1 X_i = \pi_i$$

The mean response $E(Y_i) = \beta_0 + \beta_1 X_i$ as given by the response function is therefore simply the probability that $Y_i = 1$ when the level of the independent variable is X_i. This interpretation of the mean response applies whether the response function is a simple linear one, as here, or a complex multiple regression one. The mean response, when the dependent variable is a 0,1 indicator variable, always represents the probability that $Y = 1$ for the given levels of the independent variables. Figure 10.9 illustrates a simple linear response function for a dependent indicator variable. Here, the indicator variable Y refers to whether or

FIGURE 10.9 Illustration of response function when dependent variable is binary—industrial relations department example

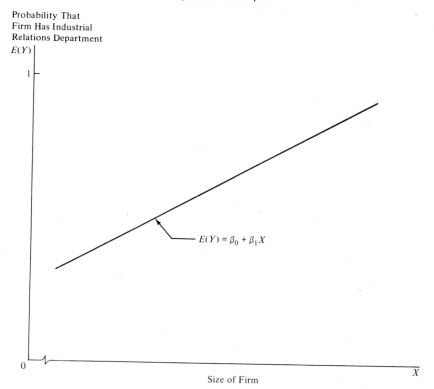

Probability That
Firm Has Industrial
Relations Department
$E(Y)$

$E(Y) = \beta_0 + \beta_1 X$

0

X

Size of Firm

not a firm has an industrial relations department, and the independent variable X is size of firm. The response function in Figure 10.9 shows the probability that firms of given size have an industrial relations department.

Special problems when dependent variable is binary

Special problems arise, unfortunately, when the dependent variable is an indicator variable. We shall consider three of these now, using a simple linear regression model as an illustration.

1. Nonnormal error terms. For a binary 0,1 dependent variable, the error terms $\varepsilon_i = Y_i - (\beta_0 + \beta_1 X_i)$ can take on only two values:

(10.43a) $\qquad$ When $Y_i = 1$: $\varepsilon_i = 1 - \beta_0 - \beta_1 X_i$

(10.43b) $\qquad$ When $Y_i = 0$: $\varepsilon_i = -\beta_0 - \beta_1 X_i$

Clearly, the normal error regression model (3.1), assuming that the ε_i are normally distributed, is not appropriate.

2. Nonconstant error variance. Another problem with the error terms ε_i is that they do not have equal variances when the dependent variable is an indicator variable. To see this, we shall obtain $\sigma^2(Y_i)$ for the simple linear regression model (10.39):

$$\sigma^2(Y_i) = E\{[Y_i - E(Y_i)]^2\} = (1 - \pi_i)^2 \pi_i + (0 - \pi_i)^2(1 - \pi_i)$$

or:

(10.44) $\qquad$ $\sigma^2(Y_i) = \pi_i(1 - \pi_i) = [E(Y_i)][1 - E(Y_i)]$

The variance of ε_i is, of course, the same as that of Y_i because $\varepsilon_i = Y_i - \pi_i$ and π_i is a constant:

(10.45) $\qquad$ $\sigma^2(\varepsilon_i) = \pi_i(1 - \pi_i) = [E(Y_i)][1 - E(Y_i)]$

or:

(10.45a) $\qquad$ $\sigma^2(\varepsilon_i) = (\beta_0 + \beta_1 X_i)(1 - \beta_0 - \beta_1 X_i)$

Note from (10.45a) that $\sigma^2(\varepsilon_i)$ depends on X_i. Hence, the error term variances will differ at different levels of X, and ordinary least squares will no longer be optimal.

3. Constraints on response function. Since the response function represents probabilities when the dependent variable is a 0,1 indicator variable, the mean responses should be constrained as follows:

(10.46) $\qquad$ $0 \le E(Y) = \pi \le 1$

Many response functions do not automatically possess this constraint. A linear response function, for instance, may fall outside the constraint limits within the range of the independent variable in the scope of the model.

These three problems create difficulties, but solutions can be found. Problem 2 concerning unequal error variances, for instance, can be handled by using weighted least squares. Problem 3 about the constraints on the response function can be handled by making sure that the mean responses for the fitted model do not fall below 0 or above 1 for levels of X within the scope of the model, or else by using a model which automatically meets the constraints.

Finally, even though the error terms are not normal when the dependent variable is binary, the method of least squares still provides unbiased estimators that are asymptotically normal under quite general conditions. Hence, when the sample size is large, inferences concerning the regression coefficients and mean responses can be made in the same fashion as when the error terms are assumed to be normally distributed.

10.7   LINEAR REGRESSION WITH DEPENDENT INDICATOR VARIABLE

We consider now the fitting of linear response functions when the dependent variable is binary. Then we shall explain the fitting of curvilinear response functions.

Illustration

A systems analyst studied the effect of computer programming experience on ability to complete a complex programming task, including debugging, within a specified time. Twenty-five persons were selected for the study. They had varying amounts of programming experience (measured in months of experience), as shown in Table 10.7, column 1. All persons were given the same programming task, and the results of their success in the task are shown in Table 10.7, column 2. The results are coded in binary fashion: if the task was completed successfully in the allotted time, $Y = 1$, and if the task was not completed successfully, $Y = 0$. Figure 10.10 contains a scatter plot of the data. This plot is not too informative because of the nature of the dependent variable, other than to indicate that ability to complete the task successfully appears to increase with amount of experience. At this point, it was decided to fit the linear regression model (10.39).

Ordinary least squares fit

One approach to fitting model (10.39) is to fit it by ordinary least squares despite the unequal error variances. The estimated regression coefficients will still be unbiased, but they will no longer have the minimum variance property among the class of unbiased linear estimators. Thus, the use of ordinary least squares may lead to inefficient estimates, i.e., estimates with larger variances than could be obtained with weighted procedures.

TABLE 10.7 Data for programming task example

Person i	(1) Months of Experience X_i	(2) Task Success Y_i	(3) $\hat{Y}_i$	(4) $\hat{w}_i$
1	14	0	.34920	4.4003
2	29	0	.82212	6.8382
3	6	0	.09697	11.4196
4	25	1	.69601	4.7263
5	18	1	.47531	4.0098
6	4	0	.03392	30.5196
7	18	0	.47531	4.0098
8	12	0	.28614	4.8956
9	22	1	.60142	4.1717
10	6	0	.09697	11.4196
11	30	1	.85365	8.0044
12	11	0	.25461	5.2691
13	30	1	.85365	8.0044
14	5	0	.06544	16.3502
15	20	1	.53837	4.0237
16	13	0	.31767	4.6135
17	9	0	.19156	6.4573
18	32	1	.91671	13.0967
19	24	0	.66448	4.4854
20	13	1	.31767	4.6135
21	19	0	.50684	4.0007
22	4	0	.03392	30.5196
23	28	1	.79059	6.0403
24	22	1	.60142	4.1717
25	8	1	.16003	7.4394

The ordinary least squares fit to the data in Table 10.7 leads to the following results:

Coefficient	Estimated Coefficient	Estimated Standard Deviation
β_0	$b_0 = -.092197$	$s(b_0) = .183272$
β_1	$b_1 = .031528$	$s(b_1) = .009606$

The estimated response function therefore is:

$$(10.47) \qquad \hat{Y} = -.092197 + .031528X$$

This response function is shown in Figure 10.10. It may be used in the ordinary manner. For instance, the estimated mean response for persons with $X_h = 14$ months experience is:

$$\hat{Y}_h = -.092197 + .031528(14) = .34920$$

Thus, we estimate that the probability is .349 that a person with 14 months experience will successfully complete the programming task.

FIGURE 10.10 Scatter plot and fitted regression line—programming task example

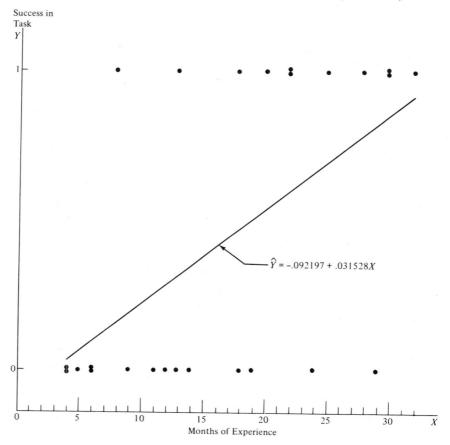

As noted earlier from the scatter plot, this probability increases with increasing experience. The coefficient b_1 indicates that the estimated probability increases by .0315 for each additional month of experience. Clearly, the model cannot be extrapolated much beyond the range of the data here or else the estimated probabilities would be negative or exceed 1, meaningless results. For $X_h = 40$ months experience, for instance, $\hat{Y}_h = 1.17$.

Weighted least squares fit

It was pointed out in Chapter 5 that weighted least squares provides efficient estimates when the error variances are unequal. Since we know from (10.45) that:

$$\sigma^2(\varepsilon_i) = \pi_i(1 - \pi_i) = [E(Y_i)][1 - E(Y_i)]$$

it would appear that using weighted least squares with weights:

$$(10.48) \qquad w_i = \frac{1}{\pi_i(1 - \pi_i)} = \frac{1}{[E(Y_i)][1 - E(Y_i)]}$$

is the proper approach to take. There exists a difficulty, however, in carrying this out, namely, that the w_i in (10.48) involve $E(Y_i)$ and hence the parameters β_0 and β_1, which are unknown and must be estimated.

A way out of this difficulty is to use a two-stage least squares procedure:

1. Stage 1—fit the regression model by ordinary least squares.
2. Stage 2—estimate the weights w_i from the results of stage 1:

$$(10.49) \qquad \hat{w}_i = \frac{1}{\hat{Y}_i(1 - \hat{Y}_i)}$$

and then use these estimated weights in the matrix $\mathbf{W}$ in (7.70) for obtaining weighted least squares estimates for the regression model by means of (7.71).

To apply this approach in our example, we first need to find the estimated weights $\hat{w}_i$ from the ordinary least squares fit. This ordinary least squares fit has been obtained earlier in (10.47). Hence, we are ready to calculate $\hat{Y}_i$ and then $1/\hat{Y}_i(1 - \hat{Y}_i)$ for each observation. For the first observation, for instance, $X_1 = 14$ so that $\hat{Y}_1 = .34920$, as found earlier. Hence, the estimated weight for the first observation is:

$$\hat{w}_1 = \frac{1}{.34920(.65080)} = 4.4003$$

In the same manner, the other weights are calculated. In Table 10.7, columns 3 and 4 contain the fitted values and the estimated weights, respectively.

To obtain the weighted least squares estimates for our example, we utilized a computer run with the weights as given in Table 10.7. This led to the following results:

Regression Coefficient	Estimated Regression Coefficient	Estimated Standard Deviation
β_0	$b_0 = -.117113$	$s(b_0) = .111841$
β_1	$b_1 = .032672$	$s(b_1) = .006644$

The fitted response function by the two-stage least squares approach therefore is:

$$(10.50) \qquad \hat{Y} = -.117113 + .032672X$$

Note that this estimated response function does not differ markedly from the one obtained by unweighted least squares in (10.47), although the estimated standard deviations of the regression coefficients are now somewhat smaller.

Comments

1. If the mean responses range between about .2 and .8 for the scope of the model, there is little to be gained from weighted least squares since the error variances will differ but little. Only if the mean responses range below .2 and/or above .8 will the error variances be sufficiently unequal to make weighted least squares worthwhile.

2. In our example, the fitted response function does not fall below 0 or above 1 within the range of the data (4 months to 32 months experience). If it did, a curvilinear response function would have to be employed. We discuss one important curvilinear model below.

3. The weighted least squares approach could employ additional stages, with the weights refined at each stage. Usually, however, the gain from additional iterations is not large. For our example, for instance, another iteration led to the following results:

$$b_0 = -.125665 \qquad s(b_0) = .084555$$
$$b_1 = .033075 \qquad s(b_1) = .005885$$

Note that there is relatively little decrease from the previous stage in the estimated standard deviations and only small changes in the regression coefficients.

4. If there are repeat observations at the different X levels, the procedure of fitting a linear response function can be simplified. Let $X_1, \ldots, X_c$ denote the X levels, p_j the proportion of 1's at X_j $(j = 1, \ldots, c)$, and n_j the number of observations at X_j. Fitting the sample proportions p_j with weights:

$$(10.51) \qquad w_j = n_j$$

leads to the identical estimated response function as ordinary least squares applied to the individual Y observations. If there are hundreds of Y observations located at a limited number of X_j levels, much computational effort can be saved by fitting the sample proportions p_j.

If weighted least squares estimates are to be obtained, only one stage is required when sample proportions p_j are available. The weights would be as follows:

$$(10.52) \qquad \hat{w}_j = \frac{n_j}{p_j(1 - p_j)}$$

where $\hat{w}_j$ denotes the estimated weight.

We discuss fitting a regression model to grouped data more fully below.

5. While we have illustrated only simple linear regression, the extension to multiple regression is straightforward.

10.8 LOGISTIC RESPONSE FUNCTION

Both theoretical and empirical considerations suggest that when the dependent variable is an indicator variable, the shape of the response function will frequently be curvilinear. Figure 10.11 contains a curvilinear response function which has been found appropriate in many instances involving a binary dependent variable. Note that this response function is shaped like a tilted S, and that it has asymptotes at 0 and 1. The latter feature assures that the constraints on $E(Y)$ in (10.46) are automatically met. The response function plotted in Figure 10.11 is called the *logistic* function and is given by:

FIGURE 10.11 Example of logistic response function

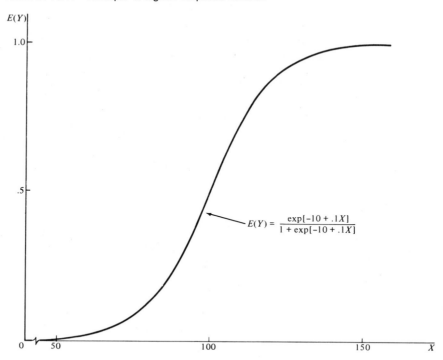

$$E(Y) = \frac{\exp[-10 + .1X]}{1 + \exp[-10 + .1X]}$$

$$(10.53) \qquad E(Y) = \frac{\exp(\beta_0 + \beta_1 X)}{1 + \exp(\beta_0 + \beta_1 X)}$$

An interesting property of the logistic function is that it can easily be linearized. Let us denote $E(Y)$ by π, since the mean response is a probability when the dependent variable is a 0,1 indicator variable. Then if we make the transformation:

$$(10.54) \qquad \pi' = \log_e\left(\frac{\pi}{1 - \pi}\right)$$

we obtain from (10.53):

$$(10.55) \qquad \pi' = \beta_0 + \beta_1 X$$

The transformation (10.54) is called the *logistic* or *logit* transformation of the probability π.

Fitting of logistic function

The fitting of the transformed logistic response function (10.55) is relatively simple when there are repeat observations at each X level. We now explain the fitting procedure for this case, which arises frequently in practice. We cite two examples:

1. A pricing experiment involves showing a new product to a consumer, providing information about it, and then asking the consumer whether he or she would buy the product at a given price. Five prices are studied, and n persons are exposed to a given price. The dependent variable is binary (would purchase, would not purchase); the independent variable is price and has five classes.

2. Four hundred heads of households are asked to indicate on a 10-point scale their intent to buy a new car within the next 12 months. A year later, each household is interviewed to determine whether a new car was purchased. The dependent variable is binary (did purchase, did not purchase). The independent variable is the measure of intent to buy and has 10 classes.

We shall denote the X levels at which observations are obtained by $X_1, \ldots, X_c$. The number of observations at level X_j will be denoted by n_j ($j = 1, \ldots, c$). It can be shown that we need consider only the total number of 1's at each X level, and not the individual Y values. Let R_j be the number of 1's at the level X_j. Hence, the proportion of 1's at the level X_j, denoted by p_j, is:

$$(10.56) \qquad p_j = \frac{R_j}{n_j}$$

Table 10.8 on page 365 contains the X_j, n_j, R_j, and p_j values for an example we shall discuss shortly. X_j refers to the price reduction offered by a coupon, n_j is the number of households which received a coupon with price reduction X_j, R_j is the number of these households that redeemed the coupon, and p_j is the proportion of households receiving a coupon with price reduction X_j that did redeem the coupon.

We fit the transformed logistic response function (10.55):

$$\pi' = \beta_0 + \beta_1 X$$

by making the logistic transformation (10.54) on the sample proportions:

$$(10.57) \qquad p_j' = \log_e\left(\frac{p_j}{1 - p_j}\right)$$

and using p_j' as the dependent variable. If transformation (10.57) is to be carried out manually, pocket calculators with $\log_e$ function keys or tables of natural logarithms can be used.

The logistic transformation, while linearizing the response function, does not eliminate the unequal variances of the error terms. Hence, weighted least squares should be used. It can be shown that when n_j is reasonably large, the approximate variance of p_j' is:

$$(10.58) \qquad \sigma^2(p_j') = \frac{1}{n_j \pi_j(1 - \pi_j)}$$

which is estimated by:

$$(10.59) \qquad s^2(p'_j) = \frac{1}{n_j p_j (1 - p_j)}$$

Hence, the estimated weights to be used in the weighted least squares computations are:

$$(10.60) \qquad \hat{w}_j = n_j p_j (1 - p_j)$$

Note that the use of the estimated weights (10.60) requires the sample sizes n_j to be reasonably large.

The fitting of a linear regression model with independent variable X and dependent variable p'_j, using weighted least squares, is straightforward. Once the fitted response function has been obtained:

$$(10.61) \qquad \hat{\pi}' = b_0 + b_1 X$$

it can be transformed back into the original units, if desired:

$$(10.62) \qquad \hat{\pi} = \frac{\exp(b_0 + b_1 X)}{1 + \exp(b_0 + b_1 X)}$$

Example

In a study of the effectiveness of coupons offering a price reduction on a given product, 1,000 homes were selected and a coupon and advertising material for the product were mailed to each. The coupons offered different price reductions (5, 10, 15, 20, and 30 cents), and 200 homes were assigned at random to each of the price reduction categories. The independent variable X in this study is the amount of price reduction, and the dependent variable Y is a binary variable indicating whether or not the coupon was redeemed within a six-month period.

It was expected that the logistic response function would be an appropriate description of the relation between price reduction and probability that the coupon is utilized. Since there were repeat observations at each X_j, and since the number of repeat observations at each X_j was large ($n_j = 200$, for all j), the procedure described earlier could be used for fitting the logistic response function.

Table 10.8 contains the basic data for this experimental study in columns 1 through 4. The transformed proportions p'_j are shown in column 5. For instance, for $X_1 = 5$, we have:

$$p'_1 = \log_e\left(\frac{.160}{1 - .160}\right) = -1.65823$$

Finally, the approximate weights $\hat{w}_j$ are shown in column 6 of Table 10.8. For instance, for $X_1 = 5$, we have:

$$\hat{w}_1 = n_1 p_1 (1 - p_1) = 200(.160)(.840) = 26.880$$

TABLE 10.8 Data for coupon effectiveness example

(1)	(2)	(3)	(4)	(5)	(6)
Price Reduction X_J	Number of Households n_J	Number of Coupons Redeemed R_J	Proportion of Coupons Redeemed p_J	Transformed Proportion p'_j	Weight $\hat{w}_j$
5	200	32	.160	-1.6582	26.880
10	200	51	.255	-1.0721	37.995
15	200	70	.350	-.6190	45.500
20	200	103	.515	.0600	49.955
30	200	148	.740	1.0460	38.480

Prior to the fitting of the logistic model, the transformed proportions p'_j were plotted against X_j. This plot is shown in Figure 10.12. It appears from there that a linear response function would fit the transformed proportions well. Hence, it was decided to proceed with fitting the transformed logistic model (10.55).

The fitting of the transformed logistic model by weighted least squares is straightforward. A computer run using weighted least squares, with the weights those in Table 10.8, column 6, led to the following fitted response function:

$$(10.63) \qquad \hat{\pi}' = -2.18506 + .108700X$$

This response function is plotted in Figure 10.12, and it appears to fit the data well.

The fitted response function (10.63) is used in the ordinary manner. To estimate the probability of a coupon redemption if the price reduction is, say, 25 cents, we first substitute in (10.63):

$$\hat{\pi}' = -2.18506 + .108700(25) = .53244$$

and then transform to the original variable:

$$\text{antilog}_e(.53244) = 1.70308 = \frac{\hat{\pi}}{1 - \hat{\pi}}$$

so that:

$$\hat{\pi} = .630$$

when $X = 25$ cents. Hence, we estimate that about 63 percent of 25-cent coupons will be redeemed within six months.

The interpretation of the slope $b_1 = .1087$ in the fitted logistic response function (10.63) is not simple, unlike the straightforward interpretation of the slope in a linear regression model. The reason is that the effect of increasing X by a unit varies for the logistic model according to the location of the starting point on the X scale. One interpretation of b_1 is found in the property of the logistic function that the "odds" $\hat{\pi}/(1 - \hat{\pi})$ are multiplied by $\exp(b_1)$ for any unit increase in X.

FIGURE 10.12 Plot of transformed data and fitted logistic response function—coupon effectiveness example

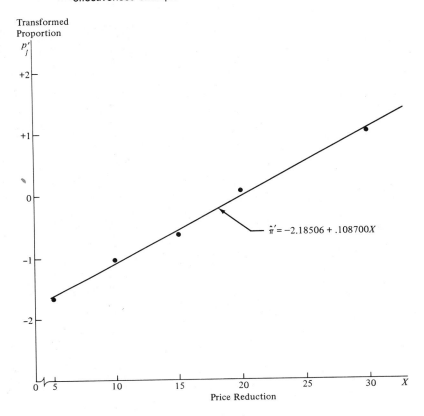

Comments

1. The procedure illustrated assumes that no $p_j = 0$ or 1. If one of these cases should occur, or if some p_j are very close to 0 or 1, modifications in the transformation (10.57) and in the weights (10.60) should be made. One approach widely used is to modify the extreme p_j as follows:

$$(10.64a) \qquad p_j = \frac{1}{2n_j} \qquad \text{if } R_j = 0$$

$$(10.64b) \qquad p_j = 1 - \frac{1}{2n_j} \qquad \text{if } R_j = n_j$$

Reference 10.1 discusses several appropriate modifications for the case when p_j equals 0 or 1.

2. A curvilinear response function of almost the same shape as the logistic function (10.53) is obtained by transforming the p_j by means of the cumulative normal distribution. This transformation is called a *probit* transformation. Reference 10.2 describes this trans-

formation. The probit transformation leads to more complex calculations than the logistic transformation, but it does have the desirable property that inferences can be made more readily with it than with the logistic transformation.

3. The transformed logistic model (10.55) can easily be extended into a multiple regression model. For, say, two independent variables, the transformed logistic response function would be:

$$(10.65) \qquad \pi' = \beta_0 + \beta_1 X_1 + \beta_2 X_2$$

where π' is the logistic transformation (10.54) of π.

4. The logistic model sometimes is motivated by threshold theory. Consider the breaking strength of concrete blocks, measured in pounds per square inch. Each block is assumed to have a threshold T_i, such that it will break if pressure equal to or greater than T_i is applied and it will not break if smaller pressure is applied. A block can be tested for only one pressure level, since some weakening of the block occurs with the first test. Thus, it is not possible to find the actual threshold of each block, but only whether or not the threshold level is above or below a particular pressure applied to the block. Of course, different pressures can be applied to different blocks in order to obtain information about the breaking strength thresholds of the blocks in the population.

Let $Y_i = 1$ if the block breaks when pressure X_i is applied, and $Y_i = 0$ if the block does not break. The earlier statement of threshold theory then implies:

$$(10.66) \qquad \begin{aligned} Y_i &= 1 \text{ whenever } T_i \leq X_i \\ Y_i &= 0 \text{ whenever } T_i > X_i \end{aligned}$$

It follows that for given pressure X_i applied to a block selected at random:

$$(10.67) \qquad \pi_i = P(Y_i = 1 \mid X_i) = P(T_i \leq X_i)$$

Now $P(T \leq X)$ is the cumulative probability distribution of the thresholds of all blocks in the population. If this cumulative probability distribution of thresholds is logistic:

$$P(T \leq X) = \frac{\exp(\beta_0 + \beta_1 X)}{1 + \exp(\beta_0 + \beta_1 X)}$$

we arrive at the logistic response model (10.53).

Incidentally, if the probability distribution of thresholds were normal, the probit response model in comment 2 above would be the relevant one according to threshold theory.

PROBLEMS

10.1. A student who used a regression model that included indicator variables was upset when receiving only the following output on the multiple regression printout: XTRANSPOSE X SINGULAR. What is a likely source of the difficulty?

10.2. Refer to regression model (10.4). Portray graphically the response curves for this model if $\beta_0 = 25.3$, $\beta_1 = .20$, and $\beta_2 = -12.1$.

10.3. In a regression study of factors affecting learning time for a certain task (measured in minutes), sex of learner was included as an independent variable (X_2) that was coded $X_2 = 1$ if male and 0 if female. It was found that $b_2 = 22.3$ and

$s(b_2) = 3.8$. An observer questioned whether the coding scheme for sex is fair because it results in a positive coefficient, leading to longer learning times for males than females. Comment.

10.4. Refer to **Calculator maintenance** Problem 2.16. The users of the desk calculators are either training institutions that use a student model, or business firms that employ a commercial model. An analyst at Tri-City wishes to fit a regression model including both number of calculators serviced (X_1) and type of calculator (X_2) as independent variables and estimate the effect of calculator model (S—student, C—commercial) on number of minutes spent on the service call. Records show that the models serviced in the 18 calls were:

i:	1	2	3	4	5	6	7	8	9
Model:	C	S	C	C	C	S	S	C	C

i:	10	11	12	13	14	15	16	17	18
Model:	C	C	S	S	C	S	C	C	C

Assume that regression model (10.4) is appropriate, and let $X_2 = 1$ if student model and 0 if commercial model.
a. Explain the meaning of all regression coefficients in the model.
b. Fit the regression model and state the estimated regression function.
c. Estimate the effect of calculator model on mean service time with a 95 percent confidence interval. Interpret your interval estimate.
d. Why would the analyst wish to include the number of calculators variable in the model when interest is in estimating the effect of type of calculator model on service time?
e. Obtain the residuals and plot them against X_1X_2. Is there any indication that an interaction term in the model would be helpful?

10.5. Refer to **Grade point average** Problem 2.15. An assistant to the director of admissions conjectures that the predictive power of the model could be improved by adding information on whether the student had chosen a major field of concentration at the time the application was submitted. Assume that regression model (10.4) is appropriate, where X_1 is entrance test score and $X_2 = 1$ if student had indicated a major field of concentration at the time of application and 0 if the major field was undecided. Students 3, 4, 5, 6, 9, and 17 had indicated a major field at the time of application.
a. Explain how each regression coefficient in model (10.4) is interpreted here.
b. Fit the regression model and state the estimated regression function.
c. Test whether the X_2 variable can be dropped from the regression model; use $\alpha = .01$. State the alternatives, decision rule, and conclusion.
d. Obtain the residuals for model (10.4) and plot them against X_1X_2. Is there any evidence in your plot that it would be helpful to include an interaction term in the model?

10.6. Refer to regression models (10.4) and (10.6). Would the conclusion that $\beta_2 = 0$ have the same implication for each of these models? Explain.

10.7. Refer to regression model (10.6). Portray graphically the response curves for this model if $\beta_0 = 25$, $\beta_1 = .30$, $\beta_2 = -12.5$, and $\beta_3 = .05$. Describe the nature of the interaction effect.

10.8. Refer to **Calculator maintenance** Problems 2.16 and 10.4. Fit regression model (10.6) and test whether the interaction term can be dropped from the model; control the α risk at .10. State the alternatives, decision rule, and conclusion. If the interaction term cannot be dropped from the model, describe the nature of the interaction effect.

10.9. Refer to **Grade point average** Problems 2.15 and 10.5.
  a. Fit regression model (10.6) and state the estimated regression function.
  b. Test whether the interaction term can be dropped from the model; use $\alpha = .05$. State the alternatives, decision rule, and conclusion. If the interaction term cannot be dropped from the model, describe the nature of the interaction effect.

10.10. In a regression analysis of on-the-job head injuries to warehouse laborers caused by falling objects, Y is a measure of severity of the injury, X_1 is an index reflecting both the weight of the object and the distance it fell, and X_2 and X_3 are indicator variables for nature of head protection worn at the time of the accident, coded as follows:

Type of Protection	X_2	X_3
Hard hat	1	0
Bump cap	0	1
None	0	0

The response function to be used in the study is $E(Y) = \beta_0 + \beta_1 X_1 + \beta_2 X_2 + \beta_3 X_3$.
  a. Develop the response function for each type-of-protection category.
  b. For each of the following questions, specify the alternatives H_0 and H_a for the appropriate test: (1) With X_1 fixed, does wearing a bump cap reduce the expected severity of injury as compared with wearing no protection? (2) With X_1 fixed, is the expected severity of injury the same when wearing a hard hat as when wearing a bump cap?

10.11. Refer to the tool wear regression model (10.12). Suppose the indicator variables had been defined as follows: $X_2 = 1$ if tool model M2 and 0 otherwise, $X_3 = 1$ if tool model M3 and 0 otherwise, $X_4 = 1$ if tool model M4 and 0 otherwise. Indicate the meaning of each of the following: (1) β_3, (2) $\beta_4 - \beta_3$, (3) β_1, (4) β_7.

10.12. Refer to the advertising expenditures regression model (10.16).
  a. How is β_4 interpreted here? What is the meaning of $\beta_3 - \beta_2$ here?
  b. State the alternatives for a test of whether the response functions are the same in incorporated and not-incorporated firms with high-quality sales management.

10.13. A marketing research trainee in the national office of a chain of shoe stores used the following response function to study seasonal (winter, spring, summer, fall) effects on sales of a certain line of shoes: $E(Y) = \beta_0 + \beta_1 X_1 + \beta_2 X_2 + \beta_3 X_3$. The X's are indicator variables defined as follows: $X_1 = 1$ if winter and 0 otherwise, $X_2 = 1$ if spring and 0 otherwise, $X_3 = 1$ if fall and 0 otherwise. After fitting the model, she tested the regression coefficients β_j ($j = 0, \ldots, 3$) and came to the following set of conclusions at an .05 family level of significance: $\beta_0 \neq 0$, $\beta_1 = 0$, $\beta_2 \neq 0$, $\beta_3 \neq 0$. In her report she then wrote: "Results of

regression analysis show that climatic and other seasonal factors have no influence in determining sales of this shoe line in the winter. Seasonal influences do
exist in the other seasons.'' Do you agree with this interpretation of the test
results?

10.14. **Assessed valuations.** A tax consultant studied the current relation between
selling price and assessed valuation of one-family residential dwellings in a large
tax district. He obtained data for a random sample of nine recent ''arm's-length''
sales transactions of one-family dwellings located on corner lots and also obtained data for a random sample of 14 recent sales of one-family dwellings not
located on corner lots. In the data that follow, both selling price (Y) and assessed
valuation (X_1) are expressed in thousand dollars. Assume that the error term
variances in the two populations are equal and that regression model (10.6) is
appropriate.

Corner Lots

i:	1	2	3	4	5	6	7	8	9
X_{i1}:	17.5	12.5	20.0	16.0	15.0	14.7	17.5	12.3	11.5
Y_i:	56.2	42.5	68.6	54.8	50.0	47.5	56.9	34.0	39.0

Noncorner Lots

i:	1	2	3	4	5	6	7	8	9	10	11	12	13	14
X_{i1}:	10.0	13.8	15.0	19.5	17.0	12.5	14.5	12.8	12.0	16.0	10.0	17.0	10.8	15.0
Y_i:	31.2	36.9	41.0	51.8	48.0	33.3	38.0	35.9	32.0	44.3	29.0	46.1	30.0	42.0

a. Plot the sample data for the two populations on one graph, using different
symbols for the two samples. Does the regression relation appear to be the
same for the two populations?

b. Test for identity of the regression functions for dwellings on corner lots and
dwellings in other locations; control the risk of Type I error at .10. State the
alternatives, decision rule, and conclusion.

c. Plot the estimated regression functions for the two populations and describe
the nature of the differences between them.

d. Estimate the difference in the slopes of the two regression functions using a
90 percent confidence interval.

e. Prepare a residual plot for each sample. Does the assumption of equal error
term variances appear to be reasonable here?

10.15. **Tire testing.** A testing laboratory with equipment that simulates highway driv-
ing studied for two makes (A, B) of a certain type of truck tire the relation
between operating cost per mile (Y) and cruising speed (X_1). The observations
are shown below (all data are coded). An engineer now wishes to decide whether
or not the regression of operating cost on cruising speed is the same for the two
makes. Assume that the error term variances for the two makes of tires are the
same and that model (10.6) is appropriate.

Make A

i:	1	2	3	4	5	6	7	8	9	10
X_{i1}:	10	20	20	30	40	40	50	60	60	70
Y_i:	9.8	12.5	14.2	14.9	19.0	16.5	20.9	22.4	24.1	25.8

Make B

i:	1	2	3	4	5	6	7	8	9	10
X_{i1}:	10	20	20	30	40	40	50	60	60	70
Y_i:	15.0	14.5	16.1	16.5	16.4	19.1	20.9	22.3	19.8	21.4

a. Plot the sample data for the two populations on one graph, using different symbols for the two samples. Does the relation between speed and cost appear to be the same for the two makes of tires?

b. Test whether or not the regression functions are the same for the two makes of tires. Control the risk of Type I error at .05. State the alternatives, decision rule, and conclusion.

c. Suppose the question of interest simply had been whether the two regression lines have equal slopes. Answer this question by setting up a 95 percent confidence interval for the difference between the two slopes. What do you find?

d. Prepare a residual plot for each make of tire. Does the assumption of equal error term variances appear to be reasonable here?

10.16. Refer to **Muscle mass** Problem 2.23. The nutritionist conjectures that the regression of muscle mass on age follows a two-piece linear relation, with the slope changing at age 60 without discontinuity.

a. State the regression model that applies if the nutritionist's conjecture is correct. What are the respective response functions when age is 60 or less and when age is over 60?

b. Fit the regression model specified in part (a) and state the estimated regression function.

c. Test whether a two-piece linear regression function is needed; use $\alpha = .01$. State the alternatives, decision rule, and conclusion. What is the P-value of the test?

10.17. **Shipment handling.** Global Electronics periodically imports shipments of a certain large part used as a component in several of its products. The size of the shipment varies depending upon production schedules. For handling and distribution to assembling plants, shipments of size 250 thousand parts or less are sent to warehouse A; larger shipments are sent to warehouse B since this warehouse has specialized equipment that provides greater economies of scale for large shipments. The data below were collected on the 10 most recent shipments; X is the size of the shipment (in thousand parts), and Y is the direct cost of handling the shipment in the warehouse (in thousand dollars).

i:	1	2	3	4	5	6	7	8	9	10
X_i:	225	350	150	200	175	180	325	290	400	125
Y_i:	11.95	14.13	8.93	10.98	10.03	10.13	13.75	13.30	15.00	7.97

A two-piecewise linear regression model with a possible discontinuity at $X = 250$ is to be fitted.

a. Specify the regression model to be used.

b. Fit this regression model. Plot the fitted response function and the data. Is there any indication that greater economies of scale are obtained in handling relatively large shipments than relatively small ones?

c. Test whether or not both the two separate slopes and the discontinuity can be dropped from the model. Control the level of significance at .025. State the alternatives, decision rule, and conclusion.

    d. For relatively small shipments, what is the point estimate of the increase in expected handling cost for each increase of one thousand in the size of the shipment? What is the corresponding estimate for relatively large shipments?

10.18. In time series analysis, the X variable representing time usually is defined to take on values 1, 2, etc., for the successive time periods. Does this represent an allocated code when the time periods are actually 1979, 1980, etc.?

10.19. An analyst wishes to include number of older siblings in family as an independent variable in a regression analysis of factors affecting maturation in eighth graders. The number of older siblings in the sample observations ranges from zero to four. Discuss whether this variable should be placed in the model as an ordinary quantitative variable or by means of four 0,1 indicator variables.

10.20. Refer to regression model (10.2) for the insurance innovation study. Suppose X_0 were dropped to eliminate the linear dependence in the **X** matrix so that the model is $Y_i = \beta_1 X_{i1} + \beta_2 X_{i2} + \beta_3 X_{i3} + \varepsilon_i$. What is the meaning here of the regression coefficients β_1, β_2, and β_3?

10.21. Refer to Figure 10.10. A student stated: "The least squares line cannot be correct as it stands because it is obvious that when Y can take on only values of 0 and 1 the least squares line has to be horizontal." Comment. What would be the implication if the least squares line were horizontal?

10.22. **Performance ability.** A psychologist made a small-scale study to examine the nature of the relation, if any, between an employee's emotional stability (X) and the employee's ability to perform in a task group (Y). Emotional stability was measured by a written test and ability to perform in a task group ($Y = 1$ if able, $Y = 0$ if unable) was evaluated by the supervisor. The results were:

i:	1	2	3	4	5	6	7	8
X_i:	474	619	584	638	399	481	624	582
Y_i:	0	1	0	1	0	1	1	1

    a. Fit a linear regression function by ordinary least squares.
    b. Now use the two-stage weighted least squares procedure to fit the linear regression function. Was use of weighted least squares helpful here in terms of the precision of the regression coefficients?
    c. Is there any evidence available from this small-scale study whether or not a linear response function is appropriate for the range of X values encountered?

10.23. **Annual dues.** The board of directors of a professional association conducted a random sample survey of 30 members to assess the effects of several possible amounts of dues increase. The sample results follow. X denotes the dollar increase in annual dues posited in the survey interview and $Y = 1$ if the interviewee indicated that the membership will not be renewed at that amount of dues increase and 0 if the membership will be renewed.

i:	1	2	3	4	5	6	7	8	9	10	11	12	13	14	15
X_i:	30	30	30	31	32	33	34	35	35	35	36	37	38	39	40
Y_i:	0	1	0	0	0	0	1	0	0	1	1	0	0	1	0

i:	16	17	18	19	20	21	22	23	24	25	26	27	28	29	30
X_i:	40	40	41	42	43	44	45	45	45	46	47	48	49	50	50
Y_i:	1	1	0	1	1	1	0	1	1	0	1	1	0	1	1

a. Fit a linear regression function by ordinary least squares. State the estimated regression function. Does the estimated regression function fall below 0 or above 1 within the scope of the fitted model?

b. Construct a 95 percent confidence interval for the slope of the response function. Interpret your interval estimate.

c. Estimate the mean response for a $40 dues increase. Use a 95 percent confidence interval. Interpret your interval estimate.

d. Fit a linear regression function by the two-stage weighted least squares procedure. Was weighted least squares helpful here in terms of the precision of the regression coefficients? Explain.

10.24. Bottle return. A carefully controlled experiment was conducted to study the effect of the size of the deposit level on the likelihood that a returnable one-liter soft-drink bottle will be returned. A bottle return was scored 1, and no return was scored 0. The data to follow show the number of bottles that were returned (R_j) out of 500 sold (n_j) at each of six deposit levels (X_j, in cents):

j:	1	2	3	4	5	6
Deposit level X_j:	2	5	10	20	25	30
Number sold n_j:	500	500	500	500	500	500
Number returned R_j:	72	103	170	296	406	449

An analyst believes that the logistic response function (10.53) is appropriate for studying the relation between size of deposit and the probability a bottle will be returned.

a. Apply transformation (10.57) and plot the transformed proportions against X. Does the plot support the use of the transformed model (10.55)?

b. Fit the logistic response function using the transformed proportions and weighted least squares, and state the fitted response function.

c. What is the estimated probability that a bottle will be returned when the deposit is 10 cents? When the deposit is 30 cents?

d. Estimate the amount of deposit for which 75 percent of the bottles will be returned.

10.25. Toxicity experiment. In an experiment testing the effect of a toxic substance, 1,500 experimental insects were divided at random into six groups of 250 each. The insects in each group were exposed to a fixed dose of the toxic substance. A day later, each insect was observed. Death from exposure was scored 1, and survival was scored 0. The results are shown below; X_j denotes the dose level (on a logarithmic scale) received by the insects in group j and R_j denotes the number of insects that died out of the 250 (n_j) in the group.

j:	1	2	3	4	5	6
X_j:	1	2	3	4	5	6
R_j:	28	53	93	126	172	197
n_j:	250	250	250	250	250	250

a. Fit the logistic response function (10.53) using transformation (10.57) and weighted least squares. State the fitted response function in the original units.

b. Plot the fitted response function in the original units and the original data. Does the fit appear to be a good one?

c. What is the estimated median dose—i.e., the dose for which 50 percent of the experimental insects would be expected to die?

EXERCISES

10.26. Refer to the instrument calibration study Example 2 in Section 10.4. Suppose that three instruments (A, B, C) had been developed to identical specifications, that the regression functions relating gauge reading (Y) to actual pressure (X_1) are second-order polynomials for each instrument, that the error term variances are the same, and that the polynomial coefficients may differ from one instrument to the next. Let X_3 denote a second indicator variable, where $X_3 = 1$ if instrument C and 0 otherwise.

a. Expand model (10.21) to cover this situation.

b. State the alternatives, define the test statistic, and give the decision rule for each of the following tests when the level of significance is .01: (1) test whether the second-order regression functions for the three instruments are identical, (2) test whether all three regression functions have the same intercept, (3) test whether both the linear and quadratic effects are the same in all three regression functions.

10.27. Refer to **Muscle mass** Problem 10.16. Specify the model for the case where the slope changes at age 40 and again at age 60 with no discontinuities.

10.28. In a regression study, three types of banks were involved, namely, commercial, mutual savings, and savings and loan. Consider the following system of indicator variables for type of bank:

Type of Bank	X_2	X_3
Commercial	1	0
Mutual savings	0	1
Savings and loan	−1	−1

a. Develop a first-order linear regression model for relating last year's profit or loss (Y) to size of bank (X_1) and type of bank (X_2, X_3).

b. State the response functions for the three types of banks.

c. Interpret each of the following quantities: (1) β_2, (2) β_3, (3) $-\beta_2 - \beta_3$.

10.29. Refer to regression model (10.17) and exclude variable X_3.

a. Obtain the $\mathbf{X'X}$ matrix for this special case of a single qualitative independent variable, for $i = 1, \ldots, n$ and n_1 firms that are not incorporated.

b. Using (7.21), find **b**.

c. Using (7.26) and (7.27), find SSR and SSE.

PROJECTS

10.30. Refer to the **SMSA** data set. The number of active physicians (Y) is to be regressed against total population (X_1), total personal income (X_2), and geographic region.

    a. Fit a first-order regression model. Let $X_3 = 1$ if NE and 0 otherwise, $X_4 = 1$ if NC and 0 otherwise, and $X_5 = 1$ if S and 0 otherwise.

    b. Examine whether the effect for the Northeastern region on number of active physicians differs from the effect for the North Central region by constructing an appropriate 90 percent confidence interval. Interpret your interval estimate.

    c. Test whether any geographic effects are present; use $\alpha = .10$. State the alternatives, decision rule, and conclusion. What is the P-value of the test?

10.31. Refer to the **SENIC** data set. Infection risk (Y) is to be regressed against length of stay (X_1), age (X_2), routine chest X-ray ratio (X_3), and medical school affiliation (X_4).

    a. Fit a first-order regression model. Let $X_4 = 1$ if hospital has medical school affiliation and 0 if not.

    b. Estimate the effect of medical school affiliation on infection risk using a 98 percent confidence interval. Interpret your interval estimate.

    c. It has been suggested that the effect of medical school affiliation on infection risk may interact with the effects of age and routine chest X-ray ratio. Add appropriate interaction terms to the model, fit the revised regression model, and test whether the interaction terms are helpful; use $\alpha = .10$. State the alternatives, decision rule, and conclusion.

10.32. Refer to the **SENIC** data set. Length of stay (Y) is to be regressed on age (X_1), routine culturing ratio (X_2), average daily census (X_3), available facilities and services (X_4), and region (X_5, X_6, X_7).

    a. Fit a first-order regression model. Let $X_5 = 1$ if NE and 0 otherwise, $X_6 = 1$ if NC and 0 otherwise, and $X_7 = 1$ if S and 0 otherwise.

    b. Test whether the routine culturing ratio can be dropped from the model; use a level of significance of .05. State the alternatives, decision rule, and conclusion.

    c. Examine whether the effect on length of stay for hospitals located in the Western region differs from that for hospitals located in the other three regions by constructing an appropriate confidence interval for each pairwise comparison. Use the Bonferroni procedure with a 95 percent family confidence coefficient. Summarize your findings.

10.33. Refer to the **SENIC** data set. A public health official has requested a study of whether medical school affiliation (Y) is related to available facilities and services (X). Let $Y = 1$ if the hospital has a medical school affiliation and 0 if not.

    a. Fit a linear response function, using ordinary least squares.

    b. Plot the data and the fitted regression line. Is the regression line consistent with probability characteristics (i.e., is it between 0 and 1) within the scope of the observations?

CITED REFERENCES

10.1 Cox, D. R. *The Analysis of Binary Data*. London: Methuen & Co. Ltd., 1970.

10.2 Finney, D. J. *Probit Analysis*. 3d ed. Cambridge: Cambridge University Press, 1971.

11

Multicollinearity, influential observations, and other topics in regression analysis—II

In this chapter, we take up selected topics in regression analysis dealing with reparameterization to improve computational accuracy, multicollinearity, and detection of influential observations.

11.1 REPARAMETERIZATION TO IMPROVE COMPUTATIONAL ACCURACY

Roundoff errors in least squares calculations

Least squares results can be sensitive to rounding of data in intermediate stages of calculations. When the number of independent variables is small—say, three or less—roundoff effects can be controlled by carrying a sufficient number of digits in intermediate calculations. Indeed, most computer regression programs use double precision arithmetic (e.g., use of 16 digits instead of 8 digits) in all computations to control roundoff effects. Still, with a large number of independent variables, serious roundoff effects can arise despite the use of many digits in intermediate calculations.

Roundoff errors tend to enter into least squares calculations primarily when the inverse of $\mathbf{X}'\mathbf{X}$ is taken. Of course, any errors in $(\mathbf{X}'\mathbf{X})^{-1}$ may be magnified when calculating $\mathbf{b}$ or making other subsequent calculations. The danger of serious roundoff errors in $(\mathbf{X}'\mathbf{X})^{-1}$ is particularly great when (1) $\mathbf{X}'\mathbf{X}$ has a determinant which is close to zero and/or (2) the elements of $\mathbf{X}'\mathbf{X}$ differ substantially in order of magnitude. The first condition arises when some or all of the independent variables are highly intercorrelated. Remedial measures for this situation will be discussed in Section 11.4. One such measure is to discard one or several of the intercorrelated independent variables in an effort to shift the determinant away from near zero so that roundoff errors will not have as severe an effect.

The second condition arises when the variables have substantially different magnitudes so that the entries in the $\mathbf{X}'\mathbf{X}$ matrix cover a wide range, say, from 15 to 49,000,000. A solution for this condition is to transform the variables and thereby reparameterize the regression model. We now address this second problem. The transformation which we take up is called the *correlation transformation*. It makes all entries in the $\mathbf{X}'\mathbf{X}$ matrix for the transformed variables fall between -1 and $+1$ inclusive, so that the calculation of the inverse matrix becomes much less subject to roundoff errors due to dissimilar orders of magnitudes than with the original variables. Many computer regression packages automatically use this transformation to obtain the basic regression results and then retransform to the original variables. In any case, users of computer programs are well advised to check that the regression package to be employed makes appropriate provisions to prevent roundoff errors from getting out of hand.

Correlation transformation

We shall illustrate the correlation transformation for the case of two independent variables. The basic regression model which we shall assume is the usual first-order one:

$$(11.1) \qquad Y_i = \beta_0 + \beta_1 X_{i1} + \beta_2 X_{i2} + \varepsilon_i$$

The first step in order to make the entries in the $\mathbf{X}'\mathbf{X}$ matrix of similar magnitude is to express all observations on the independent variables as deviations from their respective means. To illustrate this for $n = 2$, if:

$$X_{11} = 3, X_{21} = 5 \quad \text{and} \quad X_{12} = 1{,}006, X_{22} = 1{,}010$$

the deviations would be:

$$X_{11} - \bar{X}_1 = -1, X_{21} - \bar{X}_1 = 1 \quad \text{and} \quad X_{12} - \bar{X}_2 = -2, X_{22} - \bar{X}_2 = 2$$

Note that the deviations are of similar magnitude whereas the original observations were not.

To use the deviations $X_{i1} - \bar{X}_1$ and $X_{i2} - \bar{X}_2$, model (11.1) must be modified by adding and subtracting the same terms:

$$Y_i = (\beta_0 + \beta_1 \bar{X}_1 + \beta_2 \bar{X}_2) + \beta_1(X_{i1} - \bar{X}_1) + \beta_2(X_{i2} - \bar{X}_2) + \varepsilon_i$$

or:

$$(11.2) \qquad Y_i = \beta_0' + \beta_1(X_{i1} - \bar{X}_1) + \beta_2(X_{i2} - \bar{X}_2) + \varepsilon_i$$

where:

(11.2a)
$$\beta_0' = \beta_0 + \beta_1 \bar{X}_1 + \beta_2 \bar{X}_2$$

It can be shown that the least squares estimator of β_0' is always $\bar{Y}$. Hence, we can rewrite (11.2) as follows:

(11.3)
$$Y_i - \bar{Y} = \beta_1(X_{i1} - \bar{X}_1) + \beta_2(X_{i2} - \bar{X}_2) + \varepsilon_i$$

While it might appear that by eliminating one parameter from the model we have been able to increase the degrees of freedom available for MSE by one, this is not so since the dependent variable observations $Y_i - \bar{Y}$ are now subject to the restriction $\Sigma(Y_i - \bar{Y}) = 0$.

The second step in developing the correlation transformation is to express each deviation variable in units of its standard deviation:

(11.4)
$$\frac{Y_i - \bar{Y}}{s_Y} \qquad \frac{X_{i1} - \bar{X}_1}{s_1} \qquad \frac{X_{i2} - \bar{X}_2}{s_2}$$

where s_Y, s_1, and s_2 are the respective standard deviations of Y, X_1, and X_2 defined as follows:

(11.5a)
$$s_Y = \sqrt{\frac{\Sigma(Y_i - \bar{Y})^2}{n - 1}}$$

(11.5b)
$$s_1 = \sqrt{\frac{\Sigma(X_{i1} - \bar{X}_1)^2}{n - 1}}$$

(11.5c)
$$s_2 = \sqrt{\frac{\Sigma(X_{i2} - \bar{X}_2)^2}{n - 1}}$$

The final step in obtaining the correlation transformation is to use the following function of the standardized variables in (11.4):

(11.6a)
$$Y_i' = \frac{1}{\sqrt{n - 1}}\left(\frac{Y_i - \bar{Y}}{s_Y}\right)$$

(11.6b)
$$X_{i1}' = \frac{1}{\sqrt{n - 1}}\left(\frac{X_{i1} - \bar{X}_1}{s_1}\right)$$

(11.6c)
$$X_{i2}' = \frac{1}{\sqrt{n - 1}}\left(\frac{X_{i2} - \bar{X}_2}{s_2}\right)$$

Reparameterized model

The regression model with the transformed variables Y', X_1', and X_2' as defined by the correlation transformation in (11.6) is a simple extension of model (11.3):

(11.7)
$$Y_i' = \beta_1' X_{i1}' + \beta_2' X_{i2}' + \varepsilon_i'$$

It is easy to show that the new parameters β_1' and β_2' in (11.7) and the original parameters β_0, β_1, and β_2 in (11.1) are related as follows:

(11.8a)
$$\beta_1 = \left(\frac{s_Y}{s_1}\right)\beta_1'$$

(11.8b)
$$\beta_2 = \left(\frac{s_Y}{s_2}\right)\beta_2'$$

(11.8c)
$$\beta_0 = \bar{Y} - \beta_1\bar{X}_1 - \beta_2\bar{X}_2$$

Thus, the new regression coefficients β_1' and β_2' and the original regression coefficients β_1 and β_2 are related by simple scaling factors involving ratios of standard deviations.

X'X matrix for transformed variables

The **X** matrix for the transformed variables in model (11.7) is:

$$\mathbf{X} = \begin{bmatrix} X_{11}' & X_{12}' \\ X_{21}' & X_{22}' \\ \cdot & \cdot \\ \cdot & \cdot \\ \cdot & \cdot \\ X_{n1}' & X_{n2}' \end{bmatrix}$$

Remember that model (11.7) does not contain an intercept term; hence, there is no column of 1's in the **X** matrix. The **X'X** matrix then is:

(11.9)
$$\mathbf{X'X} = \begin{bmatrix} X_{11}' & X_{21}' & \cdots & X_{n1}' \\ X_{12}' & X_{22}' & \cdots & X_{n2}' \end{bmatrix} \begin{bmatrix} X_{11}' & X_{12}' \\ X_{21}' & X_{22}' \\ \cdot & \cdot \\ \cdot & \cdot \\ X_{n1}' & X_{n2}' \end{bmatrix}$$

$$= \begin{bmatrix} \Sigma(X_{i1}')^2 & \Sigma X_{i1}'X_{i2}' \\ \Sigma X_{i2}'X_{i1}' & \Sigma(X_{i2}')^2 \end{bmatrix}$$

Let us now consider the elements of this **X'X** matrix. First, we have:

$$\Sigma(X_{i1}')^2 = \Sigma\left(\frac{X_{i1} - \bar{X}_1}{\sqrt{n-1}\,s_1}\right)^2 = \frac{\Sigma(X_{i1} - \bar{X}_1)^2}{n-1} \div s_1^2 = 1$$

Similarly:

$$\Sigma(X_{i2}')^2 = 1$$

Finally:

$$\Sigma X_{i1}'X_{i2}' = \Sigma\left(\frac{X_{i1} - \bar{X}_1}{\sqrt{n-1}\,s_1}\right)\left(\frac{X_{i2} - \bar{X}_2}{\sqrt{n-1}\,s_2}\right)$$

$$= \frac{1}{n-1} \frac{\Sigma(X_{i1} - \bar{X}_1)(X_{i2} - \bar{X}_2)}{s_1 s_2}$$

$$= \frac{\Sigma(X_{i1} - \bar{X}_1)(X_{i2} - \bar{X}_2)}{[\Sigma(X_{i1} - \bar{X}_1)^2]^{1/2}[\Sigma(X_{i2} - \bar{X}_2)^2]^{1/2}}$$

But this equals r_{12}, the coefficient of correlation between X_1 and X_2 by (3.73). Since $\Sigma X'_{i1} X'_{i2} = \Sigma X'_{i2} X'_{i1}$, we find that the $\mathbf{X'X}$ matrix for the transformed variables, denoted by $\mathbf{r}_{XX}$, is:

(11.10)
$$\mathbf{r}_{XX} = \mathbf{X'X} = \begin{bmatrix} 1 & r_{12} \\ r_{12} & 1 \end{bmatrix}$$

By (3.72), r_{12} must fall between -1 and $+1$ inclusive. Hence, it follows that the elements of the $\mathbf{r}_{XX}$ matrix must have values between -1 and $+1$ inclusive. While our example dealt with the case of two independent variables, the same result follows for any number of independent variables transformed according to the principle of (11.6).

The matrix:

$$\mathbf{r}_{XX} = \begin{bmatrix} 1 & r_{12} \\ r_{12} & 1 \end{bmatrix}$$

is called the *correlation matrix of the independent variables*. For $p - 1$ independent variables, $\mathbf{r}_{XX}$ is a $(p - 1) \times (p - 1)$ matrix containing 1's on the main diagonal and the correlation coefficients r_{ij} off the main diagonal.

Regression calculations

The least squares estimators b'_1 and b'_2 for the transformed model (11.7) are obtained in the usual fashion. The inverse of the $\mathbf{r}_{XX}$ matrix is:

(11.11)
$$\mathbf{r}_{XX}^{-1} = \frac{1}{1 - r_{12}^2} \begin{bmatrix} 1 & -r_{12} \\ -r_{12} & 1 \end{bmatrix}$$

It is easy to show that the $\mathbf{X'Y}$ matrix for the transformed variables is:

(11.12)
$$\mathbf{X'Y} = \begin{bmatrix} r_{Y1} \\ r_{Y2} \end{bmatrix}$$

where r_{Y1} and r_{Y2} are the coefficients of correlation between Y and X_1 and between Y and X_2, respectively. Hence, the estimated regression coefficients for the reparameterized model (11.7) are by (7.21):

(11.13) $$\mathbf{b} = \frac{1}{1 - r_{12}^2} \begin{bmatrix} 1 & -r_{12} \\ -r_{12} & 1 \end{bmatrix} \begin{bmatrix} r_{Y1} \\ r_{Y2} \end{bmatrix} = \frac{1}{1 - r_{12}^2} \begin{bmatrix} r_{Y1} - r_{12}r_{Y2} \\ r_{Y2} - r_{12}r_{Y1} \end{bmatrix}$$

The return to the estimated regression coefficients for the original model is accomplished by employing the relations in (11.8):

(11.14a)
$$b_1 = \left(\frac{s_Y}{s_1} \right) b'_1$$

(11.14b)
$$b_2 = \left(\frac{s_Y}{s_2}\right)b_2'$$

(11.14c)
$$b_0 = \bar{Y} - b_1\bar{X}_1 - b_2\bar{X}_2$$

Comments

1. Some computer packages present both the regression coefficients b_k for the original model as well as the coefficients b_k' for the transformed model. The latter are sometimes labeled *beta coefficients* in printouts.

2. The regression coefficients b_k' for the reparameterized model are the same as the standardized regression coefficients B_k discussed in Section 7.9, as a comparison of (7.69) with (11.14) makes clear. Thus, use of the transformed variables in (11.6) automatically leads to the standardized regression coefficients.

3. Some computer printouts show the magnitude of the determinant of the correlation matrix of the independent variables. A near-zero value for this determinant implies both a high degree of linear association among the independent variables and a high potential for roundoff errors. For the case of two independent variables, this determinant is seen to be $1 - r_{12}^2$, which approaches 0 as r_{12}^2 approaches 1.

4. When the correlation matrix of the independent variables is augmented by a row and column for Y, it is called the *correlation matrix*. A correlation matrix shows the coefficients of correlation for all pairs of dependent and independent variables. This information is useful in a variety of tasks—for instance, in selecting the final independent variables to be included in the model. Many computer programs display the correlation matrix in the printout.

For the case of two independent variables, the correlation matrix is as follows:

$$\begin{bmatrix} 1 & r_{Y1} & r_{Y2} \\ r_{Y1} & 1 & r_{12} \\ r_{Y2} & r_{12} & 1 \end{bmatrix}$$

Since the correlation matrix is symmetric, the lower (or upper) triangular block of elements is frequently omitted in computer printouts.

5. It is possible to use the correlation transformation with a computer package that does not permit regression through the origin, because the intercept coefficient b_0' will always be zero for data so transformed. The other regression coefficients will also be correct, as will be all sums of squares. The degrees of freedom and mean squares will not all be correct, however, and will need to be corrected for the regression through the origin.

11.2 PROBLEMS OF MULTICOLLINEARITY

When we discussed multiple regression in Chapter 8, we noted some key problems that typically arise when the independent variables which are being considered for the model are highly correlated among themselves:

1. Adding or deleting an independent variable changes the regression coefficients.

2. The extra sum of squares associated with an independent variable varies,

depending upon which independent variables already are included in the model.
3. The estimated regression coefficients individually may not be statistically significant even though a definite statistical relation exists between the dependent variable and the set of independent variables.

These problems can also arise without substantial multicollinearity being present, but only under unusual circumstances not likely to be found in practice.

We shall now expand on the topic of multicollinearity because high intercorrelations among independent variables are frequently found in nonexperimental data in management and the social and biological sciences. For example, such pairs of independent variables as family income and liquid assets and store sales and number of employees would tend to be correlated highly.

Nature of problem

To see the essential nature of the problem of multicollinearity, we shall employ a simple example. The data in Table 11.1 refer to four sample observations on a dependent variable and two independent variables. Mr. A was asked to fit the multiple regression model:

(11.15) $$E(Y) = \beta_0 + \beta_1 X_1 + \beta_2 X_2$$

He returned in a short time with the fitted model:

(11.16) $$\hat{Y} = -87 + X_1 + 18X_2$$

He was proud of this model because it fit the data perfectly. The fitted values are shown in Table 11.1.

It so happened that Ms. B also was asked to fit model (11.15) to the same data, and she arrived at the fitted model:

(11.17) $$\hat{Y} = -7 + 9X_1 + 2X_2$$

Again, this model fits perfectly, as shown in Table 11.1.

TABLE 11.1 Example of perfectly correlated independent variables

Observation i	X_{i1}	X_{i2}	Y_i	Fitted Values	
				Model (11.16)	Model (11.17)
1	2	6	23	23	23
2	8	9	83	83	83
3	6	8	63	63	63
4	10	10	103	103	103

Model (11.16):
$$\hat{Y} = -87 + X_1 + 18X_2$$

Model (11.17):
$$\hat{Y} = -7 + 9X_1 + 2X_2$$

Indeed, it can be shown that infinitely many models will fit the data in Table 11.1 perfectly. The reason is that the independent variables X_1 and X_2 are perfectly related, according to the relation:

(11.18)
$$X_2 = 5 + .5X_1$$

Note carefully that fitted models (11.16) and (11.17) are entirely different response surfaces. The regression coefficients are different, and the fitted values will differ when X_1 and X_2 do not follow relation (11.18). For example, the fitted value for model (11.16) when $X_1 = 5$ and $X_2 = 5$ is:

$$\hat{Y} = -87 + 5 + 18(5) = 8$$

while the fitted value for model (11.17) is:

$$\hat{Y} = -7 + 9(5) + 2(5) = 48$$

Thus, when X_1 and X_2 are perfectly related and, as in our example, the data do not contain any random error component, many different response functions will lead to the same perfectly fitted values for the observations and to the same fitted values for any other (X_1, X_2) combinations following the relation between X_1 and X_2. Yet these response functions are not the same and will lead to different fitted values for (X_1, X_2) combinations that do not follow the relation between X_1 and X_2.

Two key implications of this example are:

1. The perfect relation between X_1 and X_2 did not inhibit our ability to obtain a good fit to the data.
2. Since many different models provide the same good fit, one cannot interpret any one set of regression coefficients as reflecting the effects of the different independent variables. Thus, in fitted model (11.16), $b_1 = 1$ and $b_2 = 18$ do not imply that X_2 is the key independent variable and X_1 plays little role, because model (11.17) provides an equally good fit and its regression coefficients have opposite comparative magnitudes.

Effects of multicollinearity

In actual practice, we seldom find independent variables that are perfectly related or data that do not contain some random error component. Nevertheless, the implications just noted for our idealized example still have relevance.

1. The fact that some or all independent variables are correlated among themselves does not, in general, inhibit our ability to obtain a good fit nor does it tend to affect inferences about mean responses or predictions of new observations, provided these inferences are made within the region of observations. (Figure 7.10 on p. 261 provides an illustration of the concept of the region of observations for the case of two independent variables.)

2. The counterpart in real life to the many different regression functions providing equally good fits to the data in our idealized example is that the estimated regression coefficients tend to have large sampling variability when the inde-

pendent variables are highly correlated. Thus, the estimated regression coefficients tend to vary widely from one sample to the next when the independent variables are highly correlated. As a result, only imprecise information may be available about the individual true regression coefficients. Indeed, each of the estimated regression coefficients individually may be statistically not significant even though a definite statistical relation exists between the dependent variable and the set of independent variables.

3. The common interpretation of regression coefficients as measuring the change in the expected value of the dependent variable when the corresponding independent variable is increased by one unit while all other independent variables are held constant is not fully applicable when multicollinearity exists. While it may be conceptually possible to vary one independent variable and hold the others constant, it may not be possible in practice to do so for independent variables that are highly correlated. For example, in a regression model for predicting crop yield from amount of rainfall and hours of sunshine, the relation between the two independent variables makes it unrealistic to consider varying one while holding the other constant. Therefore, the simple interpretation of the regression coefficients as measuring marginal effects is often unwarranted with highly correlated independent variables.

Example.   To illustrate these basic points, consider the data in Table 11.2 for the body fat example which was discussed earlier in Chapter 8. We shall now

TABLE 11.2 Body fat example with three independent variables, two of which are highly correlated

Subject i	Triceps Skinfold Thickness X_{i1}	Thigh Circumference X_{i2}	Midarm Circumference X_{i3}	Body Fat Y_i
1	19.5	43.1	29.1	11.9
2	24.7	49.8	28.2	22.8
3	30.7	51.9	37.0	18.7
4	29.8	54.3	31.1	20.1
5	19.1	42.2	30.9	12.9
6	25.6	53.9	23.7	21.7
7	31.4	58.5	27.6	27.1
8	27.9	52.1	30.6	25.4
9	22.1	49.9	23.2	21.3
10	25.5	53.5	24.8	19.3
11	31.1	56.6	30.0	25.4
12	30.4	56.7	28.3	27.2
13	18.7	46.5	23.0	11.7
14	19.7	44.2	28.6	17.8
15	14.6	42.7	21.3	12.8
16	29.5	54.4	30.1	23.9
17	27.7	55.3	25.7	22.6
18	30.2	58.6	24.6	25.4
19	22.7	48.2	27.1	14.8
20	25.2	51.0	27.5	21.1

consider also a third independent variable—X_3, the midarm circumference—in addition to triceps skinfold thickness (X_1) and thigh circumference (X_2), the two independent variables previously considered.

Suppose that we regress body fat (Y) on triceps skinfold thickness (X_1) only. The results of a least squares fit of the response function:

$$(11.19) \qquad E(Y) = \beta_0 + \beta_1 X_1$$

are shown in Table 11.3a. The coefficient of determination r_{Y1}^2 (the notation shows that the relation between Y and X_1 is being considered) is:

$$r_{Y1}^2 = \frac{SSR(X_1)}{SSTO} = \frac{352.27}{495.39} = .711$$

which indicates that the variability of Y is reduced by 71 percent by considering independent variable X_1. Also note that $s(b_1)$ is relatively small:

$$\frac{s(b_1)}{b_1} = \frac{.1288}{.8572} = .1503$$

Let us now add independent variable X_2 to the model. This variable is highly correlated with X_1. The coefficient of determination between the two independent variables X_1 and X_2, denoted by r_{12}^2, is $r_{12}^2 = .853$. The results of fitting the response function:

$$(11.20) \qquad E(Y) = \beta_0 + \beta_1 X_1 + \beta_2 X_2$$

are shown in Table 11.3b. Note the following:

1. The fit of model (11.20) has not been made worse in the sense of a higher error sum of squares SSE, despite the introduction of a highly correlated independent variable. Indeed, we noted earlier that SSE can never increase as the result of introducing another independent variable. (MSE can increase if the reduction in SSE is not adequate to compensate for the loss of one degree of freedom.) For our example, the coefficient of multiple determination is:

$$R^2 = \frac{SSR(X_1, X_2)}{SSTO} = \frac{385.44}{495.39} = .778$$

indicating that the variability of Y is reduced by 78 percent when both X_1 and X_2 are considered. Further, $MSE = 6.47$ now, as compared with $MSE = 7.95$ when only X_1 is included in the model.

2. The estimate of the regression coefficient β_1 for model (11.20) has larger sampling variability than before; the sampling variation of b_1 now is $s(b_1) = .3034$ as compared to $.1288$ when only X_1 is included in the model. Also, the relative sampling variation of b_2 is quite large:

$$\frac{s(b_2)}{b_2} = \frac{.2912}{.6594} = .4416$$

TABLE 11.3 Regression results for body fat example

(a) Regression of Y on X_1

$$\hat{Y} = -1.496 + .8572X_1$$

Source of Variation	SS	df	MS
Regression	352.27	1	352.27
Error	143.12	18	7.95
Total	495.39	19	

Variable	Estimated Regression Coefficient	Estimated Standard Deviation	t^*
X_1	$b_1 = .8572$	$s(b_1) = .1288$	6.66

(b) Regression of Y on X_1 and X_2

$$\hat{Y} = -19.174 + .2224X_1 + .6594X_2$$

Source of Variation	SS	df	MS
Regression	385.44	2	192.72
Error	109.95	17	6.47
Total	495.39	19	

Variable	Estimated Regression Coefficient	Estimated Standard Deviation	t^*
X_1	.2224	.3034	.73
X_2	.6594	.2912	2.26

(c) Regression of Y on X_1, X_2, and X_3

$$\hat{Y} = 117.08 + 4.334X_1 - 2.857X_2 - 2.186X_3$$

Source of Variation	SS	df	MS
Regression	396.98	3	132.33
Error	98.41	16	6.15
Total	495.39	19	

Variable	Estimated Regression Coefficient	Estimated Standard Deviation	t^*
X_1	4.334	3.016	1.44
X_2	-2.857	2.582	-1.11
X_3	-2.186	1.596	-1.37

Indeed, separate tests of β_1 and β_2, each at the level of significance .01, would lead to the conclusion that $\beta_1 = 0$ and $\beta_2 = 0$, whereas a test of the entire regression relation, based on:

$$F^* = \frac{MSR(X_1, X_2)}{MSE(X_1, X_2)} = \frac{192.72}{6.47} = 29.787$$

would lead to the conclusion, for level of significance .01, that a regression relation does exist.

Let us finally add independent variable X_3 to the model. This variable is not highly correlated with either of the other two independent variables, the coefficients of determination between X_3 and the other two independent variables being $r_{13}^2 = .210$ and $r_{23}^2 = .007$, respectively. The results of fitting the response function:

(11.21) $$E(Y) = \beta_0 + \beta_1 X_1 + \beta_2 X_2 + \beta_3 X_3$$

are shown in Table 11.3c. Note the following:

1. The fit of the model has been improved somewhat further, the coefficient of multiple determination being:

$$R^2 = \frac{SSR(X_1, X_2, X_3)}{SSTO} = \frac{396.98}{495.39} = .801$$

as compared to $R^2 = .778$ for model (11.20). Further, $MSE(X_1, X_2, X_3) = 6.15$ now, as compared with $MSE(X_1, X_2) = 6.47$ for the model with X_1 and X_2 only.

2. The estimate of the regression coefficient β_2 has actually changed signs (the estimate changed from .6594 to -2.857). In addition, the sampling variations of b_1 and b_2 in model (11.21) both increased dramatically; $s(b_1) = 3.016$ now whereas it was .3034 for model (11.20), and $s(b_2) = 2.582$ now whereas before it was .2912. Again the high degree of multicollinearity among the independent variables X_1 and X_2 is responsible for the inflated variability of the estimates of the regression coefficients. Separate tests of β_1, β_2, and β_3, each at the level of significance .01, would lead to the conclusion that $\beta_1 = 0$, $\beta_2 = 0$, and $\beta_3 = 0$, whereas a test for the entire regression relation, based on:

$$F^* = \frac{MSR(X_1, X_2, X_3)}{MSE(X_1, X_2, X_3)} = \frac{132.33}{6.15} = 21.517$$

would lead to the conclusion, at the level of significance .01, that a regression relation does exist.

3. To the extent that a change in thigh circumference (X_2) is almost always accompanied by a corresponding change in triceps skinfold thickness (X_1), the usefulness of the measures β_1 and β_2 is diminished because they reflect the effect of a change in one variable with no change in the other.

Comments

1. It was noted in Section 11.1 that a near-zero determinant of $\mathbf{X'X}$ is a potential source of serious roundoff errors in least squares results. Severe multicollinearity has the

effect of making this determinant come close to zero. Thus, under severe multicollinearity, the regression coefficients may be subject to large roundoff errors as well as large sampling variances. Hence, it is particularly advisable to employ the correlation transformation (11.6) when multicollinearity is present.

2. Just as high intercorrelations between the independent variables tend to make the estimated regression coefficients imprecise (i.e., erratic from sample to sample), so do the coefficients of partial correlation between the dependent variable and each of the independent variables tend to become erratic from sample to sample when the independent variables are highly correlated.

3. The effect of intercorrelations between the independent variables on the standard deviations of the estimated regression coefficients can be seen readily when the variables in the model are transformed by means of the correlation transformation (11.6). Consider the model with two independent variables:

$$(11.22) \qquad Y_i = \beta_0 + \beta_1 X_{i1} + \beta_2 X_{i2} + \varepsilon_i$$

This model in the transformed variables is given by (11.7) and is:

$$(11.23) \qquad Y_i' = \beta_1' X_{i1}' + \beta_2' X_{i2}' + \varepsilon_i'$$

The $(X'X)^{-1}$ matrix for this transformed model is given by (11.11):

$$(11.24) \qquad r_{XX}^{-1} = (X'X)^{-1} = \frac{1}{1 - r_{12}^2} \begin{bmatrix} 1 & -r_{12} \\ -r_{12} & 1 \end{bmatrix}$$

where r_{12} is the coefficient of correlation between X_1 and X_2. Hence, the variance-covariance matrix of the estimated regression coefficients is by (7.37):

$$(11.25) \qquad \sigma^2(\mathbf{b}) = (\sigma')^2 r_{XX}^{-1} = (\sigma')^2 \frac{1}{1 - r_{12}^2} \begin{bmatrix} 1 & -r_{12} \\ -r_{12} & 1 \end{bmatrix}$$

where $(\sigma')^2$ is the error term variance for the transformed model (11.23).

Thus, the estimated regression coefficients b_1' and b_2' have the same variance:

$$(11.26) \qquad \sigma^2(b_1') = \sigma^2(b_2') = \frac{(\sigma')^2}{1 - r_{12}^2}$$

which becomes larger as the correlation between X_1 and X_2 increases. Indeed, as X_1 and X_2 approach perfect correlation (i.e., as r_{12}^2 approaches 1), the variances of b_1' and b_2' become larger without limit.

4. We have noted that high multicollinearity is usually not a problem when the purpose of the regression analysis is to make inferences on the response function or predictions of new observations, provided these inferences are made within the range of observations. In our body fat example, for instance, the estimated mean body fat when the only independent variable included in the model is triceps skinfold thickness (X_1), together with its estimated standard deviation, are as follows for $X_{h1} = 25.0$ (calculations not shown):

$$\hat{Y}_h = 19.93 \qquad s(\hat{Y}_h) = .632$$

When the highly correlated independent variable thigh circumference (X_2) is also included in the model, the estimated mean body fat, together with its estimated standard deviation, are as follows for $X_{h1} = 25.0$ and $X_{h2} = 50.0$:

$$\hat{Y}_h = 19.36 \qquad s(\hat{Y}_h) = .624$$

Thus, the precision of the estimated mean response is equally good as before, despite the addition of the second independent variable which is highly correlated with the first one. This stability in the precision of the estimated mean response occurred despite the fact that the estimated standard deviation of b_1 became substantially larger when X_2 was added to the model (Table 11.3). The essential reason for the stability is that the covariance between b_1 and b_2 is negative, and plays a strong counteracting influence to the increase in $s^2(b_1)$ in determining the value of $s^2(\hat{Y}_h)$ as given in (7.68).

When all three independent variables are included in the model, the estimated mean body fat, together with its estimated standard deviation, are as follows for $X_{h1} = 25.0$, $X_{h2} = 50.0$, and $X_{h3} = 29.0$:

$$\hat{Y}_h = 19.19 \qquad s(\hat{Y}_h) = .621$$

Thus, the addition of the third independent variable, which is not highly correlated with the first two independent variables, does not materially affect the precision of the estimated mean response either.

11.3 VARIANCE INFLATION FACTORS AND OTHER METHODS OF DETECTING MULTICOLLINEARITY

A variety of informal and formal methods have been developed for detecting the presence of serious multicollinearity.

Informal methods

Indications of the presence of serious multicollinearity are given by the following diagnostics:

1. Large changes in the estimated regression coefficients when a variable is added or deleted, or when an observation is altered or deleted.
2. Nonsignificant results in individual tests on the regression coefficients for important independent variables.
3. Estimated regression coefficients with an algebraic sign that is the opposite of that expected from theoretical considerations or prior experience.
4. Large coefficients of correlation between pairs of independent variables in the correlation matrix $\mathbf{r}_{XX}$.
5. Wide confidence intervals for the regression coefficients representing important independent variables.

Example. In the body fat example, the independent variables triceps skinfold thickness and thigh circumference are highly correlated with each other. Also, we noted large changes in the estimated regression coefficients and their estimated standard errors when a variable was added, nonsignificant results in individual tests on anticipated important variables, and an estimated negative coefficient when a positive coefficient was expected. Therefore, serious multicollinearity among the independent variables is suspected.

Note

The informal methods just described have important limitations. They do not provide quantitative measurements of the impact of multicollinearity nor may they identify the nature of the multicollinearity. For instance, if independent variables X_1, X_2, and X_3 have low pairwise correlations, the correlation matrix $\mathbf{r}_{XX}$ will provide no indication of the presence of multicollinearity even though the three variables may be closely related as a group. Thus, examination of simple correlation coefficients will not necessarily disclose the existence of relations among groups of independent variables.

Another limitation of the informal diagnostic methods is that sometimes the observed behavior may occur without multicollinearity being present.

Variance inflation factors

One formal method of detecting the presence of multicollinearity is by means of variance inflation factors. These factors measure how much the variances of the estimated regression coefficients are inflated as compared to when the independent variables are not linearly related.

To understand the significance of variance inflation factors, we begin with the precision of least squares estimated regression coefficients, which is measured by their variances. We know from (7.37) that the variance-covariance matrix of the estimated regression coefficients is:

$$(11.27) \qquad \boldsymbol{\sigma}^2(\mathbf{b}) = \sigma^2(\mathbf{X'X})^{-1}$$

To reduce roundoff errors in calculating $(\mathbf{X'X})^{-1}$, we noted in Section 11.1 that it is desirable to first transform the variables by means of the correlation transformation (11.6). In the transformed model, the estimated regression coefficients b'_k are, as we have seen, the standardized coefficients defined in (7.69). The variance-covariance matrix of the estimated standardized regression coefficients is according to (11.25):

$$(11.28) \qquad \boldsymbol{\sigma}^2(\mathbf{b}) = (\sigma')^2 \mathbf{r}_{XX}^{-1}$$

where $\mathbf{r}_{XX}$ is the matrix of the pairwise simple correlation coefficients among the independent variables, as illustrated in (11.10) for $p - 1 = 2$ independent variables, and $(\sigma')^2$ is the error term variance for the transformed model.

Note from (11.28) that the variance of b'_k ($k = 1, \ldots, p - 1$) is equal to the product of the error term variance $(\sigma')^2$ and the kth diagonal element of the matrix $\mathbf{r}_{XX}^{-1}$. This second factor is called the *variance inflation factor (VIF)*. It can be shown that the variance inflation factor for b'_k, denoted by $(VIF)_k$, is:

$$(11.29) \qquad (VIF)_k = (1 - R_k^2)^{-1} \qquad k = 1, 2, \ldots, p - 1$$

where R_k^2 is the coefficient of multiple determination when X_k is regressed on the $p - 2$ other X variables in the model. Hence, we have:

$$(11.30) \qquad \sigma^2(b'_k) = (\sigma')^2 (VIF)_k = \frac{(\sigma')^2}{1 - R_k^2}$$

We presented in (11.26) the special results for $\sigma^2(b'_k)$ when $p - 1 = 2$, for which $R_k^2 = r_{12}^2$, the coefficient of simple determination between X_1 and X_2.

The variance inflation factor $(VIF)_k$ is equal to 1 when $R_k^2 = 0$, i.e., when X_k is not linearly related to the other X variables. When $R_k^2 \neq 0$, then $(VIF)_k$ is greater than 1, indicating an inflated variance for b_k'. This is evident from (11.30) since the denominator becomes smaller as R_k^2 becomes larger, leading to a larger variance. When X_k has a perfect linear association with the other X variables in the model so that $R_k^2 = 1$, then $(VIF)_k$ and $\sigma^2(b_k')$ are unbounded.

The largest $(VIF)_k$ among all X variables is often used as an indicator of the severity of multicollinearity. A maximum $(VIF)_k$ in excess of 10 is often taken as an indication that multicollinearity may be unduly influencing the least squares estimates.

The mean of the $(VIF)_k$'s also provides information about the severity of the multicollinearity in terms of how far the estimated standardized regression coefficients b_k' are from the true values β_k'. It can be shown that the expected value of the sum of these squared errors $(b_k' - \beta_k')^2$ is given by:

$$(11.31) \qquad E\left[\sum_{k=1}^{p-1} (b_k' - \beta_k')^2\right] = (\sigma')^2 \sum_{k=1}^{p-1} (VIF)_k$$

Thus, large $(VIF)_k$ values result, on the average, in larger differences between the estimated and true standardized regression coefficients.

When no X variable is linearly related to the others in the model, $R_k^2 \equiv 0$; hence, $(VIF)_k \equiv 1$ and:

$$(11.31a) \quad E\left[\sum_{k=1}^{p-1} (b_k' - \beta_k')^2\right] = (\sigma')^2(p-1) \qquad \text{when } (VIF)_k \equiv 1$$

A ratio of the results in (11.31) and (11.31a) provides useful information about the effect of multicollinearity on the sum of the squared errors:

$$\frac{(\sigma')^2 \Sigma(VIF)_k}{(\sigma')^2(p-1)} = \frac{\Sigma(VIF)_k}{p-1}$$

Note that this ratio is simply the mean of the $(VIF)_k$ factors, to be denoted by $(\overline{VIF})$:

$$(11.32) \qquad (\overline{VIF}) = \frac{\sum_{i=1}^{p-1} (VIF)_k}{p-1}$$

Mean VIF values considerably larger than 1 are indicative of serious multicollinearity problems.

Example. Table 11.4 contains the estimated standardized regression coefficients and the $(VIF)_k$ values for our body fat example. The maximum $(VIF)_k$ is 708.84 and $(\overline{VIF}) = 459.26$. Thus, the expected sum of the squared errors in the least squares regression coefficients is nearly 460 times as large as it would be if the X variables were uncorrelated. In addition, all three $(VIF)_k$ factors greatly exceed 10, which again indicates that serious multicollinearity problems exist.

TABLE 11.4 Variance inflation factors for body fat example

Variable	b_k'	$(VIF)_k$
X_1	4.2637	708.84
X_2	-1.5614	104.61
X_3	-2.9287	564.34
Maximum $(VIF)_k = 708.84$		$\overline{(VIF)} = 459.26$

It is interesting to note that $(VIF)_3 = 564$ despite the fact that both r_{13}^2 and r_{23}^2 are small. Here is an instance where X_3 is strongly related to X_1 and X_2 together $(R_3^2 = .998)$, even though the pairwise coefficients of simple determination are small. Examination of the correlation matrix $\mathbf{r}_{XX}$ would not have disclosed this multicollinearity.

Comments

1. A number of computer regression programs use the reciprocal of the variance inflation factor to detect instances where an X variable should not be allowed into the fitted regression model because of excessively high interdependence between this variable and the other X variables in the model. Tolerance limits for $1/(VIF)_k = 1 - R_k^2$ frequently used are .01, .001, or .0001, below which the variable is not entered into the model.

2. A limitation of variance inflation factors for detecting multicollinearities is that they cannot distinguish between several simultaneous multicollinearities.

3. A number of other methods for detecting multicollinearity have been proposed. These are more complex than variance inflation factors and are discussed in specialized texts such as Reference 11.1.

11.4 RIDGE REGRESSION AND OTHER REMEDIAL MEASURES FOR MULTICOLLINEARITY

A variety of remedial measures have been proposed for the difficulties caused by multicollinearity. Some of these leave intact the method of least squares for estimating the regression coefficients while others introduce modifications in the method of estimation.

Remedial measures with ordinary least squares

We consider first remedial measures that may be employed with ordinary least squares.

1. As we have seen, the presence of serious multicollinearity often does not affect the usefulness of the fitted model for making inferences about mean responses or making predictions, provided that the values of the independent variables for which inferences are to be made follow the same multicollinearity pattern as the data on which the regression model is based. Hence, one remedial

measure is to restrict the use of the fitted regression model to inferences for values of the independent variables which follow the same pattern of multicollinearity.

2. In polynomial regression models, as we noted in Chapter 9, expressing the independent variable(s) in the form of deviations from the mean serves to reduce substantially the multicollinearity among the first-order, second-order, and higher-order terms for any given independent variable.

3. One or several independent variables may be dropped from the model in order to lessen the multicollinearity and thereby reduce the standard errors of the estimated regression coefficients of the independent variables remaining in the model. This remedial measure has two important limitations. First, no direct information is obtained about the dropped independent variables. Second, the magnitudes of the regression coefficients for the independent variables remaining in the model are affected by the correlated independent variables not included in the model.

4. Sometimes it is possible to add some observations which break the pattern of multicollinearity. Often, however, this option is not available. In business and economics, for instance, many independent variables cannot be controlled, so that new observations will tend to show the same intercorrelation patterns as the earlier observations.

5. In some economic studies, it is possible to estimate the regression coefficients for different independent variables from different sets of data to avoid the problems of multicollinearity. Demand studies, for instance, may use both cross-section and time series data to this end. Suppose the independent variables in a demand study are price and income, and the relation to be estimated is:

$$(11.33) \qquad Y_i = \beta_0 + \beta_1 X_{i1} + \beta_2 X_{i2} + \varepsilon_i$$

where Y is demand, X_1 is income, and X_2 is price. The income coefficient β_1 may then be estimated from cross-section data. The demand variable Y is thereupon adjusted:

$$(11.34) \qquad Y_i' = Y_i - b_1 X_{i1}$$

Finally, the price coefficient β_2 is estimated by regressing the adjusted demand variable Y' on X_2.

Ridge regression

Biased estimation. Ridge regression is one of several methods that have been proposed to remedy multicollinearity problems by modifying the method of least squares to allow biased estimators of the regression coefficients. When an estimator has only a small bias and is substantially more precise than an unbiased estimator, it may well be the preferred estimator since it will have a larger probability of being close to the true parameter value. Figure 11.1 illustrates this situation. Estimator b is unbiased but imprecise, while estimator b^R is much

FIGURE 11.1   Biased estimator with small variance may be preferable to unbiased estimator with large variance

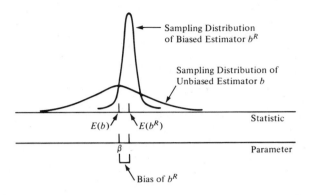

more precise but has a small bias. The probability that b^R falls near the true value β is much greater than that for the unbiased estimator b.

A measure of the combined effect of bias and sampling variation is the expected value of the squared deviation of the biased estimator b^R from the true parameter β. This measure is called the *mean squared error*, and it can be shown to equal:

$$(11.35) \qquad E(b^R - \beta)^2 = \sigma^2(b^R) + [E(b^R) - \beta]^2$$

Thus, the mean squared error equals the variance of the estimator plus the squared bias. Note that if the estimator is unbiased, the mean squared error is identical to the variance of the estimator.

Ridge estimators.   For ordinary least squares, the normal equations are given by (7.20):

$$(11.36) \qquad (\mathbf{X'X})\mathbf{b} = \mathbf{X'Y}$$

When all variables are transformed by the correlation transformation (11.6), the transformed regression model is given by:

$$(11.37) \qquad Y_i' = \beta_1' X_{i1}' + \beta_2' X_{i2}' + \cdots + \beta_{p-1}' X_{i,p-1}' + \varepsilon_i'$$

and the least squares normal equations become:

$$(11.38) \qquad \mathbf{r}_{XX}\mathbf{b} = \mathbf{r}_{YX}$$

where $\mathbf{r}_{XX}$ is a $(p - 1) \times (p - 1)$ matrix containing the pairwise coefficients of simple correlation between the independent variables:

$$(11.39a) \qquad \mathop{\mathbf{r}_{XX}}_{(p-1)\times(p-1)} = \begin{bmatrix} 1 & r_{12} & \cdots & r_{1,p-1} \\ r_{21} & 1 & \cdots & r_{2,p-1} \\ \cdot & \cdot & & \cdot \\ \cdot & \cdot & & \cdot \\ \cdot & \cdot & & \cdot \\ r_{p-1,1} & r_{p-1,2} & \cdots & 1 \end{bmatrix}$$

and $\mathbf{r}_{YX}$ is a $(p-1) \times 1$ vector containing the coefficients of simple correlation between the dependent variable and each of the independent variables:

$$(11.39b) \qquad \mathop{\mathbf{r}_{YX}}_{(p-1)\times 1} = \begin{bmatrix} r_{Y1} \\ r_{Y2} \\ \cdot \\ \cdot \\ \cdot \\ r_{Y,p-1} \end{bmatrix}$$

Formulas (11.7), (11.10), and (11.12) are illustrative of the case $p - 1 = 2$ independent variables.

The ridge standardized regression estimators are obtained by introducing into the least squares normal equations (11.38) a biasing constant $c \geq 0$, in the following form:

$$(11.40) \qquad (\mathbf{r}_{XX} + c\mathbf{I})\mathbf{b}^R = \mathbf{r}_{YX}$$

where $\mathbf{b}^R$ is the vector of the standardized ridge regression coefficients b_k^R:

$$(11.41) \qquad \mathbf{b}^R = \begin{bmatrix} b_1^R \\ b_2^R \\ \cdot \\ \cdot \\ \cdot \\ b_{p-1}^R \end{bmatrix}$$

and $\mathbf{I}$ is the $(p - 1) \times (p - 1)$ identity matrix. Solution of the normal equations (11.40) yields the ridge standardized regression coefficients:

$$(11.42) \qquad \mathbf{b}^R = (\mathbf{r}_{XX} + c\mathbf{I})^{-1}\mathbf{r}_{YX}$$

The constant c reflects the amount of bias in the estimators. When $c = 0$, (11.42) reduces to the ordinary least squares regression coefficients in standardized form, as given in (7.69). When $c > 0$, the ridge regression coefficients are biased but tend to be more stable (i.e., less variable) than ordinary least squares estimators.

Choice of biasing constant c. It can be shown that the bias component of the total mean squared error of the ridge regression estimator $\mathbf{b}^R$ increases as c gets larger (with all b_k^R tending toward zero), while at the same time the variance component becomes smaller. It can further be shown that there always exists some value c for which the ridge regression estimator $\mathbf{b}^R$ has a smaller total mean squared error than the ordinary least squares estimator $\mathbf{b}$. The difficulty is that the optimum value of c varies from one application to another and is unknown.

A commonly used method of determining the biasing constant c is based on the *ridge trace* and the variance inflation factors $(VIF)_k$ in (11.29). The ridge trace is a simultaneous plot of the values of the $p - 1$ estimated ridge standardized regression coefficients for different values of c, usually between 0 and 1.

Extensive experience has indicated that an estimated regression coefficient b_k^R may fluctuate widely as c is changed slightly from 0, and may even change signs. Gradually, however, these wide fluctuations cease and the magnitude of the regression coefficient tends to change only slowly as c is increased further. At the same time, the value of $(VIF)_k$ tends to fall rapidly as c is changed from 0, and gradually $(VIF)_k$ also tends to change only moderately as c is increased further. One therefore examines the ridge trace and the variance inflation factors and chooses the smallest value of c where it is deemed that the regression coefficients first become stable in the ridge trace and the variance inflation factors have become sufficiently small. The choice is thus a judgmental one.

Example. We noted previously the severe multicollinearity in the data for our body fat example. Indeed, in the model with three independent variables (Table 11.3c, p. 387), the estimated regression coefficient b_2 is negative even though it was expected that amount of body fat is positively related to thigh circumference. Ridge regression calculations were made for the body fat example data in Table 11.2 (calculations not shown). The ridge standardized regression coefficients for selected values of c are presented in Table 11.5, and the

TABLE 11.5 Ridge estimated standardized regression coefficients for different biasing constants c — body fat example

c	b_1^R	b_2^R	b_3^R
.000	4.264	−2.929	−1.561
.001	2.035	−.9408	−.7087
.002	1.441	−.4113	−.4813
.003	1.165	−.1661	−.3758
.004	1.006	−.0248	−.3149
.005	.9028	.0670	−.2751
.006	.8300	.1314	−.2472
.007	.7760	.1791	−.2264
.008	.7343	.2158	−.2103
.009	.7012	.2448	−.1975
.010	.6742	.2684	−.1870
.020	.5463	.3774	−.1369
.030	.5004	.4134	−.1181
.040	.4760	.4302	−.1076
.050	.4605	.4392	−.1005
.060	.4494	.4443	−.0952
.070	.4409	.4472	−.0909
.080	.4341	.4486	−.0873
.090	.4283	.4491	−.0841
.100	.4234	.4490	−.0812
.200	.3914	.4347	−.0613
.300	.3703	.4154	−.0479
.400	.3529	.3966	−.0376
.500	.3377	.3791	−.0295
.600	.3240	.3629	−.0229
.700	.3116	.3481	−.0174
.800	.3002	.3344	−.0129
.900	.2896	.3218	−.0091
1.000	.2798	.3101	−.0059

TABLE 11.6 *VIF* values for regression coefficients and R^2 for different biasing constants c— body fat example

c	$(VIF)_1$	$(VIF)_2$	$(VIF)_3$	R^2
.000	708.84	564.34	104.61	.8014
.001	125.73	100.27	19.28	.7888
.002	50.56	40.45	8.28	.7852
.003	27.18	21.84	4.86	.7832
.004	16.98	13.73	3.36	.7819
.005	11.64	9.48	2.58	.7809
.006	8.50	6.98	2.19	.7801
.007	6.50	5.38	1.82	.7794
.008	5.15	4.30	1.62	.7787
.009	4.19	3.54	1.48	.7781
.010	3.49	2.98	1.38	.7775
.020	1.10	1.08	1.01	.7726
.030	.63	.70	.92	.7682
.040	.45	.56	.88	.7639
.050	.37	.49	.85	.7597
.060	.32	.45	.83	.7556
.070	.30	.42	.81	.7515
.080	.28	.40	.79	.7475
.090	.26	.39	.78	.7436
.100	.25	.37	.76	.7397
.200	.21	.31	.63	.7031
.300	.18	.27	.54	.6702
.400	.17	.24	.46	.6405
.500	.15	.21	.40	.6134
.600	.14	.19	.35	.5887
.700	.13	.18	.31	.5659
.800	.12	.16	.28	.5449
.900	.11	.15	.25	.5254
1.000	.11	.14	.23	.5073

variance inflation factors are given in Table 11.6. The coefficients of multiple determination R^2 are also shown in the latter table. Figure 11.2 presents the ridge trace of the estimated standardized regression coefficients. To facilitate the analysis, the horizontal c scale in Figure 11.2 is logarithmic.

Note the instability in Figure 11.2 of the regression coefficients for very small values of c. The estimated regression coefficient b_2^R, in fact, changes signs. Also note the rapid decrease in the variance inflation factors in Table 11.6. It was decided to employ $c = .02$ here because for this value of the biasing constant the ridge regression coefficients have *VIF* values near 1 and the estimated regression coefficients appear to have become reasonably stable. The resulting model for $c = .02$ is:

$$\hat{Y}' = .5463X_1' + .3774X_2' - .1369X_3'$$

Transforming back to the original variables by (11.8), as extended to three independent variables, we obtain:

$$\hat{Y} = -7.3978 + .5553X_1 + .3681X_2 - .1917X_3$$

FIGURE 11.2 Ridge trace of estimated standardized regression coefficients—body fat example

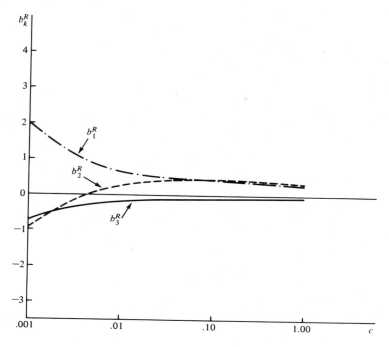

where $\bar{Y} = 20.195$, $\bar{X}_1 = 25.305$, $\bar{X}_2 = 51.170$, $\bar{X}_3 = 27.620$, $s_Y = 5.106$, $s_1 = 5.023$, $s_2 = 5.235$, and $s_3 = 3.647$.

The improper sign on the estimate for β_2 has now been eliminated, and the estimated regression coefficients are more in line with prior expectations. The sum of the squared residuals for the transformed variables, which increases with c, has only increased from .1986 at $c = 0$ to .2274 at $c = .02$ while R^2 decreased from .8014 to .7726. These changes are relatively modest. The estimated mean body fat when $X_{h1} = 25.0$, $X_{h2} = 50.0$, and $X_{h3} = 29.0$ is 19.33 for the ridge regression at $c = .02$ compared to 19.19 utilizing the ordinary least squares solution. Thus, the ridge solution at $c = .02$ appears to be quite satisfactory here and a reasonable alternative to the ordinary least squares solution.

Comments

1. The normal equations (11.40) for the ridge estimators are as follows:

(11.43)
$$
\begin{aligned}
(1 + c)b_1^R + \quad\quad r_{12}b_2^R + \cdots + r_{1,p-1}b_{p-1}^R &= r_{Y1} \\
r_{21}b_1^R + (1 + c)b_2^R + \cdots + r_{2,p-1}b_{p-1}^R &= r_{Y2} \\
\vdots \quad\quad\quad\quad\quad\quad\quad\quad\quad \vdots \quad\quad\quad\quad &\quad \vdots \\
r_{p-1,1}b_1^R + \quad r_{p-1,2}b_2^R + \cdots + (1 + c)b_{p-1}^R &= r_{Y,p-1}
\end{aligned}
$$

where r_{ij} is the coefficient of simple correlation between the ith and jth independent variables and r_{Yj} is the coefficient of simple correlation between the dependent variable Y and the jth independent variable.

2. Ridge regression estimates tend to be stable in the sense that they are usually little affected by small changes in the data on which the fitted regression is based. In contrast, ordinary least squares estimates may be highly unstable under these conditions when the independent variables are highly multicollinear. Also, the ridge estimated regression function at times will provide good estimates of mean responses or predictions of new observations for levels of the independent variables outside the region of the observations on which the regression function is based. In contrast, the estimated regression function based on ordinary least squares may perform quite poorly in such instances. Of course, any estimation or prediction well outside the region of the observations should always be made with great caution.

3. A major limitation of ridge regression is that ordinary inference procedures are not applicable and exact distributional properties are not known. Another limitation is that the choice of the biasing constant c is a judgmental one. While formal methods have been developed for making this choice, these methods have their own limitations.

4. The ridge regression procedures have been generalized to allow for differing biasing constants for the different estimated regression coefficients.

Other remedial measures

Still other approaches to remedying the problems of multicollinearity have been developed. These include regression with principal components, where the independent variables are linear combinations of the original independent variables, and Bayesian regression, where prior information about the regression coefficients is incorporated into the estimation procedure. More information about these approaches, as well as about ridge regression and generalized ridge regression, may be obtained from specialized works such as Reference 11.1.

11.5 IDENTIFICATION OF OUTLYING OBSERVATIONS

Frequently in regression analysis applications, the data set contains some observations which are outlying or extreme, i.e., observations which are well separated from the remainder of the data. These outlying observations may involve large residuals and often have dramatic effects on the fitted least squares regression function. It is therefore important to study the outlying observations carefully and decide whether they should be retained or eliminated, and if retained, whether their influence should be reduced in the fitting process and/or the regression model revised.

An observation may be outlying or extreme with respect to its Y value, its X value(s), or both. Figure 11.3 illustrates this for the case of regression with a single independent variable. In the scatter plot in Figure 11.3, observation 1 is outlying with respect to its Y value. Note that this point falls far outside the scatter, although its X value is near the middle of the range of the observations on the independent variable. Observations 2 and 3 are outlying with respect to their X values since they have much larger X values than those for the other observations; observation 3 is also outlying with respect to its Y value.

FIGURE 11.3 Scatter plot for regression with one independent variable illustrating outlying observations

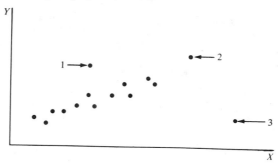

Not all outlying observations have a strong influence on the fitted regression function. Observation 1 in Figure 11.3 may not be too influential because there are a number of other observations that have similar X values, which will keep the fitted regression function from being displaced too far by the outlying observation. Likewise, observation 2 may not be too influential because its Y value is consistent with the regression relation displayed by the nonextreme observations. Observation 3, on the other hand, is likely to be very influential in affecting the fit of the regression function because it is outlying with regard to its X value, and its Y value is not consistent with the regression relation for the other observations.

In the case of regression with one or two independent variables, it is relatively simple to identify outlying observations with respect to their X or Y values and to study whether or not they are influential in affecting the fitted regression function. When more than two independent variables are included in the regression model, however, the identification of outlying observations by simple graphic means, such as scatter and residual plots, becomes difficult and more powerful methods are required. We now discuss some methods for identifying observations that are outlying with respect to their X or Y values.

Use of hat matrix for identifying outlying X observations

We encountered the hat matrix $\mathbf{H}$ in Chapter 6 where we noted in (6.93) that the least squares residuals can be expressed as a linear combination of the observations Y_i by means of the hat matrix:

(11.44) $$\mathbf{e} = (\mathbf{I} - \mathbf{H})\mathbf{Y}$$

The hat matrix $\mathbf{H}$ is given by (6.93a):

(11.45) $$\underset{n \times n}{\mathbf{H}} = \mathbf{X}(\mathbf{X}'\mathbf{X})^{-1}\mathbf{X}'$$

Similarly, the fitted values $\hat{Y}_i$ can be expressed as linear combinations of the observations Y_i through the hat matrix:

(11.46) $$\hat{\mathbf{Y}} = \mathbf{H}\mathbf{Y}$$

Further, we noted in (6.95) that the variances and covariances of the residuals involve the hat matrix:

(11.47) $$\sigma^2(\mathbf{e}) = \sigma^2(\mathbf{I} - \mathbf{H})$$

so that the variance of residual e_i, denoted by $\sigma^2(e_i)$, is:

(11.48) $$\sigma^2(e_i) = \sigma^2(1 - h_{ii})$$

where h_{ii} is the ith element on the main diagonal of the hat matrix.

The diagonal element h_{ii} of the hat matrix can be obtained directly from:

(11.49) $$h_{ii} = \mathbf{X}_i'(\mathbf{X'X})^{-1}\mathbf{X}_i$$

where $\mathbf{X}_i$ corresponds to the $\mathbf{X}_h$ vector in (7.45) except that $\mathbf{X}_i$ pertains to the ith sample observation:

(11.49a) $$\mathbf{X}_i_{p \times 1} = \begin{bmatrix} 1 \\ X_{i,1} \\ \cdot \\ \cdot \\ \cdot \\ X_{i,p-1} \end{bmatrix}$$

Note that $\mathbf{X}_i'$ is simply the ith row of the $\mathbf{X}$ matrix, pertaining to the ith sample observation.

The diagonal elements h_{ii} have some useful properties:

(11.50) $$0 \le h_{ii} \le 1 \qquad \sum_{i=1}^{n} h_{ii} = p$$

where p is the number of regression parameters in the regression function including the intercept term.

The diagonal element h_{ii} in the hat matrix is called the *leverage* (in terms of the X values) of the ith observation. It indicates whether or not the X values for the ith observation are outlying, because it can be shown that h_{ii} is a measure of the distance between the X values for the ith observation and the means of the X values for all n observations. Thus, a large leverage value h_{ii} indicates that the ith observation is distant from the center of the X observations. Figure 11.4 illustrates this for the case of two independent variables. Observation 1 is distant from the center $(\bar{X}_1, \bar{X}_2)$ and has a large leverage value $h_{11} = .812$ while observation 2 is near the center and has a small leverage value $h_{22} = .253$.

If the ith observation is an outlying X observation—i.e., one with a large leverage value h_{ii}—it exercises substantial leverage in determining the fitted value $\hat{Y}_i$. This is so for the following reasons:

1. The fitted value $\hat{Y}_i$ is a linear combination of the observed Y values, as shown in (11.46), and h_{ii} is the weight of observation Y_i in determining this fitted value. Thus, the larger is h_{ii}, the more important is Y_i in determining

FIGURE 11.4 Illustration of observations with X values near and far from center

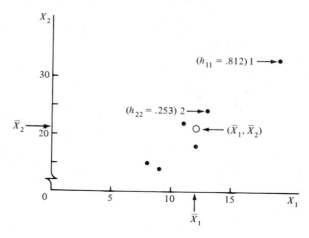

$\hat{Y}_i$. Remember that h_{ii} is a function only of the X values, so h_{ii} measures the role of the X values in determining how important Y_i is in affecting the fitted value $\hat{Y}_i$.

2. The larger is h_{ii}, the smaller is the variance of the residual e_i, as may be seen from (11.48). Hence, the larger is h_{ii}, the closer the fitted value $\hat{Y}_i$ will tend to be to the observed value Y_i. In the extreme case where $h_{ii} = 1$, $\sigma^2(e_i) = 0$ so that the fitted value $\hat{Y}_i$ is then forced to equal the observed value Y_i. Since observations with high leverage tend to have smaller residuals, it may not be possible to detect them by an examination of the residuals alone.

A leverage value h_{ii} is usually considered to be large if it is more than twice as large as the mean leverage value, denoted by $\bar{h}$, which according to (11.50) is:

(11.51)
$$\bar{h} = \frac{\sum\limits_{i=1}^{n} h_{ii}}{n} = \frac{p}{n}$$

Hence, leverage values greater than $2p/n$ are considered by this rule to indicate outlying observations with regard to the X values. Additional evidence of an extreme observation is the existence of a gap between the leverage values for most of the observations and the unusually large leverage value(s).

Example. We continue with the body fat example, using only the two independent variables triceps skinfold thickness (X_1) and thigh circumference (X_2) so that the results using the hat matrix can be compared to simple graphic plots. The data for this example were presented earlier in Table 8.3 (p. 275). Figure 11.5 contains a scatter plot of X_2 against X_1, where the observations are shown by their observation number. We note from Figure 11.5 that observations 15 and 3 appear to be outlying ones with respect to the pattern of the X values. Observa-

FIGURE 11.5 Scatter plot of thigh circumference against triceps skinfold thickness— body fat example

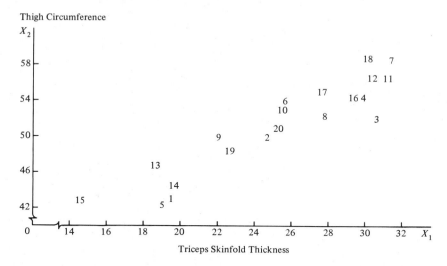

tion 15 is outlying for X_1 and at the low end of the range for X_2, while observation 3 is outlying in terms of the pattern of multicollinearity, though it is not outlying for either of the independent variables separately. Observations 1 and 5 also appear to be somewhat extreme.

Calculation of the hat matrix (11.45) confirms these impressions. Table 11.7, column 2, contains the leverage values h_{ii} for the body fat example. Note that the two largest leverage values are $h_{33} = .372$ and $h_{15,15} = .333$. Both exceed the criterion of twice the mean leverage value, $2p/n = 2(3)/20 = .30$, and are separated by a substantial gap from the next largest leverage values $h_{55} = .248$ and $h_{11} = .201$. Having identified observations 3 and 15 as outlying observations in terms of their X values, we shall need to ascertain how influential these observations are in the fitting of the regression function. We consider this question after taking up the identification of outlying Y observations.

Use of studentized deleted residuals for identifying outlying Y observations

The detection of outlying or extreme Y observations based on an examination of the residuals has been considered in earlier chapters. We utilized there either the residuals e_i:

$$(11.52) \qquad e_i = Y_i - \hat{Y}_i$$

or the standardized residuals:

$$(11.53) \qquad \frac{e_i}{\sqrt{MSE}}$$

TABLE 11.7 Residuals, diagonal elements of the hat matrix, studentized deleted residuals, and Cook's distances—body fat example

i	(1) e_i	(2) h_{ii}	(3) d_i^*	(4) D_i
1	−1.683	.201	−.730	.046
2	3.643	.059	1.534	.046
3	−3.176	.372	−1.656	.490
4	−3.158	.111	−1.348	.072
5	.000	.248	.000	.000
6	−.361	.129	−.148	.001
7	.716	.156	.298	.006
8	4.015	.096	1.760	.098
9	2.655	.115	1.117	.053
10	−2.475	.110	−1.034	.044
11	.336	.120	.137	.001
12	2.226	.109	.923	.035
13	−3.947	.178	−1.825	.211
14	3.447	.148	1.524	.125
15	.571	.333	.267	.013
16	.642	.095	.258	.002
17	−.851	.106	.344	.005
18	−.783	.197	.335	.010
19	−2.857	.067	−1.176	.032
20	1.040	.050	.409	.003

We introduce now two refinements to make the analysis of residuals more effective for identifying outlying Y observations.

When the residuals e_i have substantially different variances $\sigma^2(e_i)$, as given in (11.48), it is better to consider the magnitude of e_i relative to $\sigma(e_i)$ instead of $\sqrt{MSE}$ to give recognition to differences in their sampling errors. Since by (11.48) we have:

$$\sigma^2(e_i) = \sigma^2(1 - h_{ii})$$

an unbiased estimator of this variance is:

(11.54) $$s^2(e_i) = MSE(1 - h_{ii})$$

The ratio of e_i to $s(e_i)$ is called the *studentized residual* and will be denoted by e_i^*:

(11.55) $$e_i^* = \frac{e_i}{s(e_i)}$$

Note that the residuals e_i will have substantially different sampling variations if the leverage values h_{ii} differ markedly, but the studentized residuals have constant variance (when the model is appropriate).

The second refinement is to measure the ith residual $e_i = Y_i - \hat{Y}_i$ when the fitted regression is based on the observations excluding the ith one. In this way, the fitted value $\hat{Y}_i$ cannot be influenced by the ith observation to be close to Y_i because this observation is not part of the data set on which the fitted value is based. Thus, the ith observation is deleted, the regression function is fitted to the remaining $n - 1$ observations, and the point estimate of the expected value when the X levels are those of the ith observation, to be denoted by $\hat{Y}_{(i)}$, will be compared with the actual Y_i observed value. The notation $\hat{Y}_{(i)}$ reminds us that the ith observation was omitted when fitting the regression function. The residual:

$$(11.56) \qquad d_i = Y_i - \hat{Y}_{(i)}$$

is called a *deleted residual* and is denoted by d_i.

Note that a deleted residual corresponds to the prediction error in the numerator of (3.34) when predicting a new observation from the fitted regression function based on earlier observations, except that in (3.34) the difference considered is $\hat{Y}_{(i)} - Y_i$ and the notation differs from the present one. Hence, we know from (7.55a) that the estimated variance of d_i is:

$$(11.57) \qquad s^2(d_i) = MSE_{(i)}(1 + \mathbf{X}_i'(\mathbf{X}_{(i)}'\mathbf{X}_{(i)})^{-1}\mathbf{X}_i)$$

where $\mathbf{X}_i$ is the X observations vector (11.49a) for the ith observation, $MSE_{(i)}$ is the mean square error when the ith observation is omitted in fitting the regression function, and $\mathbf{X}_{(i)}$ is the $\mathbf{X}$ matrix with the ith observation deleted. Also, it follows from (7.55) that:

$$(11.58) \qquad \frac{d_i}{s(d_i)} \sim t(n - p - 1)$$

Remember that $n - 1$ observations are used here in predicting the ith observation; hence, the degrees of freedom are $(n - 1) - p = n - p - 1$.

Combining the two refinements, we shall use for diagnosis of outlying or extreme Y observations the deleted residual d_i in (11.56) and studentize it by dividing it by its estimated standard deviation given by (11.57). The *studentized deleted residual*, to be denoted by d_i^*, therefore is:

$$(11.59) \qquad d_i^* = \frac{d_i}{s(d_i)}$$

We know from (11.58) that each studentized deleted residual d_i^* follows the t distribution with $n - p - 1$ degrees of freedom. The d_i^*, however, are not independent.

Fortunately, the studentized deleted residuals d_i^* in (11.59) can be calculated without having to fit regression functions with the ith observation omitted. It can be shown that an algebraically equivalent expression for d_i^* is:

$$(11.59a) \qquad d_i^* = e_i \left[\frac{n - p - 1}{SSE(1 - h_{ii}) - e_i^2} \right]^{1/2}$$

Thus, the studentized deleted residual d_i^* can be calculated from the residual e_i, the error sum of squares SSE, and the leverage value h_{ii}, all for the fitted regression based on the n observations.

To identify outlying Y observations, we examine the studentized deleted residuals for large absolute values and use the appropriate t distribution to ascertain how far in the tails such outlying values fall.

Example. We illustrate the calculation of studentized deleted residuals for the first observation in the body fat example. The X values for this observation are $X_{11} = 19.5$ and $X_{12} = 43.1$. Using the fitted regression function from Table 8.4a, we obtain:

$$\hat{Y}_1 = -19.174 + .2224(19.5) + .6594(43.1) = 13.583$$

Since $Y_1 = 11.9$, the residual for this observation is $e_1 = 11.9 - 13.583 = -1.683$. We also know from Table 8.4a that $SSE = 109.95$ and from Table 11.7 that $h_{11} = .201$. Hence, by (11.59a), we find:

$$d_1^* = -1.683 \left[\frac{20 - 3 - 1}{109.95(1 - .201) - (-1.683)^2} \right]^{1/2} = -.730$$

The studentized deleted residuals for all 20 observations are shown in column 3 of Table 11.7.

Note that observations 3, 8, and 13 have the largest absolute studentized deleted residuals. If we consider tail areas of .05 on each side to be extreme, we will need to compare the absolute values of the studentized deleted residuals with $t(.95; 16) = 1.746$. Based on this comparison, we should consider observations 8 and 13 extreme enough to warrant studying whether or not they are influential observations. Incidentally, consideration of the residuals e_i (shown in Table 11.7, column 1) here would also have identified observations 8 and 13 as the most outlying ones.

11.6   IDENTIFICATION OF INFLUENTIAL OBSERVATIONS AND REMEDIAL MEASURES

After identifying outlying observations with respect to their X values and/or their Y values, the next step is to ascertain whether or not they are influential in affecting the fit of the regression function, possibly leading to serious distortion effects.

One measure of the influence of the ith observation on the fit of the regression function is the difference between the vector **b** of the estimated regression coefficients based on all n observations and the vector $\mathbf{b}_{(i)}$ based on the $n - 1$ observations with the ith observation deleted:

$$(11.60) \qquad\qquad\qquad \mathbf{b} - \mathbf{b}_{(i)}$$

Another possible measure of the influence of the ith observation is the difference between the fitted value $\hat{Y}_i$ based on the regression with all n observations and the fitted value $\hat{Y}_{(i)}$ obtained when the ith observation is deleted:

$$(11.61) \qquad\qquad\qquad \hat{Y}_i - \hat{Y}_{(i)}$$

Cook's distance measure

An overall measure of the impact of the ith observation on the estimated regression coefficients is *Cook's distance measure* D_i. Recall from (7.43) that the boundary of the confidence region for all p regression coefficients β_k ($k = 0, 1, \ldots, p - 1$) is given by:

$$(11.62) \qquad \frac{(\mathbf{b} - \boldsymbol{\beta})'\mathbf{X}'\mathbf{X}(\mathbf{b} - \boldsymbol{\beta})}{pMSE} = F(1 - \alpha; p, n - p)$$

Cook's distance measure D_i uses the same structure for measuring the combined impact of the differences in the estimated regression coefficients when the ith observation is deleted:

$$(11.63) \qquad D_i = \frac{(\mathbf{b} - \mathbf{b}_{(i)})'\mathbf{X}'\mathbf{X}(\mathbf{b} - \mathbf{b}_{(i)})}{pMSE}$$

While D_i does not follow the F distribution, it has been found useful to relate the value D_i to the corresponding F distribution according to (11.62) and ascertain the percentile value. If the percentile value is less than about 10 or 20 percent, the ith observation has little apparent influence on the fitted regression function. If, on the other hand, the percentile value is near 50 percent or more, the distance between the vectors $\mathbf{b}$ and $\mathbf{b}_{(i)}$ should be considered large, implying that the ith observation has a substantial influence on the fit of the regression function.

Fortunately, Cook's distance measure D_i can be calculated without fitting new regression functions where the ith observation is deleted. An algebraically equivalent expression is:

$$(11.63a) \qquad D_i = \frac{e_i^2}{pMSE} \left[\frac{h_{ii}}{(1 - h_{ii})^2} \right]$$

Note from (11.63a) that D_i depends on two factors: (1) the size of the residual e_i and (2) the leverage value h_{ii}. The larger is either e_i or h_{ii}, the larger is D_i. Thus, the ith observation can be influential: (1) by having a large residual e_i and only a moderate leverage value h_{ii}, or (2) by having a large leverage value h_{ii} with only a moderately sized residual e_i, or (3) by having both a large residual e_i and a large leverage value h_{ii}.

Example. In the body fat example, we had identified observations 3 and 15 as outlying X observations and observations 8 and 13 as outlying Y observations. We now calculate Cook's distance measure for each of these observations. To illustrate the calculations, we shall consider observation 3. From Table 11.7, we have $e_3 = -3.176$ and $h_{33} = .372$; and from Table 8.4a, we have $MSE = 6.47$. Since $p = 3$, we obtain using (11.63a):

$$D_3 = \frac{(-3.176)^2}{3(6.47)} \left[\frac{.372}{(1 - .372)^2} \right] = .490$$

The distance measures for all of the observations are presented in Table 11.7, column 4. We note from column 4 that observation 3 clearly is the most influential observation, with the next largest distance measure $D_{13} = .211$ being substantially smaller.

To assess the magnitude of $D_3 = .490$, we refer to the corresponding F distribution, namely, $F(p, n - p) = F(3, 17)$. It can be shown that .490 is about the 31st percentile of this distribution. Hence, it appears that observation 3 does influence the regression fit, but the extent of the influence may not be large enough to call for consideration of remedial measures.

Additional insights about the extent of the influence of observation 3 may be obtained by comparing the fitted value $\hat{Y}_3$ when all observations are utilized with the fitted value $\hat{Y}_{(3)}$ when observation 3 is deleted. It can be shown that $\hat{Y}_3 = 21.877$ and $\hat{Y}_{(3)} = 23.756$, so that omission of observation 3 increases the fitted value by 8.6 percent. This change indicates that observation 3, although it influences the fitted regression, may not play such a strong role as to require consideration of remedial measures.

Comments

1. Cook's distance measure D_i may be viewed as reflecting in the aggregate the differences between the fitted values for each observation when all n observations are used in the data base and the fitted values when the ith observation is deleted, since it can be shown that an equivalent expression for D_i is:

$$(11.64) \qquad D_i = \frac{(\hat{Y} - \hat{Y}_{(i)})'(\hat{Y} - \hat{Y}_{(i)})}{pMSE}$$

Here, $\hat{Y}$ as usual is the vector of the fitted values when all n observations are used in the data base for the regression fit and $\hat{Y}_{(i)}$ is the vector of the fitted values when the ith observation is deleted from the data base.

2. Analysis of outlying and influential observations is a necessary component of good regression analysis. However, it is neither automatic nor foolproof and requires good judgment by the analyst. The methods which have been described often work well but at other times will be ineffective. For example, if two influential outlying observations are nearly coincident, an analysis which deletes one observation at a time and estimates the change in fit will result in virtually no change for these two outlying observations. The reason is that the retained outlying observation will mask the effect of the deleted outlying observation.

Remedial measures

After using the hat matrix, studentized deleted residuals, and Cook's distances to identify outlying influential observations that have a substantial impact on the least squares regression fit, one must decide what to do about such observations. Clearly, an outlying influential observation should not be automatically discarded, because it may be entirely correct and simply represents an unlikely event. Discarding of such an outlying observation could lead to the undesirable consequence of increased variances of some of the estimated regression coefficients.

If, on the other hand, the circumstances surrounding the data provide an explanation of the unusual observation which indicates an exceptional situation not to be covered by the model, the discarding of the observation may be appropriate. Thus, when an outlying influential observation can definitely be shown to be the result of a gross measurement error, it would be appropriate to discard that observation.

When the outlying influential observation is accurate, it may not represent an unlikely event but rather a failure of the model. The failure may be either the omission of an important independent variable or the choice of an incorrect functional form, such as omission of a curvature effect for an independent variable included in the model. Often, identification of outlying influential observations leads to valuable insights for strengthening the model.

When an outlying influential observation is accurate but no explanation can be found for it, a less severe alternative than discarding the observation is to dampen its influence. We shall now discuss one method of doing this.

Method of least absolute deviations. This method is one of a variety of robust methods that have the property of being insensitive to both outlying data values and inadequacies of the model employed. The method of least absolute deviations estimates the regression coefficients by minimizing the sum of the absolute deviations of the observations from their means. The criterion to be minimized is:

$$(11.65) \qquad \sum_{i=1}^{n} |Y_i - (\beta_0 + \beta_1 X_{i1} + \cdots + \beta_{p-1} X_{i,p-1})|$$

Since absolute deviations rather than squared ones are involved here, the method of least absolute deviations places less emphasis on outlying observations than does the method of least squares.

The estimated regression coefficients according to the method of least absolute deviations can be obtained by linear programming techniques. Details about the computational aspects may be found in specialized texts, such as Reference 11.2.

Example. In the body fat example, observation 3 was identified as having a substantial influence on the fitted regression function. If observation 3 were deleted, the fitted regression function would be:

$$\hat{Y} = -12.428 + .5641X_1 + .3635X_2$$

An alternative to deleting observation 3 would be to reduce its influence. Using the method of least absolute deviations, the fitted regression function is:

$$\hat{Y} = -17.027 + .4173X_1 + .5203X_2$$

Since the fitted regression based on all observations is (Table 8.4a):

$$\hat{Y} = -19.174 + .2224X_1 + .6594X_2$$

it is seen that the method of least absolute deviations leads to more modest changes than dropping observation 3 entirely. An analysis of the residuals shows that the method of least absolute deviations resulted in reductions of those residuals that are largest absolutely with the method of least squares.

Comments

1. The residuals for the method of least absolute deviations ordinarily will not sum to zero.

2. The solution for the estimated regression coefficients with the method of least absolute deviations may not be unique.

3. The method of least absolute deviations is also called *minimum absolute deviations, minimum sum of absolute deviations,* and *minimum L_1-norm.*

4. Numerous other robust procedures besides the method of least absolute deviations have been proposed. Reference 11.3 discusses a number of these procedures.

PROBLEMS

11.1. Refer to the example of perfectly correlated independent variables in Table 11.1.
   a. Develop another model, like models (11.16) and (11.17), that fits the data perfectly.
   b. What is the intersection of the infinitely many response surfaces that fit the data perfectly?

11.2. The progress report of a research analyst to the supervisor stated: "All the estimated regression coefficients in our model with three independent variables to predict sales are statistically significant. Our new preliminary model with seven independent variables, which includes the three variables of our smaller model, is less satisfactory because only two of the seven regression coefficients are statistically significant. Yet in some initial trials the expanded model is giving more precise sales predictions than the smaller model. The reasons for this anomaly are now being investigated." Comment.

11.3. Two authors wrote as follows: "Our research utilized a multiple regression model. Two of the independent variables important in our theory turned out to be highly correlated in our observations. This made it difficult to assess the individual effects of each of these variables separately. We retained both variables in our model, however, because the high coefficient of multiple determination made this difficulty unimportant." Comment.

11.4. A student asked: "Why is it necessary to perform diagnostic checks of the fit when R^2 is large?" Comment.

11.5. **Cosmetics sales.** An assistant in the district sales office of a national cosmetics firm obtained data, shown below, on advertising expenditures and sales last year in the district's 14 territories. X_1 denotes expenditures for point-of-sale displays in beauty salons and department stores (in thousand dollars) while X_2 and X_3 represent the corresponding expenditures for local media advertising and prorated share of national media advertising, respectively. Y denotes sales (in thousand cases). The assistant was instructed to estimate the increase in expected sales when X_1 is increased by one thousand dollars and X_2 and X_3 are held constant, and was told to use an ordinary multiple regression model with linear terms for the independent variables and with independent normal error terms.

i:	1	2	3	4	5	6	7
X_{i1}:	4.2	6.5	3.0	2.1	2.9	7.2	4.8
X_{i2}:	4.0	6.5	3.5	2.0	3.0	7.0	5.0
X_{i3}:	3.0	5.0	4.0	3.0	4.0	3.0	4.5
Y_i:	8.26	14.70	9.73	5.62	7.84	12.18	8.56

i:	8	9	10	11	12	13	14
X_{i1}:	4.3	2.6	3.1	6.2	5.5	2.2	3.0
X_{i2}:	4.0	2.5	3.0	6.0	5.5	2.0	2.8
X_{i3}:	5.0	5.0	4.0	4.5	5.0	4.0	3.0
Y_i:	10.77	7.56	8.90	12.51	10.46	7.15	6.74

a. State the regression model to be employed and fit it to the data.

b. Test whether there is a regression relation between sales and the three independent variables. Use a level of significance of .05. State the alternatives, decision rule, and conclusion.

c. Test for each of the regression coefficients β_k ($k = 1, 2, 3$) individually whether or not $\beta_k = 0$. Use a level of significance of .05 each time. Do the conclusions of these tests correspond to that obtained in part (b)?

d. Obtain the correlation matrix.

e. What do the results in parts (b), (c), and (d) suggest about the suitability of the data for the research objective?

11.6. Refer to **Cosmetics sales** Problem 11.5.

a. Verify that the variance inflation factor for variable X_1 is $(VIF)_1 = 66.29$. The other variance inflation factors are $(VIF)_2 = 66.99$ and $(VIF)_3 = 1.09$. What do these suggest about the effects of multicollinearity here?

b. The assistant eventually decided to drop variables X_2 and X_3 from the model "to clear up the picture." Fit the assistant's revised model. Is the assistant now in a better position to achieve the research objective?

c. Why would an experiment here be more effective in providing suitable data to meet the research objective? How would you design such an experiment? What model would you employ?

11.7. Refer to **Patient satisfaction** Problem 7.17.

a. Obtain the correlation matrix. What does it show about pairwise linear associations among the independent variables?

b. The variance inflation factors are $(VIF)_1 = 1.35$, $(VIF)_2 = 2.76$, and $(VIF)_3 = 2.87$. What do these results suggest about the effects of multicollinearity here? Are these results more revealing than those in part (a)?

11.8. Refer to **Brand preference** Problem 7.8.

a. Obtain the correlation matrix. What does it show about the linear association between the two independent variables?

b. Find the two variance inflation factors. Why are they both equal to 1?

11.9. Refer to **Mathematicians salaries** Problem 7.20.

a. Obtain the correlation matrix. What does it show about the pairwise linear associations among the independent variables?

b. Obtain the variance inflation factors. Do they indicate that a serious multicollinearity problem exists here?

11.10. Refer to **Cosmetics sales** Problem 11.5. Given below are the estimated ridge standardized regression coefficients, the variance inflation factors, and R^2 for selected biasing constants c.

c:	.000	.005	.01	.02	.03	.04	.05	.06
b_1^R:	.273	.327	.349	.368	.376	.380	.382	.383
b_2^R:	.549	.494	.470	.447	.435	.427	.422	.417
b_3^R:	.260	.260	.260	.259	.257	.256	.254	.253
$(VIF)_1$:	66.29	24.11	12.45	5.20	2.92	1.91	1.38	1.07
$(VIF)_2$:	66.99	24.36	12.57	5.25	2.94	1.92	1.39	1.07
$(VIF)_3$:	1.09	1.06	1.04	1.01	.99	.97	.95	.93
R^2:	.840	.838	.836	.832	.828	.824	.821	.816

a. Make a ridge trace plot for the given c values. Do the ridge regression coefficients exhibit substantial changes near $c = 0$?

b. Suggest a reasonable value for the biasing constant c based on the ridge trace, the (VIF)'s, and R^2.

c. Transform the estimated standardized regression coefficients selected in part (b) back to the original variables and obtain the fitted values for the 14 observations. How similar are these fitted values to those obtained with the ordinary least squares fit in Problem 11.5a?

11.11. Refer to **Chemical shipment** Problem 7.12. Given below are the estimated ridge standardized regression coefficients, variance inflation factors, and R^2 for selected biasing constants c.

c:	.000	.005	.01	.05	.07	.09	.10	.20
b_1^R:	.451	.453	.455	.460	.460	.459	.458	.444
b_2^R:	.561	.556	.552	.526	.517	.508	.504	.473
$(VIF)_1 = (VIF)_2$:	7.03	6.20	5.51	2.65	2.03	1.61	1.46	.71
R^2:	.987	.984	.982	.962	.952	.943	.940	.894

a. Make a ridge trace plot for the given c values. Do the regression coefficients exhibit substantial changes near $c = 0$?

b. Why are the $(VIF)_1$ values the same as the $(VIF)_2$ values here?

c. Suggest a reasonable value for the biasing constant c based on the ridge trace in part (a), the (VIF)'s, and R^2.

d. Transform the estimated standardized regression coefficients selected in part (c) back to the original variables and obtain the fitted values for the 20 observations. How similar are these fitted values to those obtained with the ordinary least squares fit in Problem 7.12a?

11.12. Refer to **Brand preference** Problem 7.8. The diagonal elements of the hat matrix are: $h_{55} = h_{66} = h_{77} = h_{88} = h_{99} = h_{10,10} = h_{11,11} = h_{12,12} = .137$ and $h_{11} = h_{22} = h_{33} = h_{44} = h_{13,13} = h_{14,14} = h_{15,15} = h_{16,16} = .237$.

a. Explain the reason for the pattern in the diagonal elements of the hat matrix.

b. According to the rule of thumb stated in the chapter, are any of the observations outlying with regard to their X values?

c. Obtain the studentized deleted residuals and identify any outlying Y observations.

d. Calculate Cook's distance D_i for each observation. Are any observations influential according to this measure?

11.13. Refer to **Chemical shipment** Problem 7.12. The diagonal elements of the hat matrix are:

i:	1	2	3	4	5	6	7	8	9	10
h_{ii}:	.091	.194	.131	.268	.149	.141	.429	.067	.135	.165

i:	11	12	13	14	15	16	17	18	19	20
h_{ii}:	.179	.051	.110	.156	.095	.128	.097	.230	.112	.073

a. Identify any outlying X observations using the rule of thumb presented in the chapter.

b. Obtain the studentized deleted residuals and identify any outlying Y observations.

c. Calculate Cook's distance D_i for each observation. Are any observations influential according to this measure? How much is the fitted value for observation $i = 7$ changed when all observations are included and when observation 7 is omitted from the fit? [*Hint:* $\hat{Y}_{(i)} = Y_i - e_i/(1 - h_{ii})$.]

11.14. Refer to **Patient satisfaction** Problem 7.17. The diagonal elements of the hat matrix are:

i:	1	2	3	4	5	6	7	8	9	10	11	12
h_{ii}:	.134	.193	.070	.235	.204	.319	.060	.174	.339	.104	.209	.143

i:	13	14	15	16	17	18	19	20	21	22	23
h_{ii}:	.078	.231	.158	.238	.059	.057	.254	.137	.313	.072	.230

a. Identify any outlying X observations.

b. Obtain the studentized deleted residuals and identify any outlying Y observations.

c. Calculate Cook's distance D_i for each observation. Are any observations influential according to this measure?

11.15. Refer to **Mathematicians salaries** Problem 7.20. The diagonal elements of the hat matrix are:

i:	1	2	3	4	5	6	7	8	9	10	11	12
h_{ii}:	.184	.059	.132	.071	.214	.146	.115	.179	.241	.288	.083	.128

i:	13	14	15	16	17	18	19	20	21	22	23	24
h_{ii}:	.320	.098	.186	.151	.267	.146	.198	.206	.118	.136	.225	.110

a. Identify any outlying X observations.

b. Obtain the studentized deleted residuals and identify any outlying Y observations.

c. Calculate Cook's distance D_i for each observation. Are any observations influential according to this measure?

EXERCISES

11.16. Refer to the work crew productivity example data in Table 8.1.

a. For the variables transformed according to (11.6), obtain: (1) $\mathbf{X'X}$, (2) $\mathbf{X'Y}$, (3) $\mathbf{b}$, (4) $s^2(\mathbf{b})$.

b. Show that the regression coefficients obtained in part (a3) are the same as the standardized regression coefficients according to (7.69).

11.17. Show that the least squares estimator of β_0' in (11.2a) is $\bar{Y}$.

11.18. Derive the relations between the β_k and β_k' ($k = 1, 2$) in (11.8a) and (11.8b).

11.19. Derive the expression for $\mathbf{X'Y}$ in (11.12) for the transformed model (11.7).

11.20. Derive the mean squared error in (11.35).

11.21. Refer to the least absolute deviations estimates for the body fat example on page 410—namely, $b_0 = -17.027$, $b_1 = .4173$, and $b_2 = .5203$.
   a. Find the sum of the absolute deviations of the sample observations from the fitted values based on the least absolute deviations estimates.
   b. For the least squares estimated regression coefficients $b_0 = -19.174$, $b_1 = .2224$, and $b_2 = .6594$, find the sum of the absolute deviations. Is this sum larger than the sum obtained in part (a)?

PROJECTS

11.22. Refer to **Patient satisfaction** Problem 7.17.
   a. Obtain the estimated ridge standardized regression coefficients, variance inflation factors, and R^2 for the following biasing constants: $c = .000, .005, .01, .02, .03, .04, .05$.
   b. Make a ridge trace plot for the given c values. Do the ridge regression coefficients exhibit substantial changes near $c = 0$?
   c. Suggest a reasonable value for the biasing constant c based on the ridge trace, the (VIF)'s, and R^2.
   d. Transform the estimated standardized regression coefficients selected in part (c) back to the original variables and obtain the fitted values for the 23 observations. How similar are these fitted values to those obtained with the ordinary least squares fit in Problem 7.17a?

11.23. Refer to **Mathematicians salaries** Problem 7.20.
   a. Obtain the estimated ridge standardized regression coefficients, variance inflation factors, and R^2 for the following biasing constants: $c = .000, .005, .01, .02, .03, .04, .05$.
   b. Make a ridge trace plot for the given c values. Do the ridge regression coefficients exhibit substantial changes near $c = 0$?
   c. Suggest a reasonable value for the biasing constant c based on the ridge trace, the (VIF)'s, and R^2.
   d. Transform the estimated standardized regression coefficients selected in part (c) back to the original variables and obtain the fitted values for the 24 observations. How similar are these fitted values to those obtained with the ordinary least squares fit in Problem 7.20a?

11.24. Refer to the **SENIC** data set.
   a. Regress the logarithm of length of stay (Y') on infection risk (X_1), number of beds (X_2), and average daily census (X_3).
   b. Obtain the residuals and identify outliers.

    c. Obtain the correlation matrix and the variance inflation factors. What do these suggest about the effects of multicollinearity?

    d. Obtain the estimated ridge regression coefficients, variance inflation factors, and R^2 for the values of the biasing constant c given in Table 11.6.

    e. Make a ridge trace plot and determine a reasonable value for the biasing constant c based on this plot, the (VIF)'s, and R^2.

11.25. Refer to the **SMSA** data set.

    a. Regress number of active physicians (Y) on number of hospital beds (X_1), total personal income (X_2), and total serious crimes (X_3).

    b. Obtain the residuals and identify outliers.

    c. Obtain the correlation matrix and the variance inflation factors. What do these suggest about the effects of multicollinearity?

    d. Obtain the estimated ridge regression coefficients, variance inflation factors, and R^2 for the values of the biasing constant c in Table 11.6.

    e. Make a ridge trace plot and determine a reasonable value for the biasing constant c based on this plot, the (VIF)'s, and R^2.

CITED REFERENCES

11.1 Belsley, David A.; Edwin Kuh; and Roy E. Welsch. *Regression Diagnostics: Identifying Influential Data and Sources of Collinearity*. New York: John Wiley & Sons, 1980.

11.2 Kennedy, William J., Jr., and James E. Gentle. *Statistical Computing*. New York: Marcel Dekker, 1980.

11.3 Hogg, Robert V. "Statistical Robustness: One View of Its Use in Applications Today." *The American Statistician* 33 (1979), pp. 108–15.

12

Selection of independent variables

One of the most difficult problems in regression analysis often is the selection of the set of independent variables to be employed in the model. In this chapter, we take up several computer-assisted search methods for helping to identify one or a number of possible sets of independent variables to be included in the regression model.

12.1 NATURE OF PROBLEM

As we have seen in previous chapters, regression analysis has three major uses: (1) description, (2) control, and (3) prediction. For each of these uses, the investigator must specify the set of independent variables to be employed for describing, controlling, and/or predicting the dependent variable.

In some fields, theory can aid in selecting the independent variables to be employed and in specifying the functional form of the regression relation. Often in these fields, controlled experiments can be undertaken to furnish data on the basis of which the regression parameters can be estimated and the theoretical form of the regression function tested.

In many other subject matter fields, however, including the social and behavioral sciences and management, serviceable theoretical models are relatively rare. To complicate matters further, the available theoretical models may involve

417

independent variables that are not directly measurable, such as a family's future earnings over the next 10 years. Under these conditions, investigators are often forced to prospect for independent variables that could conceivably be related to the dependent variable under study. Obviously, such a set of independent variables will be large. For example, a company's sales of portable dishwashers in a district may be affected by population size, per capita income, percent of population in urban areas, percent of population under 50 years old, percent of families with children at home, etc. etc.!

After such a lengthy list has been compiled, some of the independent variables can be screened out. An independent variable (1) may not be fundamental to the problem, (2) may be subject to large measurement errors, and/or (3) may effectively duplicate another independent variable in the list. Other independent variables that cannot be measured may either be deleted or replaced by proxy variables that are highly correlated with them.

Typically, the number of independent variables that remain after this initial screening is still large. Further, many of these variables will be highly inter-correlated. Hence, the investigator usually will wish to reduce the number of independent variables to be used in the final model. There are several reasons for this. A regression model with a large number of independent variables is expensive to maintain. Further, regression models with a limited number of independent variables are easier to analyze and understand. Finally, the presence of many highly intercorrelated independent variables may add little to the predictive power of the model while substantially increasing the sampling variation of the regression coefficients, detracting from the model's descriptive abilities, and increasing the problem of roundoff errors (as we have seen in Chapter 11).

The investigator must be careful, however, not to eliminate key explanatory variables because that could seriously damage the explanatory power of the model and lead to biased estimates of regression coefficients, mean responses, and predictions of new observations.

The problem then is how to shorten the list of independent variables so as to obtain, in some sense, a "good" selection of independent variables. This subset of independent variables needs to be small enough so that maintenance costs are manageable and analysis is facilitated, yet it must be large enough so that adequate description, control, or prediction is possible.

Since the purposes of regression analysis vary, no one subset of independent variables is usually "best" for all uses. For instance, descriptive uses of a regression model typically will emphasize precise estimation of the regression coefficients, while predictive uses will focus on the prediction errors. Often, different subsets of the potential independent variables will best serve these varying purposes. Even for a given purpose, it is often found that several subsets are about equally "good" according to a given criterion, and the choice of the subset of independent variables to be employed in the regression model needs to be made on the basis of additional considerations. The entire selection process is, and should be, pragmatic, with large doses of subjective judgment. The selection procedures to be discussed in this chapter are aids to the investigator's judgment and should not be used in a purely mechanical fashion. A mechanical approach, for instance, might omit an important independent variable just because it occurred in the sample within a narrow range of values and therefore turned out to be statistically nonsignificant.

We shall discuss in this chapter two approaches to the selection of independent variables. The first approach considers all possible regression models that can be developed from the pool of potential independent variables and identifies subsets of the independent variables which are "good" according to a criterion specified by the investigator. The second approach employs automatic search procedures to arrive at a single subset of the independent variables.

The choice of the independent variables to be employed in the regression model does not, of course, fully determine the model to be utilized. Other determinations must also be made, such as whether an independent variable appears in linear form, in a transformed fashion, or with a quadratic term added, and whether interaction terms should be included in the model. Our discussion of the choice of independent variables in this chapter assumes that the investigator has already considered the functional form of the regression relation (whether given variables are to appear in linear form, quadratic form, etc.), whether the independent variables or the dependent variable are first transformed (e.g., by a logarithmic transformation), and whether any interaction terms are to be included. At this point, a selection procedure is employed to reduce the number of X variables, which include not only the potential independent variables in first-order form but also quadratic and other curvature terms and interaction terms.

Note

All too often, unwary investigators will screen the set of independent variables by fitting the regression model containing the entire set of potential X variables and then simply dropping those for which the t^* statistic (8.25):

(12.1)
$$t_k^* = \frac{b_k}{s(b_k)}$$

has a small absolute value. As we know from Chapter 11, this procedure can lead to the dropping of important intercorrelated independent variables. Clearly, a good search procedure must be able to handle important intercorrelated independent variables in such a way that not all of them will be dropped.

12.2 EXAMPLE

In order to illustrate the selection procedures to be discussed in the following sections, we shall use a relatively simple example which has four potential independent variables. By limiting the number of potential independent variables, we shall be able to explain the selection procedures without overwhelming the reader with masses of computer printouts.

A hospital surgical unit was interested in predicting survival in patients under-

TABLE 12.1 Potential independent variables and dependent variable—surgical unit example

Case Number	Blood Clotting Score X_1	Prognostic Index X_2	Enzyme Function Test X_3	Liver Function Test X_4	Survival Time Y	$Y' = \log_{10} Y$
1	6.7	62	81	2.59	200	2.3010
2	5.1	59	66	1.70	101	2.0043
3	7.4	57	83	2.16	204	2.3096
4	6.5	73	41	2.01	101	2.0043
5	7.8	65	115	4.30	509	2.7067
6	5.8	38	72	1.42	80	1.9031
7	5.7	46	63	1.91	80	1.9031
8	3.7	68	81	2.57	127	2.1038
9	6.0	67	93	2.50	202	2.3054
10	3.7	76	94	2.40	203	2.3075
11	6.3	84	83	4.13	329	2.5172
12	6.7	51	43	1.86	65	1.8129
13	5.8	96	114	3.95	830	2.9191
14	5.8	83	88	3.95	330	2.5185
15	7.7	62	67	3.40	168	2.2253
16	7.4	74	68	2.40	217	2.3365
17	6.0	85	28	2.98	87	1.9395
18	3.7	51	41	1.55	34	1.5315
19	7.3	68	74	3.56	215	2.3324
20	5.6	57	87	3.02	172	2.2355
21	5.2	52	76	2.85	109	2.0374
22	3.4	83	53	1.12	136	2.1335
23	6.7	26	68	2.10	70	1.8451
24	5.8	67	86	3.40	220	2.3424
25	6.3	59	100	2.95	276	2.4409
26	5.8	61	73	3.50	144	2.1584
27	5.2	52	86	2.45	181	2.2577
28	11.2	76	90	5.59	574	2.7589
29	5.2	54	56	2.71	72	1.8573
30	5.8	76	59	2.58	178	2.2504
31	3.2	64	65	0.74	71	1.8513
32	8.7	45	23	2.52	58	1.7634
33	5.0	59	73	3.50	116	2.0645
34	5.8	72	93	3.30	295	2.4698
35	5.4	58	70	2.64	115	2.0607
36	5.3	51	99	2.60	184	2.2648
37	2.6	74	86	2.05	118	2.0719
38	4.3	8	119	2.85	120	2.0792
39	4.8	61	76	2.45	151	2.1790
40	5.4	52	88	1.81	148	2.1703
41	5.2	49	72	1.84	95	1.9777
42	3.6	28	99	1.30	75	1.8751
43	8.8	86	88	6.40	483	2.6840
44	6.5	56	77	2.85	153	2.1847
45	3.4	77	93	1.48	191	2.2810
46	6.5	40	84	3.00	123	2.0899
47	4.5	73	106	3.05	311	2.4928
48	4.8	86	101	4.10	398	2.5999
49	5.1	67	77	2.86	158	2.1987
50	3.9	82	103	4.55	310	2.4914
51	6.6	77	46	1.95	124	2.0934
52	6.4	85	40	1.21	125	2.0969
53	6.4	59	85	2.33	198	2.2967
54	8.8	78	72	3.20	313	2.4955

going a particular type of liver operation. A random selection of 54 patients was available for analysis. From each patient record, the following information was extracted from the preoperational evaluation:

X_1 blood clotting score
X_2 prognostic index, which includes the age of patient
X_3 enzyme function test score
X_4 liver function test score

These constitute the potential independent variables for a predictive regression model. The dependent variable was survival time, which was ascertained in a follow-up study. The data on the potential independent variables and the dependent variable are presented in Table 12.1. Since the survival time distribution is substantially skewed to the right, the logarithm of the survival time $Y' = \log_{10} Y$ was taken as the dependent variable.

The surgical unit wished to obtain a subset of independent variables for predicting Y'. The task was assigned to an analyst, who first obtained the correlation matrix for all the variables from a computer run. This matrix provides valuable basic information on the nature of the problems to be encountered. Table 12.2 contains the correlation matrix based on a computer run, omitting the duplicate terms below the main diagonal.

Table 12.2 indicates that all of the independent variables are linearly associated with Y', X_4 showing the highest degree of association and X_1 the lowest. The correlation matrix further shows intercorrelations among the potential independent variables. In particular, the individual pairwise correlations between X_4 and X_1, X_2, and X_3 are moderately high. The task now is to determine whether all four independent variables are required for the prediction model.

TABLE 12.2   Correlation matrix for surgical unit example

	Y'	X_1	X_2	X_3	X_4
Y'	1.000	.346	.593	.665	.726
X_1		1.000	.090	−.150	.502
X_2			1.000	−.024	.369
X_3				1.000	.416
X_4					1.000

12.3 ALL POSSIBLE REGRESSION MODELS

The *all-possible-regressions selection procedure* calls for an examination of all possible regression models involving the potential X variables and identifying "good" subsets according to some criterion. When this selection procedure is used with our surgical unit example, for instance, 16 different regression models are to be considered, as shown in Table 12.3. First, there is the regression model with no X variables, i.e., the model $Y_i = \beta_0 + \varepsilon_i$. Then there are the regression models with one X variable (X_1, X_2, X_3, X_4), with two X variables (X_1 and X_2, X_1 and X_3, X_1 and X_4, X_2 and X_3, X_2 and X_4, X_3 and X_4), and so on.

Different criteria for comparing the various regression models may be used with the all-possible-regressions selection procedure. We shall discuss three— R_p^2, MSE_p, and C_p. Before doing so, we need to develop some notation. Let us denote the number of potential X variables in the pool by $P - 1$. We assume throughout this chapter that all regression models contain an intercept term β_0. Hence, the regression function containing all potential X variables contains P parameters, and the function with no X variables contains one parameter (β_0).

The number of X variables in a subset will be denoted by $p - 1$, as always, so that there are p parameters in the regression function for this subset of X variables. Thus, we have:

$$(12.2) \qquad\qquad 1 \leq p \leq P$$

The all-possible-regressions approach assumes that the number of observations n exceeds the maximum number of potential parameters:

$$(12.3) \qquad\qquad n > P$$

and, indeed, it is highly desirable that n be substantially larger than P so that sound results can be obtained.

R_p^2 criterion

The R_p^2 criterion calls for an examination of the coefficient of multiple determination R^2, defined in (7.31), in order to select one or several subsets of X variables. We show the number of parameters in the regression model as a subscript of R^2. Thus, R_p^2 indicates that there are p parameters, or $p - 1$ predictor variables, in the regression equation on which R_p^2 is based.

Since R_p^2 is a ratio of sums of squares:

$$(12.4) \qquad\qquad R_p^2 = \frac{SSR_p}{SSTO} = 1 - \frac{SSE_p}{SSTO}$$

and the denominator is constant for all possible regressions, R_p^2 varies inversely with the error sums of squares SSE_p. But we know that SSE_p can never increase as additional independent variables are included in the model. Thus, R_p^2 will be a maximum when all $P - 1$ potential X variables are included in the regression model. The reason for using the R_p^2 criterion with the all-possible-regressions approach therefore cannot be to maximize R_p^2. Rather, the intent is to find the point where adding more X variables is not worthwhile because it leads to a very small increase in R_p^2. Often, this point is reached when only a limited number of X variables is included in the regression model. Clearly, the determination of where diminishing returns set in is a judgmental one.

Example. Table 12.3, column 4, contains the R_p^2 values for all possible regression models for our surgical unit example. The data were obtained from a series of computer runs. For instance, when X_4 is the only X variable in the

TABLE 12.3 R_p^2, MSE_p, and C_p values for all possible regression models—surgical unit example

X Variables in Model	(1) p	(2) df	(3) SSE_p	(4) R_p^2	(5) MSE_p	(6) C_p
None	1	53	3.9728	0	.0750	1,721.6
X_1	2	52	3.4960	.120	.0672	1,510.7
X_2	2	52	2.5762	.352	.0495	1,100.1
X_3	2	52	2.2154	.442	.0426	939.0
X_4	2	52	1.8777	.527	.0361	788.3
X_1, X_2	3	51	2.2324	.438	.0438	948.6
X_1, X_3	3	51	1.4073	.646	.0276	580.3
X_1, X_4	3	51	1.8759	.528	.0368	789.5
X_2, X_3	3	51	.7431	.813	.0146	283.7
X_2, X_4	3	51	1.3922	.650	.0273	573.5
X_3, X_4	3	51	1.2455	.687	.0244	508.0
X_1, X_2, X_3	4	50	.1099	.972	.00220	3.1
X_1, X_2, X_4	4	50	1.3905	.650	.0278	574.8
X_1, X_3, X_4	4	50	1.1157	.719	.0223	452.1
X_2, X_3, X_4	4	50	.4653	.883	.00931	161.7
X_1, X_2, X_3, X_4	5	49	.1098	.972	.00224	5.0

regression model, we obtain:

$$R_2^2 = 1 - \frac{SSE(X_4)}{SSTO} = 1 - \frac{1.8777}{3.9728} = .527$$

Note that $SSTO = SSE_1 = 3.9728$.

The R_p^2 values are plotted in Figure 12.1. The maximum R_p^2 value for the possible subsets of $p - 1$ predictor variables, denoted by $\max(R_p^2)$, appears at the top of the graph for each p. These points are connected by dashed lines to show the impact of adding additional X variables. Figure 12.1 makes it clear that little increase in $\max(R_p^2)$ takes place after three X variables are included in the model. Hence, the use of the subset (X_1, X_2, X_3) in the regression model appears to be reasonable according to the R_p^2 criterion.

Note that variable X_4, which singly correlates most highly with the dependent variable, is not in the $\max(R_p^2)$ models for $p = 3$ and $p = 4$, indicating that X_2 and X_3 contain much of the information presented by X_4. If it were desired that X_4 be retained in the model and that the subset model be limited to three X variables, the subset (X_2, X_3, X_4) should then be considered as next best according to the R_p^2 criterion for $p = 4$. The coefficient of multiple determination associated with this subset, $R_4^2 = .883$, would be somewhat smaller than $R_4^2 = .972$ for the subset (X_1, X_2, X_3).

MSE_p or R_a^2 criterion

Since R_p^2 does not take account of the number of parameters in the model, and since $\max(R_p^2)$ can never decrease as p increases, the use of the adjusted coeffi-

FIGURE 12.1 R_p^2 plot for surgical unit example

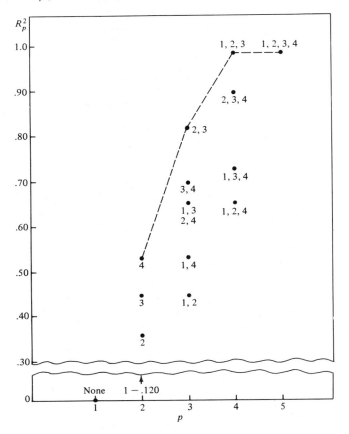

cient of multiple determination R_a^2 in (7.33):

$$(12.5) \qquad R_a^2 = 1 - \left(\frac{n-1}{n-p}\right)\frac{SSE}{SSTO} = 1 - \frac{MSE}{\dfrac{SSTO}{n-1}}$$

has been suggested as a criterion which takes the number of parameters in the model into account through the degrees of freedom. It can be seen from (12.5) that R_a^2 increases if and only if MSE decreases since $SSTO/(n-1)$ is fixed for the given Y observations. Hence, R_a^2 and MSE are equivalent criteria. We shall consider here the criterion MSE_p. Min(MSE_p) can, indeed, increase as p increases when the reduction in SSE_p becomes so small that it is not sufficient to offset the loss of an additional degree of freedom. Users of the MSE_p criterion either seek to find the subset of X variables that minimizes MSE_p, or one or several subsets for which MSE_p is so close to the minimum that adding more variables is not worthwhile.

Example. The MSE_p values for all possible regression models for our surgical unit example are shown in Table 12.3, column 5. For instance, if the regression model contains only X_4, we have:

$$MSE_2 = \frac{SSE(X_4)}{n-2} = \frac{1.8777}{52} = .0361$$

Figure 12.2 contains the MSE_p plot for our example. We have connected the $\min(MSE_p)$ values for each p by dashed lines. The story which Figure 12.2 tells is very similar to that told by Figure 12.1. The subset (X_1, X_2, X_3) appears to be best. Indeed, the mean square error achieved with this subset is practically the same as that with (X_1, X_2, X_3, X_4), which uses all potential X variables.

If X_4 were to be included in the model with $p = 4$, the subset (X_2, X_3, X_4) would be best, involving $MSE_4 = .009$ which is somewhat higher than $MSE_4 = .002$ for subset (X_1, X_2, X_3).

FIGURE 12.2 MSE_p plot for surgical unit example

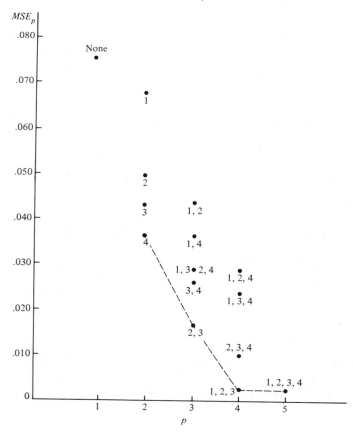

C_p criterion

This criterion is concerned with the *total mean squared error* of the n fitted values for each of the various subset regression models. The mean squared error concept involves a bias component and a random error component. The mean squared error for an estimated regression coefficient was defined in (11.35). Here, the mean squared error pertains to the fitted values $\hat{Y}_i$ for the regression model employed. The bias component for the ith fitted value $\hat{Y}_i$ is:

$$(12.6) \qquad E(\hat{Y}_i) - E(Y_i)$$

where $E(\hat{Y}_i)$ is the expectation of the ith fitted value for the given regression model and $E(Y_i)$ is the true mean response. The random error component for $\hat{Y}_i$ is simply $\sigma^2(\hat{Y}_i)$, its variance. The mean squared error for $\hat{Y}_i$ is then the sum of the squared bias and the variance:

$$(12.7) \qquad [E(\hat{Y}_i) - E(Y_i)]^2 + \sigma^2(\hat{Y}_i)$$

The total mean squared error for all n fitted values $\hat{Y}_i$ is the sum of the n individual mean squared errors:

$$(12.8) \qquad \sum_{i=1}^{n} [E(\hat{Y}_i) - E(Y_i)]^2 + \sum_{i=1}^{n} \sigma^2(\hat{Y}_i)$$

The criterion measure, denoted by Γ_p, is simply the total mean squared error divided by σ^2, the true error variance:

$$(12.9) \qquad \Gamma_p = \frac{1}{\sigma^2} \left\{ \sum_{i=1}^{n} [E(\hat{Y}_i) - E(Y_i)]^2 + \sum_{i=1}^{n} \sigma^2(\hat{Y}_i) \right\}$$

The model which includes all $P - 1$ potential X variables is assumed to have been carefully chosen so that $MSE(X_1, \ldots, X_{P-1})$ is an unbiased estimator of σ^2. It can then be shown that an estimator of Γ_p is C_p:

$$(12.10) \qquad C_p = \frac{SSE_p}{MSE(X_1, \ldots, X_{P-1})} - (n - 2p)$$

where SSE_p is the error sum of squares for the fitted subset regression model with p parameters (i.e., with $p - 1$ predictor variables).

When there is no bias in the regression model with $p - 1$ predictor variables so that $E(\hat{Y}_i) \equiv E(Y_i)$, the expected value of C_p is approximately p:

$$(12.11) \qquad E[C_p \mid E(\hat{Y}_i) \equiv E(Y_i)] \approx p$$

Thus, when the C_p values for all possible regression models are plotted against p, those models with little bias will tend to fall near the line $C_p = p$. Models with substantial bias will tend to fall considerably above this line.

In using the C_p criterion, one seeks to identify subsets of X variables for which (1) the C_p value is small and (2) the C_p value is near p. Sets of X variables with small C_p values have a small total mean squared error, and when the C_p value is also near p, the bias of the regression model is small. It may sometimes occur that the regression model based on the subset of X variables with the smallest C_p value involves substantial bias. In that case, one may at times prefer a regression model based on a somewhat larger subset of X variables for which the C_p value is slightly larger but which does not involve a substantial bias component. Reference 12.1 contains extended discussions of applications of the C_p criterion.

Example. Table 12.3, column 6, contains the C_p values for all possible regression models for our surgical unit example. For instance, when X_4 is the only X variable in the regression model, the C_p value is:

$$C_2 = \frac{SSE(X_4)}{MSE(X_1, X_2, X_3, X_4)} - [n - 2(2)]$$

$$= \frac{1.8777}{.00224} - (54 - 4) = 788.3$$

The C_p values for all possible regression models are plotted in Figure 12.3. We

FIGURE 12.3 C_p plot for surgical unit example

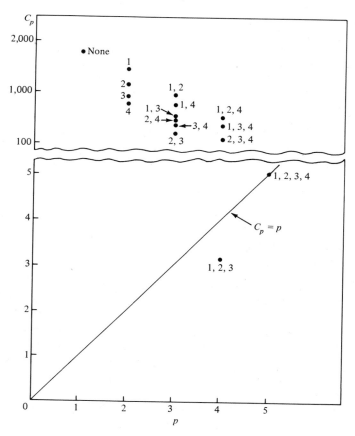

find again that subset (X_1, X_2, X_3) is suggested. This subset has the smallest C_p value, with no indication of any bias in the regression model. The fact that the C_p measure for this model, $C_4 = 3.1$, is below $p = 4$ is the result of random variation in the C_p measure.

Note that use of all potential X variables (X_1, X_2, X_3, X_4) would involve a larger total mean squared error. Also, use of subset (X_2, X_3, X_4) with C_p value $C_4 = 161.7$ would be poor because of the substantial bias with that model.

Comments

1. Effective use of the C_p criterion requires careful development of the pool of $P - 1$ potential X variables, with the independent variables expressed in appropriate form (linear, quadratic, transformed) and useless variables excluded so that $MSE(X_1, \ldots, X_{P-1})$ provides an unbiased estimate of the error variance σ^2.

2. The C_p criterion places major emphasis on the fit of the subset model for the n sample observations. At times, a modification of the C_p criterion which emphasizes new observations to be predicted might be preferable.

3. To see why C_p as defined in (12.10) is an estimator of Γ_p, we need to utilize two results that we shall simply state. First, it can be shown that:

$$(12.12) \qquad \sum_{i=1}^{n} \sigma^2(\hat{Y}_i) = p\sigma^2$$

Thus, the total random error of the n fitted values $\hat{Y}_i$ increases as the number of variables in the regression model increases.

Further, it can be shown that:

$$(12.13) \qquad E(SSE_p) = \Sigma[E(\hat{Y}_i) - E(Y_i)]^2 + (n - p)\sigma^2$$

Hence, Γ_p in (12.9) can be expressed as follows:

$$(12.14) \qquad \Gamma_p = \frac{1}{\sigma^2}[E(SSE_p) - (n - p)\sigma^2 + p\sigma^2]$$

$$= \frac{E(SSE_p)}{\sigma^2} - (n - 2p)$$

Replacing $E(SSE_p)$ by the estimator SSE_p and using $MSE(X_1, \ldots, X_{P-1})$ as an estimator of σ^2 yields C_p in (12.10).

Identification of "best" subsets by use of algorithm

A major disadvantage of the all-possible-regressions selection procedure is the amount of computation required. Since each potential independent variable either can be included or excluded, there are $2^{(P-1)}$ possible regression models when there are $P - 1$ potential X variables. When $P - 1 = 10$, for instance, there are 1,024 possible regression models. With the availability of large computers today, running all possible regression models for as many as 10 potential X variables is not too time consuming. Beyond that, however, the development of all possible regression models becomes inefficient.

Even when all possible regression models can easily be calculated, as when there are eight variables in the pool of potential X variables, it may be difficult for the investigator to evaluate carefully all of the regression models fitted. In the case of 8 potential X variables, for instance, there would be 256 models to consider, a major task for an investigator.

Thus, whenever the pool of potential X variables is not very small, it is highly desirable that the investigator be able to concentrate on a limited number of regression models which are the "best" ones according to a specified criterion. This limited number might consist of the "best" 5 or 10 subsets according to the criterion employed, so that the investigator can then carefully study these regression models for choosing the final model to be employed.

Time-saving algorithms have been developed in which the "best" subsets according to a specified criterion are identified without requiring the fitting of most of the possible subset regression models. Thus, if, say, the C_p criterion is to be employed and the five "best" subsets according to this criterion are to be identified, these algorithms search for the five subsets of X variables with the smallest C_p values using much less computational effort than when all possible subsets are evaluated. These algorithms are called *"best" subsets algorithms*. Not only do these algorithms provide the best subsets according to the specified criterion, but they often will also provide a number of "good" subsets for each possible number of X variables in the model to give the investigator helpful information in making the final selection of the subset of X variables to be employed in the regression model.

Example. In our surgical unit example, use of one of the "best" subsets algorithms will provide a portion of the information in Table 12.3. Suppose that the C_p criterion is to be employed and that the "best" three subsets are to be identified. The algorithm will then identify subsets (X_1, X_2, X_3), (X_1, X_2, X_3, X_4), and (X_2, X_3, X_4) as the three subsets with the smallest C_p values. In addition, information about three "good" subsets for each level of p may be provided.

Some final comments

The all-possible-regressions selection approach or a "best" subsets algorithm leads to the identification of a small number of subsets which are "good" according to a specified criterion. While in our surgical unit example, each of the three criteria pointed to the same "best" subset, this is not always the case. It therefore may be desirable at times to consider more than one criterion in evaluating possible subsets of X variables.

Once the investigator has identified a few subsets as "good" ones, a final choice of the model variables must be made. This choice is aided by residual analyses and examinations of influential observations for each of the competing models. Information gained by these analyses, together with knowledge by the invesigator about the phenomenon under study, will be helpful in choosing the final regression model to be employed.

12.4 STEPWISE REGRESSION

In those occasional cases when the pool of potential X variables contains 40 to 60 or even more variables, use of a "best" subsets algorithm may not be feasible. An automatic search procedure that develops sequentially the subset of X variables to be included in the regression model may be helpful in those cases. The *stepwise regression procedure* is probably the most widely used of the automatic search methods. It was developed to economize on computational efforts, as compared with the all-possible-regressions approach, while arriving at a reasonably "good" subset of independent variables. Essentially, this search method develops a sequence of regression models, at each step adding or deleting an X variable. The criterion for adding or deleting an X variable can be stated equivalently in terms of error sum of squares reduction, coefficient of partial correlation, or F^* statistic.

Search algorithm

We shall describe the stepwise regression search algorithm in terms of the F^* statistic for the partial F test.

1. The stepwise regression routine first fits a simple regression model for each of the $P - 1$ potential X variables. For each simple regression model, the F^* statistic (3.59) for testing whether or not the slope is zero is obtained:

$$(12.15) \qquad F_k^* = \frac{MSR(X_k)}{MSE(X_k)}$$

Recall that $MSR(X_k) = SSR(X_k)$ measures the reduction in the total variation of Y associated with the use of the variable X_k. The X variable with the largest F^* value is the candidate for first addition. If this F^* value exceeds a predetermined level, the X variable is added. Otherwise, the program terminates with no X variable considered sufficiently helpful to enter the regression model.

2. Assume X_7 is the variable entered at step 1. The stepwise regression routine now fits all regression models with two X variables, where X_7 is one of the pair. For each such regression model, the partial F test statistic (8.24):

$$(12.16) \qquad F_k^* = \frac{MSR(X_k \mid X_7)}{MSE(X_7, X_k)} = \left[\frac{b_k}{s(b_k)} \right]^2$$

is obtained. This is the statistic for testing whether or not $\beta_k = 0$ when X_7 and X_k are the variables in the model. The X variable with the largest F^* value is the candidate for addition at the second stage. If this F^* value exceeds a predetermined level, the second X variable is added. Otherwise the program terminates.

3. Suppose X_3 is added at the second stage. Now the stepwise regression routine examines whether any of the other X variables already in the model should be dropped. For our illustration, there is at this stage only one other X variable in the model, X_7, so that only one partial F test statistic is obtained:

$$(12.17) \qquad F_7^* = \frac{MSR(X_7 | X_3)}{MSE(X_3, X_7)}$$

At later stages, there would be a number of these F^* statistics, for each of the variables in the model besides the one last added. The variable for which this F^* value is smallest is the candidate for deletion. If this F^* value falls below a predetermined limit, the variable is dropped from the model; otherwise, it is retained.

4. Suppose X_7 is retained so that both X_3 and X_7 are now in the model. The stepwise regression routine now examines which X variable is the next candidate for addition, then examines whether any of the variables already in the model should now be dropped, and so on until no further X variables can either be added or deleted, at which point the search terminates.

It should be noted that the stepwise regression algorithm allows an X variable, brought into the model at an earlier stage, to be dropped subsequently if it is no longer helpful in conjunction with variables added at later stages.

Example

Figure 12.4 shows the computer printout obtained when a particular stepwise regression routine (BMDP2R, Ref. 12.2) was applied to our surgical unit example. The minimum acceptable F limit for adding a variable and the maximum acceptable F limit for removing a variable were specified to be 4.0 and 3.9, respectively, as shown at the top left of Figure 12.4. Since the degrees of freedom associated with MSE vary, depending on the number of X variables in the model, and since repeated tests on the same data are undertaken, fixed F limits for adding or deleting a variable have no precise probabilistic meaning. Note, however, that $F(.95; 1, 60) = 4.00$, so that the specified F limits of 4.0 and 3.9 would correspond roughly to a level of significance of .05 for any single test based on approximately 50 degrees of freedom.

The minimum acceptable tolerance of .01 shown in the upper left of Figure 12.4 is a specification to guard against the entry of a variable that is highly correlated with the other X variables already in the model. As explained in Section 11.3, the tolerance is defined as $1 - R_k^2$, where R_k^2 is the coefficient of multiple determination when X_k is regressed on the other X variables in the regression model. The tolerance specification of .01 in Figure 12.4 provides that no variable is to be added to the model which has a coefficient of multiple determination with the other X variables already in the model which exceeds $1 - .01 = .99$ or which would cause the R_k^2 for any variable in the model to exceed .99.

We shall now follow through the steps.

1. At step 0, no X variable is in the model so that the model to be fitted is $Y_i = \beta_0 + \varepsilon_i$. The residual or error sum of squares shown in the ANOVA table in Figure 12.4 for step 0 is therefore $\Sigma(Y_i - \bar{Y})^2 = SSTO = 3.9728$. For each potential X variable, the F^* statistic (12.15) is calculated. In Figure 12.4, these F_k^*

FIGURE 12.4 Stepwise regression for surgical unit example (BMDP2R, Ref. 12.2)

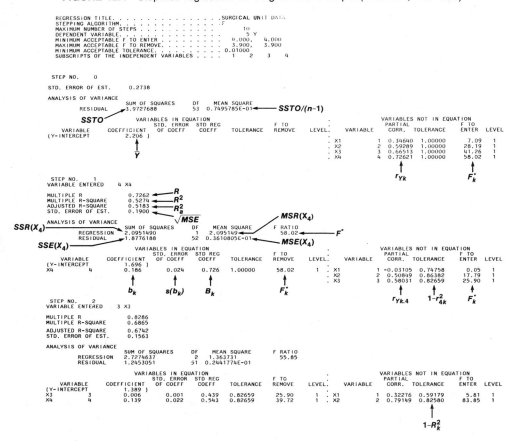

values are shown under the heading "Variables not in equation" and are called "F to enter" values. We see that F_4^* is the largest one:

$$F_4^* = \frac{MSR(X_4)}{MSE(X_4)} = \frac{2.09515}{.03611} = 58.02$$

Since this value exceeds the minimum acceptable F-to-enter value 4.0, X_4 is added to the model.

The column headed "Level" refers to a package option which permits the user to give different priorities to the various potential X variables. Note that in the present example all X variables have the same priority.

2. At this stage, step 1 has been completed. The current regression model contains X_4, and the estimated regression coefficients, the analysis of variance table, and selected other information about the current model are provided.

FIGURE 12.4   *(concluded)*

```
STEP NO.     3
VARIABLE ENTERED      2 X2

MULTIPLE R              0.9396
MULTIPLE R-SQUARE      0.8829
ADJUSTED R-SQUARE     0.8759
STD. ERROR OF EST.     0.0965

ANALYSIS OF VARIANCE
                      SUM OF SQUARES     DF    MEAN SQUARE      F RATIO
          REGRESSION    3.5075836         3    1.169194        125.67
          RESIDUAL      0.46518236       50  0.9303644E-02

                        VARIABLES IN EQUATION                              .          VARIABLES NOT IN EQUATION
                              STD. ERROR  STD REG                 F TO      .                  PARTIAL              F TO
       VARIABLE      COEFFICIENT  OF COEFF    COEFF   TOLERANCE  REMOVE  LEVEL.  VARIABLE   CORR.  TOLERANCE   ENTER  LEVEL
(Y-INTERCEPT          0.942 )                                             .
X2        2           0.008      0.001       0.488    0.82580    83.85    1  . X1         1  0.87411  0.55586   158.69    1
X3        3           0.007      0.001       0.543    0.79021    99.63    1  .
X4        4           0.082      0.015       0.320    0.68298    29.86    1  .


STEP NO.     4
VARIABLE ENTERED      1 X1

MULTIPLE R              0.9861
MULTIPLE R-SQUARE      0.9724
ADJUSTED R-SQUARE     0.9701
STD. ERROR OF EST.     0.0473

ANALYSIS OF VARIANCE
                      SUM OF SQUARES     DF    MEAN SQUARE      F RATIO
          REGRESSION    3.8630199         4    0.9657550       431.19
          RESIDUAL      0.10974741       49  0.2239743E-02

                        VARIABLES IN EQUATION                              .          VARIABLES NOT IN EQUATION
                              STD. ERROR  STD REG                 F TO      .                  PARTIAL              F TO
       VARIABLE      COEFFICIENT  OF COEFF    COEFF   TOLERANCE  REMOVE  LEVEL.  VARIABLE   CORR.  TOLERANCE   ENTER  LEVEL
(Y-INTERCEPT          0.489 )                                             .
X1        1           0.069      0.005       0.401    0.55586   158.69    1  .
X2        2           0.009      0.000       0.571    0.77567   449.08    1  .
X3        3           0.009      0.000       0.736    0.59594   571.83    1  .
X4        4           0.002      0.010       0.008    0.39134     0.04    1  .


STEP NO.     5
VARIABLE REMOVED      4 X4

MULTIPLE R              0.9861
MULTIPLE R-SQUARE      0.9724
ADJUSTED R-SQUARE     0.9707
STD. ERROR OF EST.     0.0469

ANALYSIS OF VARIANCE
                      SUM OF SQUARES     DF    MEAN SQUARE      F RATIO
          REGRESSION    3.8629322         3    1.287643        586.17
          RESIDUAL      0.10983557       50  0.2196711E-02

                        VARIABLES IN EQUATION                              .          VARIABLES NOT IN EQUATION
                              STD. ERROR  STD REG                 F TO      .                  PARTIAL              F TO
       VARIABLE      COEFFICIENT  OF COEFF    COEFF   TOLERANCE  REMOVE  LEVEL.  VARIABLE   CORR.  TOLERANCE   ENTER  LEVEL
(Y-INTERCEPT          0.484 )                                             .
X1        1           0.069      0.004       0.405    0.97011   288.23    1  . X4         4  0.02833  0.39134     0.04    1
X2        2           0.009      0.000       0.574    0.99178   590.58    1  .
X3        3           0.010      0.000       0.739    0.97751   966.28    1  .

* * * * * F-LEVELS(   4.000,    3.900) OR TOLERANCE INSUFFICIENT FOR FURTHER STEPPING
```

Next, all regression models containing X_4 and another independent variable are fitted, and the F^* statistics calculated. They are now:

$$F_k^* = \frac{MSR(X_k \mid X_4)}{MSE(X_4, X_k)}$$

These statistics are shown in step 1 under the heading "Variables not in equation." X_3 has the highest F^* value, which exceeds 4.0, so that X_3 now enters the model.

3. Step 2 in Figure 12.4 summarizes the situation at this point. X_3 and X_4 are now in the model, and information about this model is provided. Next, a test whether X_4 should be dropped is undertaken. The F^* statistic is shown under the heading "Variables in equation" and is called "F to remove":

$$F_4^* = \frac{MSR(X_4 \mid X_3)}{MSE(X_3, X_4)} = 39.72$$

Since this F^* value exceeds the maximum acceptable F-to-remove value 3.9, X_4 is not dropped.

4. Next, all regression models containing X_3, X_4, and one of the remaining potential X variables are fitted. The appropriate F^* statistics now are:

$$F_k^* = \frac{MSR(X_k \mid X_3, X_4)}{MSE(X_3, X_4, X_k)}$$

These statistics are shown in step 2 under the heading "Variables not in equation." X_2 has the highest F^* value, which exceeds 4.0, so that X_2 now enters the model.

5. Step 3 in Figure 12.4 summarizes the situation at this point. X_2, X_3, and X_4 are now in the model. Next, a test is undertaken whether X_3 or X_4 should be dropped. The F^* statistics to remove a variable are shown under the heading "Variables in equation" in step 3. F_4^* is smallest:

$$F_4^* = \frac{MSR(X_4 \mid X_2, X_3)}{MSE(X_2, X_3, X_4)} = 29.86$$

Since its value exceeds 3.9, X_4 is not dropped from the model.

6. At this point, only X_1 remains in the potential pool. Its F^* value to enter exceeds 4.0 (see "Variables not in equation" under step 3), so X_1 is entered into the model.

7. Step 4 in Figure 12.4 summarizes the addition of variable X_1 into the model containing variables X_2, X_3, and X_4. Next, a test is undertaken to determine whether either X_2, X_3, or X_4 should be dropped. The F^* statistics are shown under the heading "Variables in equation" in step 4. Note that:

$$F_4^* = \frac{MSR(X_4 \mid X_1, X_2, X_3)}{MSE(X_1, X_2, X_3, X_4)} = .04$$

is smallest, and that its value is less than 3.9; hence, X_4 is deleted.

8. Step 5 summarizes the dropping of X_4 from the model. Since the only potential variable remaining is X_4, which has just been dropped, it cannot enter the regression now. The algorithm therefore next considers the "F to remove" values in step 5 which indicate that F_1^* is smallest. However, since its value exceeds 3.9, X_1 is not dropped from the model and the search process is terminated.

Thus, the stepwise search algorithm identifies (X_1, X_2, X_3) as the "best" subset of X variables, a result that happens to be consistent with our previous analyses based on the all-possible-regressions approach.

Comments

1. In the surgical unit example, the tolerance requirement was always met; hence, no variable was excluded from the model as a result of too high a correlation with the other X variables in the model.

2. Variations of the rules for entering and removing variables illustrated in the example are possible. For instance, different F-to-enter and F-to-remove values can be employed in accordance with the degrees of freedom associated with MSE in the F^* statistic. However, this refinement often is not utilized, and fixed values are employed instead since the repeated testing in the search procedure does not permit precise probabilistic interpretations.

3. The minimum acceptable F-to-enter value should never be smaller than the maximum acceptable F-to-remove value; otherwise cycling is possible where a variable is continually entered and removed.

4. The order in which variables enter the regression model does not reflect their importance. In our surgical unit example, for instance, X_4 was the first variable to enter the model yet it was eventually dropped.

5. The stepwise regression routine we employed prints out the partial correlation coefficients at each stage. These could be used equivalently to the F^* values for screening the X variables, and indeed some routines actually use the partial correlation coefficients for screening.

6. The F limits for adding and deleting a variable need not be selected in terms of approximate significance levels, but may be determined descriptively in terms of error reduction. For instance, an F limit of 2.0 for adding a variable may be specified with the thought that the marginal error reduction associated with the added variable should be at least twice as great as the remaining error mean square once that variable has been added.

7. A limitation of the stepwise regression search approach is that it presumes there is a single "best" subset of X variables and seeks to identify it. As noted earlier, there is often no unique "best" subset. Hence, some statisticians suggest that all possible regression models with a similar number of X variables as in the stepwise regression solution be fitted subsequently to study whether some other subsets of X variables might be better.

Another limitation of the stepwise regression routine is that it sometimes arrives at an unreasonable "best" subset when the X variables are very highly correlated.

Other automatic search procedures

There are a number of other automatic search procedures which have been proposed to find a "best" subset of independent variables. We mention two of these. Neither of the two methods, however, has gained the acceptance of the stepwise search procedure.

Forward selection. This search procedure is a simplified version of stepwise regression, omitting the test whether a variable once entered into the model should be dropped.

Backward elimination. This search procedure is the opposite of forward selection. It begins with the model containing all potential X variables and identifies the one with the smallest F^* value. For instance, the F^* value for X_1 is:

$$(12.18) \qquad F_1^* = \frac{MSR(X_1 \mid X_2, \ldots, X_{P-1})}{MSE(X_1, \ldots, X_{P-1})}$$

If the minimum F_k^* value is less than a predetermined limit, that independent variable is dropped. The model with the remaining $P - 2$ predictor variables is then fitted, and the next candidate for dropping is identified. This process continues until no further independent variables can be dropped.

The backward elimination procedure requires more computations than the forward selection method since it starts with the biggest possible model. However, it does have the advantage of showing users the implications of models with many variables.

12.5 SELECTION OF VARIABLES WITH RIDGE REGRESSION

In Section 11.4, we discussed the use of ridge regression for helping to overcome problems related to multicollinearities among the X variables. The ridge trace mentioned there (Figure 11.2, p. 399) can also be used to identify variables which might be dropped from the regression model. It has been suggested that variables be dropped whose ridge trace is unstable, with the coefficient tending toward the value of zero. Also, variables should be dropped whose ridge trace is stable but at a very small value. Finally, variables with unstable ridge traces that do not tend toward zero should be considered as candidates for dropping.

12.6 IMPLEMENTATION OF SELECTION PROCEDURES

Options and refinements

Our discussion of the major selection procedures for identifying "good" sets of X variables has focused on the main conceptual issues and not on options, variations, and refinements available with particular computer packages. It is essential that the specific features of the package to be employed are fully understood so that intelligent use of the package can be made. In some packages, there is an option for regression models through the origin. Some packages permit variables to be brought into the model and tested in pairs or other groupings instead of singly, to save computing time or for other reasons. Some packages, once a "best" regression model is identified, will fit all the possible regression models in the same number of variables and will develop information for each model so that a final choice can be made by the user. Some stepwise programs have options for forcing variables into the regression model; such variables are not removed even if their F^* values become too low.

The diversity of these options and special features serves to emphasize a point made earlier: there is no unique way of searching for "good" subsets of X variables, and subjective elements must play an important role in the search process.

Completion of model building process

The screening of variables by a computerized selection process is only one step in the building of a regression model. Once the set of X variables has been identified, the resulting model needs to be studied for its aptness by the methods of Chapters 4 and 11. If repeat observations are available, a formal test for lack of fit can be made. In any case, a variety of residual plots and analyses can be employed to identify the nature of lack of fit, outliers, and influential observations. When the original set of $P - 1$ potential X variables excludes cross-product terms and powers of the independent variables to keep the selection problem within reasonable bounds, residual plots against such ''missing'' variables, or augmenting the model of ''best'' independent variables by adding cross-product and/or power terms, can be useful in identifying ways in which the model fit can be improved further.

Cautions in use of final model

The model-building process, as we have just noted, requires repeated analyses on the same set of data in order to arrive at a model which fits the data well. A consequence is that the model may be subject to *prediction bias*, i.e., the indicated predictive ability of the model for the data on which the model is based may be greater than the model's predictive ability for new data. The prediction bias arises because the choice of the final model is so uniquely related to the observations at hand. The prediction bias may be particularly large when the effects of independent variables are small.

It is good statistical practice to measure the prediction bias by observing the predictive power of the model on a new set of data. If necessary, some of the original data can be kept aside for this calibration of predictive power and the model derived only from the remaining data.

Often, a predictive model is desired for values of the independent variables which cover only a portion of the entire observation space. In that case, it is good practice to test the stability of the regression model by fitting it to that portion of the observations which fall in the space of future interest and comparing the regression results with those for the model based on all observations. Similarly, if the data are time series, it is often desirable to study the stability of the regression model over time by fitting the model also to the most recent data alone, and comparing results.

In this connection, it is worthwhile repeating an earlier caution. When the independent variables are highly intercorrelated, use of the model for prediction for values of the independent variables that do not follow the past pattern of multicollinearity becomes highly suspect.

PROBLEMS

12.1. A speaker stated: ''In a well-designed experiment involving quantitative independent variables, a procedure for screening the independent variables after the observations are obtained is not necessary.'' Discuss.

12.2. An educational researcher wishes to predict the grade point average in graduate work for applicants to the Graduate School. List a dozen variables that might be useful independent variables here.

12.3. Agency revenues. An economic consultant was retained by a large employment agency in a metropolitan area to develop a regression model for predicting monthly agency revenues (Y). She decided that three economic indicators for the area were potentially useful as independent variables, namely, average weekly overtime hours of production workers in manufacturing (X_1), number of job vacancies in manufacturing (X_2), and index of help wanted advertising in newspapers (X_3). Monthly observations on agency revenues and the three independent variables (all seasonally adjusted) were obtained for the past 25 months. The ANOVA table for the model $Y_i = \beta_0 + \beta_1 X_{i1} + \beta_2 X_{i2} + \beta_3 X_{i3} + \varepsilon_i$ is as follows:

Source of Variation	SS	df	MS
Regression	5,409.89	3	1,803.30
Error	16.35	21	.78
Total	5,426.24	24	

a. Test to determine whether a regression relation exists. Use a level of significance of .05. State the alternatives, decision rule, and conclusion.

b. If a regression relation had not existed, what would this imply about screening the independent variables?

12.4. Refer to **Agency revenues** Problem 12.3. The consultant decided to screen the independent variables to determine the best set for predicting agency revenues. The regression sums of squares for all possible regression models were found to be as follows:

Independent Variables in Model	SSR	Independent Variables in Model	SSR
X_1	2,970.64	X_1, X_2	5,123.80
X_2	3,654.85	X_1, X_3	5,409.59
X_3	3,584.54	X_2, X_3	3,741.30
		X_1, X_2, X_3	5,409.89

a. Indicate which subset of independent variables is best for predicting Y according to each of the following criteria: (1) R_p^2, (2) MSE_p, (3) C_p. Support your recommendations with appropriate graphs.

b. Did the three criteria in part (a) identify the same best subset? Does this always happen?

c. Would stepwise regression have any advantages here as a screening procedure over all possible regressions?

d. An observer states: "There are only three variables, so why screen? You might as well use all three." Discuss.

12.5. Refer to **Patient satisfaction** Problem 7.17. The ANOVA table for the model $Y_i = \beta_0 + \beta_1 X_{i1} + \beta_2 X_{i2} + \beta_3 X_{i3} + \varepsilon_i$ is as follows:

Source of Variation	SS	df	MS
Regression	4,133.62	3	1,377.87
Error	2,011.59	19	105.87
Total	6,145.21	22	

The hospital administrator decided to screen the independent variables to determine the best subset for predicting patient satisfaction. The regression sums of squares for all possible regression models are as follows:

Independent Variables in Model	SSR	Independent Variables in Model	SSR
X_1	3,678.44	X_1, X_2	4,081.21
X_2	2,120.61	X_1, X_3	4,063.98
X_3	2,229.32	X_2, X_3	2,426.92
		X_1, X_2, X_3	4,133.62

a. Indicate which subset of independent variables you would recommend as best for predicting Y according to each of the following criteria: (1) R_p^2, (2) MSE_p, (3) C_p. Support your recommendations with appropriate graphs.

b. Did the three criteria in part (a) identify the same best subset? Does this always happen?

c. Would stepwise regression have any advantages here as a screening procedure over all possible regressions?

12.6. Roofing shingles. Data on sales last year (Y, in thousand squares) in 20 sales districts are given below for a maker of asphalt roofing shingles. Shown also are promotional expenditures (X_1, in thousand dollars), number of active accounts (X_2), number of competing brands (X_3), and district potential (X_4, coded) for each of the districts.

District i	X_{i1}	X_{i2}	X_{i3}	X_{i4}	Y_i
1	5.5	31	10	8	79.3
2	2.5	55	8	6	200.1
3	8.0	67	12	9	163.2
4	3.0	50	7	16	200.1
5	3.0	38	8	15	146.0
6	2.9	71	12	17	177.7
7	8.0	30	12	8	30.9
8	9.0	56	5	10	291.9
9	4.0	42	8	4	160.0
10	6.5	73	5	16	339.4
11	5.5	60	11	7	159.6
12	5.0	44	12	12	86.3
13	6.0	50	6	6	237.5
14	5.0	39	10	4	107.2
15	3.5	55	10	4	155.0
16	8.0	70	6	14	291.4
17	6.0	40	11	6	100.2
18	4.0	50	11	8	135.8
19	7.5	62	9	13	223.3
20	7.0	59	9	11	195.0

It is believed that a regression model containing only first-order terms and no interaction terms will be appropriate.

a. Find the three best subsets according to the C_p criterion. Is there relatively little bias in the subset model with the smallest C_p value?

b. For the subset model with the smallest C_p value, obtain the residuals and plot them against $\hat{Y}$ and each of the independent variables in the subset model on separate graphs. Also prepare a normal probability plot. On the basis of your plots, do you suggest any modifications in the model to be employed?

12.7. **Job proficiency.** A personnel officer in a governmental agency administered four newly developed aptitude tests to each of 25 applicants for entry-level clerical positions in the agency. For purposes of the study, all 25 applicants were accepted for positions irrespective of their test scores. After a probationary period, each applicant was rated for proficiency on the job. The scores on the four tests (X_1, X_2, X_3, X_4) and the job proficiency score (Y) for the 25 employees were as follows:

Subject	Test Score				Job Proficiency Score
	X_1	X_2	X_3	X_4	Y
1	86	110	100	87	88
2	62	97	99	100	80
3	110	107	103	103	96
4	101	117	93	95	76
5	100	101	95	88	80
6	78	85	95	84	73
7	120	77	80	74	58
8	105	122	116	102	116
9	112	119	106	105	104
10	120	89	105	97	99
11	87	81	90	88	64
12	133	120	113	108	126
13	140	121	96	89	94
14	84	113	98	78	71
15	106	102	109	109	111
16	109	129	102	108	109
17	104	83	100	102	100
18	150	118	107	110	127
19	98	125	108	95	99
20	120	94	95	90	82
21	74	121	91	85	67
22	96	114	114	103	109
23	104	73	93	80	78
24	94	121	115	104	115
25	91	129	97	83	83

It is expected that a regression model containing only first-order terms and no interaction terms will be appropriate.

a. Find the three best subsets according to the C_p criterion. Is there relatively little bias in the subset model with the smallest C_p value?

b. For the subset model with the smallest C_p value, obtain the residuals and plot them against $\hat{Y}$ and each of the independent variables in the subset model on separate graphs. Also prepare a normal probability plot. On the basis of your plots, do you suggest any modifications in the model to be employed?

12.8. Two researchers investigated factors affecting summer attendance at privately operated beaches on Lake Ontario, and collected information on attendance and 11 explanatory variables for 42 beaches. Two summers were studied, of relatively hot and relatively cool weather, respectively. A "best" subsets algorithm now is to be used to screen the potential independent variables.

a. Should the screening be done for both summers combined or should it be done separately for each summer? Explain the problems involved and how you might handle them.

b. Will the "best" subsets screening procedure select those independent variables that are most important in a causal sense for determining beach attendance?

12.9. In stepwise regression, what advantage is there in using a relatively large F limit for adding variables? What advantage is there in using a smaller F limit for adding variables?

12.10. In stepwise regression, why should the F limit for deleting variables never exceed the F limit for adding variables?

12.11. Draw a flowchart of each of the following selection methods: (1) stepwise regression, (2) forward selection, (3) backward elimination.

12.12. Refer to **Agency revenues** Problems 12.3 and 12.4. The consultant was interested to learn how the stepwise selection procedure and some of its variations would perform in this application.

a. Determine the subset of variables that is selected as best by the stepwise regression procedure using F limits of 4.2 and 4.1 to add or delete a variable, respectively. Show your steps.

b. To what level of significance in any individual test is the F limit of 4.2 for adding a variable approximately equivalent here?

c. Determine the subset of variables that is selected as best by the forward selection procedure using an F limit of 4.2 to add a variable. Show your steps.

d. Determine the subset of variables that is selected as best by the backward elimination procedure using an F limit of 4.1 to delete a variable. Show your steps.

e. Compare the results of the three selection procedures. How consistent are these results? How do the results compare with those for all possible regressions in Problem 12.4?

12.13. Refer to **Agency revenues** Problem 12.12a. Suppose the consultant "forced" X_2 into the best subset for administrative reasons by arbitrarily entering it first and not removing it even if its F^* value becomes too low. Which subset of variables (including X_2) is now selected as best by the stepwise regression procedure if F limits of 4.2 and 4.1 are used to add or delete a variable, respectively? Did the forced inclusion of X_2 affect the selection of the other variables included in the best subset? Will this always happen?

12.14. Refer to **Patient satisfaction** Problems 7.17 and 12.5. The hospital administrator was interested to learn how the stepwise selection procedure and some of its variations would perform here.

a. Determine the subset of variables that is selected as best by the stepwise regression procedure using F limits of 3.0 and 2.9 to add or delete a variable, respectively. Show your steps.

b. To what level of significance in any individual test is the F limit of 3.0 for adding a variable approximately equivalent here?

c. Determine the subset of variables that is selected as best by the forward selection procedure using an F limit of 3.0 to add a variable. Show your steps.

d. Determine the subset of variables that is selected as best by the backward elimination procedure using an F limit of 2.9 to delete a variable. Show your steps.

e. Compare the results of the three selection procedures. How consistent are these results? How do the results compare with those for all possible regressions in Problem 12.5?

12.15. Refer to **Roofing shingles** Problem 12.6.

a. Using stepwise regression, find the best subset of independent variables to predict sales. Use F limits for adding or deleting a variable of 4.0 and 3.9, respectively.

b. How does the best subset according to stepwise regression compare with the best subset according to the C_p criterion obtained in Problem 12.6a?

12.16. Refer to **Job proficiency** Problem 12.7.

a. Using stepwise regression, find the best subset of independent variables to predict job proficiency. Use F limits of 3.5 and 3.4 for adding or deleting a variable, respectively.

b. How does the best subset according to stepwise regression compare with the best subset according to the C_p criterion obtained in Problem 12.7a?

12.17. An engineer has stated: "Screening of variables should always be done using the objective stepwise regression procedure." Discuss.

EXERCISES

12.18. The true quadratic regression function is $E(Y) = 15 + 20X + 3X^2$. The fitted linear regression function is $\hat{Y} = 13 + 40X$, for which $E(b_0) = 10$ and $E(b_1) = 45$. What are the bias and sampling error components of the mean squared error for $X_i = 10$? For $X_i = 20$?

12.19. Prove (12.12). [*Hint:* Use Exercise 6.33 and (11.50).]

12.20. Refer to (12.16). Show that the same variable X_k that maximizes the test statistic F_k^* also maximizes the coefficient of partial determination $r_{Yk.7}^2$.

PROJECTS

12.21. Refer to the **SENIC** data set. Length of stay (Y) is to be predicted, and the pool of potential independent variables includes all other variables in the data set except medical school affiliation and region. It is believed that a model with $\log_{10} Y$ as the dependent variable and the independent variables in first-order terms with no interaction terms will be appropriate.

a. Using the C_p criterion, obtain the three best subsets. Which of these subset models appears to have the smallest bias? Which of the three models would you recommend as best?

b. Divide the data set into two halves by considering the first 56 observations as one half and the remaining 57 observations as the other half. Fit the regression model recommended as best in part (a) using the first 56 observations. Then obtain the deviations of the remaining 57 observations from their respective "predicted" values, i.e., obtain $Y_i - \hat{Y}_i$. How well does the model perform on the hold-out (validation) sample? Calculate $\Sigma(Y_i - \hat{Y}_i)^2/n$ for the last 57 observations and compare it with MSE for the first 56 observations. Is there any evidence of a large prediction bias?

c. For the recommended subset model in part (a), obtain the residuals and plot them against the fitted values and each of the independent variables in the subset model. Also prepare a normal probability plot. Do these plots suggest any modifications in the model?

d. Would stepwise regression with F limits of 3.5 and 3.4 for adding or deleting a variable, respectively, lead to the same best model as the all-possible-regressions approach?

12.22. Refer to the **SMSA** data set. A public safety official wishes to predict the rate of serious crimes in an SMSA (Y, total number of serious crimes per 100,000 population). The pool of potential independent variables includes all other variables in the data set except region.

a. Using the C_p criterion, obtain the three best subsets. Which of these subset models appears to have the smallest bias? Which of the three models would you recommend as best?

b. Divide the data set into two halves by considering the odd-numbered observations as one half and the even-numbered observations as the other half. Fit the regression model recommended as best in part (a) using the odd-numbered observations. Then obtain the deviations of the even-numbered observations from their respective "predicted" values, i.e., obtain $Y_i - \hat{Y}_i$. How well does the model perform on the hold-out (validation) sample? Calculate $\Sigma(Y_i - \hat{Y}_i)^2/n$ for the even-numbered observations and compare it with MSE for the odd-numbered observations. Is there any evidence of a large prediction bias?

c. For the recommended subset model in part (a), obtain the residuals and plot them against the fitted values and each of the independent variables in the subset model. Also prepare a normal probability plot. Do these plots suggest any modifications in the model?

d. Would stepwise regression with F-to-enter and F-to-remove values of 4.0 and 3.9, respectively, lead to the same best model as the all-possible-regressions approach?

CITED REFERENCES

12.1 Daniel, Cuthbert, and Fred S. Wood. *Fitting Equations to Data*. 2d ed. New York: Wiley-Interscience, 1980.

12.2 Dixon, W. J., and M. B. Brown, eds. *BMDP-81, Biomedical Computer Programs, P-Series*. Berkeley, Calif.: University of California Press, 1981.

13

Autocorrelation in time series data

The basic regression models considered so far have assumed that the random error terms ε_i are either uncorrelated random variables or independent normal random variables. In business and economics, many regression applications involve time series data. For such data, the assumption of uncorrelated or independent error terms is often not appropriate; rather, the error terms are frequently correlated positively over time. Error terms correlated over time are said to be *autocorrelated* or *serially correlated*.

A major cause of positively autocorrelated error terms in business and economic regression applications involving time series data is the omission of one or several key variables from the model. When time-ordered effects of such "missing" key variables are positively correlated, the error terms in the regression model will tend to be positively autocorrelated since the error terms include effects of missing variables. Suppose, for example, that annual sales of a product are regressed against average yearly price over a period of 30 years. If population size has an important effect on sales, its omission from the model may lead to the error terms being positively autocorrelated because the effect of population size on sales likely is positively correlated over time.

Another common cause of positively autocorrelated error terms in economic data is systematic coverage errors in the dependent variable time series, which errors often tend to be positively correlated over time.

13.1 PROBLEMS OF AUTOCORRELATION

If the error terms in the regression model are positively autocorrelated, the use of ordinary least squares procedures has a number of important consequences. We summarize these first, and then discuss them in more detail:

1. The ordinary least squares regression coefficients are still unbiased, but they no longer have the minimum variance property and may be quite inefficient.
2. *MSE* may seriously underestimate the variance of the error terms.
3. $s(b_k)$ calculated according to ordinary least squares procedures may seriously underestimate the true standard deviation of the estimated regression coefficient with that procedure.
4. The confidence intervals and tests using the t and F distributions, discussed earlier, are no longer strictly applicable.

To illustrate these problems intuitively, we shall consider the simple linear regression model with time series data:

$$Y_t = \beta_0 + \beta_1 X_t + \varepsilon_t$$

Here, Y_t and X_t are observations for period t. Let us assume that the error terms ε_t are positively autocorrelated as follows:

$$\varepsilon_t = \varepsilon_{t-1} + u_t$$

The u_t, called *disturbances*, are independent normal random variables. Thus, any error term ε_t is the sum of the previous error term ε_{t-1} and a new disturbance term u_t. We shall assume here that the u_t have mean 0 and variance 1.

In Table 13.1, column 1, we show 10 random observations on the normal variable u_t with mean 0 and variance 1, obtained from a standard normal random numbers generator. Suppose now that $\varepsilon_0 = 3.0$; we obtain then:

$$\varepsilon_1 = \varepsilon_0 + u_1 = 3.0 + .5 = 3.5$$
$$\varepsilon_2 = \varepsilon_1 + u_2 = 3.5 - .7 = 2.8$$

etc.

TABLE 13.1 Example of positively autocorrelated error terms

t	(1) u_t	(2) $\varepsilon_{t-1} + u_t = \varepsilon_t$
0	—	3.0
1	+.5	3.0 + .5 = 3.5
2	−.7	3.5 − .7 = 2.8
3	+.3	2.8 + .3 = 3.1
4	0	3.1 + 0 = 3.1
5	−2.3	3.1 − 2.3 = .8
6	−1.9	.8 − 1.9 = −1.1
7	+.2	−1.1 + .2 = − .9
8	−.3	− .9 − .3 = −1.2
9	+.2	−1.2 + .2 = −1.0
10	−.1	−1.0 − .1 = −1.1

The error terms ε_t are shown in Table 13.1, column 2, and they are plotted in Figure 13.1. Note the systematic pattern in these error terms. Their positive relation over time is shown by the fact that adjacent error terms tend to be of the same magnitude.

Suppose that X_t in the regression model represents time, such that $X_1 = 1$, $X_2 = 2$, etc. Further, suppose we know that $\beta_0 = 2$ and $\beta_1 = .5$. Figure 13.2a contains the true regression line and the observed Y values based on the error terms in Figure 13.1. Figure 13.2b contains the estimated regression line, fitted by ordinary least squares methods, and the observed Y values. Notice that the fitted regression line differs sharply from the true regression line because the initial ε_0 value was large and the succeeding positively autocorrelated error terms tended to be large for some time. This persistency pattern in the positively autocorrelated error terms leads to a fitted regression line far from the true one. Had the initial ε_0 value been small, say, $\varepsilon_0 = -.2$, and the disturbances different, a sharply different fitted regression line might have been obtained because of the persistency pattern, as shown in Figure 13.2c. This variation from sample to sample in the fitted regression lines because of the positively autocorrelated error terms may be so substantial as to lead to large variances of the estimated regression coefficients when ordinary least squares methods are used.

Another key problem with applying ordinary least squares methods when the error terms are positively autocorrelated, as mentioned before, is that MSE may

FIGURE 13.1 Example of positively autocorrelated error terms

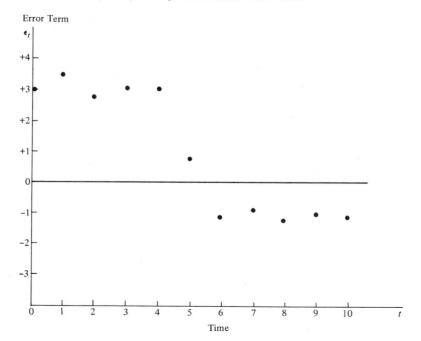

FIGURE 13.2 Regression with positively autocorrelated error terms

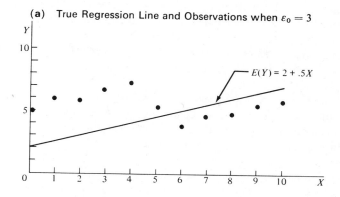

(a) True Regression Line and Observations when $\varepsilon_0 = 3$

$E(Y) = 2 + .5X$

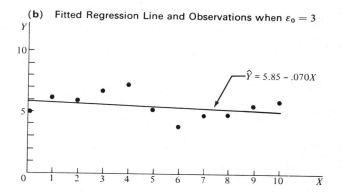

(b) Fitted Regression Line and Observations when $\varepsilon_0 = 3$

$\hat{Y} = 5.85 - .070X$

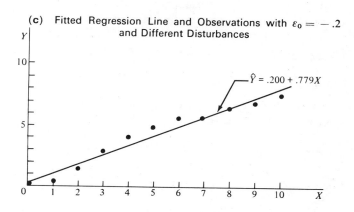

(c) Fitted Regression Line and Observations with $\varepsilon_0 = -.2$ and Different Disturbances

$\hat{Y} = .200 + .779X$

seriously underestimate the variance of the ε_t. Figure 13.2 makes this clear. Note that the variability of the Y values around the fitted regression line in Figure 13.2b is substantially smaller than the variability of the Y's around the true regression line in Figure 13.2a. This is one of the factors leading to an indication of greater precision of the regression coefficients than is actually the case when ordinary least squares methods are used in the presence of positively autocorrelated errors.

In view of the seriousness of the problems created by autocorrelated errors, it is important that their presence be detected. A plot of residuals against time is an effective, though subjective, means of detecting autocorrelated errors. Formal statistical tests have also been developed. A widely used test is based on the first-order autoregressive error model, which we take up next. This model is a simple one, yet experience suggests that it is frequently applicable in business and economics when the error terms are serially correlated.

13.2 FIRST-ORDER AUTOREGRESSIVE ERROR MODEL

Simple linear regression

The simple linear regression model for one independent variable with the random error terms following a first-order autoregressive process is:

$$(13.1) \qquad Y_t = \beta_0 + \beta_1 X_t + \varepsilon_t$$
$$\varepsilon_t = \rho\varepsilon_{t-1} + u_t$$

where:

ρ is a parameter such that $|\rho| < 1$
u_t are independent $N(0, \sigma^2)$

Note that (13.1) is identical to the simple linear regression model (3.1) except for the structure of the error terms. Each error term in model (13.1) consists of a fraction of the previous error term (when $\rho > 0$) plus a new disturbance term u_t. The parameter ρ is called the *autocorrelation parameter*.

Multiple regression

The multiple regression model with the random error terms following a first-order autoregressive process is:

$$(13.2) \qquad Y_t = \beta_0 + \beta_1 X_{t1} + \beta_2 X_{t2} + \cdots + \beta_{p-1} X_{t,p-1} + \varepsilon_t$$
$$\varepsilon_t = \rho\varepsilon_{t-1} + u_t$$

where:

$|\rho| < 1$
u_t are independent $N(0, \sigma^2)$

Thus, we see again that the multiple regression model (13.2) is identical to the earlier multiple regression model (7.7) except for the structure of the error terms.

Properties of error terms

It is instructive to expand the definition of the first-order autoregressive error term ε_t:

$$\varepsilon_t = \rho\varepsilon_{t-1} + u_t$$

Since $\varepsilon_{t-1} = \rho\varepsilon_{t-2} + u_{t-1}$, we obtain:

$$\varepsilon_t = \rho(\rho\varepsilon_{t-2} + u_{t-1}) + u_t = \rho^2\varepsilon_{t-2} + \rho u_{t-1} + u_t$$

Replacing now ε_{t-2} by $\rho\varepsilon_{t-3} + u_{t-2}$, we obtain:

$$\varepsilon_t = \rho^3\varepsilon_{t-3} + \rho^2 u_{t-2} + \rho u_{t-1} + u_t$$

Continuing in this fashion, we find:

$$(13.3) \qquad \varepsilon_t = \sum_{s=0}^{\infty} \rho^s u_{t-s}$$

Thus, the error term ε_t in period t is a linear combination of the current and preceding disturbance terms. When $0 < \rho < 1$, (13.3) indicates that the further the period is in the past, the smaller is the weight of that disturbance term in determining ε_t.

Mean. Since $E(u_t) = 0$ according to models (13.1) and (13.2) for all t, it follows from (13.3) that:

$$(13.4) \qquad E(\varepsilon_t) = 0$$

Thus, ε_t has expectation zero, just as for regression models with uncorrelated error terms.

Variance. Since according to models (13.1) and (13.2) the u_t are independent with variance σ^2, it follows from (13.3) that:

$$\sigma^2(\varepsilon_t) = \sum_{s=0}^{\infty} \rho^{2s}\sigma^2(u_{t-s}) = \sigma^2 \sum_{s=0}^{\infty} \rho^{2s}$$

Now for $|\rho| < 1$, it is known that:

$$\sum_{s=0}^{\infty} \rho^{2s} = \frac{1}{1 - \rho^2}$$

Hence, we have:

$$(13.5) \qquad \sigma^2(\varepsilon_t) = \frac{\sigma^2}{1 - \rho^2}$$

We thus see that the error terms have constant variance, just as for regression models with uncorrelated error terms.

Covariance. To find the covariance of ε_t and ε_{t-1}, we need to recognize that:

$$\sigma^2(\varepsilon_t) = E(\varepsilon_t^2)$$
$$\sigma(\varepsilon_t, \varepsilon_{t-1}) = E(\varepsilon_t\varepsilon_{t-1})$$

These results follow from theorems (1.14a) and (1.19a), respectively, since $E(\varepsilon_t) = 0$ by (13.4).

By (13.3), we have:

$$E(\varepsilon_t\varepsilon_{t-1}) = E[(u_t + \rho u_{t-1} + \rho^2 u_{t-2} + \cdots)(u_{t-1} + \rho u_{t-2} + \rho^2 u_{t-3} + \cdots)]$$

which can be rewritten:

$$E(\varepsilon_t\varepsilon_{t-1}) = E\{[u_t + \rho(u_{t-1} + \rho u_{t-2} + \cdots)][u_{t-1} + \rho u_{t-2} + \rho^2 u_{t-3} + \cdots]\}$$
$$= E[u_t(u_{t-1} + \rho u_{t-2} + \rho^2 u_{t-3} + \cdots)]$$
$$+ E[\rho(u_{t-1} + \rho u_{t-2} + \rho^2 u_{t-3} + \cdots)^2]$$

Since $E(u_t u_{t-s}) = 0$ for all $s \neq 0$ by the assumed independence of the u_t, the first term drops out and we obtain:

$$E(\varepsilon_t\varepsilon_{t-1}) = \rho E(\varepsilon_{t-1}^2) = \rho\sigma^2(\varepsilon_{t-1})$$

Hence, by (13.5), which holds for all t, we have:

$$(13.6) \qquad \sigma(\varepsilon_t, \varepsilon_{t-1}) = \rho\left(\frac{\sigma^2}{1 - \rho^2}\right)$$

In general, it can be shown that:

$$(13.7) \qquad \sigma(\varepsilon_t, \varepsilon_{t-s}) = \rho^s\left(\frac{\sigma^2}{1 - \rho^2}\right) \qquad s \neq 0$$

Thus, the error terms for models (13.1) and (13.2) are autocorrelated unless the autocorrelation parameter ρ equals zero.

Note

It follows directly from (13.5) and (13.6) that the autocorrelation parameter ρ is the coefficient of correlation between ε_t and ε_{t-1}, as defined in (15.3).

13.3 DURBIN-WATSON TEST FOR AUTOCORRELATION

The Durbin-Watson test assumes the first-order autoregressive error models (13.1) or (13.2), with the values of the independent variable(s) fixed. The test consists of determining whether or not the autocorrelation parameter ρ in (13.1) or (13.2) is zero. Note that if $\rho = 0$, $\varepsilon_t = u_t$. Hence, the error terms ε_t are then independent since the u_t are independent.

Because correlated error terms in business and economic applications tend to show positive serial correlation, the usual test alternatives considered are:

(13.8)
$$H_0: \rho = 0$$
$$H_a: \rho > 0$$

The test statistic D is obtained by first fitting the ordinary least squares regression function and calculating the residuals:

(13.9)
$$e_t = Y_t - \hat{Y}_t$$

and then calculating the statistic:

(13.10)
$$D = \frac{\sum_{t=2}^{n} (e_t - e_{t-1})^2}{\sum_{t=1}^{n} e_t^2}$$

where n is the number of observations.

An exact test procedure is not available, but Durbin and Watson have obtained lower and upper bounds d_L and d_U such that a value of D outside these bounds leads to a definite decision. The decision rule for testing between the alternatives in (13.8) is:

(13.11)
If $D > d_U$, conclude H_0
If $D < d_L$, conclude H_a
If $d_L \leq D \leq d_U$, the test is inconclusive

Small values of D lead to the conclusion that $\rho > 0$ because the adjacent error terms ε_t and ε_{t-1} tend to be of the same magnitude when they are positively autocorrelated. Hence, the differences in the residuals, $e_t - e_{t-1}$, would tend to be small when $\rho > 0$, leading to a small numerator in D and hence to a small test statistic D.

Table A–6 contains the bounds d_L and d_U for various sample sizes (n), for two levels of significance (.05 and .01), and for various numbers of X variables ($p - 1$) in the regression model.

Example

The Blaisdell Company wished to predict its sales by using industry sales as a predictor variable. (Accurate predictions of industry sales are available from the industry's trade association.) In Table 13.2, columns 1 and 2 contain seasonally adjusted quarterly data on company sales and industry sales, respectively, for the period 1977–81. A scatter plot (not shown) suggested that a linear regression model is appropriate. The market research analyst was, however, concerned whether or not the error terms were positively autocorrelated. He therefore used the Durbin-Watson test with the alternatives:

$$H_0: \rho = 0$$
$$H_a: \rho > 0$$

TABLE 13.2 Durbin-Watson test calculations for Blaisdell Company example (company and industry sales data are seasonally adjusted)

Year and Quarter	t	(1) Company Sales ($ millions) Y_t	(2) Industry Sales ($ millions) X_t	(3) Residual e_t	(4) $e_t - e_{t-1}$	(5) $(e_t - e_{t-1})^2$	(6) e_t^2
1977: 1	1	20.96	127.3	−.026052	—	—	.0006787
2	2	21.40	130.0	−.062015	−.035963	.0012933	.0038459
3	3	21.96	132.7	.022021	.084036	.0070620	.0004849
4	4	21.52	129.4	.163754	.141733	.0200882	.0268154
1978: 1	5	22.39	135.0	.046570	−.117184	.0137321	.0021688
2	6	22.76	137.1	.046377	−.000193	.0000000	.0021508
3	7	23.48	141.2	.043617	−.002760	.0000076	.0019024
4	8	23.66	142.8	−.058435	−.102052	.0104146	.0034146
1979: 1	9	24.10	145.5	−.094399	−.035964	.0012934	.0089112
2	10	24.01	145.3	−.149142	−.054743	.0029968	.0222433
3	11	24.54	148.3	−.147991	.001151	.0000013	.0219013
4	12	24.30	146.4	−.053054	.094937	.0090130	.0028147
1980: 1	13	25.00	150.2	−.022928	.030126	.0009076	.0005257
2	14	25.64	153.1	.105852	.128780	.0165843	.0112046
3	15	26.36	157.3	.085464	−.020388	.0004157	.0073041
4	16	26.98	160.7	.106102	.020638	.0004259	.0112576
1981: 1	17	27.52	164.2	.029112	−.076990	.0059275	.0008475
2	18	27.78	165.6	.042316	.013204	.0001743	.0017906
3	19	28.24	168.7	−.044160	−.086476	.0074781	.0019501
4	20	28.78	171.7	−.033009	.011151	.0001243	.0010896
Total						.0979400	.1333018

He fitted an ordinary least squares regression line to the data in Table 13.2. The results are shown in Table 13.3a. He then calculated the residuals e_t, which are shown in column 3 of Table 13.2 and which are plotted against time in Figure 13.3. Note how the residuals consistently are above or below the fitted values for extended periods. Positive autocorrelation in the error terms is suggested by such a pattern when an appropriate regression function has been employed.

FIGURE 13.3 Residuals plotted against time—Blaisdell Company example

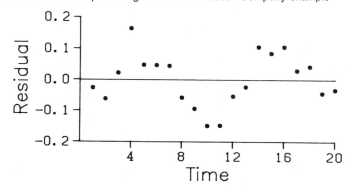

TABLE 13.3 Regression results for Blaisdell Company example

(a) Original Variables Y_t and X_t

Regression Coefficient	Estimated Regression Coefficient	Estimated Standard Deviation
β_0	$b_0 = -1.45475$	$s(b_0) = .21415$
β_1	$b_1 = .17628$	$s(b_1) = .00144$

$$\hat{Y} = -1.45475 + .17628X$$

(b) Transformed Variables $Y'_t = Y_t - rY_{t-1}$
and $X'_t = X_t - rX_{t-1}$

Regression Coefficient	Estimated Regression Coefficient	Estimated Standard Deviation
β'_0	$b'_0 = -.39396$	$s(b'_0) = .16704$
$\beta'_1 = \beta_1$	$b'_1 = .17376$	$s(b'_1) = .00296$

$$b_0 = \frac{b'_0}{1-r} = \frac{-.39396}{1-.631166} = -1.06811$$

$$s(b_0) = \frac{s(b'_0)}{1-r} = \frac{.16704}{1-.631166} = .45288$$

(c) First Differences $Y'_t = Y_t - Y_{t-1}$ and $X'_t = X_t - X_{t-1}$

Regression Coefficient	Estimated Regression Coefficient	Estimated Standard Deviation
$\beta'_1 = \beta_1$	$b_1 = .16849$	$s(b_1) = .005096$

Columns 4, 5, and 6 of Table 13.2 contain the necessary calculations for the test statistic D. The analyst then obtained:

$$D = \frac{\sum_{t=2}^{20} (e_t - e_{t-1})^2}{\sum_{t=1}^{20} e_t^2} = \frac{.09794}{.13330} = .735$$

Using a level of significance of .01, he found in Table A–6 for $n = 20$ and $p - 1 = 1$:

$$d_L = .95$$
$$d_U = 1.15$$

Since $D = .735$ falls below $d_L = .95$, decision rule (13.11) indicates that the appropriate conclusion is H_a, namely, that the error terms are positively auto-correlated.

Comments

1. If a test for negative autocorrelation is required, the test statistic to be used is $4 - D$, where D is defined as above. The test is then conducted in the same manner

described for testing for positive autocorrelation. That is, if the quantity $4 - D$ falls below d_L, we conclude $\rho < 0$, that negative autocorrelation exists, and so on.

2. A two-sided test for H_0: $\rho = 0$ versus H_a: $\rho \neq 0$ can be made by employing both one-sided tests separately. The Type I risk with the two-sided test is 2α, where α is the Type I risk with each one-sided test.

3. When the Durbin-Watson test employing the bounds d_L and d_U gives indeterminate results, in principle more observations are required. Of course, with time series data it may be impossible to obtain more observations, or additional observations may lie in the future and be obtainable only with great delay. Durbin and Watson (Ref. 13.1) do give an approximate test which may be used when the bounds test is indeterminate, but the degrees of freedom should be larger than about 40 before this approximate test will give more than a rough indication of whether autocorrelation exists.

A reasonable procedure is to treat indeterminate results as suggesting the presence of autocorrelated errors and employ one of the remedial actions to be discussed next. If such an action does not lead to substantially different regression results, the assumption of uncorrelated error terms would appear to be satisfactory. When the remedial action does lead to substantially different regression results (such as larger estimated standard errors for the regression coefficients or the elimination of autocorrelated errors), the results obtained by means of the remedial action are probably the more useful ones.

4. The Durbin-Watson test is not robust against misspecifications of the model. For example, the Durbin-Watson test may not disclose the presence of autocorrelated errors that follow a second-order autoregressive pattern, where an error term in period t is directly related to the error term in period $t - 2$.

5. While the Durbin-Watson test is widely used, other tests for autocorrelation are available. One such alternative test, due to Theil and Nagar, is found in Reference 13.2.

13.4 REMEDIAL MEASURES FOR AUTOCORRELATION

The two principal remedial measures when autocorrelated error terms are present are to add one or more independent variables to the model or to use transformed variables.

Addition of independent variables

As noted earlier, one major cause of autocorrelated error terms is the omission from the model of one or more key independent variables that have time-ordered effects on the dependent variable. When autocorrelated error terms are found to be present, the first remedial action should always be to search for missing key independent variables. The missing variable, population size, was mentioned previously for a regression of annual sales of a product on average yearly price of the product during a 30-year period. Sometimes, use of a simple linear trend variable or use of indicator variables for seasonal effects can be helpful in eliminating or reducing autocorrelation in the error terms.

Use of transformed variables

Only when use of additional independent variables is not helpful in eliminating the problem of autocorrelated errors should a remedial action based on transformed variables be employed. We now explain two methods of transforming the variables. One is based on an iterative approach and the other uses first differences. We shall explain these methods for simple linear regression, but the same principles apply for multiple regression.

Iterative approach. The iterative approach for simple linear regression is motivated by an interesting property of model (13.1). Consider the transformed dependent variable:

$$Y'_t = Y_t - \rho Y_{t-1}$$

Substituting in this expression for Y_t and Y_{t-1} according to model (13.1), we obtain:

$$Y'_t = (\beta_0 + \beta_1 X_t + \varepsilon_t) - \rho(\beta_0 + \beta_1 X_{t-1} + \varepsilon_{t-1})$$
$$= \beta_0(1 - \rho) + \beta_1(X_t - \rho X_{t-1}) + (\varepsilon_t - \rho\varepsilon_{t-1})$$

But by (13.1), $\varepsilon_t - \rho\varepsilon_{t-1} = u_t$. Hence:

$$Y'_t = \beta_0(1 - \rho) + \beta_1(X_t - \rho X_{t-1}) + u_t$$

where the u_t are independent error terms. Thus, when we use the transformed variables:

(13.12a) $$Y'_t = Y_t - \rho Y_{t-1}$$

(13.12b) $$X'_t = X_t - \rho X_{t-1}$$

the transformed regression model becomes:

(13.13) $$Y'_t = \beta'_0 + \beta'_1 X'_t + u_t$$

where:

$$\beta'_0 = \beta_0(1 - \rho)$$
$$\beta'_1 = \beta_1$$

Note that the transformed regression model (13.13) has independent error terms. This means that ordinary least squares methods have their usual optimum properties with this model.

The parameters in the original model (13.1) are related to the parameters in the transformed model (13.13) as follows:

(13.14a) $$\beta_0 = \frac{\beta'_0}{1 - \rho}$$

(13.14b) $$\beta_1 = \beta'_1$$

The transformed model (13.13) cannot be used directly because the autocorrelation parameter ρ needed to obtain the transformed variables in (13.12) is unknown. We can, however, estimate ρ. Note that the autoregressive error process assumed in model (13.1) can be viewed as a regression through the origin:

$$\varepsilon_t = \rho\varepsilon_{t-1} + u_t$$

where ε_t is the dependent variable, ε_{t-1} the independent variable, u_t the error term, and ρ the slope of the line through the origin. Since the ε_t and ε_{t-1} are unknown we use the residuals e_t and e_{t-1}, obtained by ordinary least squares methods, as the dependent and independent variables, respectively, and estimate ρ by fitting a straight line through the origin. From our previous discussion of regression through the origin, we know by (5.17) that the estimate of the slope ρ, denoted by r, is:

$$(13.15) \qquad r = \frac{\displaystyle\sum_{t=2}^{n} e_{t-1} e_t}{\displaystyle\sum_{t=2}^{n} e_{t-1}^2}$$

We now obtain the transformed variables:

$$(13.16a) \qquad Y_t' = Y_t - rY_{t-1}$$

$$(13.16b) \qquad X_t' = X_t - rX_{t-1}$$

and use ordinary least squares with these transformed variables.

The Durbin-Watson test is then employed to test whether the error terms for the transformed model are uncorrelated. If the test indicates that they are uncorrelated, the procedure terminates. Otherwise, the parameter ρ is reestimated from the new residuals for the regression model with the original variables, using the regression coefficients derived from the fit of the regression model with the transformed variables. A new set of transformed variables is then obtained with the new r. This process may be continued for several iterations until the Durbin-Watson test suggests that the error terms in the transformed model are uncorrelated.

Example. For our Blaisdell Company example, the necessary calculations for estimating the autocorrelation parameter ρ, based on the residuals obtained with ordinary least squares applied to the original variables, appear in Table 13.4. Column 1 repeats the residuals from Table 13.2. Column 2 contains the residuals e_{t-1}, and columns 3 and 4 contain the necessary calculations. Hence, we estimate:

$$r = \frac{.0834478}{.1322122} = .631166$$

We now obtain the transformed variables Y_t' and X_t' in (13.16a) and (13.16b):

$$Y_t' = Y_t - .631166Y_{t-1}$$
$$X_t' = X_t - .631166X_{t-1}$$

These are shown in Table 13.5. Ordinary least squares fitting of linear regression is now used with these transformed variables. The results are shown in Table 13.3b.

TABLE 13.4 Calculations for estimating ρ for Blaisdell Company example

t	(1) e_t	(2) e_{t-1}	(3) $e_t e_{t-1}$	(4) e_{t-1}^2
1	−.026052	——	——	——
2	−.062015	−.026052	.0016156	.0006787
3	.022021	−.062015	−.0013656	.0038459
4	.163754	.022021	.0036060	.0004849
5	.046570	.163754	.0076260	.0268154
6	.046377	.046570	.0021598	.0021688
7	.043617	.046377	.0020228	.0021508
8	−.058435	.043617	−.0025488	.0019024
9	−.094399	−.058435	.0055162	.0034146
10	−.149142	−.094399	.0140789	.0089112
11	−.147991	−.149142	.0220718	.0222433
12	−.053054	−.147991	.0078515	.0219013
13	−.022928	−.053054	.0012164	.0028147
14	.105852	−.022928	−.0024270	.0005257
15	.085464	.105852	.0090465	.0112046
16	.106102	.085464	.0090679	.0073041
17	.029112	.106102	.0030889	.0112576
18	.042316	.029112	.0012319	.0008475
19	−.044160	.042316	−.0018687	.0017906
20	−.033009	−.044160	.0014577	.0019501
Total			.0834478	.1322122

TABLE 13.5 Transformed variables for first iteration—Blaisdell Company example

t	(1) Y_t	(2) X_t	(3) $Y_t' = Y_t - .631166Y_{t-1}$	(4) $X_t' = X_t - .631166X_{t-1}$
1	20.96	127.3	——	——
2	21.40	130.0	8.1708	49.653
3	21.96	132.7	8.4530	50.648
4	21.52	129.4	7.6596	45.644
5	22.39	135.0	8.8073	53.327
6	22.76	137.1	8.6282	51.893
7	23.48	141.2	9.1147	54.667
8	23.66	142.8	8.8402	53.679
9	24.10	145.5	9.1666	55.369
10	24.01	145.3	8.7989	53.465
11	24.54	148.3	9.3857	56.592
12	24.30	146.4	8.8112	52.798
13	25.00	150.2	9.6627	57.797
14	25.64	153.1	9.8608	58.299
15	26.36	157.3	10.1769	60.668
16	26.98	160.7	10.3425	61.418
17	27.52	164.2	10.4911	62.772
18	27.78	165.6	10.4103	61.963
19	28.24	168.7	10.7062	64.179
20	28.78	171.7	10.9559	65.222

Based on the fitted regression for the transformed variables in Table 13.3b, residuals were obtained and the Durbin-Watson statistic calculated. The result was (calculations not shown) $D = 1.65$. From Table A–6, we find for $\alpha = .01$, $p - 1 = 1$, and $n = 19$:

$$d_L = .93 \qquad d_U = 1.13$$

Since $D = 1.65 > d_U = 1.13$, we conclude that the autocorrelation coefficient for the error terms in the model with the transformed variables is zero.

Having successfully handled the problem of autocorrelated error terms, we now transform the estimated regression coefficients and standard deviations back to the model with the original variables (Table 13.3b):

$$b_0 = -1.06811 \qquad s(b_0) = .45288$$
$$b_1 = \quad .17376 \qquad s(b_1) = .00296$$

Note that the estimated regression coefficients $b_0 = -1.06811$ and $b_1 = .17376$ obtained with the iterative method are close to those obtained with ordinary least squares (Table 13.3a), but that the estimated standard errors $s(b_0) = .45288$ and $s(b_1) = .00296$ with the iterative method are larger than their ordinary-least-squares counterparts. The larger standard errors are to be expected, since we noted earlier that positive autocorrelation may lead to estimated standard deviations $s(b_k)$ calculated according to ordinary least squares that seriously underestimate the true standard deviations $\sigma(b_k)$.

Comments

1. The iterative approach does not always work properly. A major reason is that when the error terms are positively autocorrelated, the estimate r in (13.15) tends to underestimate the autocorrelation parameter ρ. When this bias is serious, it can significantly reduce the effectiveness of the iterative approach.

2. There exists an approximate relation between the Durbin-Watson test statistic D in (13.10) and the estimated autocorrelation parameter r in (13.15):

(13.17) $$D \simeq 2(1 - r)$$

This relation indicates that the Durbin-Watson statistic ranges approximately between 0 and 4 since r takes on values between -1 and 1, and that D is approximately 2 when $r = 0$. Note that for the Blaisdell Company example, $D = .735$, $r = .631$, and $2(1 - r) = .738$.

First differences approach. Some economists and statisticians have suggested that instead of iterative estimation of ρ, which is not always successful, the autocorrelation parameter be assumed to equal 1. If $\rho = 1$, $\beta_0' = \beta_0(1 - \rho) = 0$, and the transformed model (13.13) becomes:

(13.18) $$Y_t' = \beta_1' X_t' + u_t$$

where:

(13.18a) $$Y_t' = Y_t - Y_{t-1}$$
$$X_t' = X_t - X_{t-1}$$

Thus, the regression coefficient $\beta_1' = \beta_1$ can be directly estimated by ordinary least squares methods for regression through the origin. Note that the transformed variables (13.18a) are ordinary first differences. It has been found that this first differences approach is effective in a variety of applications in reducing the autocorrelations of the error terms, and of course it is much simpler than the iterative approach.

Example. Table 13.6 contains the transformed variables Y_t' and X_t', based on the first differences transformations in (13.18a) for our Blaisdell Company example. Application of ordinary least squares for estimating a linear regression through the origin led to the results shown in Table 13.3c. Note that the estimated regression coefficient $b_1 = .16849$ is similar to that obtained with ordinary least squares applied to the original variables ($b_1 = .17628$), but it has a larger standard error, again as expected.

TABLE 13.6 First differences for Blaisdell Company data

t	(1) Y_t	(2) X_t	(3) $Y_t' = Y_t - Y_{t-1}$	(4) $X_t' = X_t - X_{t-1}$
1	20.96	127.3	—	—
2	21.40	130.0	.44	2.7
3	21.96	132.7	.56	2.7
4	21.52	129.4	−.44	−3.3
5	22.39	135.0	.87	5.6
6	22.76	137.1	.37	2.1
7	23.48	141.2	.72	4.1
8	23.66	142.8	.18	1.6
9	24.10	145.5	.44	2.7
10	24.01	145.3	−.09	−.2
11	24.54	148.3	.53	3.0
12	24.30	146.4	−.24	−1.9
13	25.00	150.2	.70	3.8
14	25.64	153.1	.64	2.9
15	26.36	157.3	.72	4.2
16	26.98	160.7	.62	3.4
17	27.52	164.2	.54	3.5
18	27.78	165.6	.26	1.4
19	28.24	168.7	.46	3.1
20	28.78	171.7	.54	3.0

Note

Sometimes the first differences approach can overcorrect, leading to negative autocorrelations in the error terms. Hence, it may be appropriate to use a two-sided Durbin-Watson test when testing for autocorrelation with first differences data. One complication arises here. The first differences model (13.18) has no intercept term, but the Durbin-Watson test requires a fitted regression with an intercept term. A valid test for autocorrelation in a no-intercept model can be carried out by fitting for this purpose a regression function with an intercept term. Of course, the fitted no-intercept model is still the model of basic interest. In our Blaisdell Company example, the Durbin-Watson statistic for the fitted first differences regression model with an intercept term is $D = 1.75$ (calculation not shown). This indicates uncorrelated error terms for either a one-sided test (with $\alpha = .01$) or a two-sided test (with $\alpha = .02$).

Comments

1. The autoregressive error structure can also be used to advantage in situations where predictions of the dependent variable are to be made. Johnston (Ref. 13.3) discusses this problem.

2. The first-order autoregressive error process in models (13.1) and (13.2) is the simplest kind. A second-order process would be:

(13.19) $$\varepsilon_t = \rho_1 \varepsilon_{t-1} + \rho_2 \varepsilon_{t-2} + u_t$$

Still higher-order processes could be postulated. Specialized approaches have been developed for complex autoregressive error processes. These are discussed in treatments of time series procedures and forecasting, such as in Reference 13.4.

PROBLEMS

13.1. Refer to Table 13.1.
   a. Plot ε_t against ε_{t-1} for $t = 1, \ldots, 10$ on a graph. How is the positive first-order autocorrelation in the error terms shown by the plot?
   b. If you plotted u_t against ε_{t-1} for $t = 1, \ldots, 10$, what pattern would you expect?

13.2. Refer to **Plastic hardness** Problem 2.18. If the same test item were measured at 12 different points in time, would the error terms in the regression model likely be autocorrelated? Discuss.

13.3. A student stated that the first-order autoregressive error models (13.1) and (13.2) are too simple for business time series data because the error term in period t in such data is also influenced by random effects that occurred more than one period in the past. Comment.

13.4. A student writing a term paper used ordinary least squares in fitting a simple linear regression model to some time series data containing positively autocorrelated errors, and found that the 90 percent confidence interval for β_1 was too wide to be useful. She then decided to employ model (13.1) to improve the precision of the estimate. Comment.

13.5. For each of the following tests concerning the autocorrelation parameter ρ in model (13.2) with three independent variables, state the appropriate decision rule based on the Durbin-Watson statistic for a sample of size 38: (1) $H_0: \rho = 0$, $H_a: \rho \neq 0$, $\alpha = .02$; (2) $H_0: \rho = 0$, $H_a: \rho < 0$, $\alpha = .05$; (3) $H_0: \rho = 0$, $H_a: \rho > 0$, $\alpha = .01$.

13.6. Refer to **Calculator maintenance** Problem 2.16. The observations are listed in time order. Assume that regression model (13.1) is appropriate. Test whether or not positive autocorrelation is present; use $\alpha = .01$. State the alternatives, decision rule, and conclusion.

13.7. Refer to **Chemical shipment** Problem 7.12. The observations are listed in time order. Assume that regression model (13.2) is appropriate. Test whether or not positive autocorrelation is present; use $\alpha = .05$. State the alternatives, decision rule, and conclusion.

13.8. Refer to **Crop yield** Problem 9.13. The observations are listed in time order. Assume that regression model (13.2) with first- and second-order terms for the two independent variables and no interaction term is appropriate. Test whether or not positive autocorrelation is present; use $\alpha = .01$. State the alternatives, decision rule, and conclusion.

13.9. **Microcomputer components.** A staff analyst for a manufacturer of microcomputer components has compiled monthly data for the past 16 months on the value of industry production of processing units that use these components (X, in million dollars) and the value of the firm's components used (Y, in thousand dollars). The analyst believes that a simple linear regression relation is appropriate but anticipates positive autocorrelation. The data follow.

t:	1	2	3	4	5	6	7	8
X_t:	2.052	2.026	2.002	1.949	1.942	1.887	1.986	2.053
Y_t:	102.9	101.5	100.8	98.0	97.3	93.5	97.5	102.2

t:	9	10	11	12	13	14	15	16
X_t:	2.102	2.113	2.058	2.060	2.035	2.080	2.102	2.150
Y_t:	105.0	107.2	105.1	103.9	103.0	104.8	105.0	107.2

a. Fit a simple linear regression model by ordinary least squares and obtain the residuals. Also obtain $s(b_0)$ and $s(b_1)$.

b. Plot the residuals against time and explain whether you find any evidence of positive autocorrelation.

c. Conduct a formal test for positive autocorrelation using a significance level of .05. State the alternatives, decision rule, and conclusion. Is the residual analysis in part (b) in accord with the test result?

13.10. Refer to **Microcomputer components** Problem 13.9. The analyst has decided to employ regression model (13.1) and use the iterative approach to fit the model.

a. Obtain a point estimate of the autocorrelation parameter. How well does the approximate relationship (13.17) hold here between this point estimate and the Durbin-Watson test statistic?

b. Use one iteration to obtain the estimates b_0' and b_1' of the regression coefficients β_0' and β_1' in transformed model (13.13) and state the estimated regression function. Also obtain $s(b_0')$ and $s(b_1')$.

c. Test whether any positive autocorrelation remains after the first iteration using a significance level of .05. State the alternatives, decision rule, and conclusion.

d. Restate the estimated regression function obtained in part (b) in terms of the original variables. Also obtain $s(b_0)$ and $s(b_1)$. Compare the estimated regression coefficients obtained with the iterative method and their standard errors with those obtained with ordinary least squares in Problem 13.9a.

e. Based on the results in parts (c) and (d), does the iterative method appear to have been effective here?

13.11. Refer to **Microcomputer components** Problems 13.9 and 13.10. The staff analyst wishes to try also the first differences approach.
  a. Estimate the regression coefficient β_1 by the first differences approach, and obtain the estimated standard deviation of this estimate.
  b. Compare the results obtained in part (a) with those in Problem 13.10d. Summarize your findings.
  c. Test whether or not the error terms with the first differences approach are autocorrelated using a two-sided test and a level of significance of .10. State the alternatives, decision rule, and conclusion. Why is a two-sided test meaningful here?

13.12. **Advertising agency.** The managing partner of an advertising agency has become concerned about possible inefficiencies in the handling of client accounts. Monthly data on amount of billings (Y, in thousands of constant dollars) and on number of hours of staff time (X, in thousand hours) for the 20 most recent months follow. A simple linear regression model is believed to be appropriate, but positively autocorrelated error terms may be present.

t:	1	2	3	4	5	6	7	8	9	10
X_t:	2.521	2.171	2.234	2.524	2.305	2.523	3.020	3.014	3.532	3.461
Y_t:	220.4	203.9	207.2	221.9	211.3	222.7	247.6	247.6	272.9	269.1

t:	11	12	13	14	15	16	17	18	19	20
X_t:	3.737	3.801	3.576	3.586	3.447	2.723	3.019	3.117	3.623	3.618
Y_t:	283.9	287.0	275.4	275.1	269.1	232.8	248.1	252.4	278.6	278.5

  a. Fit a simple linear regression model by ordinary least squares and obtain the residuals. Also obtain $s(b_0)$ and $s(b_1)$.
  b. Plot the residuals against time and explain whether you find any evidence of positive autocorrelation.
  c. Conduct a formal test for positive autocorrelation using a significance level of .01. State the alternatives, decision rule, and conclusion. Is the residual analysis in part (b) in accord with the test result?

13.13. Refer to **Advertising agency** Problem 13.12. Assume that regression model (13.1) is applicable.
  a. Obtain a point estimate of the autocorrelation parameter. How well does the approximate relationship (13.17) hold here between the point estimate and the Durbin-Watson test statistic?
  b. Use one iteration to obtain the estimates b_0' and b_1' of the regression coefficients β_0' and β_1' in transformed model (13.13) and state the estimated regression function. Also obtain $s(b_0')$ and $s(b_1')$.
  c. Test whether any positive autocorrelation remains after the first iteration using a significance level of .01. State the alternatives, decision rule, and conclusion.
  d. Restate the estimated regression function obtained in part (b) in terms of the original variables. Also obtain $s(b_0)$ and $s(b_1)$. Compare the estimated regression coefficients obtained with the iterative method and their standard errors with those obtained with ordinary least squares in Problem 13.12a.
  e. Based on the results in parts (c) and (d), does the iterative method appear to have been effective here?

13.14. Refer to **Advertising agency** Problems 13.12 and 13.13.

    a.   Estimate the regression coefficient β_1 using the first differences approach by means of a 95 percent confidence interval. Interpret your interval estimate.

    b.   How does the estimated standard deviation of b_1 for the first differences estimate obtained in part (a) compare with that for the iterative method estimate in Problem 13.13d?

13.15. **McGill Company sales.** The data below show seasonally adjusted quarterly sales for the McGill Company (Y, in million dollars) and for the entire industry (X, in million dollars), for the most recent 20 quarters.

t:	1	2	3	4	5	6	7
X_t:	127.3	130.0	132.7	129.4	135.0	137.1	141.1
Y_t:	20.96	21.40	21.96	21.52	22.39	22.76	23.48

t:	8	9	10	11	12	13	14
X_t:	142.8	145.5	145.3	148.3	146.4	150.2	153.1
Y_t:	23.66	24.10	24.01	24.54	24.28	25.00	25.64

t:	15	16	17	18	19	20
X_t:	157.3	160.7	164.2	165.6	168.7	172.0
Y_t:	26.46	26.98	27.52	27.78	28.24	28.78

The first-order autoregressive error model (13.1) is to be employed.

    a.   Would you expect the autocorrelation parameter ρ to be positive, negative, or zero here?

    b.   Fit the linear regression model by ordinary least squares, obtain the residuals, and plot them against time. What do you find?

    c.   Conduct a formal test for positive autocorrelation using $\alpha = .05$. State the alternatives, decision rule, and conclusion.

13.16. Refer to **McGill Company sales** Problem 13.15.

    a.   Use one iteration with the iterative method to estimate the parameters β_0 and β_1 in regression model (13.1). Also obtain the estimated standard deviations of these estimates.

    b.   Test whether any positive autocorrelation remains after the first iteration; use $\alpha = .05$. State the alternatives, decision rule, and conclusion. Does the iterative approach appear to have been effective here?

    c.   Estimate β_1 with a 90 percent confidence interval. Interpret your interval estimate.

13.17. Refer to **McGill Company sales** Problems 13.15 and 13.16.

    a.   Estimate the regression coefficient β_1 in model (13.1) by the first differences approach using a 90 percent confidence interval.

    b.   Compare your result in part (a) with that in Problem 13.16c. State your findings.

    c.   Test whether or not the error terms with the first differences approach are positively autocorrelated using $\alpha = .05$. State the alternatives, decision rule, and conclusion.

13.18. A student applying the first differences transformation (13.18a) found that several X_t' values equaled zero but that the corresponding Y_t' values were nonzero. Does this signify that the first differences transformation is not apt for the data?

EXERCISES

13.19. Derive (13.7) for $s = 2$.

13.20. Refer to the first-order autoregressive error model (13.1). Suppose Y_t is company's percent share of the market, X_t is company's selling price as a percent of average competitive selling price, $\beta_0 = 100$, $\beta_1 = -.35$, $\rho = .6$, $\sigma^2 = 1$, and $\varepsilon_0 = 2.403$. Let X_t and u_t be as follows for $t = 1, \ldots, 10$:

t:	1	2	3	4	5	6	7	8	9	10
X_t:	100	115	120	90	85	75	70	95	105	110
u_t:	.764	.509	−.242	−1.808	−.485	.501	−.539	.434	−.299	.030

a. Plot the true regression line. Generate the observations Y_t ($t = 1, \ldots, 10$), and plot these on the same graph. Fit a least squares regression line to the observations and plot it also on the same graph. How does your fitted regression line relate to the true line?

b. Repeat the steps in part (a) but this time let $\rho = 0$. In which of the two cases does the fitted regression line come closer to the true line? Is this the expected outcome?

c. Generate the observations Y_t for $\rho = -.7$. For each of the cases $\rho = .6$, $\rho = 0$, and $\rho = -.7$, obtain the successive error term differences $\varepsilon_t - \varepsilon_{t-1}$ ($t = 1, \ldots, 10$).

d. For which of the three cases in part (c) is $\Sigma(\varepsilon_t - \varepsilon_{t-1})^2$ smallest? For which is it largest? What generalization does this suggest?

13.21. Suppose the autoregressive error process for the model $Y_t = \beta_0 + \beta_1 X_t + \varepsilon_t$ is that given by (13.19).

a. What would be the transformed variables Y_t' and X_t' to be used with the iterative method?

b. How would you estimate the parameters ρ_1 and ρ_2 for use with the iterative method?

PROJECTS

13.22. The true regression model is $Y_t = 10 + 24X_t + \varepsilon_t$, where $\varepsilon_t = .8\varepsilon_{t-1} + u_t$ and u_t are independent $N(0, 25)$.

a. Generate 11 independent random numbers from $N(0, 25)$. Use the first random number as ε_0, obtain the 10 error terms $\varepsilon_1, \ldots, \varepsilon_{10}$, and then calculate the 10 observations $Y_1, \ldots, Y_{10}$ corresponding to $X_1 = 1$, $X_2 = 2, \ldots$, $X_{10} = 10$. Fit a linear regression function by ordinary least squares and calculate MSE.

b. Repeat part (a) 100 times, using new random numbers each time.

c. Calculate the mean of the 100 estimates b_1. Does it appear that b_1 is an unbiased estimator of β_1 despite the presence of positive autocorrelation?

d. Calculate the mean of the 100 estimates MSE. Does it appear that MSE is a biased estimator of σ^2? If so, does the magnitude of the bias appear to be small or large?

CITED REFERENCES

13.1 Durbin, J., and G. S. Watson. "Testing for Serial Correlation in Least Squares Regression. II." *Biometrika* 38 (1951), pp. 159–78.

13.2 Theil, H., and A. L. Nagar. "Testing the Independence of Regression Disturbances." *Journal of the American Statistical Association* 56 (1961), pp. 793–806.

13.3 Johnston, J. *Econometric Methods.* 2d ed. New York: McGraw-Hill, 1971.

13.4 Box, G. E. P., and G. M. Jenkins. *Time Series Analysis, Forecasting and Control.* Rev. ed. San Francisco: Holden-Day, 1976.

14

Nonlinear regression

The linear regression models considered up to this point are satisfactory for most regression applications. There are occasions, however, when a nonlinear regression model is most suitable. We shall consider in this chapter nonlinear regression models, how to obtain least squares estimates of the regression parameters in such models, and how to make inferences about these regression parameters. The analysis of nonlinear regression models is numerically tedious and is therefore heavily computer-oriented.

14.1 LINEAR, INTRINSICALLY LINEAR, AND NONLINEAR REGRESSION MODELS

Linear regression models

In previous chapters, we considered linear regression models, i.e., models which are linear in the parameters. Such models can be represented by:

$$(14.1) \qquad Y_i = \beta_0 + \beta_1 X_{i1} + \beta_2 X_{i2} + \cdots + \beta_{p-1} X_{i,p-1} + \varepsilon_i$$

Linear regression models, as we have seen, include not only first-order models in $p - 1$ independent variables but also more complex models. For instance, a polynomial regression model in one or more independent variables is linear in the parameters, such as the following model in two independent variables with linear, quadratic, and interaction terms:

$$(14.2) \quad Y_i = \beta_0 + \beta_1 X_{i1} + \beta_2 X_{i1}^2 + \beta_3 X_{i2} + \beta_4 X_{i2}^2 + \beta_5 X_{i1} X_{i2} + \varepsilon_i$$

Also, models with transformed variables that are linear in the parameters belong to the class of linear regression models, such as the following model:

$$(14.3) \quad \log_{10} Y_i = \beta_0 + \beta_1 \sqrt{X_{i1}} + \beta_2 \exp(X_{i2}) + \varepsilon_i$$

Intrinsically linear regression models

In addition to the multitude of models that are linear in the parameters, there are other models that, though not linear in the parameters, can be transformed so that the parameters appear in linear fashion. For example, the exponential model:

$$(14.4) \qquad Y_i = \gamma_0 [\exp(\gamma_1 X_i)] \varepsilon_i$$

is nonlinear in the parameters γ_0 and γ_1. However, this model can be transformed into the linear form (14.1) by using the logarithmic transformation:

$$(14.5) \qquad \log_e Y_i = \log_e \gamma_0 + \gamma_1 X_i + \log_e \varepsilon_i$$

Letting:

$$\log_e Y_i = Y_i'$$
$$\log_e \gamma_0 = \beta_0$$
$$\gamma_1 = \beta_1$$
$$\log_e \varepsilon_i = \varepsilon_i'$$

we can write model (14.5) in the usual form for a linear model:

$$(14.5a) \qquad Y_i' = \beta_0 + \beta_1 X_i + \varepsilon_i'$$

We say that model (14.4) is an *intrinsically linear model* because it can be expressed in the linear form (14.1) by a suitable transformation. Another intrinsically linear regression model is:

$$(14.6) \qquad Y_i = [\exp(\gamma)] X_i + \varepsilon_i$$

If we let $\exp(\gamma) = \beta_1$, we then have a model with the regression through the origin:

$$(14.6a) \qquad Y_i = \beta_1 X_i + \varepsilon_i$$

Note

When an intrinsically linear model has been transformed into the linear model form, such as model (14.4) transformed into model (14.5a), it is important to study the linearized model for aptness. For instance, if the error terms ε_i in (14.4) are normally distributed, the transformed error terms ε_i' in (14.5a) will not be normally distributed.

Nonlinear regression models

Nonlinear regression models are not linear in the parameters and cannot be made so by transformation. For example, *exponential model* (14.4) with an additive error term:

$$(14.7) \qquad Y_i = \gamma_0 \exp(\gamma_1 X_i) + \varepsilon_i$$

is intrinsically nonlinear because no transformation exists which transforms this model into the linear form (14.1). A more general nonlinear exponential model in one independent variable with an additive error term is:

$$(14.8) \qquad Y_i = \gamma_0 + \gamma_1 \exp(\gamma_2 X_i) + \varepsilon_i$$

This model is commonly used in growth studies where the rate of growth at a given time X is proportional to the amount of growth remaining as time increases, with γ_0 representing the maximum growth value. Model (14.8) is often employed to relate the concentration of a substance (Y) to elapsed time (X). Figure 14.1a shows the response function for exponential model (14.8), for $\gamma_0 = 100$, $\gamma_1 = -50$, and $\gamma_2 = -2$.

Another nonlinear regression model is the *general logistic model* with additive error term:

$$(14.9) \qquad Y_i = \frac{\gamma_0}{1 + \gamma_1 \exp(\gamma_2 X_i)} + \varepsilon_i$$

This model has been used in population studies to relate number of species (Y) to time (X). Recall that logistic response function (10.53) was used in Chapter 10 for situations where the dependent variable is a 0, 1 indicator variable. Logistic response function (10.53) is a special type of logistic function. Figure 14.1b

FIGURE 14.1 Graphs of nonlinear regression response functions

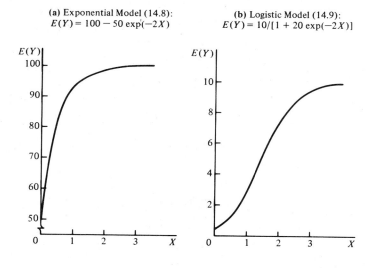

(a) Exponential Model (14.8):
$E(Y) = 100 - 50 \exp(-2X)$

(b) Logistic Model (14.9):
$E(Y) = 10/[1 + 20 \exp(-2X)]$

shows the response function for the general logistic model, for $\gamma_0 = 10$, $\gamma_1 = 20$, and $\gamma_2 = -2$.

Note

When nonlinear growth models are used for time series data, it is important to ascertain whether or not the error terms are uncorrelated, just as when linear regression models are applied to time series data.

14.2 EXAMPLE

In order to illustrate the analysis of nonlinear regression models, we shall use a relatively simple two-parameter example. In so doing, we shall be able to explain the concepts and procedures without overwhelming the reader with details.

A hospital administrator wished to develop a regression model for predicting the degree of long-term recovery after discharge from the hospital for severely injured patients. The predictor variable to be utilized is number of days of hospitalization (X), and the dependent variable is a prognosis index for long-term recovery (Y), with large values of the index reflecting a good prognosis. Data for 15 patients were studied; these are presented in Table 14.1.

Initially, the intrinsically linear exponential model (14.4) was fitted to the data, employing the logarithmic transformation model in (14.5). A residual analysis for this model suggested, however, that the error variance increases with X. It was therefore decided to use the two-parameter exponential model (14.7) with an additive error term, which is an intrinsically nonlinear model:

$$(14.10) \qquad Y_i = \gamma_0 \exp(\gamma_1 X_i) + \varepsilon_i$$

TABLE 14.1 Data for severely injured patients example

Patient i	Days Hospitalized X_i	Prognosis Index Y_i
1	2	54
2	5	50
3	7	45
4	10	37
5	14	35
6	19	25
7	26	20
8	31	16
9	34	18
10	38	13
11	45	8
12	52	11
13	53	8
14	60	4
15	65	6

It is now desired to estimate the regression parameters γ_0 and γ_1 for this model and to study the fit of the model.

14.3 LEAST SQUARES ESTIMATION IN NONLINEAR REGRESSION

We noted in Chapter 2 that the method of least squares requires the minimization of the criterion Q which for simple linear regression is:

$$Q = \sum_{i=1}^{n} (Y_i - \beta_0 - \beta_1 X_i)^2$$

Those estimates of β_0 and β_1 which minimize Q for the given sample observations (X_i, Y_i) are the least squares estimates and are denoted by b_0 and b_1. We also noted that one can search for the least squares estimates by trying various values of b_0 and b_1 and evaluating Q each time until the minimum value of Q is found. Alternatively, one can find analytically the least squares estimates by differentiating Q with respect to β_0 and β_1, setting the derivatives equal to 0, and solving the normal equations.

The same two basic approaches apply in nonlinear regression. We shall first consider the use of the normal equations and then the use of direct search procedures.

Normal equations

In linear regression, we represent an observation Y_i as the sum of the mean response and an error term:

(14.11) $$Y_i = E(Y_i) + \varepsilon_i$$

where:

$$E(Y_i) = \beta_0 + \beta_1 X_{i1} + \cdots + \beta_{p-1} X_{i,p-1}$$

Since nonlinear regression models take many different forms, as we saw earlier, we shall now simply indicate that $E(Y_i)$ is a function of p regression parameters $\gamma_0, \gamma_1, \ldots, \gamma_{p-1}$ and the ith observations on the X variables:

(14.12) $$E(Y_i) = f(\mathbf{X}_i, \boldsymbol{\gamma})$$

where:

(14.12a)
$$\underset{q \times 1}{\mathbf{X}_i} = \begin{bmatrix} X_{i1} \\ X_{i2} \\ \cdot \\ \cdot \\ \cdot \\ X_{iq} \end{bmatrix} \qquad \underset{p \times 1}{\boldsymbol{\gamma}} = \begin{bmatrix} \gamma_0 \\ \gamma_1 \\ \cdot \\ \cdot \\ \cdot \\ \gamma_{p-1} \end{bmatrix}$$

We denote the number of X variables by q, since the number of X variables in nonlinear regression is not directly related to the number of parameters, unlike linear regression. Thus, for model (14.10), we have:

$$f(\mathbf{X}_i, \boldsymbol{\gamma}) = \gamma_0 \exp(\gamma_1 X_i)$$

where the two ($p = 2$) parameters are γ_0 and γ_1 and $\mathbf{X}_i$ consists of one ($q = 1$) value X_i for the ith observation.

With this notation, we express a nonlinear regression model as follows:

(14.13) $$Y_i = f(\mathbf{X}_i, \boldsymbol{\gamma}) + \varepsilon_i$$

The least squares criterion Q can then be written:

(14.14) $$Q = \sum_{i=1}^{n} [Y_i - f(\mathbf{X}_i, \boldsymbol{\gamma})]^2$$

The partial derivative of Q with respect to γ_k is:

(14.15) $$\frac{\partial Q}{\partial \gamma_k} = \sum_{i=1}^{n} -2[Y_i - f(\mathbf{X}_i, \boldsymbol{\gamma})] \left[\frac{\partial f(\mathbf{X}_i, \boldsymbol{\gamma})}{\partial \gamma_k} \right]$$

When the p partial derivatives are each set equal to 0 and the parameters γ_k are replaced by the least squares estimates g_k, we obtain after some simplification the p normal equations:

(14.16) $$\sum_{i=1}^{n} Y_i \left[\frac{\partial f(\mathbf{X}_i, \boldsymbol{\gamma})}{\partial \gamma_k} \right]_{\boldsymbol{\gamma}=\mathbf{g}} - \sum_{i=1}^{n} f(\mathbf{X}_i, \mathbf{g}) \left[\frac{\partial f(\mathbf{X}_i, \boldsymbol{\gamma})}{\partial \gamma_k} \right]_{\boldsymbol{\gamma}=\mathbf{g}} = 0$$

$$k = 0, 1, \ldots, p - 1$$

where $\mathbf{g}$ is the vector of the least squares estimates g_k:

(14.16a) $$\mathbf{g} = \begin{bmatrix} g_0 \\ g_1 \\ \cdot \\ \cdot \\ \cdot \\ g_{p-1} \end{bmatrix}$$

and the terms in brackets in (14.16) are partial derivatives with the parameters γ_k replaced by the least squares estimates g_k.

The normal equations (14.16) for nonlinear regression models are nonlinear in the parameter estimates g_k and are usually difficult to solve, even in the simplest of cases. Hence, numerical procedures are ordinarily required to obtain a solution iteratively. To make things still more difficult, multiple solutions may be present.

Example. In our severely injured patients example, we are employing the response function:

(14.17) $$f(\mathbf{X}_i, \boldsymbol{\gamma}) = \gamma_0 \exp(\gamma_1 X_i)$$

Hence, the partial derivatives of $f(\mathbf{X}_i, \boldsymbol{\gamma})$ are:

(14.18a)
$$\frac{\partial f(\mathbf{X}_i, \boldsymbol{\gamma})}{\partial \gamma_0} = \exp(\gamma_1 X_i)$$

(14.18b)
$$\frac{\partial f(\mathbf{X}_i, \boldsymbol{\gamma})}{\partial \gamma_1} = \gamma_0 X_i \exp(\gamma_1 X_i)$$

Replacing γ_0 and γ_1 by the respective least squares estimates g_0 and g_1 in (14.17), (14.18a), and (14.18b), the normal equations (14.16) therefore are:

$$\Sigma Y_i \exp(g_1 X_i) \qquad - \Sigma g_0 \exp(g_1 X_i) \exp(g_1 X_i) \qquad = 0$$
$$\Sigma Y_i g_0 X_i \exp(g_1 X_i) - \Sigma g_0 \exp(g_1 X_i) g_0 X_i \exp(g_1 X_i) = 0$$

Upon simplification, the normal equations become:

$$\Sigma Y_i \exp(g_1 X_i) \qquad - g_0 \Sigma \exp(2g_1 X_i) \qquad = 0$$
$$\Sigma Y_i X_i \exp(g_1 X_i) - g_0 \Sigma X_i \exp(2g_1 X_i) = 0$$

These normal equations are not linear in g_0 and g_1, and no closed-form solution exists. Thus, numerical methods will be required to find the least squares estimates iteratively.

Note

When the error terms in a nonlinear regression model are independent $N(0, \sigma^2)$, the least squares estimates are the same as the maximum likelihood estimates.

Gauss-Newton method

In many nonlinear regression problems, it is more practical to find the least squares estimates by direct search procedures rather than by first obtaining the normal equations and then using numerical methods to solve these equations iteratively. The major statistical computer packages employ one or more direct search procedures for solving nonlinear regression problems.

The *Gauss-Newton method*, also called the *linearization method*, uses a Taylor series expansion to approximate the nonlinear regression model with linear terms and then employs ordinary least squares to estimate the parameters. Iteration of these steps generally leads to a solution to the nonlinear regression problem.

The Gauss-Newton method begins with initial or starting values for the regression parameters $\gamma_0, \gamma_1, \dots, \gamma_{p-1}$. We shall denote these by $g_0^{(0)}, g_1^{(0)}, \dots,$ $g_{p-1}^{(0)}$, where the superscript in parentheses denotes the iteration number. The starting values $g_k^{(0)}$ may be obtained from previous or related studies, theoretical expectations, or a preliminary search for parameter values that lead to a comparatively low criterion value Q in (14.14). We shall later discuss in more detail the choice of the starting values.

Once the starting values for the parameters have been obtained, we approximate the mean responses $f(\mathbf{X}_i, \boldsymbol{\gamma})$ for the n observations by the linear terms in the Taylor series expansion around the starting values $g_k^{(0)}$. We obtain for the ith observation:

$$(14.19) \quad f(\mathbf{X}_i, \boldsymbol{\gamma}) \simeq f(\mathbf{X}_i, \mathbf{g}^{(0)}) + \sum_{k=0}^{p-1} \left[\frac{\partial f(\mathbf{X}_i, \boldsymbol{\gamma})}{\partial \gamma_k} \right]_{\boldsymbol{\gamma} = \mathbf{g}^{(0)}} (\gamma_k - g_k^{(0)})$$

where:

$$(14.19a) \qquad\qquad \mathbf{g}^{(0)} = \begin{bmatrix} g_0^{(0)} \\ g_1^{(0)} \\ \cdot \\ \cdot \\ \cdot \\ g_{p-1}^{(0)} \end{bmatrix}$$

is the vector of the parameter starting values. The terms in brackets in (14.19) are the same partial derivatives of the regression function we encountered earlier in the normal equations (14.16), but here they are evaluated at $\gamma_k = g_k^{(0)}$ for $k = 0, 1, \ldots, p - 1$.

Let us now simplify the notation as follows:

$$(14.20a) \qquad\qquad f_i^{(0)} = f(\mathbf{X}_i, \mathbf{g}^{(0)})$$

$$(14.20b) \qquad\qquad \beta_k^{(0)} = \gamma_k - g_k^{(0)}$$

$$(14.20c) \qquad\qquad D_{ik}^{(0)} = \left[\frac{\partial f(\mathbf{X}_i, \boldsymbol{\gamma})}{\partial \gamma_k} \right]_{\boldsymbol{\gamma} = \mathbf{g}^{(0)}}$$

The Taylor approximation (14.19) for the ith observation mean response then becomes in this notation:

$$f(\mathbf{X}_i, \boldsymbol{\gamma}) \simeq f_i^{(0)} + \sum_{k=0}^{p-1} D_{ik}^{(0)} \beta_k^{(0)}$$

and an approximation to the nonlinear regression model (14.13):

$$Y_i = f(\mathbf{X}_i, \boldsymbol{\gamma}) + \varepsilon_i$$

is:

$$(14.21) \qquad\qquad Y_i \simeq f_i^{(0)} + \sum_{k=0}^{p-1} D_{ik}^{(0)} \beta_k^{(0)} + \varepsilon_i$$

When we shift the $f_i^{(0)}$ term to the left and denote the difference $Y_i - f_i^{(0)}$ by $Y_i^{(0)}$, we obtain a linear regression model approximation:

$$(14.22) \qquad\qquad Y_i^{(0)} \simeq \sum_{k=0}^{p-1} D_{ik}^{(0)} \beta_k^{(0)} + \varepsilon_i \qquad i = 1, \ldots, n$$

where:

(14.22a)
$$Y_i^{(0)} = Y_i - f_i^{(0)}$$

We shall represent this approximation in matrix form as follows:

(14.23)
$$\mathbf{Y}^{(0)} \simeq \mathbf{D}^{(0)}\boldsymbol{\beta}^{(0)} + \boldsymbol{\varepsilon}$$

where:

(14.23a)
$$\mathbf{Y}^{(0)}_{\substack{n \times 1}} = \begin{bmatrix} Y_1 - f_1^{(0)} \\ \vdots \\ Y_n - f_n^{(0)} \end{bmatrix}$$

(14.23b)
$$\mathbf{D}^{(0)}_{\substack{n \times p}} = \begin{bmatrix} D_{10}^{(0)} & \cdots & D_{1,p-1}^{(0)} \\ \vdots & & \vdots \\ D_{n0}^{(0)} & \cdots & D_{n,p-1}^{(0)} \end{bmatrix}$$

(14.23c)
$$\boldsymbol{\beta}^{(0)}_{\substack{p \times 1}} = \begin{bmatrix} \beta_0^{(0)} \\ \vdots \\ \beta_{p-1}^{(0)} \end{bmatrix}$$

Note that the approximate model (14.23) is precisely in the form of the general linear regression model (7.18), with the $\mathbf{D}$ matrix of partial derivatives now playing the role of the $\mathbf{X}$ matrix. We can therefore estimate the parameters $\boldsymbol{\beta}^{(0)}$ by means of the normal equations for the ordinary linear regression model and obtain according to (7.21):

(14.24)
$$\mathbf{b}^{(0)} = (\mathbf{D}^{(0)\prime}\mathbf{D}^{(0)})^{-1}\mathbf{D}^{(0)\prime}\mathbf{Y}^{(0)}$$

where $\mathbf{b}^{(0)}$ is the vector of the least squares estimated regression coefficients. We use these least squares estimates to obtain revised estimated regression coefficients $g_k^{(1)}$ by means of (14.20b):

$$g_k^{(1)} = g_k^{(0)} + b_k^{(0)}$$

where $g_k^{(1)}$ denotes the revised estimate of γ_k at the end of the first iteration. In matrix form, we represent the revision process as follows:

(14.25)
$$\mathbf{g}^{(1)} = \mathbf{g}^{(0)} + \mathbf{b}^{(0)}$$

At this point, we can examine whether the revised regression coefficients represent adjustments in the proper direction. The least squares criterion measure Q in (14.14) for the starting regression coefficients $\mathbf{g}^{(0)}$, to be denoted by $SSE^{(0)}$, is:

(14.26)
$$SSE^{(0)} = \sum_{i=1}^{n} [Y_i - f(\mathbf{X}_i, \mathbf{g}^{(0)})]^2 = \sum_{i=1}^{n} (Y_i - f_i^{(0)})^2$$

At the end of the first iteration, the estimated regression coefficients are $\mathbf{g}^{(1)}$, and the least squares criterion measure, now denoted by $SSE^{(1)}$, is:

(14.27) $$SSE^{(1)} = \sum_{i=1}^{n} [Y_i - f(\mathbf{X}_i, \mathbf{g}^{(1)})]^2 = \sum_{i=1}^{n} (Y_i - f_i^{(1)})^2$$

If the Gauss-Newton method is working effectively in the first iteration, $SSE^{(1)}$ should be smaller than $SSE^{(0)}$ since the revised estimated regression coefficients $\mathbf{g}^{(1)}$ should be better estimates.

Note that the nonlinear regression functions $f(\mathbf{X}_i, \mathbf{g}^{(0)})$ and $f(\mathbf{X}_i, \mathbf{g}^{(1)})$ are used in calculating $SSE^{(0)}$ and $SSE^{(1)}$, and not the linear approximations from the Taylor series expansion.

The revised regression coefficients $\mathbf{g}^{(1)}$ are not, of course, the least squares estimates for the nonlinear regression problem because the fitted model (14.23) is only an approximation of the nonlinear model. The Gauss-Newton method therefore repeats the procedure just described, with $\mathbf{g}^{(1)}$ now as the starting values. This produces a new set of revised estimates, denoted by $\mathbf{g}^{(2)}$, and a new least squares criterion measure $SSE^{(2)}$. The iterative process is continued until the difference between successive coefficient estimates $\mathbf{g}^{(s+1)} - \mathbf{g}^{(s)}$ and/or the difference between successive least squares criterion measures $SSE^{(s+1)} - SSE^{(s)}$ become negligible. We shall denote the final estimates of the regression coefficients simply by $\mathbf{g}$ and the final least squares criterion measure, which is the error sum of squares, by SSE.

The Gauss-Newton method works effectively in many nonlinear regression applications. In some instances, however, the method may require numerous iterations before converging, and in a few cases it may not converge at all.

Example.   In our severely injured patients example, initial values of the parameters γ_0 and γ_1 were taken to be the estimates of these parameters when the logarithmic transformation model (14.5) of the intrinsically linear exponential model (14.4) was fitted. These initial values are $g_0^{(0)} = 56.6646$ and $g_1^{(0)} = -.03797$ (calculations not shown). The least squares criterion measure at this stage requires evaluation of the nonlinear regression function (14.17) for each observation, utilizing the starting parameter values $g_0^{(0)}$ and $g_1^{(0)}$. For instance, for the first observation, for which $X_1 = 2$, we obtain:

$$\begin{aligned} f(\mathbf{X}_1, \mathbf{g}^{(0)}) = f_1^{(0)} &= g_0^{(0)} \exp(g_1^{(0)} X_1) \\ &= (56.6646) \exp[-.03797(2)] \\ &= 52.5208 \end{aligned}$$

Since $Y_1 = 54$, the deviation from the mean response is:

$$Y_1^{(0)} = Y_1 - f_1^{(0)} = 54 - 52.5208 = 1.4792$$

We see that the deviation $Y_1^{(0)}$ is actually the residual for observation 1 at the initial fitting stage since $f_1^{(0)}$ is the estimated mean response when the initial estimates $\mathbf{g}^{(0)}$ of the parameters are employed. The stage 0 residuals for this and the other sample observations are presented in Table 14.2 and constitute the $\mathbf{Y}^{(0)}$ vector.

The least squares criterion measure at this initial stage then is simply the sum of the squared stage 0 residuals:

$$SSE^{(0)} = \Sigma(Y_i - f_i^{(0)})^2 = \Sigma(Y_i^{(0)})^2$$
$$= (1.4792)^2 + \cdots + (1.1977)^2 = 56.0869$$

To revise the initial values for the parameters, we require the $\mathbf{D}^{(0)}$ matrix and the $\mathbf{Y}^{(0)}$ vector. The latter was already obtained in the process of calculating the least squares criterion measure at the start. To obtain the $\mathbf{D}^{(0)}$ matrix, we need the partial derivatives of the regression function (14.17) evaluated at $\gamma = \mathbf{g}^{(0)}$. The partial derivatives are given in (14.18). Table 14.2 shows the $\mathbf{D}^{(0)}$ matrix entries in symbolic form and also the numerical values. To illustrate the calculations for observation $i = 1$, we know from Table 14.1 that $X_1 = 2$. Hence, evaluating the partial derivatives at $\mathbf{g}^{(0)}$, we find:

$$\left[\frac{\partial f(\mathbf{X}_1, \gamma)}{\partial \gamma_0}\right]_{\gamma=\mathbf{g}^{(0)}} = \exp(g_1^{(0)}X_1)$$
$$= \exp[-.03797(2)] = .92687$$

$$\left[\frac{\partial f(\mathbf{X}_1, \gamma)}{\partial \gamma_1}\right]_{\gamma=\mathbf{g}^{(0)}} = g_0^{(0)}X_1\exp(g_1^{(0)}X_1)$$
$$= 56.6646(2)\exp[-.03797(2)] = 105.0416$$

TABLE 14.2 $\mathbf{Y}^{(0)}$ and $\mathbf{D}^{(0)}$ matrices for severely injured patients example

$$\mathbf{Y}^{(0)}_{15\times1} = \begin{bmatrix} Y_1 - f_1^{(0)} \\ \\ \\ \\ \\ \cdot \\ \\ \cdot \\ \\ \\ \cdot \\ \\ \\ Y_{15} - f_{15}^{(0)} \end{bmatrix} = \begin{bmatrix} Y_1 - g_0^{(0)}\exp(g_1^{(0)}X_1) \\ \\ \\ \\ \\ \cdot \\ \\ \cdot \\ \\ \\ \cdot \\ \\ \\ Y_{15} - g_0^{(0)}\exp(g_1^{(0)}X_{15}) \end{bmatrix} = \begin{bmatrix} 1.4792 \\ 3.1337 \\ 1.5609 \\ -1.7624 \\ 1.6996 \\ -2.5422 \\ -1.1139 \\ -1.4629 \\ 2.4172 \\ -.3871 \\ -2.2625 \\ 3.1327 \\ .4259 \\ -1.8063 \\ 1.1977 \end{bmatrix}$$

$$\mathbf{D}^{(0)}_{15\times2} = \begin{bmatrix} \exp(g_1^{(0)}X_1) & g_0^{(0)}X_1\exp(g_1^{(0)}X_1) \\ & \\ \cdot & \cdot \\ & \\ \cdot & \cdot \\ & \\ \exp(g_1^{(0)}X_{15}) & g_0^{(0)}X_{15}\exp(g_1^{(0)}X_{15}) \end{bmatrix} = \begin{bmatrix} .92687 & 105.0416 \\ .82708 & 234.3317 \\ .76660 & 304.0736 \\ .68407 & 387.6236 \\ .58768 & 466.2057 \\ .48606 & 523.3020 \\ .37261 & 548.9603 \\ .30818 & 541.3505 \\ .27500 & 529.8162 \\ .23625 & 508.7088 \\ .18111 & 461.8140 \\ .13884 & 409.0975 \\ .13367 & 401.4294 \\ .10247 & 348.3801 \\ .08475 & 312.1510 \end{bmatrix}$$

We are now ready to obtain the least squares estimates $\mathbf{b}^{(0)}$ by solving (14.24):

$$\mathbf{b}^{(0)} = (\mathbf{D}^{(0)\prime}\mathbf{D}^{(0)})^{-1}\mathbf{D}^{(0)\prime}\mathbf{Y}^{(0)}$$

We find (calculations not shown):

$$\mathbf{D}^{(0)\prime}\mathbf{D}^{(0)} = \begin{bmatrix} 3.63328 & 2{,}211.285 \\ 2{,}211.285 & 2{,}694{,}343.69 \end{bmatrix}$$

and:

$$\mathbf{D}^{(0)\prime}\mathbf{Y}^{(0)} = \begin{bmatrix} 3.4229 \\ -24.1911 \end{bmatrix}$$

so that:

$$\mathbf{b}^{(0)} = \begin{bmatrix} 3.63328 & 2{,}211.285 \\ 2{,}211.285 & 2{,}694{,}343.69 \end{bmatrix}^{-1} \begin{bmatrix} 3.4229 \\ -24.1911 \end{bmatrix}$$

$$= \begin{bmatrix} 5.499E-1 & -4.513E-4 \\ -4.513E-4 & 7.416E-7 \end{bmatrix} \begin{bmatrix} 3.4229 \\ -24.1911 \end{bmatrix} = \begin{bmatrix} 1.8932 \\ -1.563E-3 \end{bmatrix}$$

By (14.25), we obtain the revised least squares estimates $\mathbf{g}^{(1)}$:

$$\mathbf{g}^{(1)} = \mathbf{g}^{(0)} + \mathbf{b}^{(0)} = \begin{bmatrix} 56.6646 \\ -.03797 \end{bmatrix} + \begin{bmatrix} 1.8932 \\ -.001563 \end{bmatrix}$$

$$= \begin{bmatrix} 58.5578 \\ -.03953 \end{bmatrix}$$

Hence, $g_0^{(1)} = 58.5578$ and $g_1^{(1)} = -.03953$ are the revised parameter estimates at the end of the first iteration. Note that the estimated regression coefficients have been revised moderately from the initial values, as can be seen from Table 14.3a, which presents the estimated regression coefficients as well as the least squares criterion measures for the first three iterations. Note also that the least squares criterion measure has been reduced in the first iteration.

While iteration 1 led to moderate revisions in the estimated regression coefficients and a substantially better fit according to the least squares criterion, Table 14.3a indicates that iteration 2 resulted only in minor revisions of the estimated regression coefficients and little improvement in the fit. Iteration 3 led to no change in either the estimates of the coefficients or the least squares criterion measure.

Hence, the search procedure was terminated after three iterations. The final regression coefficient estimates therefore are $g_0 = 58.6065$ and $g_1 = -.03959$, and the fitted regression model is:

(14.28) $$\hat{Y} = (58.6065)\exp(-.03959X)$$

The error sum of squares for this fitted model is $SSE = 49.4593$. Figure 14.2 presents a scatter plot of the data and the estimated regression function.

TABLE 14.3 Gauss-Newton method iterations to obtain nonlinear least squares estimates—severely injured patients example

(a) Estimates of Parameters and Least Squares Criterion Measure

Iteration	g_0	g_1	SSE
0	56.6646	−.03797	56.0869
1	58.5578	−.03953	49.4638
2	58.6055	−.03959	49.4593
3	58.6065	−.03959	49.4593

(b) Final Least Squares Estimates

k	g_k	$s(g_k)$
0	58.6065	1.472
1	−.03959	.00171

$$MSE = \frac{49.4593}{13} = 3.80456$$

(c) Estimated Approximate Variance-Covariance Matrix of Estimated Regression Coefficients

$$\mathbf{s}^2(\mathbf{g}) = MSE(\mathbf{D'D})^{-1} = 3.80456 \begin{bmatrix} 5.696E - 1 & -4.682E - 4 \\ -4.682E - 4 & 7.697E - 7 \end{bmatrix}$$

$$= \begin{bmatrix} 2.1672 & -1.781E - 3 \\ -1.781E - 3 & 2.928E - 6 \end{bmatrix}$$

FIGURE 14.2 Plot of data and fitted nonlinear regression function—severely injured patients example

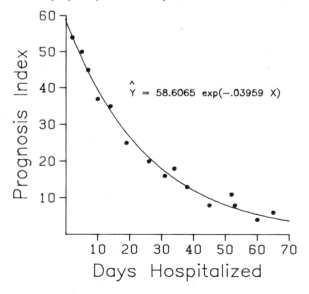

$\hat{Y} = 58.6065\ \exp(-.03959\ X)$

A plot of the residuals against the fitted values $\hat{Y}$ (not shown) did not suggest any serious departures from the model assumptions, and thus exponential model (14.17) and the estimated regression function (14.28) were accepted for the prognosis analysis for severely injured patients.

Comments

1. The choice of initial starting values is very important with the Gauss-Newton method because a poor choice may result in slow convergence, convergence to a local minimum, or even divergence. Good starting values will generally result in faster convergence, and if multiple minima exist, will lead to a solution that is the global minimum rather than a local minimum.

2. A variety of methods are available for obtaining starting values for the regression parameters. Often, experience can be utilized to provide good starting values for the regression parameters. Another possibility is to select p representative observations, set the regression function $f(\mathbf{X}_i, \boldsymbol{\gamma})$ equal to Y_i for each of the p observations (thereby ignoring the random error), solve the p equations for the p parameters, and use the solutions as the starting values, provided they lead to reasonably good fits of the observed data. Still another possibility is to do a grid search in the parameter space by selecting in a grid fashion various trial choices of $\mathbf{g}$, evaluate the least squares criterion Q for each of these choices, and use as the starting values that $\mathbf{g}$ vector for which Q is smallest.

3. When using the Gauss-Newton or some other direct search procedure, it is often desirable to try other sets of starting values after a solution has been obtained to make sure that the same solution will be found.

4. Some computer packages for nonlinear regression require that the user specify the starting values for the regression parameters. Others do a grid search to obtain starting values.

5. Some nonlinear regression computer programs using the Gauss-Newton method require the user to input the partial derivatives of the regression function, while others numerically calculate estimated partial derivatives from the regression function. In addition, most nonlinear computer programs have a library of commonly used regression functions which need only be specified by the user.

6. The Gauss-Newton method may produce iterations which oscillate widely or result in increases in the error sum of squares. Sometimes, these are only temporary but occasionally serious convergence problems exist. Various modifications of the Gauss-Newton method have been suggested to improve its performance, such as the Hartley modification (Ref. 14.1).

Other direct search procedures

A number of other direct search procedures besides the Gauss-Newton method are frequently used. One is the *method of steepest descent*. This method searches for the minimum least squares criterion measure Q by iteratively determining the direction in which the regression coefficients $\mathbf{g}$ should be changed. The method of steepest descent is particularly effective when the starting values $\mathbf{g}^{(0)}$ are not good and are far from the final values $\mathbf{g}$.

The *Marquardt algorithm* seeks to utilize the best features of the Gauss-

Newton method and the method of steepest descent, and occupies a middle ground between these two methods.

Additional information about direct search procedures can be found in specialized sources, such as References 14.2 and 14.3.

14.4 INFERENCES ABOUT NONLINEAR REGRESSION PARAMETERS

Estimated variances and covariances

Inferences about nonlinear regression parameters require an estimate of the error term variance σ^2. This estimate is the same as for linear regression:

$$(14.29) \qquad MSE = \frac{SSE}{n-p} = \frac{\Sigma(Y_i - \hat{Y}_i)^2}{n-p} = \frac{\Sigma[Y_i - f(\mathbf{X}_i, \mathbf{g})]^2}{n-p}$$

where $\mathbf{g}$ is the vector of the final parameter estimates. For nonlinear regression, MSE is not an unbiased estimator of σ^2, but the bias is small when the sample size is large.

When the error terms are independent and normally distributed and the sample size is reasonably large, the following theorem is helpful:

(14.30)   When the error terms ε_i are independent $N(0, \sigma^2)$ and the sample size n is reasonably large, the sampling distribution of $\mathbf{g}$ is approximately normal with:

$$\mathbf{E(g)} \cong \mathbf{\gamma}$$

Thus, when the sample size is large the least squares estimators $\mathbf{g}$ for nonlinear regression are approximately normally distributed and unbiased. An estimate of the approximate variance-covariance matrix of the regression coefficients is:

$$(14.31) \qquad\qquad \mathbf{s}^2(\mathbf{g}) = MSE(\mathbf{D'D})^{-1}$$

where $\mathbf{D}$ is the matrix of partial derivatives evaluated at the final least squares estimates $\mathbf{g}$, just as $\mathbf{D}^{(0)}$ in (14.23a) is the matrix of partial derivatives evaluated at $\mathbf{g}^{(0)}$.

Note that the estimated approximate variance-covariance matrix $\mathbf{s}^2(\mathbf{g})$ is of exactly the same form as the one for linear regression in (7.39), with $\mathbf{D}$ again playing the role of the $\mathbf{X}$ matrix.

Example. For our severely injured patients example, we know from Table 14.3a that the final error sum of squares is $SSE = 49.4593$. Since $p = 2$ parameters are present in regression model (14.17), we have:

$$MSE = \frac{SSE}{n-p} = \frac{49.4593}{15-2} = 3.80456$$

Table 14.3b presents this mean square, and Table 14.3c contains the estimated variance-covariance matrix of the regression coefficients. The matrix $(\mathbf{D'D})^{-1}$ is based on the final regression coefficient estimates **g** and is shown without computational details.

We see from Table 14.3c that $s^2(g_0) = 2.1672$ and $s^2(g_1) = .000002928$. The estimated standard deviations of the regression coefficients are given in Table 14.3b.

Interval estimation of a single γ_k

When the error terms in the nonlinear regression model (14.13) are independent and normally distributed, the following approximate result holds when the sample size is large:

$$(14.32) \qquad \frac{g_k - \gamma_k}{s(g_k)} \sim t(n - p) \qquad k = 0, 1, \ldots, p - 1$$

Hence, approximate $1 - \alpha$ confidence limits for any single γ_k are the usual ones:

$$(14.33) \qquad g_k \pm t(1 - \alpha/2; n - p)s(g_k)$$

Example.   For our severely injured patients example, it is desired to estimate γ_1 with a 95 percent confidence interval. We require $t(.975; 13) = 2.160$, and find from Table 14.3b that $g_1 = -.03959$ and $s(g_1) = .00171$. Hence:

$$-.03959 - 2.160(.00171) \leq \gamma_1 \leq -.03959 + 2.160(.00171)$$
$$-.0433 \leq \gamma_1 \leq -.0359$$

Thus, we can conclude with 95 percent confidence that γ_1 is between $-.0433$ and $-.0359$.

Simultaneous interval estimation of several γ_k

Approximate joint confidence regions for the regression parameters in nonlinear regression can be developed, but they are difficult to interpret except when $p - 1 = 2$. Bonferroni joint confidence intervals, on the other hand, are easy to obtain and interpret, as in linear regression. If m parameters are to be estimated with family confidence coefficient $1 - \alpha$, the joint Bonferroni confidence limits are:

$$(14.34) \qquad g_k \pm Bs(g_k)$$

where:

$$(14.34a) \qquad B = t(1 - \alpha/2m; n - p)$$

Example.   In our severely injured patients example, it is desired to obtain simultaneous interval estimates for γ_0 and γ_1 with a 90 percent family confidence coefficient. With the Bonferroni procedure we therefore require separate confidence intervals for the two parameters, each with a 95 percent statement confidence coefficient. We have already obtained a confidence interval for γ_1 with a 95 percent statement confidence coefficient. A 95 percent statement confidence interval for γ_0 is:

$$58.6065 - 2.160(1.472) \le \gamma_0 \le 58.6065 + 2.160(1.472)$$
$$55.43 \le \gamma_0 \le 61.79$$

Hence, the joint confidence intervals with family confidence coefficient of 90 percent are:

$$55.4 \le \gamma_0 \le 61.8$$
$$-.0433 \le \gamma_1 \le -.0359$$

Test concerning a single γ_k

A test concerning a single γ_k is set up in the usual fashion. To test:

(14.35a)
$$H_0\colon \gamma_k = \gamma_{k0}$$
$$H_a\colon \gamma_k \neq \gamma_{k0}$$

where γ_{k0} is the specified value of γ_k, we may use the usual t^* test statistic when n is reasonably large:

(14.35b)
$$t^* = \frac{g_k - \gamma_{k0}}{s(g_k)}$$

and the decision rule:

(14.35c)
$$\text{If } |t^*| \le t(1 - \alpha/2; n - p), \text{ conclude } H_0$$
$$\text{Otherwise conclude } H_a$$

Example. In our severely injured patients example, we wish to test:

$$H_0\colon \gamma_0 = 50$$
$$H_a\colon \gamma_0 \neq 50$$

The test statistic (14.35b) here is:

$$t^* = \frac{58.6065 - 50}{1.472} = 5.85$$

For $\alpha = .01$, we require $t(.995; 13) = 3.012$. Since $|t^*| = 5.85 > 3.012$, we conclude H_a, that $\gamma_0 \neq 50$.

Test concerning several γ_k

When a test is desired concerning several γ_k simultaneously, we use the same approach as for the general linear test, first fitting the full model and obtaining $SSE(F)$, then fitting the reduced model and obtaining $SSE(R)$, and finally calculating the same test statistic as for linear regression:

$$(14.36) \qquad F^* = \frac{SSE(R) - SSE(F)}{df_R - df_F} \div MSE(F)$$

For large n, this test statistic is distributed approximately as $F(df_R - df_F, df_F)$ when H_0 holds.

14.5 LEARNING CURVE EXAMPLE

We shall now present a second example to provide an additional illustration of the nonlinear regression concepts developed in this chapter. An electronic products manufacturer undertook the production of a new product in two locations (location A: coded $X_1 = 1$, location B: coded $X_1 = 0$). Location B has more modern facilities and hence was expected to be more efficient than location A, even after the initial learning period. An industrial engineer calculated the expected unit production cost for a modern facility after learning has occurred. Weekly unit production costs for each location were then expressed as a fraction of this expected cost. The reciprocal of this fraction is a measure of relative efficiency, and this measure was utilized as the efficiency measure in this study.

It is well known that efficiency increases over time when a new product is produced, and that the improvements eventually slow down and the process stabilizes. Hence, it was decided to employ an exponential model with an upper asymptote for expressing the relation between relative efficiency (Y) and time (X_2), and to incorporate a constant effect for the difference in the two production locations. The model decided on was:

$$(14.37) \qquad Y_i = \gamma_0 + \gamma_1 X_{i1} + \gamma_3 \exp(\gamma_2 X_{i2}) + \varepsilon_i$$

Here, γ_0 is the upper asymptote for location B as X_2 gets large, and $\gamma_0 + \gamma_1$ is the upper asymptote for location A. The parameters γ_2 and γ_3 reflect the speed of learning, which was expected to be the same in the two locations.

While weekly data on relative production efficiency for each location were available, we shall only use observations for selected weeks during the first 90 weeks of production to simplify the presentation. The data on location, week, and relative efficiency are presented in Table 14.4. Note that learning was relatively rapid in both locations, and that the relative efficiency in location B toward the end of the 90-week period even exceeded 1.0, i.e., the actual unit costs then were lower than the industrial engineer's expected unit cost.

Model (14.37) is nonlinear in the parameters γ_2 and γ_3. Hence, a direct search estimation procedure was to be employed, for which starting values for the parameters are needed. These were developed partly from past experience, partly from analysis of the data. Previous studies indicated that γ_3 should be in the neighborhood of $-.5$, so $g_3^{(0)} = -.5$ was used. Since the difference in the relative efficiencies between locations A and B for a given week tended to average $-.0459$ during the 90-week period, a starting value $g_1^{(0)} = -.0459$ was specified. The largest observed relative efficiency for location B was 1.028, so

TABLE 14.4 Data for learning curve example

Observation i	Location X_{i1}	Week X_{i2}	Relative Efficiency Y_i
1	1	1	.483
2	1	2	.539
3	1	3	.618
4	1	5	.707
5	1	7	.762
6	1	10	.815
7	1	15	.881
8	1	20	.919
9	1	30	.964
10	1	40	.959
11	1	50	.968
12	1	60	.971
13	1	70	.960
14	1	80	.967
15	1	90	.975
16	0	1	.517
17	0	2	.598
18	0	3	.635
19	0	5	.750
20	0	7	.811
21	0	10	.848
22	0	15	.943
23	0	20	.971
24	0	30	1.012
25	0	40	1.015
26	0	50	1.007
27	0	60	1.022
28	0	70	1.028
29	0	80	1.017
30	0	90	1.023

that a starting value $g_0^{(0)} = 1.025$ was felt to be reasonable. Only a starting value for γ_2 remains to be found. This was chosen by selecting a typical relative efficiency observation in the middle of the time period, $Y_{24} = 1.012$, equating it to the response function with $X_{24,1} = 0$, $X_{24,2} = 30$, and the previous starting values for the other regression coefficients (thus ignoring the error term):

$$1.012 = 1.025 - (.5)\exp(30\gamma_2)$$

and solving for γ_2. Thereby the starting value $g_2^{(0)} = -.122$ was obtained. Tests for several other representative observations yielded similar starting values, and $g_2^{(0)} = -.122$ was therefore considered to be a reasonable initial value.

With the four starting values $g_0^{(0)} = 1.025$, $g_1^{(0)} = -.0459$, $g_2^{(0)} = -.122$, and $g_3^{(0)} = -.5$, a computer package direct search program was utilized to obtain the least squares estimates. The resulting least squares regression function was:

(14.38) $\hat{Y} = 1.0156 - .04727X_1 - (.5524)\exp(-.1348X_2)$

and the error sum of squares was $SSE = .00329$, with $30 - 4 = 26$ degrees of

FIGURE 14.3 Plot of data and fitted nonlinear regression functions— learning curve example

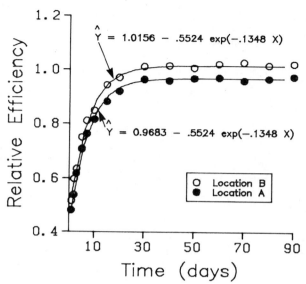

freedom. Figure 14.3 presents the scatter plot and the fitted regression functions for the two locations. Residual plots did not indicate any noticeable departures from the assumed model.

Special interest existed in the parameter γ_1, reflecting the effect of location. A 95 percent confidence interval is to be constructed. We require $t(.975; 26) = 2.056$. The computer printout contained the estimated variance-covariance matrix $s^2(\mathbf{g})$ from which it was found that $s(g_1) = \sqrt{.000016885} = .00411$. Hence, the 95 percent confidence interval for γ_1 is:

$$-.04727 - 2.056(.00411) \leq \gamma_1 \leq -.04727 + 2.056(.00411)$$
$$-.0557 \leq \gamma_1 \leq -.0388$$

Since γ_1 is seen to be negative, this confidence interval confirms that location A with its less modern facilities tends to be less efficient.

Note

When growth or learning curve models are fitted to data constituting repeated observations on the same unit, such as efficiency data for the same production unit at different points in time, the error terms may be correlated. Hence, in these situations it is important to ascertain whether or not a model assuming uncorrelated error terms is reasonable. In the learning curve example, a plot of the residuals against time order did not suggest any serious correlations among the error terms.

PROBLEMS

14.1. For each of the following models, indicate whether it is a linear regression model, an intrinsically linear model, or a nonlinear model. In the case of an intrinsically linear model, state how it can be expressed in the form of (14.1) by a suitable transformation:

a. $Y_i = \exp(\gamma_0 + \gamma_1 X_i + \varepsilon_i)$

b. $Y_i = \exp(\gamma_0 + \gamma_1 X_i) + \varepsilon_i$

c. $Y_i = \gamma_0 + \dfrac{\gamma_1}{\gamma_0} X_i + \varepsilon_i$

14.2. For each of the following models, indicate whether it is a linear regression model, an intrinsically linear model, or a nonlinear model. In the case of an intrinsically linear model, state how it can be expressed in the form of (14.1) by a suitable transformation:

a. $\log_e Y_i = \gamma_0 + \gamma_1 \log_e X_i + \varepsilon_i$

b. $Y_i = \gamma_0 X_{i1}^{\gamma_1} X_{i2}^{\gamma_2} \varepsilon_i$

c. $Y_i = \gamma_0 - \gamma_1 \gamma_2^{X_i} + \varepsilon_i$

14.3. a. Plot the logistic response function:

$$E(Y) = \frac{300}{1 + (30)\exp(-1.5X)} \qquad X \geq 0$$

b. What is the asymptote of this response function? For what value of X does the response function reach 90 percent of its asymptote?

14.4. a. Plot the exponential response function:

$$E(Y) = 49 - (30)\exp(-1.1X) \qquad X \geq 0$$

b. What is the asymptote of this response function? For what value of X does the response function reach 95 percent of its asymptote?

14.5. **Home computers.** A computer manufacturer hired a market research firm to investigate the relationship between the likelihood a family will purchase a home computer and the price of the home computer. The data below are based on a survey of 1,000 heads of households who were asked if they are likely to purchase a home computer at a given price. Ten prices (X, in hundred dollars) were studied, and 100 heads randomly selected were assigned to a given price. The proportion likely to purchase at a given price is denoted by Y.

i:	1	2	3	4	5	6	7	8	9	10
X_i:	1	2.5	5	10	20	30	40	50	75	100
Y_i:	.95	.85	.58	.46	.31	.28	.19	.11	.06	.03

The following exponential model with independent normal error terms was deemed to be appropriate:

$$Y_i = \gamma_0 + \gamma_2 \exp(-\gamma_1 X_i) + \varepsilon_i$$

a. To obtain initial estimates of γ_0, γ_1, and γ_2, note that Y approaches a lower asymptote γ_0 as X increases without bound. Hence, let $g_0^{(0)} = 0$ and observe that when we ignore the error term a logarithmic transformation then yields $Y_i' = \beta_0 + \beta_1 X_i$, where $Y_i' = \log_e Y_i$, $\beta_0 = \log_e \gamma_2$, and $\beta_1 = -\gamma_1$. Therefore, fit a linear regression function based on the transformed data and use as initial estimates $g_0^{(0)} = 0$, $g_1^{(0)} = -b_1$, and $g_2^{(0)} = \exp(b_0)$.

b. Using the starting values obtained in part (a), find the least squares estimates of the parameters γ_0, γ_1, and γ_2.

c. Assume that the number of observations is reasonably large. Obtain approximate joint confidence intervals for the parameters γ_0, γ_1, and γ_2 using the Bonferroni procedure and a 90 percent family confidence coefficient.

14.6. Refer to **Home computers** Problem 14.5.

a. Plot the estimated nonlinear regression function and the data. Does the fit appear to be adequate?

b. Obtain the residuals and plot them against the fitted values and against X on separate graphs. Also obtain a normal probability plot. Does the model appear to be adequate?

14.7. **Enzyme kinetics.** In an enzyme kinetics study the velocity of a reaction (Y) is expected to be related to the concentration (X) as follows:

$$Y_i = \frac{\gamma_0 X_i}{\gamma_1 + X_i} + \varepsilon_i$$

Eleven concentrations have been studied and the results follow:

i:	1	2	3	4	5	6	7	8	9	10	11
X_i:	1	2	3	4	5	7.5	10	15	20	30	40
Y_i:	2.1	4.9	6.5	7.0	8.4	10.2	12.5	14.6	16.1	19.7	23.2

a. To obtain starting values for γ_0 and γ_1, observe that when the error term is ignored we have $Y_i' = \beta_0 + \beta_1 X_i'$, where $Y_i' = 1/Y_i$, $\beta_0 = 1/\gamma_0$, $\beta_1 = \gamma_1/\gamma_0$, and $X_i' = 1/X_i$. Therefore fit a linear regression function to the transformed data to obtain initial estimates $g_0^{(0)} = 1/b_0$ and $g_1^{(0)} = b_1/b_0$.

b. Using the starting values obtained in part (a), find the least squares estimates of the parameters γ_0 and γ_1.

c. Assume that the number of observations is reasonably large. (1) Obtain an approximate 95 percent confidence interval for γ_0. (2) Test whether or not $\gamma_1 = 20$; use $\alpha = .05$. State the alternatives, decision rule, and conclusion.

14.8. Refer to **Enzyme kinetics** Problem 14.7.

a. Plot the estimated nonlinear regression function and the data. Does the fit appear to be adequate?

b. Obtain the residuals and plot them against the fitted values and against X on separate graphs. Also obtain a normal probability plot. What do your plots show?

14.9. **Drug responsiveness.** A pharmacologist modeled the responsiveness to a drug using the following nonlinear regression model:

$$Y_i = \gamma_0 - \frac{\gamma_0}{1 + \left(\dfrac{X_i}{\gamma_2}\right)^{\gamma_1}} + \varepsilon_i$$

X denotes the dose level, in coded form, and Y the responsiveness expressed as a percent of the maximum possible responsiveness. In the model, γ_0 is the expected response at saturation, γ_2 is the concentration that produces a half maximal response, and γ_1 is related to the slope. The data for nine dose levels follow.

i:	1	2	3	4	5	6	7	8	9
X_i:	1	2	3	4	5	6	7	8	9
Y_i:	.5	2.3	3.4	24.0	54.7	82.1	94.8	96.2	96.4

a. Obtain least squares estimates of the parameters γ_0, γ_1, and γ_2 using starting values $g_0^{(0)} = 100$, $g_1^{(0)} = 5$, and $g_2^{(0)} = 4.8$.

b. Assume that the number of observations is reasonably large. Obtain approximate joint confidence intervals for the parameters γ_0, γ_1, and γ_2 using the Bonferroni procedure with a 91 percent family confidence coefficient. Interpret your results.

14.10. Refer to **Drug responsiveness** Problem 14.9.

a. Plot the estimated nonlinear regression function and the data. Does the fit appear to be adequate?

b. Obtain the residuals and plot them against the fitted values and against X on separate graphs. Also obtain a normal probability plot. What do your plots show about the adequacy of the regression model?

14.11. **Process yield.** The yield (Y) of a chemical process depends on the temperature (X_1) and pressure (X_2). The following nonlinear regression model is expected to be applicable:

$$Y_i = \gamma_0 X_{i1}^{\gamma_1} X_{i2}^{\gamma_2} + \varepsilon_i$$

Prior to beginning full-scale production, 18 tests were undertaken to study the process yield for various temperature and pressure combinations. The results follow.

i:	1	2	3	4	5	6	7	8	9
X_{i1}:	1	10	100	1	10	100	1	10	100
X_{i2}:	1	1	1	10	10	10	100	100	100
Y_i:	12	32	103	20	61	198	38	133	406

i:	10	11	12	13	14	15	16	17	18
X_{i1}:	1	10	100	1	10	100	1	10	100
X_{i2}:	1	1	1	10	10	10	100	100	100
Y_i:	8	38	98	14	56	205	43	128	398

a. To obtain starting values for γ_0, γ_1, and γ_2, note that when we ignore the random error term, a logarithmic transformation yields $Y_i' = \beta_0 + \beta_1 X_{i1}' + \beta_2 X_{i2}'$, where $Y_i' = \log_{10} Y_i$, $\beta_0 = \log_{10} \gamma_0$, $\beta_1 = \gamma_1$, $X_{i1}' = \log_{10} X_{i1}$, $\beta_2 = \gamma_2$, and $X_{i2}' = \log_{10} X_{i2}$. Fit an ordinary first-order multiple regression model to the transformed data and use as starting values $g_0^{(0)} = $ antilog$_{10} b_0$, $g_1^{(0)} = b_1$, and $g_2^{(0)} = b_2$.

b. Using the starting values obtained in part (a), find the least squares estimates of the parameters γ_0, γ_1, and γ_2.

14.12. Refer to **Process yield** Problem 14.11. Assume that the number of observations is reasonably large so that large-sample theory is applicable.

a. Test the hypotheses H_0: $\gamma_1 = \gamma_2$ against H_a: $\gamma_1 \neq \gamma_2$ using the .05 level of significance. State the alternatives, decision rule, and conclusion.

b. Obtain approximate joint confidence intervals for the parameters γ_1 and γ_2 using the Bonferroni procedure and a 95 percent family confidence coefficient.

c. What do you conclude about the parameters γ_1 and γ_2 based on the results in parts (a) and (b)?

14.13. Refer to **Process yield** Problem 14.11.

a. Plot the estimated nonlinear regression function and the data. Does the fit appear to be adequate?

b. Obtain the residuals and plot them against $\hat{Y}$, X_1, and X_2 on separate graphs. Also obtain a normal probability plot. What do your plots show about the adequacy of the model?

14.14. Refer to **Process yield** Problem 14.11. Conduct a formal approximate test for lack of fit of the nonlinear regression function. Use $\alpha = .05$ and assume that the number of observations is reasonably large. State the alternatives, decision rule, and conclusion.

EXERCISES

14.15. (Calculus needed.) Refer to **Home computers** Problem 14.5. Obtain the least squares normal equations and show that they are nonlinear in the estimated regression coefficients g_0, g_1, and g_2.

14.16. (Calculus needed.) Refer to **Enzyme kinetics** Problem 14.7. Obtain the least squares normal equations and show that they are nonlinear in the estimated regression coefficients g_0 and g_1.

14.17. (Calculus needed.) Refer to **Process yield** Problem 14.11. Obtain the least squares normal equations and show that they are nonlinear in the estimated regression coefficients g_0, g_1, and g_2.

14.18. Refer to **Drug responsiveness** Problem 14.9.

a. Assuming that $E(\varepsilon_i) = 0$, show that:

$$E(Y) = \gamma_0 \left(\frac{A}{1 + A} \right)$$

where:

$$A = \exp[\gamma_1 (\log_e X - \log_e \gamma_2)] = \exp(\beta_0 + \beta_1 X')$$

and $\beta_0 = -\gamma_1 \log_e \gamma_2$, $\beta_1 = \gamma_1$, and $X' = \log_e X$.

b. Assuming γ_0 is known, show that:

$$\frac{E(Y')}{1 - E(Y')} = \exp(\beta_0 + \beta_1 X')$$

where $Y' = Y/\gamma_0$.

c. What transformation do these results suggest for obtaining a simple linear regression function in the transformed variables?

d. How can starting values for finding the least squares estimates of the nonlinear regression parameters be obtained from the estimates of the linear regression coefficients?

PROJECTS

14.19. Refer to **Enzyme kinetics** Problem 14.7. Starting values for finding the least squares estimates of the parameters of the nonlinear regression model are to be obtained by a grid search. The following bounds for the two parameters have been specified:

$$5 \le \gamma_0 \le 65$$
$$5 \le \gamma_1 \le 65$$

Obtain 49 grid points by using all possible combinations of the boundary values and five other equally spaced points for each parameter range. Evaluate the least squares criterion (14.14) for each grid point and identify the point providing the best fit. Does this point give reasonable starting values here?

14.20. Refer to **Process yield** Problem 14.11. Starting values for finding the least squares estimates of the nonlinear regression model coefficients are to be obtained by a grid search. The following bounds for the parameters have been postulated:

$$1 \le \gamma_0 \le 21$$
$$.2 \le \gamma_1 \le .8$$
$$.1 \le \gamma_2 \le .7$$

Obtain 27 gridpoints by using all possible combinations of the boundary values and the midpoint for each of the parameter ranges. Evaluate the least squares criterion (14.14) for each gridpoint and identify the point providing the best fit. Does this point give reasonable starting values here?

CITED REFERENCES

14.1 Hartley, H. O. "The Modified Gauss-Newton Method for the Fitting of Non-linear Regression Functions by Least Squares." *Technometrics* 3 (1961), pp. 269–80.

14.2 Gallant, A. R. "Nonlinear Regression." *The American Statistician* 29 (1975), pp. 73–81.

14.3 Kennedy, W. J., Jr., and J. E. Gentle. *Statistical Computing.* New York: Marcel Dekker, 1980.

15

Normal correlation models

The purpose of this chapter is to indicate the relation between regression models and their uses, discussed in Chapters 2–14, and normal correlation models. We first take up bivariate normal correlation models, and then consider multivariate normal models.

15.1 DISTINCTION BETWEEN REGRESSION AND CORRELATION MODELS

As we know, the basic regression models taken up in this book assume that the independent variables $X_1, \ldots, X_{p-1}$ are fixed constants, and primary interest exists in making inferences about the dependent variable Y on the basis of the independent variables.

We saw in Chapter 3, for the case of a single independent variable, that the regression analysis for a normal error regression model is applicable even when X is a random variable, provided that the conditional distributions of Y follow certain specifications and the marginal distribution of X does not involve the regression model parameters β_0, β_1, and σ^2. Thus, in the case where X is a random variable, only the conditional distributions of Y were specified, and a restriction was placed on the marginal distribution of X. We did not, however, seek to completely specify the joint distribution of X and Y. While the discussion

491

in Chapter 3 dealt with only a single independent variable, all of the points apply to multiple regression models containing a number of independent variables which are random.

Correlation models, like regression models with random independent variables, consist of variables all of which are random. Correlation models differ from regression models by specifying the joint distribution of the variables completely. Furthermore, the variables in a correlation model play a symmetrical role, with no one variable automatically designated as the dependent variable. Correlation models are employed to study the nature of the relations between the variables, and also may be used for making inferences about any one of the variables on the basis of the others.

Thus, an analyst may use a correlation model for the two variables "height of person" and "weight of person" in a study of a sample of persons, each variable being taken as random. He or she might wish to study the relation between the two variables or might be interested in making inferences about weight of a person on the basis of the person's height, in making inferences about height on the basis of weight, or in both.

Other examples where a correlation model may be appropriate are:

1. To study the relations between service station sales of gasoline, auxiliary products, and repair services.
2. To study the relation between company net income determined by generally accepted accounting principles and net income according to tax regulations.
3. To study the relations between a person's blood pressure, body temperature, and weight.

The correlation model most widely employed is the normal correlation model. We discuss it now for the case of two variables.

15.2 BIVARIATE NORMAL DISTRIBUTION

The normal correlation model for the case of two variables is based on the *bivariate normal distribution*. Let us denote the two variables as Y_1 and Y_2. (We do not use the notation X and Y in this chapter because both variables play a symmetrical role in correlation analysis.) We say that Y_1 and Y_2 are *jointly normally distributed* if their joint probability distribution is the bivariate normal distribution.

Density function

The density function for the bivariate normal distribution is as follows:

$$(15.1) \quad f(Y_1, Y_2) = \frac{1}{2\pi\sigma_1\sigma_2\sqrt{1-\rho_{12}^2}} \exp\left\{ -\frac{1}{2(1-\rho_{12}^2)}\left[\left(\frac{Y_1-\mu_1}{\sigma_1}\right)^2 \right.\right.$$
$$\left.\left. -2\rho_{12}\left(\frac{Y_1-\mu_1}{\sigma_1}\right)\left(\frac{Y_2-\mu_2}{\sigma_2}\right) + \left(\frac{Y_2-\mu_2}{\sigma_2}\right)^2 \right] \right\}$$

Note that this density function involves five parameters: $\mu_1, \mu_2, \sigma_1, \sigma_2, \rho_{12}$. We shall explain the meaning of these parameters shortly. First, let us consider a graphic representation of the bivariate normal distribution.

Graphic representation

Figure 15.1 contains a graphic representation of a bivariate normal distribution. It is a surface in three-dimensional space. For every pair of (Y_1, Y_2) values, there is a density $f(Y_1, Y_2)$ represented by the height of the surface at that point. The surface is continuous, and probability corresponds to volume under the surface.

FIGURE 15.1 Example of bivariate normal distribution

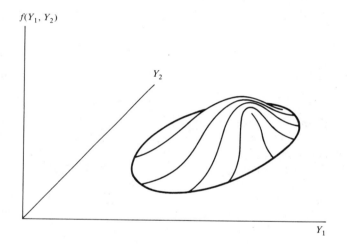

Marginal distributions

If Y_1 and Y_2 are jointly normally distributed, it can be shown that their marginal distributions have the following characteristics:

(15.2a)   The marginal distribution of Y_1 is normal with mean μ_1 and standard deviation σ_1:

$$f_1(Y_1) = \frac{1}{\sqrt{2\pi}\,\sigma_1} \exp\left[-\frac{1}{2}\left(\frac{Y_1 - \mu_1}{\sigma_1} \right)^2 \right]$$

(15.2b)   The marginal distribution of Y_2 is normal with mean μ_2 and standard deviation σ_2:

$$f_2(Y_2) = \frac{1}{\sqrt{2\pi}\,\sigma_2} \exp\left[-\frac{1}{2}\left(\frac{Y_2 - \mu_2}{\sigma_2} \right)^2 \right]$$

Thus, when Y_1 and Y_2 are jointly normally distributed, each of the two variables by itself is normally distributed. It is not generally true, however, that if Y_1 and Y_2 are each normally distributed, they must be jointly normally distributed in accord with (15.1).

Meaning of parameters

The five parameters of the bivariate normal density function (15.1) have the following meaning:

1. μ_1 and σ_1 are, respectively, the mean and standard deviation of the marginal distribution of Y_1.
2. μ_2 and σ_2 are, respectively, the mean and standard deviation of the marginal distribution of Y_2.
3. ρ_{12} is the *coefficient of correlation* between the random variables Y_1 and Y_2. It is defined as follows:

$$(15.3) \qquad \rho_{12} = \frac{\sigma_{12}}{\sigma_1 \sigma_2}$$

where σ_{12} is the covariance between Y_1 and Y_2, as defined in (1.19):

$$(15.4) \qquad \sigma_{12} = E[(Y_1 - \mu_1)(Y_2 - \mu_2)]$$

If Y_1 and Y_2 are independent, $\sigma_{12} = 0$ according to (1.23) so that $\rho_{12} = 0$ then. If Y_1 and Y_2 are positively related—i.e., Y_1 tends to be large when Y_2 is large, and small when Y_2 is small—σ_{12} is positive and so is ρ_{12}. On the other hand, if Y_1 and Y_2 are negatively related—i.e., Y_1 tends to be large when Y_2 is small, and vice versa—σ_{12} is negative and so is ρ_{12}. The coefficient of correlation ρ_{12} is a pure number, and can take on any value between -1 and $+1$ inclusive. It assumes $+1$ if Y_1 and Y_2 are perfectly positively related in a linear fashion, and -1 if the perfect linear relation is a negative one.

Contour representation

Bivariate normal distributions frequently are portrayed in terms of a contour diagram. A contour curve on such a diagram is composed of all the points on the surface that are equidistant from the $Y_1 Y_2$ plane. To put this another way, a contour curve is composed of all (Y_1, Y_2) outcomes which have constant density $f(Y_1, Y_2)$. Thus we can picture a contour as the cross section obtained by slicing a bivariate normal surface horizontally at a fixed distance above the $Y_1 Y_2$ plane, as in Figure 15.2.

Figure 15.3 presents a contour diagram for the bivariate normal surface of Figure 15.1. It is a property of the bivariate normal distribution that all contour curves are ellipses except when $\rho_{12} = 0$ and $\sigma_1 = \sigma_2$. Note that the ellipses have a common center at (μ_1, μ_2), and have common major and minor axes. Also note that the higher the horizontal cross section of the surface is above the $Y_1 Y_2$ plane, the smaller is the corresponding contour ellipse.

FIGURE 15.2 Contour ellipse for bivariate normal surface

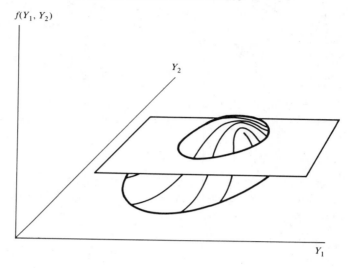

$f(Y_1, Y_2)$

Y_2

Y_1

FIGURE 15.3 Contour diagram for bivariate normal surface in Figure 15.1

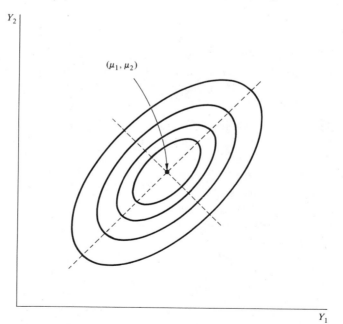

Y_2

(μ_1, μ_2)

Y_1

Figure 15.4 illustrates the effects of different parameter values on the location and shape of the bivariate normal surface. Note that when Y_1 and Y_2 are positively related so that $\rho_{12} > 0$, the principal axis has positive slope, implying that the surface tends to run along a line with positive slope. When Y_1 and Y_2 are negatively related so that $\rho_{12} < 0$, the principal axis has a negative slope, implying that the surface tends to run along a line with negative slope.

Figure 15.4 also demonstrates how the mean values μ_1 and μ_2 affect the location of the surface, and how the standard deviations σ_1 and σ_2, together with the correlation coefficient ρ_{12}, affect the shape of the surface.

15.3 CONDITIONAL INFERENCES

As noted, one principal use of bivariate correlation models is to make conditional inferences regarding one variable, given the other variable. Suppose Y_1 represents a service station's gasoline sales and Y_2 its sales of auxiliary products and services. We may then wish to predict a service station's sales of auxiliary products and services Y_2, given that its gasoline sales are $Y_1 = \$5,500$.

Such conditional inferences require the use of conditional probability distributions, which we discuss next.

Conditional probability distributions of Y_1

The conditional density function of Y_1 for any given value of Y_2 is denoted by $f(Y_1|Y_2)$ and defined as follows:

$$(15.5) \qquad f(Y_1|Y_2) = \frac{f(Y_1, Y_2)}{f_2(Y_2)}$$

where $f(Y_1, Y_2)$ is the joint density function of Y_1 and Y_2, and $f_2(Y_2)$ is the marginal density function of Y_2. When Y_1 and Y_2 are jointly normally distributed according to (15.1) so that the marginal density function $f_2(Y_2)$ is given by (15.2b), it can be shown that:

(15.6)   The conditional probability distribution of Y_1 for any given value of Y_2 is normal with mean $\alpha_{1.2} + \beta_{12}Y_2$ and standard deviation $\sigma_{1.2}$:

$$f(Y_1|Y_2) = \frac{1}{\sqrt{2\pi}\,\sigma_{1.2}} \exp\left[-\frac{1}{2}\left(\frac{Y_1 - \alpha_{1.2} - \beta_{12}Y_2}{\sigma_{1.2}} \right)^2 \right]$$

The parameters $\alpha_{1.2}$, β_{12}, and $\sigma_{1.2}$ of the conditional probability distributions of Y_1 are functions of the parameters of the joint probability distribution (15.1), as follows:

$$(15.7a) \qquad \alpha_{1.2} = \mu_1 - \mu_2 \rho_{12}\frac{\sigma_1}{\sigma_2}$$

$$(15.7b) \qquad \beta_{12} = \rho_{12}\frac{\sigma_1}{\sigma_2}$$

FIGURE 15.4 Effects of parameter values on location and shape of bivariate normal distribution

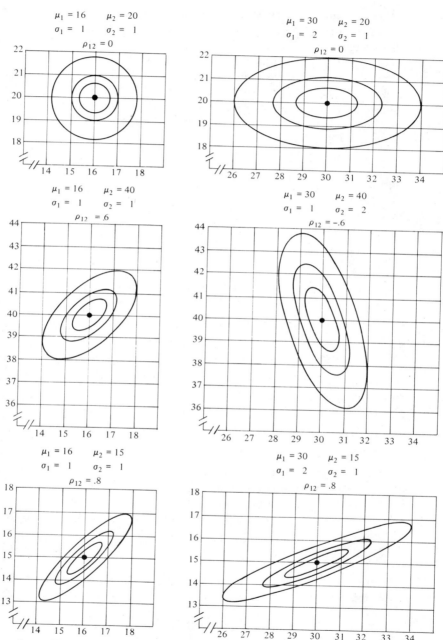

Source: Adapted, with permission, from Helen M. Walker and Joseph Lev, *Statistical Inference* (New York: Holt, Rinehart & Winston, 1953), p. 250.

(15.7c)
$$\sigma_{1.2}^2 = \sigma_1^2(1 - \rho_{12}^2)$$

Important characteristics of conditional distributions. Three important characteristics of the conditional probability distributions of Y_1 are normality, linear regression, and constant variance. We take up each of these in turn.

1. The conditional probability distribution of Y_1 for any given value of Y_2 is normal. Imagine that we slice a bivariate normal distribution vertically at a given value of Y_2, say, at Y_{h2}. That is, we slice it parallel to the Y_1 axis. This slicing is shown in Figure 15.5. The exposed cross section has the shape of a normal distribution, and after being scaled so that its area is 1, it portrays the conditional probability distribution of Y_1, given that $Y_2 = Y_{h2}$.

FIGURE 15.5 Cross section of bivariate normal distribution at Y_{h2}

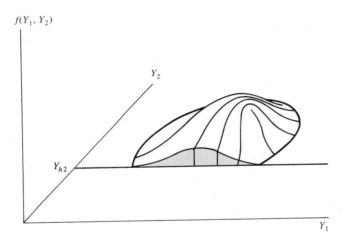

This property of normality holds no matter what the value Y_{h2} is. Thus, whenever we slice the bivariate normal distribution parallel to the Y_1 axis, we obtain (after proper scaling) a normal conditional probability distribution.

2. The means of the conditional probability distributions of Y_1 fall on a straight line, and hence are a linear function of Y_2:

(15.8)
$$E(Y_1|Y_2) = \alpha_{1.2} + \beta_{12}Y_2$$

Here $\alpha_{1.2}$ is the intercept parameter and β_{12} the slope parameter. Thus, the relation between the conditional means and Y_2 is given by a linear regression function.

3. All conditional probability distributions of Y_1 have the same standard deviation $\sigma_{1.2}$. Thus, no matter where we slice the bivariate normal distribution parallel to the Y_1 axis, the resulting conditional probability distribution (after

scaling to have an area of 1) has the same standard deviation. Hence, constant variances characterize the conditional probability distributions of Y_1.

Equivalence to normal error regression model. Suppose that we select a random sample of observations (Y_1, Y_2) from a bivariate normal population and wish to make conditional inferences about Y_1, given Y_2. The preceding discussion makes it clear that the normal error regression model (2.25) is entirely applicable because:

1. The Y_1 observations are independent.
2. The Y_1 observations when Y_2 is considered given or fixed are normally distributed with mean $E(Y_1|Y_2) = \alpha_{1.2} + \beta_{12}Y_2$ and constant variance $\sigma_{1.2}^2$.

Conditional probability distributions of Y_2

The random variables Y_1 and Y_2 play symmetrical roles in the bivariate normal probability distribution (15.1). Hence, it follows:

(15.9) The conditional probability distribution of Y_2 for any given value of Y_1 is normal with mean $\alpha_{2.1} + \beta_{21}Y_1$ and standard deviation $\sigma_{2.1}$:

$$f(Y_2|Y_1) = \frac{1}{\sqrt{2\pi}\,\sigma_{2.1}} \exp\left[-\frac{1}{2}\left(\frac{Y_2 - \alpha_{2.1} - \beta_{21}Y_1}{\sigma_{2.1}} \right)^2 \right]$$

The parameters $\alpha_{2.1}$, β_{21}, and $\sigma_{2.1}$ of the conditional probability distributions of Y_2 are functions of the parameters of the joint probability distribution (15.1), as follows:

(15.10a) $$\alpha_{2.1} = \mu_2 - \mu_1\rho_{12}\frac{\sigma_2}{\sigma_1}$$

(15.10b) $$\beta_{21} = \rho_{12}\frac{\sigma_2}{\sigma_1}$$

(15.10c) $$\sigma_{2.1}^2 = \sigma_2^2(1 - \rho_{12}^2)$$

The parameter $\alpha_{2.1}$ is the intercept of the line of regression of Y_2 on Y_1, and the parameter β_{21} is the slope of this line.

Again, we find that the conditional correlation model of Y_2 for given Y_1 is the equivalent of the normal error regression model (2.25).

Comments

1. The notation for the parameters of the conditional correlation models departs somewhat from our previous notation for regression models. The symbol α is now used to denote the regression intercept. The subscript 1.2 to α indicates that Y_1 is regressed on Y_2. Similarly, the subscript 2.1 to α indicates that Y_2 is regressed on Y_1. The symbol β_{12} indicates that it is the slope in the regression of Y_1 on Y_2, while β_{21} is the slope in the regression of Y_2 on Y_1. Finally, $\sigma_{2.1}$ is the standard deviation of the conditional probability

distributions of Y_2 for any given Y_1, while $\sigma_{1.2}$ is the standard deviation of the conditional probability distributions of Y_1 for any given Y_2. This notation can be extended straightforwardly for multivariate correlation models.

2. Two distinct regressions are involved in a bivariate normal model, that of Y_1 on Y_2 when Y_2 is fixed and that of Y_2 on Y_1 when Y_1 is fixed. In general, the two regression lines are not the same. For instance, the two slopes β_{12} and β_{21} are the same only if $\sigma_1 = \sigma_2$, as can be seen from (15.7b) and (15.10b).

3. Figure 15.6 illustrates the relation of the two regression lines to the contour ellipses. Note that both regression lines go through the point (μ_1, μ_2). If $\rho_{12} = 0$, the two regression lines intersect at right angles. The larger absolutely is ρ_{12}, the more the two regression lines come together.

FIGURE 15.6 Illustration of relation between lines of regression and contour ellipses

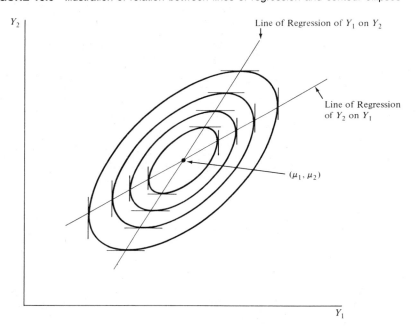

Use of regression analysis

In view of the equivalence of each of the conditional bivariate normal correlation models (15.6) and (15.9) with the normal error regression model (2.25), all conditional inferences with these correlation models can be made by means of the usual regression methods. Thus, if a researcher has data which can be appropriately described as having been generated from a bivariate normal distribution and wishes to make inferences about Y_2, given a particular value of Y_1, the ordinary regression techniques would be applicable. Thus, the regression equation of Y_2 on Y_1 would be estimated by means of (2.12), the slope of the regression equation would be estimated by means of the interval estimate (3.15), a new observation Y_2, given the value of Y_1, would be predicted by means of (3.35), and so on. Computer regression packages can be used in the usual manner. To avoid notational problems, it may be helpful to relabel the variables according to regression usage: $Y = Y_2$, $X = Y_1$. Of course, if conditional inferences on Y_1 for given values of Y_2 are desired, the notation correspondences would be: $Y = Y_1$, $X = Y_2$.

Note

When obtaining interval estimates for the conditional correlation models, the confidence coefficient refers to repeated samples where pairs of observations (Y_1, Y_2) are obtained from the bivariate normal population. We noted a similar point for regression models where the independent variables are random.

15.4 INFERENCES ON ρ_{12}

A principal use of correlation models is to study the relationships between the variables. In a bivariate normal model, the parameter ρ_{12} and its square, ρ_{12}^2, provide information about the degree of relationship between the two variables Y_1 and Y_2. Of the two measures, ρ_{12}^2 is the more meaningful one.

Coefficient of determination

The square of the coefficient of correlation ρ_{12} is called the *coefficient of determination*. We noted earlier in (15.7c) and (15.10c) that:

(15.11a) $$\sigma_{1.2}^2 = \sigma_1^2(1 - \rho_{12}^2)$$

(15.11b) $$\sigma_{2.1}^2 = \sigma_2^2(1 - \rho_{12}^2)$$

We can rewrite these expressions as follows:

(15.12a) $$\rho_{12}^2 = \frac{\sigma_1^2 - \sigma_{1.2}^2}{\sigma_1^2}$$

(15.12b) $$\rho_{12}^2 = \frac{\sigma_2^2 - \sigma_{2.1}^2}{\sigma_2^2}$$

The meaning of ρ_{12}^2 is now clear. Consider first (15.12a). ρ_{12}^2 measures how much smaller relatively is the variability in any conditional distribution of Y_1, for a given level of Y_2, than is the variability in the marginal distribution of Y_1. Thus, ρ_{12}^2 measures the relative reduction in the variability of Y_1 associated with the use of the variable Y_2. Correspondingly, (15.12b) shows that ρ_{12}^2 also measures the relative reduction in the variability of Y_2 associated with the use of the variable Y_1.

It can be shown that:

(15.13) $$0 \le \rho_{12}^2 \le 1$$

$\rho_{12}^2 = 0$ if Y_1 and Y_2 are independent, so that the variances of each variable in the conditional probability distributions are then no smaller than the variance in the marginal distribution. $\rho_{12}^2 = 1$ if there is no variability in the conditional probability distributions for each variable, so that perfect predictions of either variable can be made from the other.

Note

The interpretation of ρ_{12}^2 as measuring the relative reduction in the conditional variances as compared with the marginal variance is valid for the case of a bivariate normal population, but not for many other bivariate populations. Of course, the interpretation implies nothing in a causal sense.

Point estimators of ρ_{12} and ρ_{12}^2

The maximum likelihood estimator of ρ_{12}, denoted by r_{12}, is given by:

$$(15.14) \qquad r_{12} = \frac{\Sigma(Y_{i1} - \bar{Y}_1)(Y_{i2} - \bar{Y}_2)}{[\Sigma(Y_{i1} - \bar{Y}_1)^2]^{1/2}[\Sigma(Y_{i2} - \bar{Y}_2)^2]^{1/2}}$$

This estimator is biased (unless $\rho_{12} = 0$ or 1), but the bias is small if n is large.

The coefficient of determination ρ_{12}^2 is estimated by the square of the sample coefficient of correlation, r_{12}^2.

Note

The maximum likelihood estimator (15.14) of the population correlation coefficient is the same as the descriptive coefficient of correlation in (3.73) for the regression model. With a correlation model, r_{12} is an estimator of a parameter. With a regression model, in contrast, r_{12} is only a descriptive measure which reflects the proportion of the total sum of squares that is partitioned into the regression sum of squares. Another difference is that in the standard regression case where the X_i are fixed, the magnitude of r_{12} can be arbitrarily affected by the spacing pattern chosen for the X_i. For the bivariate normal model, on the other hand, the magnitude of r_{12} cannot be so affected since neither variable is under the control of the investigator.

Test whether $\rho_{12} = 0$

When the population is a bivariate normal one, it is frequently desired to test between:

$$(15.15) \qquad \begin{aligned} H_0 &: \rho_{12} = 0 \\ H_a &: \rho_{12} \neq 0 \end{aligned}$$

The reason for interest in this test is that in the case where Y_1 and Y_2 are jointly normally distributed, $\rho_{12} = 0$ implies that Y_1 and Y_2 are independent.

We can use regression procedures for the test since (15.7b) implies that the following alternatives are equivalent to those in (15.15):

$$(15.15a) \qquad \begin{aligned} H_0 &: \beta_{12} = 0 \\ H_a &: \beta_{12} \neq 0 \end{aligned}$$

and (15.10b) implies that the following alternatives are also equivalent to the ones in (15.15):

(15.15b)
$$H_0: \beta_{21} = 0$$
$$H_a: \beta_{21} \neq 0$$

It can be shown that the statistics for testing either (15.15a) or (15.15b) can be expressed directly in terms of r_{12}:

(15.16)
$$t^* = \frac{r_{12}\sqrt{n-2}}{\sqrt{1-r_{12}^2}}$$

If H_0 holds, t^* follows the $t(n-2)$ distribution. The appropriate decision rule to control the Type I error at α is:

(15.17)
$$\text{If } |t^*| \leq t(1 - \alpha/2; n - 2), \text{ conclude } H_0$$
$$\text{If } |t^*| > t(1 - \alpha/2; n - 2), \text{ conclude } H_a$$

Test statistic (15.16) is identical to the regression t^* test statistic (3.17).

Interval estimation of ρ_{12}

Because the sampling distribution of r_{12} is complicated when $\rho_{12} \neq 0$, interval estimation of ρ_{12} is usually done by means of a transformation.

z' **transformation.** This transformation, due to R. A. Fisher, is as follows:

(15.18)
$$z' = \frac{1}{2}\log_e\left(\frac{1 + r_{12}}{1 - r_{12}}\right)$$

When n is large (25 or more is a useful rule of thumb), the distribution of z' is approximately normal with mean and variance:

(15.19)
$$E(z') = \zeta = \frac{1}{2}\log_e\left(\frac{1 + \rho_{12}}{1 - \rho_{12}}\right)$$

(15.20)
$$\sigma^2(z') = \frac{1}{n-3}$$

Note that the transformation from r_{12} to z' in (15.18) is the same as the relation in (15.19) between ρ_{12} and $E(z') = \zeta$. Also note that the variance of z' is a known constant, depending only on the sample size n.

Table A–7 gives paired values for the left and right sides of (15.18) and (15.19), thus eliminating the need for calculations. For instance, if r_{12} or ρ_{12} equals .25, Table A–7 indicates that z' or ζ equals .2554, and vice versa. The values on the two sides of the transformation always have the same sign. Thus, if r_{12} or ρ_{12} is negative, a minus sign is attached to the value in Table A–7. For instance, if $r_{12} = -.25$, $z' = -.2554$.

Interval estimate. Since z' is approximately normally distributed for large n, it follows that the standardized statistic:

(15.21)
$$\frac{z' - \zeta}{\sigma(z')}$$

is approximately a standard normal variable when n is large. Therefore, the $1 - \alpha$ confidence limits for ζ are:

(15.22)
$$z' \pm z(1 - \alpha/2)\sigma(z')$$

where $z(1 - \alpha/2)$ is the $(1 - \alpha/2)100$ percentile of the standard normal distribution. These percentiles are given in Table A-1. The $1 - \alpha$ confidence limits for ρ_{12} are then obtained by transforming the limits on ζ by means of (15.19).

Comments

1. As usual, a confidence interval for ρ_{12} can be employed to test whether or not ρ_{12} has a specified value—say, .5—by noting whether or not the specified value falls within the confidence limits.

2. Confidence limits for ρ_{12}^2 can be obtained by squaring the respective confidence limits for ρ_{12}.

Example

An economist investigated food purchasing patterns by households in a midwestern city. He selected 200 households with family incomes between $7,500 and $17,500, and ascertained from each household, among other things, the proportions of the food budget expended for beef and poultry, respectively. He expected these to be negatively related, and wished to estimate the coefficient of correlation with a 95 percent confidence interval. The economist had some supporting evidence which suggested that the joint distribution of the two variables does not depart markedly from a bivariate normal one.

The point estimate of ρ_{12} was $r_{12} = -.61$ (data and calculations not shown). To obtain a 95 percent confidence interval estimate, we require:

$$z' = -.7089 \text{ when } r_{12} = -.61 \qquad \text{(from Table A-7)}$$

$$\sigma(z') = \frac{1}{\sqrt{200 - 3}} = .07125$$

$$z(.975) = 1.96$$

Hence, the confidence interval for ζ, by (15.22), is:

$$-.849 = -.7089 - 1.96(.07125) \leq \zeta \leq -.7089 + 1.96(.07125) = -.569$$

Using Table A-7 to transform back to ρ_{12}, we obtain:

$$-.69 \leq \rho_{12} \leq -.51$$

This confidence interval was sufficiently precise to be useful to the economist, confirming the negative relation and indicating that the degree of linear association is moderately high.

Caution

Correlation models, like regression models, do not express any causal relations. Earlier cautions about drawing conclusions as to causality from regression findings apply equally to correlation studies. Correlation findings can be useful in analyzing causal relationships, but they do not by themselves establish causal patterns.

15.5 MULTIVARIATE NORMAL DISTRIBUTION

The normal correlation model for the case of p variables $Y_1, \ldots, Y_p$ is based on the *multivariate normal distribution*. This distribution is an extension of the bivariate normal distribution, and has corresponding properties. In particular, if $Y_1, \ldots, Y_p$ are jointly normally distributed (i.e., they follow the multivariate normal distribution), the marginal probability distribution of each variable Y_k is normal, with mean μ_k and standard deviation σ_k.

Conditional inferences

One major use of multivariate correlation models is to make conditional inferences on one variable when the other variables have given values. It can be shown in the case of the multivariate normal distribution that:

(15.23)   The conditional probability distribution of Y_k for any given set of values for the other variables is normal with mean given by a linear regression function and constant variance.

Suppose Y_1, Y_2, Y_3, and Y_4 are jointly normally distributed. The conditional probability distribution of, say, Y_3, when Y_1, Y_2, and Y_4 are fixed at any specified levels, has the characteristics:

1. It is normal.
2. It has a mean given by:

$$E(Y_3 | Y_1, Y_2, Y_4) = \alpha_{3.124} + \beta_{31.24}Y_1 + \beta_{32.14}Y_2 + \beta_{34.12}Y_4$$

3. It has constant variance $\sigma^2_{3.124}$.

The notation is a straightforward extension of that for the bivariate correlation case. Thus, $\beta_{31.24}$ denotes the regression coefficient of Y_1 when Y_3 is regressed on Y_1, Y_2, and Y_4.

It is clear from the above that the conditional multivariate correlation models are equivalent to the normal error multiple regression model (7.7). Hence, in the case where the variables are jointly normally distributed, all inferences on one variable conditional on the other variables being fixed are carried out by the usual multiple regression techniques. For instance, an interval estimate of a conditional mean would be obtained by (7.51), or a prediction of a new observation on the variable of interest, given the values of the other variables, would be obtained by (7.55). To facilitate use of the regression formulas, it may be helpful to relabel the variable of interest Y and the other variables X's. Computer multiple regression packages can be used in ordinary fashion.

Coefficients of multiple correlation and determination

A major use of multivariate correlation models is to study the relationships between the variables. One set of measures useful to this end consists of the *coefficients of multiple determination* and the *coefficients of multiple correlation*.

Meaning of coefficients. A coefficient of multiple determination is associated with each variable. Suppose that Y_1, Y_2, Y_3, and Y_4 are included in the correlation model. The coefficient of multiple determination associated with, say, Y_1 is denoted by $\rho_{1.234}^2$, and defined as follows:

$$(15.24) \qquad \rho_{1.234}^2 = \frac{\sigma_1^2 - \sigma_{1.234}^2}{\sigma_1^2}$$

where $\sigma_{1.234}^2$ is the variance of the conditional distributions of Y_1 when the other variables are fixed. Thus, $\rho_{1.234}^2$ measures how much smaller, relatively, is the variability in the conditional distributions of Y_1, when the other variables are fixed at given values, than is the variability in the marginal distribution of Y_1. The other coefficients of multiple determination are defined and interpreted in similar fashion.

It can be shown that coefficients of multiple determination take on values between 0 and 1 inclusive. Thus:

$$(15.25) \qquad 0 \le \rho_{1.234}^2 \le 1$$

Let us consider the significance of the limiting values. If $\rho_{1.234}^2 = 0$, no reduction in the variability of Y_1 takes place by considering the other variables. Equivalently, $\rho_{1.234}^2 = 0$ in the case of the normal correlation model implies that Y_1 is independent of Y_2, Y_3, and Y_4 so that $\rho_{12} = \rho_{13} = \rho_{14} = 0$.

At the other extreme, if $\rho_{1.234}^2 = 1$, the conditional distributions of Y_1, given Y_2, Y_3, and Y_4, have no variability so that perfect predictions can be made of Y_1 from knowledge of the other variables.

The positive square root of a coefficient of multiple determination is the coefficient of multiple correlation. Thus, for Y_1, we have in our example:

$$(15.26) \qquad \rho_{1.234} = \sqrt{\rho_{1.234}^2}$$

$\rho_{1.234}$ can be viewed as a simple correlation coefficient, namely, between Y_1 and $\alpha_{1.234} + \beta_{12.34} Y_2 + \beta_{13.24} Y_3 + \beta_{14.23} Y_4$.

Estimation of coefficients. Coefficients of multiple determination and correlation are estimated by (7.31) and (7.34), respectively, the descriptive coefficients in the regression case. Thus, to estimate $\rho_{1.234}^2$, we need the total sum of squares for Y_1, denoted by $SSTO(Y_1)$. Then we require the regression sum of squares when Y_1 is regressed on Y_2, Y_3, and Y_4, denoted by $SSR(Y_2, Y_3, Y_4)$.

The estimator, denoted by $R^2_{1.234}$, is:

$$(15.27) \qquad R^2_{1.234} = \frac{SSR(Y_2, Y_3, Y_4)}{SSTO(Y_1)}$$

The positive square root of $R^2_{1.234}$, denoted by $R_{1.234}$, is the estimated coefficient of multiple correlation for Y_1.

Testing of coefficients. To test, say:

$$(15.28) \qquad \begin{array}{l} H_0: \rho_{1.234} = 0 \\ H_a: \rho_{1.234} \neq 0 \end{array}$$

we can utilize the equivalent test on the regression coefficients:

$$(15.28a) \qquad \begin{array}{l} H_0: \beta_{12.34} = \beta_{13.24} = \beta_{14.23} = 0 \\ H_a: \text{not all regression coefficients are zero} \end{array}$$

It turns out that the F^* statistic for this test, given in (7.30b), can be expressed directly in terms of $R^2_{1.234}$, as follows:

$$(15.29) \qquad F^* = \frac{R^2_{1.234}}{1 - R^2_{1.234}} \frac{n - 4}{3}$$

In general, the factor on the right is $(n - q - 1)/q$ when there are q predictor variables.

If H_0 holds, F^* follows the $F(q, n - q - 1)$ distribution so that the decision rule to control the Type I error risk at α is set up in the usual fashion:

$$(15.30) \qquad \begin{array}{l} \text{If } F^* \leq F(1 - \alpha; q, n - q - 1), \text{ conclude } H_0 \\ \text{If } F^* > F(1 - \alpha; q, n - q - 1), \text{ conclude } H_a \end{array}$$

Coefficients of partial correlation and determination

Meaning of coefficients. Suppose again that four variables Y_1, Y_2, Y_3, and Y_4 are included in a multivariate normal correlation model. Consider now the correlation betweeen Y_1 and Y_2 in the conditional joint distribution when each of the variables Y_3 and Y_4 is fixed at a given level. When all variables are jointly normally distributed, this correlation does not depend on the levels where Y_3 and Y_4 are fixed and is given by:

$$(15.31) \qquad \rho_{12.34} = \frac{\sigma_{12.34}}{\sigma_{1.34}\sigma_{2.34}}$$

$\rho_{12.34}$ is called the *coefficient of partial correlation* between Y_1 and Y_2 when Y_3 and Y_4 are fixed.

The square of $\rho_{12.34}$ is called the *coefficient of partial determination* and is denoted by $\rho^2_{12.34}$. The following relation holds:

$$(15.32) \qquad \rho^2_{12.34} = \frac{\sigma^2_{1.34} - \sigma^2_{1.234}}{\sigma^2_{1.34}}$$

Thus, $\rho_{12.34}^2$ measures how much smaller, relatively, is the variability in the conditional distributions of Y_1, given Y_2, Y_3, and Y_4, than it is in the conditional distributions of Y_1, given Y_3 and Y_4 only.

The other partial correlation coefficients are defined and interpreted in similar fashion. For instance, $\rho_{12.3}$ measures the correlation between Y_1 and Y_2 when only Y_3 is fixed.

A coefficient of partial correlation $\rho_{12.3}$ is called a *first-order* coefficient, a coefficient $\rho_{12.34}$ a *second-order* coefficient, and so on. All partial correlation coefficients measure the correlation between two variables; the order of the coefficient simply indicates how many other variables are fixed in the conditional bivariate distribution.

Point estimation of coefficients. Point estimators of the coefficients of partial determination and correlation are the descriptive regression measures encountered earlier; see, for instance, (8.16). Thus, to estimate $\rho_{12.3}^2$, we regress Y_1 on Y_3 and obtain the error sum of squares $SSE(Y_3)$. Next, we regress Y_1 on Y_2 and Y_3 and obtain the error sum of squares $SSE(Y_2, Y_3)$. The estimator of $\rho_{12.3}^2$ then is:

$$(15.33) \qquad r_{12.3}^2 = \frac{SSE(Y_3) - SSE(Y_2, Y_3)}{SSE(Y_3)} = \frac{SSR(Y_2 \mid Y_3)}{SSE(Y_3)}$$

Tests concerning coefficients. Researchers often wish to test whether two variables are correlated in the conditional probability distributions when other variables are fixed. Thus, when four variables Y_1, Y_2, Y_3, and Y_4 are being considered and the correlation between Y_1 and Y_2, when Y_3 and Y_4 are fixed, is of interest, the following test may be desired to see if Y_1 and Y_2 are correlated in the conditional joint distributions:

$$(15.34) \qquad \begin{aligned} H_0&: \rho_{12.34} = 0 \\ H_a&: \rho_{12.34} \neq 0 \end{aligned}$$

Such tests concerning partial correlation coefficients can be carried out via the test statistic (15.16) for the bivariate case, with r_{12} replaced by the estimated partial correlation coefficient and n replaced by $n - q$, where q is the number of variables that are held fixed.

Interval estimation of coefficients. Interval estimates of partial correlation coefficients are obtained via the z' transformation (15.18) in identical fashion to that for simple correlation coefficients. Only the standard deviation $\sigma(z')$ need be modified. It is, for q variables being held fixed:

$$(15.35) \qquad \sigma(z') = \frac{1}{\sqrt{n - q - 3}}$$

Example

An operations analyst, wishing to study the relations between three types of test scores made by applicants for entry-level clerical positions in a large insurance company, drew a sample of 250 such applicants from recent records and ascertained their scores. The variables were:

Y_1 verbal aptitude test score
Y_2 reading aptitude test score
Y_3 personal interview score

The multivariate normal model was considered to be applicable for the study.

The analyst's initial interest was in the partial correlation coefficient $\rho_{23.1}$. Since her computer program did not have an option for calculating partial correlation coefficients by a single command (many programs have such an option), the analyst ran two separate regressions, Y_2 on Y_1 and Y_2 on Y_1 and Y_3. She obtained in turn $SSE(Y_1) = 6,340$ and $SSE(Y_1, Y_3) = 6,086$. The point estimate of $\rho_{23.1}^2$ then was calculated corresponding to (15.33):

$$r_{23.1}^2 = \frac{SSE(Y_1) - SSE(Y_1, Y_3)}{SSE(Y_1)} = \frac{6,340 - 6,086}{6,340} = .040$$

Hence $r_{23.1} = .20$. (The sign of $r_{23.1}$ here is positive because the regression coefficient for Y_3 when Y_2 is regressed on Y_1 and Y_3 is positive.) Desiring a confidence interval with a 95 percent confidence coefficient, the analyst required:

$$z' = .2027 \qquad \text{when } r_{23.1} = .20$$

$$\sigma(z') = \frac{1}{\sqrt{250 - 1 - 3}} = \frac{1}{\sqrt{246}} = .06376$$

$$z(.975) = 1.96$$

The confidence interval for ζ by (15.22) is:

$$.0777 = .2027 - 1.96(.06376) \leq \zeta \leq .2027 + 1.96(.06376) = .3277$$

In transforming from ζ to ρ, the analyst obtained:

$$.08 \leq \rho_{23.1} \leq .32$$

Thus, the coefficient of partial correlation between reading aptitude score and interview score, with verbal aptitude score fixed, was at best relatively low and could, indeed, be close to zero.

Next the analyst ascertained from a printout of the regression of Y_2 on Y_3 that $r_{23} = .83$. For a 95 percent confidence interval, she then obtained:

$$.79 \leq \rho_{23} \leq .86$$

Thus, ρ_{23} turned out to be substantially larger than $\rho_{23.1}$.

The comparative magnitudes of $\rho_{23.1}$ and ρ_{23} suggested to the analyst (and further investigation verified) that in the company's interviewing procedure the interview scores (Y_3) depend in good part on verbal skills. Also, the reading aptitude scores (Y_2) obtained with the test used by the company tend to be heavily influenced by verbal aptitude. Thus, when verbal aptitude is not considered, the degree of relation between Y_2 and Y_3 is relatively high since applicants with relatively good (poor) verbal aptitude tend to score well (poorly) in both the reading aptitude test and the personal interview. However, among applicants at any given level of verbal aptitude score the degree of relationship between reading score and interview score tends to be low.

PROBLEMS

15.1. A management trainee in a production department wished to study the relation between weight of rough casting and machining time to produce the finished block. He selected castings so that the weights would be spaced equally apart in the sample and observed the corresponding machining times. Would you recommend that a regression or a correlation model be used? Explain.

15.2. A social scientist stated: "The conditions for the bivariate and multivariate normal distributions are so rarely met in my experience that I feel much safer using a regression model." Comment.

15.3. A student was investigating from a large sample whether variables Y_1 and Y_2 follow a bivariate normal distribution. She obtained the residuals when regressing Y_1 on Y_2, and also obtained the residuals when regressing Y_2 on Y_1. She then prepared a normal probability plot for each set of residuals. Do these two normal probability plots provide sufficient information for finding whether the two variables follow a bivariate normal distribution? Explain.

15.4. Refer to Figures 15.1 and 15.3. Where in the $Y_1 Y_2$ plane is the height of the bivariate normal surface greatest? How can this point be ascertained from inspection of the density function (15.1)?

15.5. Plot a contour diagram for the bivariate normal distribution with parameters $\mu_1 = 50$, $\mu_2 = 100$, $\sigma_1 = 3$, $\sigma_2 = 4$, and $\rho_{12} = .80$.

15.6. Refer to Problem 15.5.
   a. State the characteristics of the marginal distribution of Y_1.
   b. State the characteristics of the conditional distribution of Y_2 when $Y_1 = 55$.
   c. State the characteristics of the conditional distribution of Y_1 when $Y_2 = 95$.

15.7. Refer to Problem 15.5.
   a. Give the two regression lines for this model.
   b. Why are there two regression lines for a bivariate normal distribution and not just one? What is the meaning of each?
   c. Must β_{12} and β_{21} have the same sign?

15.8. Refer to Figure 15.6. For any specified value Y_{h2}, can you identify where the conditional density $f(Y_1 | Y_{h2})$ is maximized? How can this result be ascertained also from inspection of the conditional density function (15.6)?

15.9. a. Plot a contour diagram for the bivariate normal distribution with parameters $\mu_1 = 14$, $\mu_2 = 350$, $\sigma_1 = 2$, $\sigma_2 = 25$, and $\rho_{12} = .90$.

b. Give the two regression lines for this model. Why are there two regression lines and not just one?

15.10. Refer to Figure 15.6. If the two regression lines were not labeled, could you determine by inspection which line pertains to the regression of Y_2 on Y_1 and which to the regression of Y_1 on Y_2? Explain.

15.11. Explain whether any of the following would be affected if the bivariate normal model (15.1) were employed instead of the normal error regression model (3.1) with fixed levels of the independent variable: (1) point estimates of the regression coefficients, (2) confidence limits for the regression coefficients, (3) interpretation of the confidence coefficient.

15.12. Refer to **Plastic hardness** Problem 2.18. A student was analyzing these data and received the following standard query from the interactive regression and correlation computer package: CALCULATE CONFIDENCE INTERVAL FOR POPULATION CORRELATION COEFFICIENT RHO? ANSWER Y OR N. Would a "yes" response provide meaningful information here? Explain.

15.13. **Property assessments.** The observations that follow show assessed value for property tax purposes (Y_1, in thousand dollars) and sales price (Y_2, in thousand dollars) for a sample of 15 parcels of land for industrial development sold recently in "arm's-length" transactions in a tax district. Assume that bivariate normal model (15.1) is appropriate here.

i:	1	2	3	4	5	6	7	8
Y_{i1}:	13.9	16.0	10.3	11.8	16.7	12.5	10.0	11.4
Y_{i2}:	28.6	34.7	21.0	25.5	36.8	24.0	19.1	22.5

i:	9	10	11	12	13	14	15
Y_{i1}:	13.9	12.2	15.4	14.8	14.9	12.9	15.8
Y_{i2}:	28.3	25.0	31.1	29.6	35.1	30.0	36.2

a. Plot the observations in a scatter diagram. Does the bivariate normal model appear to be appropriate here? Discuss.

b. Calculate r_{12}. What parameter is estimated by r_{12}? What is the interpretation of this parameter?

c. Test whether or not Y_1 and Y_2 are statistically independent in the population, using test statistic (15.16) and level of significance .01. State the alternatives, decision rule, and conclusion.

d. To test $\rho_{12} = .6$ versus $\rho_{12} \neq .6$, would it be appropriate to use test statistic (15.16)?

15.14. **Contract profitability.** A cost analyst for a drilling and blasting contractor examined 84 contracts handled in the last two years and found that the coefficient of correlation between value of contract (Y_1) and profit contribution generated by the contract (Y_2) is $r_{12} = .61$. Assume that the bivariate normal model (15.1) applies.

a. Test whether or not Y_1 and Y_2 are statistically independent in the population; use $\alpha = .05$. State the alternatives, decision rule, and conclusion.

b. Estimate ρ_{12} with a 95 percent confidence interval.

c. Convert the confidence interval in part (b) to a 95 percent confidence interval for ρ_{12}^2. Interpret this interval estimate.

15.15. Bid preparation. A building construction consultant studied the relationship between cost of bid preparation (Y_1) and amount of bid (Y_2) for the consulting firm's clients. In a sample of 103 bids prepared by clients, $r_{12} = .87$. Assume that the bivariate normal model (15.1) applies.

a. Test whether or not $\rho_{12} = 0$; control the risk of Type I error at .10. State the alternatives, decision rule, and conclusion. What would be the implication if $\rho_{12} = 0$?

b. Obtain a 90 percent confidence interval for ρ_{12}. Interpret this interval estimate.

c. Convert the confidence interval in part (b) to a 90 percent confidence interval for ρ_{12}^2.

15.16. Water flow. An engineer, desiring to estimate the coefficient of correlation ρ_{12} between rate of water flow at point A in a stream (Y_1) and concurrent rate of flow at point B (Y_2), obtained $r_{12} = .83$ in a sample of 147 observations. Assume that the bivariate normal model (15.1) is appropriate.

a. Obtain a 99 percent confidence interval for ρ_{12}.

b. Convert the confidence interval in part (a) to a 99 percent confidence interval for ρ_{12}^2.

15.17. Pharmaceutical study. A marketing research analyst in a pharmaceutical firm wishes to study the relation between the following quantified variables in the target population:

Y_1 socioeconomic status of homemaker
Y_2 magazine media exposure to homemaker
Y_3 awareness of homemaker to firm's new product

She believes it is reasonable to assume that the variables follow a multivariate normal distribution.

a. Explain what is measured by each of the following: (1) $\sigma_{3.12}^2$, (2) $\sigma_{31.2}$, (3) $\sigma_{1.3}^2$.

b. Explain the meaning of each of the following: (1) $\rho_{2.13}^2$, (2) $\rho_{2.3}^2$, (3) $\rho_{12.3}^2$.

15.18. Household survey. In a random sample of 250 households, the coefficient of mulitiple correlation between purchases of soft drinks (Y_1) and purchases of snack foods (Y_2) and purchases of dairy products (Y_3) was $R_{1.23} = .892$. The coefficient of partial correlation $r_{13.2}$ was found to be $-.413$.

a. Test whether or not $\rho_{1.23} = 0$; use $\alpha = .05$. State the alternatives, decision rule, and conclusion.

b. Test whether or not $\rho_{13.2} = 0$; use $\alpha = .05$. State the alternatives, decision rule, and conclusion.

c. Estimate $\rho_{13.2}$ with a 95 percent confidence interval. Convert this confidence interval into a 95 percent confidence interval for $\rho_{13.2}^2$. Interpret this latter interval estimate.

15.19. Microcomputer assembly. A consultant studied the relation between visual acuity (Y_1), manual dexterity (Y_2), and tactile reflex (Y_3) in persons who had proven to be proficient as assemblers of microcomputer components. Observations on these variables were obtained for a random sample of 30 persons. Sums of squares for regressions run by the consultant follow. Assume that the multivariate normal correlation model applies.

Regression

Sum of Squares	Y_1 on Y_2	Y_1 on Y_3	Y_1 on Y_2, Y_3	Y_2 on Y_1	Y_2 on Y_3
Regression	7,436	66	7,439	6,900	42
Error	392	7,762	389	363	7,221
Total	7,828	7,828	7,828	7,263	7,263

Regression

Sum of Squares	Y_2 on Y_1, Y_3	Y_3 on Y_1	Y_3 on Y_2	Y_3 on Y_1, Y_2
Regression	6,901	46	32	64
Error	362	5,327	5,341	5,309
Total	7,263	5,373	5,373	5,373

   a.   What does $\rho_{1.23}^2$ denote here? Estimate this parameter by a point estimate. What information is provided by this estimate?

   b.   Test whether or not $\rho_{1.23} = 0$ using a level of significance of .01. State the alternatives, decision rule, and conclusion.

   c.   Test whether or not $\rho_{12.3} = 0$; use $\alpha = .01$. State the alternatives, decision rule, and conclusion.

   d.   Estimate $\rho_{12.3}$ with a 99 percent confidence interval. Interpret your interval estimate.

15.20. Refer to **Microcomputer assembly** Problem 15.19. Obtain r_{12}, r_{13}, and r_{23}. Also obtain $r_{12.3}$, $r_{13.2}$, and $r_{23.1}$. Summarize the information provided by these coefficients.

EXERCISES

15.21. The random variables Y_1 and Y_2 follow the bivariate normal distribution (15.1). Show that if $\rho_{12} = 0$, Y_1 and Y_2 are independent random variables.

15.22. (Calculus needed.)

   a.   Obtain the maximum likelihood estimators of the parameters of the bivariate normal distribution (15.1).

   b.   Using the results in part (a), obtain the maximum likelihood estimators of the parameters of the conditional probability distribution of Y_1 for any value of Y_2, as given in (15.7).

   c.   Show that the maximum likelihood estimators of $\alpha_{1.2}$ and β_{12} obtained in part (b) are the same as the least squares estimators (2.10) for the regression coefficients in the simple linear regression model.

15.23. Show that test statistics (3.17) and (15.16) are equivalent.

15.24. Show that test statistic (7.30b) for $p = 4$ is equivalent to test statistic (15.29).

15.25. Show that the ratio $SSR/SSTO$ is the same whether Y_1 is regressed on Y_2 or Y_2 is regressed on Y_1. [*Hint*: Use (3.50) and (2.10a).]

PROJECT

15.26. Refer to **Pharmaceutical study** Problem 15.17. The following observations were obtained in a pilot study involving a random sample of 42 homemakers:

Home-maker i	Y_{i1}	Y_{i2}	Y_{i3}	Home-maker i	Y_{i1}	Y_{i2}	Y_{i3}
1	76	81	53	22	57	51	32
2	92	89	78	23	49	52	28
3	85	84	66	24	71	72	54
4	63	70	46	25	70	66	55
5	50	48	38	26	91	86	67
6	89	93	68	27	95	91	78
7	35	29	20	28	81	79	60
8	67	68	41	29	40	38	15
9	94	98	77	30	41	43	18
10	83	79	61	31	88	91	72
11	24	28	12	32	63	63	40
12	31	37	14	33	59	54	34
13	42	36	17	34	24	26	11
14	71	71	53	35	96	91	79
15	87	89	71	36	74	80	52
16	28	23	3	37	30	38	14
17	36	36	12	38	81	85	65
18	61	58	35	39	96	92	78
19	74	70	47	40	38	40	16
20	66	60	40	41	52	46	35
21	81	86	66	42	76	68	42

a. The analyst first wished to test whether or not $\rho_{3.12} = 0$. Perform the test using a level of significance of .01. State the alternatives, decision rule, and conclusion. What is the implication for the analyst if $\rho_{3.12} = 0$?

b. Estimate each coefficient of simple correlation and each coefficient of partial correlation with a separate 99 percent confidence interval. What is the family confidence coefficient for the entire set of estimates?

c. Analyze the results obtained in part (b) and summarize your findings.

APPENDIX TABLES

TABLE A–1 Cumulative probabilities of the standard normal distribution

Entry is area A under the standard normal curve from $-\infty$ to $z(A)$

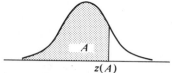

z	.00	.01	.02	.03	.04	.05	.06	.07	.08	.09
.0	.5000	.5040	.5080	.5120	.5160	.5199	.5239	.5279	.5319	.5359
.1	.5398	.5438	.5478	.5517	.5557	.5596	.5636	.5675	.5714	.5753
.2	.5793	.5832	.5871	.5910	.5948	.5987	.6026	.6064	.6103	.6141
.3	.6179	.6217	.6255	.6293	.6331	.6368	.6406	.6443	.6480	.6517
.4	.6554	.6591	.6628	.6664	.6700	.6736	.6772	.6808	.6844	.6879
.5	.6915	.6950	.6985	.7019	.7054	.7088	.7123	.7157	.7190	.7224
.6	.7257	.7291	.7324	.7357	.7389	.7422	.7454	.7486	.7517	.7549
.7	.7580	.7611	.7642	.7673	.7704	.7734	.7764	.7794	.7823	.7852
.8	.7881	.7910	.7939	.7967	.7995	.8023	.8051	.8078	.8106	.8133
.9	.8159	.8186	.8212	.8238	.8264	.8289	.8315	.8340	.8365	.8389
1.0	.8413	.8438	.8461	.8485	.8508	.8531	.8554	.8577	.8599	.8621
1.1	.8643	.8665	.8686	.8708	.8729	.8749	.8770	.8790	.8810	.8830
1.2	.8849	.8869	.8888	.8907	.8925	.8944	.8962	.8980	.8997	.9015
1.3	.9032	.9049	.9066	.9082	.9099	.9115	.9131	.9147	.9162	.9177
1.4	.9192	.9207	.9222	.9236	.9251	.9265	.9279	.9292	.9306	.9319
1.5	.9332	.9345	.9357	.9370	.9382	.9394	.9406	.9418	.9429	.9441
1.6	.9452	.9463	.9474	.9484	.9495	.9505	.9515	.9525	.9535	.9545
1.7	.9554	.9564	.9573	.9582	.9591	.9599	.9608	.9616	.9625	.9633
1.8	.9641	.9649	.9656	.9664	.9671	.9678	.9686	.9693	.9699	.9706
1.9	.9713	.9719	.9726	.9732	.9738	.9744	.9750	.9756	.9761	.9767
2.0	.9772	.9778	.9783	.9788	.9793	.9798	.9803	.9808	.9812	.9817
2.1	.9821	.9826	.9830	.9834	.9838	.9842	.9846	.9850	.9854	.9857
2.2	.9861	.9864	.9868	.9871	.9875	.9878	.9881	.9884	.9887	.9890
2.3	.9893	.9896	.9898	.9901	.9904	.9906	.9909	.9911	.9913	.9916
2.4	.9918	.9920	.9922	.9925	.9927	.9929	.9931	.9932	.9934	.9936
2.5	.9938	.9940	.9941	.9943	.9945	.9946	.9948	.9949	.9951	.9952
2.6	.9953	.9955	.9956	.9957	.9959	.9960	.9961	.9962	.9963	.9964
2.7	.9965	.9966	.9967	.9968	.9969	.9970	.9971	.9972	.9973	.9974
2.8	.9974	.9975	.9976	.9977	.9977	.9978	.9979	.9979	.9980	.9981
2.9	.9981	.9982	.9982	.9983	.9984	.9984	.9985	.9985	.9986	.9986
3.0	.9987	.9987	.9987	.9988	.9988	.9989	.9989	.9989	.9990	.9990
3.1	.9990	.9991	.9991	.9991	.9992	.9992	.9992	.9992	.9993	.9993
3.2	.9993	.9993	.9994	.9994	.9994	.9994	.9994	.9995	.9995	.9995
3.3	.9995	.9995	.9995	.9996	.9996	.9996	.9996	.9996	.9996	.9997
3.4	.9997	.9997	.9997	.9997	.9997	.9997	.9997	.9997	.9997	.9998

Selected Percentiles

Cumulative probability A:	.90	.95	.975	.98	.99	.995	.999
$z(A)$:	1.282	1.645	1.960	2.054	2.326	2.576	3.090

TABLE A–2 Percentiles of the t distribution

Entry is $t(A;v)$ where $P\{t(v) \leqslant t(A;v)\} = A$

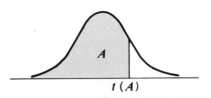

$t(A)$

				A			
v	.60	.70	.80	.85	.90	.95	.975
1	0.325	0.727	1.376	1.963	3.078	6.314	12.706
2	0.289	0.617	1.061	1.386	1.886	2.920	4.303
3	0.277	0.584	0.978	1.250	1.638	2.353	3.182
4	0.271	0.569	0.941	1.190	1.533	2.132	2.776
5	0.267	0.559	0.920	1.156	1.476	2.015	2.571
6	0.265	0.553	0.906	1.134	1.440	1.943	2.447
7	0.263	0.549	0.896	1.119	1.415	1.895	2.365
8	0.262	0.546	0.889	1.108	1.397	1.860	2.306
9	0.261	0.543	0.883	1.100	1.383	1.833	2.262
10	0.260	0.542	0.879	1.093	1.372	1.812	2.228
11	0.260	0.540	0.876	1.088	1.363	1.796	2.201
12	0.259	0.539	0.873	1.083	1.356	1.782	2.179
13	0.259	0.537	0.870	1.079	1.350	1.771	2.160
14	0.258	0.537	0.868	1.076	1.345	1.761	2.145
15	0.258	0.536	0.866	1.074	1.341	1.753	2.131
16	0.258	0.535	0.865	1.071	1.337	1.746	2.120
17	0.257	0.534	0.863	1.069	1.333	1.740	2.110
18	0.257	0.534	0.862	1.067	1.330	1.734	2.101
19	0.257	0.533	0.861	1.066	1.328	1.729	2.093
20	0.257	0.533	0.860	1.064	1.325	1.725	2.086
21	0.257	0.532	0.859	1.063	1.323	1.721	2.080
22	0.256	0.532	0.858	1.061	1.321	1.717	2.074
23	0.256	0.532	0.858	1.060	1.319	1.714	2.069
24	0.256	0.531	0.857	1.059	1.318	1.711	2.064
25	0.256	0.531	0.856	1.058	1.316	1.708	2.060
26	0.256	0.531	0.856	1.058	1.315	1.706	2.056
27	0.256	0.531	0.855	1.057	1.314	1.703	2.052
28	0.256	0.530	0.855	1.056	1.313	1.701	2.048
29	0.256	0.530	0.854	1.055	1.311	1.699	2.045
30	0.256	0.530	0.854	1.055	1.310	1.697	2.042
40	0.255	0.529	0.851	1.050	1.303	1.684	2.021
60	0.254	0.527	0.848	1.045	1.296	1.671	2.000
120	0.254	0.526	0.845	1.041	1.289	1.658	1.980
∞	0.253	0.524	0.842	1.036	1.282	1.645	1.960

TABLE A–2 *(concluded)* Percentiles of the *t* distribution

ν				A			
	.98	.985	.99	.9925	.995	.9975	.9995
1	15.895	21.205	31.821	42.434	63.657	127.322	636.590
2	4.849	5.643	6.965	8.073	9.925	14.089	31.598
3	3.482	3.896	4.541	5.047	5.841	7.453	12.924
4	2.999	3.298	3.747	4.088	4.604	5.598	8.610
5	2.757	3.003	3.365	3.634	4.032	4.773	6.869
6	2.612	2.829	3.143	3.372	3.707	4.317	5.959
7	2.517	2.715	2.998	3.203	3.499	4.029	5.408
8	2.449	2.634	2.896	3.085	3.355	3.833	5.041
9	2.398	2.574	2.821	2.998	3.250	3.690	4.781
10	2.359	2.527	2.764	2.932	3.169	3.581	4.587
11	2.328	2.491	2.718	2.879	3.106	3.497	4.437
12	2.303	2.461	2.681	2.836	3.055	3.428	4.318
13	2.282	2.436	2.650	2.801	3.012	3.372	4.221
14	2.264	2.415	2.624	2.771	2.977	3.326	4.140
15	2.249	2.397	2.602	2.746	2.947	3.286	4.073
16	2.235	2.382	2.583	2.724	2.921	3.252	4.015
17	2.224	2.368	2.567	2.706	2.898	3.222	3.965
18	2.214	2.356	2.552	2.689	2.878	3.197	3.922
19	2.205	2.346	2.539	2.674	2.861	3.174	3.883
20	2.197	2.336	2.528	2.661	2.845	3.153	3.849
21	2.189	2.328	2.518	2.649	2.831	3.135	3.819
22	2.183	2.320	2.508	2.639	2.819	3.119	3.792
23	2.177	2.313	2.500	2.629	2.807	3.104	3.768
24	2.172	2.307	2.492	2.620	2.797	3.091	3.745
25	2.167	2.301	2.485	2.612	2.787	3.078	3.725
26	2.162	2.296	2.479	2.605	2.779	3.067	3.707
27	2.158	2.291	2.473	2.598	2.771	3.057	3.690
28	2.154	2.286	2.467	2.592	2.763	3.047	3.674
29	2.150	2.282	2.462	2.586	2.756	3.038	3.659
30	2.147	2.278	2.457	2.581	2.750	3.030	3.646
40	2.123	2.250	2.423	2.542	2.704	2.971	3.551
60	2.099	2.223	2.390	2.504	2.660	2.915	3.460
120	2.076	2.196	2.358	2.468	2.617	2.860	3.373
∞	2.054	2.170	2.326	2.432	2.576	2.807	3.291

TABLE A-3 Percentiles of the χ^2 distribution

Entry is $\chi^2(A; \nu)$ where $P\{\chi^2(\nu) \le \chi^2(A; \nu)\} = A$

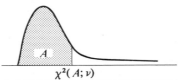

$\chi^2(A; \nu)$

| | | | | | | A | | | | |
ν	.005	.010	.025	.050	.100	.900	.950	.975	.990	.995
1	0.0^4393	0.0^3157	0.0^3982	0.0^2393	0.0158	2.71	3.84	5.02	6.63	7.88
2	0.0100	0.0201	0.0506	0.103	0.211	4.61	5.99	7.38	9.21	10.60
3	0.072	0.115	0.216	0.352	0.584	6.25	7.81	9.35	11.34	12.84
4	0.207	0.297	0.484	0.711	1.064	7.78	9.49	11.14	13.28	14.86
5	0.412	0.554	0.831	1.145	1.61	9.24	11.07	12.83	15.09	16.75
6	0.676	0.872	1.24	1.64	2.20	10.64	12.59	14.45	16.81	18.55
7	0.989	1.24	1.69	2.17	2.83	12.02	14.07	16.01	18.48	20.28
8	1.34	1.65	2.18	2.73	3.49	13.36	15.51	17.53	20.09	21.96
9	1.73	2.09	2.70	3.33	4.17	14.68	16.92	19.02	21.67	23.59
10	2.16	2.56	3.25	3.94	4.87	15.99	18.31	20.48	23.21	25.19
11	2.60	3.05	3.82	4.57	5.58	17.28	19.68	21.92	24.73	26.76
12	3.07	3.57	4.40	5.23	6.30	18.55	21.03	23.34	26.22	28.30
13	3.57	4.11	5.01	5.89	7.04	19.81	22.36	24.74	27.69	29.82
14	4.07	4.66	5.63	6.57	7.79	21.06	23.68	26.12	29.14	31.32
15	4.60	5.23	6.26	7.26	8.55	22.31	25.00	27.49	30.58	32.80
16	5.14	5.81	6.91	7.96	9.31	23.54	26.30	28.85	32.00	34.27
17	5.70	6.41	7.56	8.67	10.09	24.77	27.59	30.19	33.41	35.72
18	6.26	7.01	8.23	9.39	10.86	25.99	28.87	31.53	34.81	37.16
19	6.84	7.63	8.91	10.12	11.65	27.20	30.14	32.85	36.19	38.58
20	7.43	8.26	9.59	10.85	12.44	28.41	31.41	34.17	37.57	40.00
21	8.03	8.90	10.28	11.59	13.24	29.62	32.67	35.48	38.93	41.40
22	8.64	9.54	10.98	12.34	14.04	30.81	33.92	36.78	40.29	42.80
23	9.26	10.20	11.69	13.09	14.85	32.01	35.17	38.08	41.64	44.18
24	9.89	10.86	12.40	13.85	15.66	33.20	36.42	39.36	42.98	45.56
25	10.52	11.52	13.12	14.61	16.47	34.38	37.65	40.65	44.31	46.93
26	11.16	12.20	13.84	15.38	17.29	35.56	38.89	41.92	45.64	48.29
27	11.81	12.88	14.57	16.15	18.11	36.74	40.11	43.19	46.96	49.64
28	12.46	13.56	15.31	16.93	18.94	37.92	41.34	44.46	48.28	50.99
29	13.12	14.26	16.05	17.71	19.77	39.09	42.56	45.72	49.59	52.34
30	13.79	14.95	16.79	18.49	20.60	40.26	43.77	46.98	50.89	53.67
40	20.71	22.16	24.43	26.51	29.05	51.81	55.76	59.34	63.69	66.77
50	27.99	29.71	32.36	34.76	37.69	63.17	67.50	71.42	76.15	79.49
60	35.53	37.48	40.48	43.19	46.46	74.40	79.08	83.30	88.38	91.95
70	43.28	45.44	48.76	51.74	55.33	85.53	90.53	95.02	100.4	104.2
80	51.17	53.54	57.15	60.39	64.28	96.58	101.9	106.6	112.3	116.3
90	59.20	61.75	65.65	69.13	73.29	107.6	113.1	118.1	124.1	128.3
100	67.33	70.06	74.22	77.93	82.36	118.5	124.3	129.6	135.8	140.2

Source: Reprinted, with permission, from C. M. Thompson, "Table of Percentage Points of the Chi-Square Distribution," *Biometrika* 32 (1941), pp. 188–89.

TABLE A–4 Percentiles of the F distribution

Entry is $F(A; \nu_1, \nu_2)$ where $P\{F(\nu_1, \nu_2) \le F(A; \nu_1, \nu_2)\} = A$

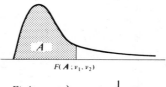

$$F(A; \nu_1, \nu_2) = \frac{1}{F(1 - A; \nu_2, \nu_1)}$$

TABLE A–4 *(continued)* Percentiles of the F distribution

Den. df	A	\multicolumn{9}{c}{Numerator df}

Den. df	A	1	2	3	4	5	6	7	8	9
1	.50	1.00	1.50	1.71	1.82	1.89	1.94	1.98	2.00	2.03
	.90	39.9	49.5	53.6	55.8	57.2	58.2	58.9	59.4	59.9
	.95	161	200	216	225	230	234	237	239	241
	.975	648	800	864	900	922	937	948	957	963
	.99	4,052	5,000	5,403	5,625	5,764	5,859	5,928	5,981	6,022
	.995	16,211	20,000	21,615	22,500	23,056	23,437	23,715	23,925	24,091
	.999	405,280	500,000	540,380	562,500	576,400	585,940	592,870	598,140	602,280
2	.50	0.667	1.00	1.13	1.21	1.25	1.28	1.30	1.32	1.33
	.90	8.53	9.00	9.16	9.24	9.29	9.33	9.35	9.37	9.38
	.95	18.5	19.0	19.2	19.2	19.3	19.3	19.4	19.4	19.4
	.975	38.5	39.0	39.2	39.2	39.3	39.3	39.4	39.4	39.4
	.99	98.5	99.0	99.2	99.2	99.3	99.3	99.4	99.4	99.4
	.995	199	199	199	199	199	199	199	199	199
	.999	998.5	999.0	999.2	999.2	999.3	999.3	999.4	999.4	999.4
3	.50	0.585	0.881	1.00	1.06	1.10	1.13	1.15	1.16	1.17
	.90	5.54	5.46	5.39	5.34	5.31	5.28	5.27	5.25	5.24
	.95	10.1	9.55	9.28	9.12	9.01	8.94	8.89	8.85	8.81
	.975	17.4	16.0	15.4	15.1	14.9	14.7	14.6	14.5	14.5
	.99	34.1	30.8	29.5	28.7	28.2	27.9	27.7	27.5	27.3
	.995	55.6	49.8	47.5	46.2	45.4	44.8	44.4	44.1	43.9
	.999	167.0	148.5	141.1	137.1	134.6	132.8	131.6	130.6	129.9
4	.50	0.549	0.828	0.941	1.00	1.04	1.06	1.08	1.09	1.10
	.90	4.54	4.32	4.19	4.11	4.05	4.01	3.98	3.95	3.94
	.95	7.71	6.94	6.59	6.39	6.26	6.16	6.09	6.04	6.00
	.975	12.2	10.6	9.98	9.60	9.36	9.20	9.07	8.98	8.90
	.99	21.2	18.0	16.7	16.0	15.5	15.2	15.0	14.8	14.7
	.995	31.3	26.3	24.3	23.2	22.5	22.0	21.6	21.4	21.1
	.999	74.1	61.2	56.2	53.4	51.7	50.5	49.7	49.0	48.5
5	.50	0.528	0.799	0.907	0.965	1.00	1.02	1.04	1.05	1.06
	.90	4.06	3.78	3.62	3.52	3.45	3.40	3.37	3.34	3.32
	.95	6.61	5.79	5.41	5.19	5.05	4.95	4.88	4.82	4.77
	.975	10.0	8.43	7.76	7.39	7.15	6.98	6.85	6.76	6.68
	.99	16.3	13.3	12.1	11.4	11.0	10.7	10.5	10.3	10.2
	.995	22.8	18.3	16.5	15.6	14.9	14.5	14.2	14.0	13.8
	.999	47.2	37.1	33.2	31.1	29.8	28.8	28.2	27.6	27.2
6	.50	0.515	0.780	0.886	0.942	0.977	1.00	1.02	1.03	1.04
	.90	3.78	3.46	3.29	3.18	3.11	3.05	3.01	2.98	2.96
	.95	5.99	5.14	4.76	4.53	4.39	4.28	4.21	4.15	4.10
	.975	8.81	7.26	6.60	6.23	5.99	5.82	5.70	5.60	5.52
	.99	13.7	10.9	9.78	9.15	8.75	8.47	8.26	8.10	7.98
	.995	18.6	14.5	12.9	12.0	11.5	11.1	10.8	10.6	10.4
	.999	35.5	27.0	23.7	21.9	20.8	20.0	19.5	19.0	18.7
7	.50	0.506	0.767	0.871	0.926	0.960	0.983	1.00	1.01	1.02
	.90	3.59	3.26	3.07	2.96	2.88	2.83	2.78	2.75	2.72
	.95	5.59	4.74	4.35	4.12	3.97	3.87	3.79	3.73	3.68
	.975	8.07	6.54	5.89	5.52	5.29	5.12	4.99	4.90	4.82
	.99	12.2	9.55	8.45	7.85	7.46	7.19	6.99	6.84	6.72
	.995	16.2	12.4	10.9	10.1	9.52	9.16	8.89	8.68	8.51
	.999	29.2	21.7	18.8	17.2	16.2	15.5	15.0	14.6	14.3

TABLE A–4 *(continued)*   Percentiles of the *F* distribution

Den. df	A	Numerator df								
		10	12	15	20	24	30	60	120	∞
1	.50	2.04	2.07	2.09	2.12	2.13	2.15	2.17	2.18	2.20
	.90	60.2	60.7	61.2	61.7	62.0	62.3	62.8	63.1	63.3
	.95	242	244	246	248	249	250	252	253	254
	.975	969	977	985	993	997	1,001	1,010	1,014	1,018
	.99	6,056	6,106	6,157	6,209	6,235	6,261	6,313	6,339	6,366
	.995	24,224	24,426	24,630	24,836	24,940	25,044	25,253	25,359	25,464
	.999	605,620	610,670	615,760	620,910	623,500	626,100	631,340	633,970	636,620
2	.50	1.34	1.36	1.38	1.39	1.40	1.41	1.43	1.43	1.44
	.90	9.39	9.41	9.42	9.44	9.45	9.46	9.47	9.48	9.49
	.95	19.4	19.4	19.4	19.4	19.5	19.5	19.5	19.5	19.5
	.975	39.4	39.4	39.4	39.4	39.5	39.5	39.5	39.5	39.5
	.99	99.4	99.4	99.4	99.4	99.5	99.5	99.5	99.5	99.5
	.995	199	199	199	199	199	199	199	199	200
	.999	999.4	999.4	999.4	999.4	999.5	999.5	999.5	999.5	999.5
3	.50	1.18	1.20	1.21	1.23	1.23	1.24	1.25	1.26	1.27
	.90	5.23	5.22	5.20	5.18	5.18	5.17	5.15	5.14	5.13
	.95	8.79	8.74	8.70	8.66	8.64	8.62	8.57	8.55	8.53
	.975	14.4	14.3	14.3	14.2	14.1	14.1	14.0	13.9	13.9
	.99	27.2	27.1	26.9	26.7	26.6	26.5	26.3	26.2	26.1
	.995	43.7	43.4	43.1	42.8	42.6	42.5	42.1	42.0	41.8
	.999	129.2	128.3	127.4	126.4	125.9	125.4	124.5	124.0	123.5
4	.50	1.11	1.13	1.14	1.15	1.16	1.16	1.18	1.18	1.19
	.90	3.92	3.90	3.87	3.84	3.83	3.82	3.79	3.78	3.76
	.95	5.96	5.91	5.86	5.80	5.77	5.75	5.69	5.66	5.63
	.975	8.84	8.75	8.66	8.56	8.51	8.46	8.36	8.31	8.26
	.99	14.5	14.4	14.2	14.0	13.9	13.8	13.7	13.6	13.5
	.995	21.0	20.7	20.4	20.2	20.0	19.9	19.6	19.5	19.3
	.999	48.1	47.4	46.8	46.1	45.8	45.4	44.7	44.4	44.1
5	.50	1.07	1.09	1.10	1.11	1.12	1.12	1.14	1.14	1.15
	.90	3.30	3.27	3.24	3.21	3.19	3.17	3.14	3.12	3.11
	.95	4.74	4.68	4.62	4.56	4.53	4.50	4.43	4.40	4.37
	.975	6.62	6.52	6.43	6.33	6.28	6.23	6.12	6.07	6.02
	.99	10.1	9.89	9.72	9.55	9.47	9.38	9.20	9.11	9.02
	.995	13.6	13.4	13.1	12.9	12.8	12.7	12.4	12.3	12.1
	.999	26.9	26.4	25.9	25.4	25.1	24.9	24.3	24.1	23.8
6	.50	1.05	1.06	1.07	1.08	1.09	1.10	1.11	1.12	1.12
	.90	2.94	2.90	2.87	2.84	2.82	2.80	2.76	2.74	2.72
	.95	4.06	4.00	3.94	3.87	3.84	3.81	3.74	3.70	3.67
	.975	5.46	5.37	5.27	5.17	5.12	5.07	4.96	4.90	4.85
	.99	7.87	7.72	7.56	7.40	7.31	7.23	7.06	6.97	6.88
	.995	10.2	10.0	9.81	9.59	9.47	9.36	9.12	9.00	8.88
	.999	18.4	18.0	17.6	17.1	16.9	16.7	16.2	16.0	15.7
7	.50	1.03	1.04	1.05	1.07	1.07	1.08	1.09	1.10	1.10
	.90	2.70	2.67	2.63	2.59	2.58	2.56	2.51	2.49	2.47
	.95	3.64	3.57	3.51	3.44	3.41	3.38	3.30	3.27	3.23
	.975	4.76	4.67	4.57	4.47	4.42	4.36	4.25	4.20	4.14
	.99	6.62	6.47	6.31	6.16	6.07	5.99	5.82	5.74	5.65
	.995	8.38	8.18	7.97	7.75	7.65	7.53	7.31	7.19	7.08
	.999	14.1	13.7	13.3	12.9	12.7	12.5	12.1	11.9	11.7

TABLE A–4 *(continued)* Percentiles of the *F* distribution

Den. df	A	Numerator df								
		1	2	3	4	5	6	7	8	9
8	.50	0.499	0.757	0.860	0.915	0.948	0.971	0.988	1.00	1.01
	.90	3.46	3.11	2.92	2.81	2.73	2.67	2.62	2.59	2.56
	.95	5.32	4.46	4.07	3.84	3.69	3.58	3.50	3.44	3.39
	.975	7.57	6.06	5.42	5.05	4.82	4.65	4.53	4.43	4.36
	.99	11.3	8.65	7.59	7.01	6.63	6.37	6.18	6.03	5.91
	.995	14.7	11.0	9.60	8.81	8.30	7.95	7.69	7.50	7.34
	.999	25.4	18.5	15.8	14.4	13.5	12.9	12.4	12.0	11.8
9	.50	0.494	0.749	0.852	0.906	0.939	0.962	0.978	0.990	1.00
	.90	3.36	3.01	2.81	2.69	2.61	2.55	2.51	2.47	2.44
	.95	5.12	4.26	3.86	3.63	3.48	3.37	3.29	3.23	3.18
	.975	7.21	5.71	5.08	4.72	4.48	4.32	4.20	4.10	4.03
	.99	10.6	8.02	6.99	6.42	6.06	5.80	5.61	5.47	5.35
	.995	13.6	10.1	8.72	7.96	7.47	7.13	6.88	6.69	6.54
	.999	22.9	16.4	13.9	12.6	11.7	11.1	10.7	10.4	10.1
10	.50	0.490	0.743	0.845	0.899	0.932	0.954	0.971	0.983	0.992
	.90	3.29	2.92	2.73	2.61	2.52	2.46	2.41	2.38	2.35
	.95	4.96	4.10	3.71	3.48	3.33	3.22	3.14	3.07	3.02
	.975	6.94	5.46	4.83	4.47	4.24	4.07	3.95	3.85	3.78
	.99	10.0	7.56	6.55	5.99	5.64	5.39	5.20	5.06	4.94
	.995	12.8	9.43	8.08	7.34	6.87	6.54	6.30	6.12	5.97
	.999	21.0	14.9	12.6	11.3	10.5	9.93	9.52	9.20	8.96
12	.50	0.484	0.735	0.835	0.888	0.921	0.943	0.959	0.972	0.981
	.90	3.18	2.81	2.61	2.48	2.39	2.33	2.28	2.24	2.21
	.95	4.75	3.89	3.49	3.26	3.11	3.00	2.91	2.85	2.80
	.975	6.55	5.10	4.47	4.12	3.89	3.73	3.61	3.51	3.44
	.99	9.33	6.93	5.95	5.41	5.06	4.82	4.64	4.50	4.39
	.995	11.8	8.51	7.23	6.52	6.07	5.76	5.52	5.35	5.20
	.999	18.6	13.0	10.8	9.63	8.89	8.38	8.00	7.71	7.48
15	.50	0.478	0.726	0.826	0.878	0.911	0.933	0.949	0.960	0.970
	.90	3.07	2.70	2.49	2.36	2.27	2.21	2.16	2.12	2.09
	.95	4.54	3.68	3.29	3.06	2.90	2.79	2.71	2.64	2.59
	.975	6.20	4.77	4.15	3.80	3.58	3.41	3.29	3.20	3.12
	.99	8.68	6.36	5.42	4.89	4.56	4.32	4.14	4.00	3.89
	.995	10.8	7.70	6.48	5.80	5.37	5.07	4.85	4.67	4.54
	.999	16.6	11.3	9.34	8.25	7.57	7.09	6.74	6.47	6.26
20	.50	0.472	0.718	0.816	0.868	0.900	0.922	0.938	0.950	0.959
	.90	2.97	2.59	2.38	2.25	2.16	2.09	2.04	2.00	1.96
	.95	4.35	3.49	3.10	2.87	2.71	2.60	2.51	2.45	2.39
	.975	5.87	4.46	3.86	3.51	3.29	3.13	3.01	2.91	2.84
	.99	8.10	5.85	4.94	4.43	4.10	3.87	3.70	3.56	3.46
	.995	9.94	6.99	5.82	5.17	4.76	4.47	4.26	4.09	3.96
	.999	14.8	9.95	8.10	7.10	6.46	6.02	5.69	5.44	5.24
24	.50	0.469	0.714	0.812	0.863	0.895	0.917	0.932	0.944	0.953
	.90	2.93	2.54	2.33	2.19	2.10	2.04	1.98	1.94	1.91
	.95	4.26	3.40	3.01	2.78	2.62	2.51	2.42	2.36	2.30
	.975	5.72	4.32	3.72	3.38	3.15	2.99	2.87	2.78	2.70
	.99	7.82	5.61	4.72	4.22	3.90	3.67	3.50	3.36	3.26
	.995	9.55	6.66	5.52	4.89	4.49	4.20	3.99	3.83	3.69
	.999	14.0	9.34	7.55	6.59	5.98	5.55	5.23	4.99	4.80

TABLE A–4 *(continued)* Percentiles of the *F* distribution

Den. df	A	10	12	15	20	24	30	60	120	∞
8	.50	1.02	1.03	1.04	1.05	1.06	1.07	1.08	1.08	1.09
	.90	2.54	2.50	2.46	2.42	2.40	2.38	2.34	2.32	2.29
	.95	3.35	3.28	3.22	3.15	3.12	3.08	3.01	2.97	2.93
	.975	4.30	4.20	4.10	4.00	3.95	3.89	3.78	3.73	3.67
	.99	5.81	5.67	5.52	5.36	5.28	5.20	5.03	4.95	4.86
	.995	7.21	7.01	6.81	6.61	6.50	6.40	6.18	6.06	5.95
	.999	11.5	11.2	10.8	10.5	10.3	10.1	9.73	9.53	9.33
9	.50	1.01	1.02	1.03	1.04	1.05	1.05	1.07	1.07	1.08
	.90	2.42	2.38	2.34	2.30	2.28	2.25	2.21	2.18	2.16
	.95	3.14	3.07	3.01	2.94	2.90	2.86	2.79	2.75	2.71
	.975	3.96	3.87	3.77	3.67	3.61	3.56	3.45	3.39	3.33
	.99	5.26	5.11	4.96	4.81	4.73	4.65	4.48	4.40	4.31
	.995	6.42	6.23	6.03	5.83	5.73	5.62	5.41	5.30	5.19
	.999	9.89	9.57	9.24	8.90	8.72	8.55	8.19	8.00	7.81
10	.50	1.00	1.01	1.02	1.03	1.04	1.05	1.06	1.06	1.07
	.90	2.32	2.28	2.24	2.20	2.18	2.16	2.11	2.08	2.06
	.95	2.98	2.91	2.84	2.77	2.74	2.70	2.62	2.58	2.54
	.975	3.72	3.62	3.52	3.42	3.37	3.31	3.20	3.14	3.08
	.99	4.85	4.71	4.56	4.41	4.33	4.25	4.08	4.00	3.91
	.995	5.85	5.66	5.47	5.27	5.17	5.07	4.86	4.75	4.64
	.999	8.75	8.45	8.13	7.80	7.64	7.47	7.12	6.94	6.76
12	.50	0.989	1.00	1.01	1.02	1.03	1.03	1.05	1.05	1.06
	.90	2.19	2.15	2.10	2.06	2.04	2.01	1.96	1.93	1.90
	.95	2.75	2.69	2.62	2.54	2.51	2.47	2.38	2.34	2.30
	.975	3.37	3.28	3.18	3.07	3.02	2.96	2.85	2.79	2.72
	.99	4.30	4.16	4.01	3.86	3.78	3.70	3.54	3.45	3.36
	.995	5.09	4.91	4.72	4.53	4.43	4.33	4.12	4.01	3.90
	.999	7.29	7.00	6.71	6.40	6.25	6.09	5.76	5.59	5.42
15	.50	0.977	0.989	1.00	1.01	1.02	1.02	1.03	1.04	1.05
	.90	2.06	2.02	1.97	1.92	1.90	1.87	1.82	1.79	1.76
	.95	2.54	2.48	2.40	2.33	2.29	2.25	2.16	2.11	2.07
	.975	3.06	2.96	2.86	2.76	2.70	2.64	2.52	2.46	2.40
	.99	3.80	3.67	3.52	3.37	3.29	3.21	3.05	2.96	2.87
	.995	4.42	4.25	4.07	3.88	3.79	3.69	3.48	3.37	3.26
	.999	6.08	5.81	5.54	5.25	5.10	4.95	4.64	4.48	4.31
20	.50	0.966	0.977	0.989	1.00	1.01	1.01	1.02	1.03	1.03
	.90	1.94	1.89	1.84	1.79	1.77	1.74	1.68	1.64	1.61
	.95	2.35	2.28	2.20	2.12	2.08	2.04	1.95	1.90	1.84
	.975	2.77	2.68	2.57	2.46	2.41	2.35	2.22	2.16	2.09
	.99	3.37	3.23	3.09	2.94	2.86	2.78	2.61	2.52	2.42
	.995	3.85	3.68	3.50	3.32	3.22	3.12	2.92	2.81	2.69
	.999	5.08	4.82	4.56	4.29	4.15	4.00	3.70	3.54	3.38
24	.50	0.961	0.972	0.983	0.994	1.00	1.01	1.02	1.02	1.03
	.90	1.88	1.83	1.78	1.73	1.70	1.67	1.61	1.57	1.53
	.95	2.25	2.18	2.11	2.03	1.98	1.94	1.84	1.79	1.73
	.975	2.64	2.54	2.44	2.33	2.27	2.21	2.08	2.01	1.94
	.99	3.17	3.03	2.89	2.74	2.66	2.58	2.40	2.31	2.21
	.995	3.59	3.42	3.25	3.06	2.97	2.87	2.66	2.55	2.43
	.999	4.64	4.39	4.14	3.87	3.74	3.59	3.29	3.14	2.97

TABLE A–4 *(continued)*   Percentiles of the F distribution

Den. df	A	\multicolumn{9}{c}{Numerator df}
		1	2	3	4	5	6	7	8	9
30	.50	0.466	0.709	0.807	0.858	0.890	0.912	0.927	0.939	0.948
	.90	2.88	2.49	2.28	2.14	2.05	1.98	1.93	1.88	1.85
	.95	4.17	3.32	2.92	2.69	2.53	2.42	2.33	2.27	2.21
	.975	5.57	4.18	3.59	3.25	3.03	2.87	2.75	2.65	2.57
	.99	7.56	5.39	4.51	4.02	3.70	3.47	3.30	3.17	3.07
	.995	9.18	6.35	5.24	4.62	4.23	3.95	3.74	3.58	3.45
	.999	13.3	8.77	7.05	6.12	5.53	5.12	4.82	4.58	4.39
60	.50	0.461	0.701	0.798	0.849	0.880	0.901	0.917	0.928	0.937
	.90	2.79	2.39	2.18	2.04	1.95	1.87	1.82	1.77	1.74
	.95	4.00	3.15	2.76	2.53	2.37	2.25	2.17	2.10	2.04
	.975	5.29	3.93	3.34	3.01	2.79	2.63	2.51	2.41	2.33
	.99	7.08	4.98	4.13	3.65	3.34	3.12	2.95	2.82	2.72
	.995	8.49	5.80	4.73	4.14	3.76	3.49	3.29	3.13	3.01
	.999	12.0	7.77	6.17	5.31	4.76	4.37	4.09	3.86	3.69
120	.50	0.458	0.697	0.793	0.844	0.875	0.896	0.912	0.923	0.932
	.90	2.75	2.35	2.13	1.99	1.90	1.82	1.77	1.72	1.68
	.95	3.92	3.07	2.68	2.45	2.29	2.18	2.09	2.02	1.96
	.975	5.15	3.80	3.23	2.89	2.67	2.52	2.39	2.30	2.22
	.99	6.85	4.79	3.95	3.48	3.17	2.96	2.79	2.66	2.56
	.995	8.18	5.54	4.50	3.92	3.55	3.28	3.09	2.93	2.81
	.999	11.4	7.32	5.78	4.95	4.42	4.04	3.77	3.55	3.38
∞	.50	0.455	0.693	0.789	0.839	0.870	0.891	0.907	0.918	0.927
	.90	2.71	2.30	2.08	1.94	1.85	1.77	1.72	1.67	1.63
	.95	3.84	3.00	2.60	2.37	2.21	2.10	2.01	1.94	1.88
	.975	5.02	3.69	3.12	2.79	2.57	2.41	2.29	2.19	2.11
	.99	6.63	4.61	3.78	3.32	3.02	2.80	2.64	2.51	2.41
	.995	7.88	5.30	4.28	3.72	3.35	3.09	2.90	2.74	2.62
	.999	10.8	6.91	5.42	4.62	4.10	3.74	3.47	3.27	3.10

TABLE A-4 *(concluded)* Percentiles of the *F* distribution

Den. df	A	Numerator df 10	12	15	20	24	30	60	120	∞
30	.50	0.955	0.966	0.978	0.989	0.994	1.00	1.01	1.02	1.02
	.90	1.82	1.77	1.72	1.67	1.64	1.61	1.54	1.50	1.46
	.95	2.16	2.09	2.01	1.93	1.89	1.84	1.74	1.68	1.62
	.975	2.51	2.41	2.31	2.20	2.14	2.07	1.94	1.87	1.79
	.99	2.98	2.84	2.70	2.55	2.47	2.39	2.21	2.11	2.01
	.995	3.34	3.18	3.01	2.82	2.73	2.63	2.42	2.30	2.18
	.999	4.24	4.00	3.75	3.49	3.36	3.22	2.92	2.76	2.59
60	.50	0.945	0.956	0.967	0.978	0.983	0.989	1.00	1.01	1.01
	.90	1.71	1.66	1.60	1.54	1.51	1.48	1.40	1.35	1.29
	.95	1.99	1.92	1.84	1.75	1.70	1.65	1.53	1.47	1.39
	.975	2.27	2.17	2.06	1.94	1.88	1.82	1.67	1.58	1.48
	.99	2.63	2.50	2.35	2.20	2.12	2.03	1.84	1.73	1.60
	.995	2.90	2.74	2.57	2.39	2.29	2.19	1.96	1.83	1.69
	.999	3.54	3.32	3.08	2.83	2.69	2.55	2.25	2.08	1.89
120	.50	0.939	0.950	0.961	0.972	0.978	0.983	0.994	1.00	1.01
	.90	1.65	1.60	1.55	1.48	1.45	1.41	1.32	1.26	1.19
	.95	1.91	1.83	1.75	1.66	1.61	1.55	1.43	1.35	1.25
	.975	2.16	2.05	1.95	1.82	1.76	1.69	1.53	1.43	1.31
	.99	2.47	2.34	2.19	2.03	1.95	1.86	1.66	1.53	1.38
	.995	2.71	2.54	2.37	2.19	2.09	1.98	1.75	1.61	1.43
	.999	3.24	3.02	2.78	2.53	2.40	2.26	1.95	1.77	1.54
∞	.50	0.934	0.945	0.956	0.967	0.972	0.978	0.989	0.994	1.00
	.90	1.60	1.55	1.49	1.42	1.38	1.34	1.24	1.17	1.00
	.95	1.83	1.75	1.67	1.57	1.52	1.46	1.32	1.22	1.00
	.975	2.05	1.94	1.83	1.71	1.64	1.57	1.39	1.27	1.00
	.99	2.32	2.18	2.04	1.88	1.79	1.70	1.47	1.32	1.00
	.995	2.52	2.36	2.19	2.00	1.90	1.79	1.53	1.36	1.00
	.999	2.96	2.74	2.51	2.27	2.13	1.99	1.66	1.45	1.00

Source: Reprinted from Table 5 of Pearson and Hartley, *Biometrika Tables for Statisticians,* Volume 2, 1972, published by the Cambridge University Press, on behalf of The Biometrika Society, by permission of the authors and publishers.

TABLE A–5 Power function for two-sided t test

$$\alpha = .05$$

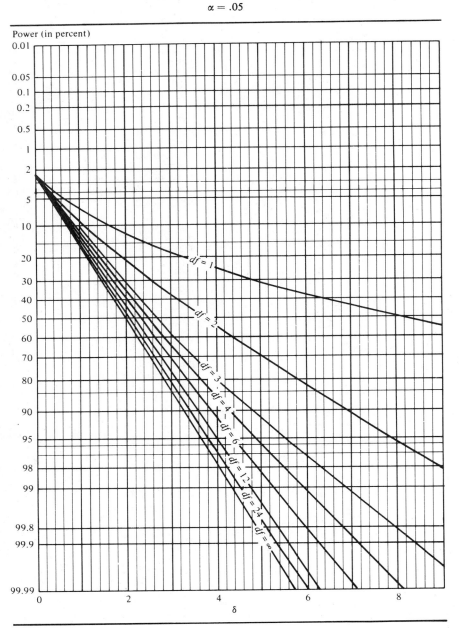

TABLE A–5 *(concluded)* Power function for two-sided *t* test

$$\alpha = .01$$

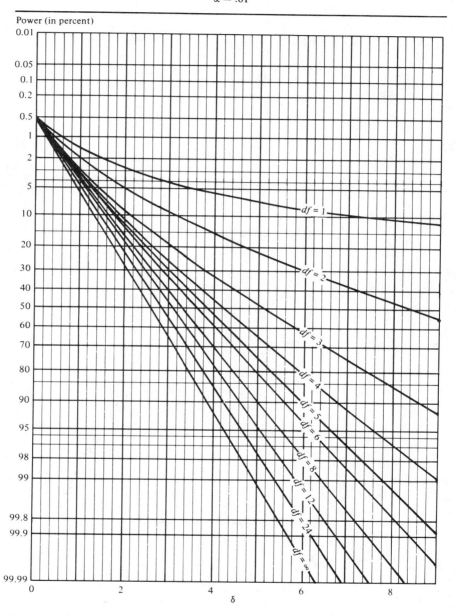

Power (in percent)

δ

Source: Reprinted, with permission, from D. B. Owen, *Handbook of Statistical Tables* (Reading, Mass.: Addison-Wesley Publishing, 1962), pp. 32, 34. Courtesy of U.S. Atomic Energy Commission.

TABLE A-6 Durbin-Watson test bounds

Level of Significance $\alpha = .05$

	$p-1=1$		$p-1=2$		$p-1=3$		$p-1=4$		$p-1=5$	
n	d_L	d_U	d_L	d_U	d_L	d_U	d_L	d_U	d_L	d_U
15	1.08	1.36	0.95	1.54	0.82	1.75	0.69	1.97	0.56	2.21
16	1.10	1.37	0.98	1.54	0.86	1.73	0.74	1.93	0.62	2.15
17	1.13	1.38	1.02	1.54	0.90	1.71	0.78	1.90	0.67	2.10
18	1.16	1.39	1.05	1.53	0.93	1.69	0.82	1.87	0.71	2.06
19	1.18	1.40	1.08	1.53	0.97	1.68	0.86	1.85	0.75	2.02
20	1.20	1.41	1.10	1.54	1.00	1.68	0.90	1.83	0.79	1.99
21	1.22	1.42	1.13	1.54	1.03	1.67	0.93	1.81	0.83	1.96
22	1.24	1.43	1.15	1.54	1.05	1.66	0.96	1.80	0.86	1.94
23	1.26	1.44	1.17	1.54	1.08	1.66	0.99	1.79	0.90	1.92
24	1.27	1.45	1.19	1.55	1.10	1.66	1.01	1.78	0.93	1.90
25	1.29	1.45	1.21	1.55	1.12	1.66	1.04	1.77	0.95	1.89
26	1.30	1.46	1.22	1.55	1.14	1.65	1.06	1.76	0.98	1.88
27	1.32	1.47	1.24	1.56	1.16	1.65	1.08	1.76	1.01	1.86
28	1.33	1.48	1.26	1.56	1.18	1.65	1.10	1.75	1.03	1.85
29	1.34	1.48	1.27	1.56	1.20	1.65	1.12	1.74	1.05	1.84
30	1.35	1.49	1.28	1.57	1.21	1.65	1.14	1.74	1.07	1.83
31	1.36	1.50	1.30	1.57	1.23	1.65	1.16	1.74	1.09	1.83
32	1.37	1.50	1.31	1.57	1.24	1.65	1.18	1.73	1.11	1.82
33	1.38	1.51	1.32	1.58	1.26	1.65	1.19	1.73	1.13	1.81
34	1.39	1.51	1.33	1.58	1.27	1.65	1.21	1.73	1.15	1.81
35	1.40	1.52	1.34	1.58	1.28	1.65	1.22	1.73	1.16	1.80
36	1.41	1.52	1.35	1.59	1.29	1.65	1.24	1.73	1.18	1.80
37	1.42	1.53	1.36	1.59	1.31	1.66	1.25	1.72	1.19	1.80
38	1.43	1.54	1.37	1.59	1.32	1.66	1.26	1.72	1.21	1.79
39	1.43	1.54	1.38	1.60	1.33	1.66	1.27	1.72	1.22	1.79
40	1.44	1.54	1.39	1.60	1.34	1.66	1.29	1.72	1.23	1.79
45	1.48	1.57	1.43	1.62	1.38	1.67	1.34	1.72	1.29	1.78
50	1.50	1.59	1.46	1.63	1.42	1.67	1.38	1.72	1.34	1.77
55	1.53	1.60	1.49	1.64	1.45	1.68	1.41	1.72	1.38	1.77
60	1.55	1.62	1.51	1.65	1.48	1.69	1.44	1.73	1.41	1.77
65	1.57	1.63	1.54	1.66	1.50	1.70	1.47	1.73	1.44	1.77
70	1.58	1.64	1.55	1.67	1.52	1.70	1.49	1.74	1.46	1.77
75	1.60	1.65	1.57	1.68	1.54	1.71	1.51	1.74	1.49	1.77
80	1.61	1.66	1.59	1.69	1.56	1.72	1.53	1.74	1.51	1.77
85	1.62	1.67	1.60	1.70	1.57	1.72	1.55	1.75	1.52	1.77
90	1.63	1.68	1.61	1.70	1.59	1.73	1.57	1.75	1.54	1.78
95	1.64	1.69	1.62	1.71	1.60	1.73	1.58	1.75	1.56	1.78
100	1.65	1.69	1.63	1.72	1.61	1.74	1.59	1.76	1.57	1.78

TABLE A–6 *(concluded)* Durbin-Watson test bounds

Level of Significance $\alpha = .01$

n	$p-1=1$ d_L	d_U	$p-1=2$ d_L	d_U	$p-1=3$ d_L	d_U	$p-1=4$ d_L	d_U	$p-1=5$ d_L	d_U
15	0.81	1.07	0.70	1.25	0.59	1.46	0.49	1.70	0.39	1.96
16	0.84	1.09	0.74	1.25	0.63	1.44	0.53	1.66	0.44	1.90
17	0.87	1.10	0.77	1.25	0.67	1.43	0.57	1.63	0.48	1.85
18	0.90	1.12	0.80	1.26	0.71	1.42	0.61	1.60	0.52	1.80
19	0.93	1.13	0.83	1.26	0.74	1.41	0.65	1.58	0.56	1.77
20	0.95	1.15	0.86	1.27	0.77	1.41	0.68	1.57	0.60	1.74
21	0.97	1.16	0.89	1.27	0.80	1.41	0.72	1.55	0.63	1.71
22	1.00	1.17	0.91	1.28	0.83	1.40	0.75	1.54	0.66	1.69
23	1.02	1.19	0.94	1.29	0.86	1.40	0.77	1.53	0.70	1.67
24	1.04	1.20	0.96	1.30	0.88	1.41	0.80	1.53	0.72	1.66
25	1.05	1.21	0.98	1.30	0.90	1.41	0.83	1.52	0.75	1.65
26	1.07	1.22	1.00	1.31	0.93	1.41	0.85	1.52	0.78	1.64
27	1.09	1.23	1.02	1.32	0.95	1.41	0.88	1.51	0.81	1.63
28	1.10	1.24	1.04	1.32	0.97	1.41	0.90	1.51	0.83	1.62
29	1.12	1.25	1.05	1.33	0.99	1.42	0.92	1.51	0.85	1.61
30	1.13	1.26	1.07	1.34	1.01	1.42	0.94	1.51	0.88	1.61
31	1.15	1.27	1.08	1.34	1.02	1.42	0.96	1.51	0.90	1.60
32	1.16	1.28	1.10	1.35	1.04	1.43	0.98	1.51	0.92	1.60
33	1.17	1.29	1.11	1.36	1.05	1.43	1.00	1.51	0.94	1.59
34	1.18	1.30	1.13	1.36	1.07	1.43	1.01	1.51	0.95	1.59
35	1.19	1.31	1.14	1.37	1.08	1.44	1.03	1.51	0.97	1.59
36	1.21	1.32	1.15	1.38	1.10	1.44	1.04	1.51	0.99	1.59
37	1.22	1.32	1.16	1.38	1.11	1.45	1.06	1.51	1.00	1.59
38	1.23	1.33	1.18	1.39	1.12	1.45	1.07	1.52	1.02	1.58
39	1.24	1.34	1.19	1.39	1.14	1.45	1.09	1.52	1.03	1.58
40	1.25	1.34	1.20	1.40	1.15	1.46	1.10	1.52	1.05	1.58
45	1.29	1.38	1.24	1.42	1.20	1.48	1.16	1.53	1.11	1.58
50	1.32	1.40	1.28	1.45	1.24	1.49	1.20	1.54	1.16	1.59
55	1.36	1.43	1.32	1.47	1.28	1.51	1.25	1.55	1.21	1.59
60	1.38	1.45	1.35	1.48	1.32	1.52	1.28	1.56	1.25	1.60
65	1.41	1.47	1.38	1.50	1.35	1.53	1.31	1.57	1.28	1.61
70	1.43	1.49	1.40	1.52	1.37	1.55	1.34	1.58	1.31	1.61
75	1.45	1.50	1.42	1.53	1.39	1.56	1.37	1.59	1.34	1.62
80	1.47	1.52	1.44	1.54	1.42	1.57	1.39	1.60	1.36	1.62
85	1.48	1.53	1.46	1.55	1.43	1.58	1.41	1.60	1.39	1.63
90	1.50	1.54	1.47	1.56	1.45	1.59	1.43	1.61	1.41	1.64
95	1.51	1.55	1.49	1.57	1.47	1.60	1.45	1.62	1.42	1.64
100	1.52	1.56	1.50	1.58	1.48	1.60	1.46	1.63	1.44	1.65

Source: Reprinted, with permission, from J. Durbin and G. S. Watson, "Testing for Serial Correlation in Least Squares Regression. II," *Biometrika* 38 (1951), pp. 159–78.

TABLE A–7 Table of z' transformation of correlation coefficient

r ρ	z' ζ	r ρ	z' ζ	r ρ	z' ζ	r ρ	z' ζ
.00	.0000	.25	.2554	.50	.5493	.75	.973
.01	.0100	.26	.2661	.51	.5627	.76	.996
.02	.0200	.27	.2769	.52	.5763	.77	1.020
.03	.0300	.28	.2877	.53	.5901	.78	1.045
.04	.0400	.29	.2986	.54	.6042	.79	1.071
.05	.0500	.30	.3095	.55	.6184	.80	1.099
.06	.0601	.31	.3205	.56	.6328	.81	1.127
.07	.0701	.32	.3316	.57	.6475	.82	1.157
.08	.0802	.33	.3428	.58	.6625	.83	1.188
.09	.0902	.34	.3541	.59	.6777	.84	1.221
.10	.1003	.35	.3654	.60	.6931	.85	1.256
.11	.1104	.36	.3769	.61	.7089	.86	1.293
.12	.1206	.37	.3884	.62	.7250	.87	1.333
.13	.1307	.38	.4001	.63	.7414	.88	1.376
.14	.1409	.39	.4118	.64	.7582	.89	1.422
.15	.1511	.40	.4236	.65	.7753	.90	1.472
.16	.1614	.41	.4356	.66	.7928	.91	1.528
.17	.1717	.42	.4477	.67	.8107	.92	1.589
.18	.1820	.43	.4599	.68	.8291	.93	1.658
.19	.1923	.44	.4722	.69	.8480	.94	1.738
.20	.2027	.45	.4847	.70	.8673	.95	1.832
.21	.2132	.46	.4973	.71	.8872	.96	1.946
.22	.2237	.47	.5101	.72	.9076	.97	2.092
.23	.2342	.48	.5230	.73	.9287	.98	2.298
.24	.2448	.49	.5361	.74	.9505	.99	2.647

Source: Abridged from Table 14 of Pearson and Hartley, *Biometrika Tables for Statisticians,* Volume 1, 1966, published by the Cambridge University Press, on behalf of The Biometrika Society, by permission of the authors and publishers.

SENIC data set

The primary objective of the Study on the Efficacy of Nosocomial Infection Control (SENIC Project) was to determine whether infection surveillance and control programs have reduced the rates of nosocomial (hospital-acquired) infection in United States hospitals. This data set consists of a random sample of 113 hospitals selected from the original 338 hospitals surveyed.

Each line of the data set has an identification number and provides information on 11 other variables for a single hospital. The data presented here are for the 1975–76 study period. The 12 variables are:

Variable Number	Variable Name	Description
1	Identification number	1-113
2	Length of stay	Average length of stay of all patients in hospital (in days)
3	Age	Average age of patients (in years)
4	Infection risk	Average estimated probability of acquiring infection in hospital (in percent)
5	Routine culturing ratio	Ratio of number of cultures performed to number of patients without signs or symptoms of hospital-acquired infection, times 100

SENIC data set *(continued)*

Variable Number	Variable Name	Description
6	Routine chest X-ray ratio	Ratio of number of X rays performed to number of patients without signs or symptoms of pneumonia, times 100
7	Number of beds	Average number of beds in hospital during study period
8	Medical school affiliation	1 = Yes 2 = No
9	Region	Geographic region, where: 1 = NE, 2 = NC, 3 = S, 4 = W
10	Average daily census	Average number of patients in hospital per day during study period
11	Number of nurses	Average number of full-time equivalent registered and licensed practical nurses during study period (number full time plus one half the number part time)
12	Available facilities and services	Percent of 35 potential facilities and services that are provided by the hospital

Reference: Special Issue, "The SENIC Project," *American Journal of Epidemiology* 111 (1980), pp. 465–653.

Data obtained from: Robert W. Haley, M.D., Hospital Infections Program, Center for Infectious Diseases, Centers for Disease Control, Atlanta, Georgia 30333.

SENIC data set *(continued)*

1	2	3	4	5	6	7	8	9	10	11	12
1	7.13	55.7	4.1	9.0	39.6	279	2	4	207	241	60.0
2	8.82	58.2	1.6	3.8	51.7	80	2	2	51	52	40.0
3	8.34	56.9	2.7	8.1	74.0	107	2	3	82	54	20.0
4	8.95	53.7	5.6	18.9	122.8	147	2	4	53	148	40.0
5	11.20	56.5	5.7	34.5	88.9	180	2	1	134	151	40.0
6	9.76	50.9	5.1	21.9	97.0	150	2	2	147	106	40.0
7	9.68	57.8	4.6	16.7	79.0	186	2	3	151	129	40.0
8	11.18	45.7	5.4	60.5	85.8	640	1	2	399	360	60.0
9	8.67	48.2	4.3	24.4	90.8	182	2	3	130	118	40.0
10	8.84	56.3	6.3	29.6	82.6	85	2	1	59	66	40.0
11	11.07	53.2	4.9	28.5	122.0	768	1	1	591	656	80.0
12	8.30	57.2	4.3	6.8	83.8	167	2	3	105	59	40.0
13	12.78	56.8	7.7	46.0	116.9	322	1	1	252	349	57.1
14	7.58	56.7	3.7	20.8	88.0	97	2	2	59	79	37.1
15	9.00	56.3	4.2	14.6	76.4	72	2	3	61	38	17.1
16	11.08	50.2	5.5	18.6	63.6	387	2	3	326	405	57.1
17	8.28	48.1	4.5	26.0	101.8	108	2	4	84	73	37.1
18	11.62	53.9	6.4	25.5	99.2	133	2	1	113	101	37.1
19	9.06	52.8	4.2	6.9	75.9	134	2	2	103	125	37.1
20	9.35	53.8	4.1	15.9	80.9	833	2	3	547	519	77.1
21	7.53	42.0	4.2	23.1	98.9	95	2	4	47	49	17.1
22	10.24	49.0	4.8	36.3	112.6	195	2	2	163	170	37.1
23	9.78	52.3	5.0	17.6	95.9	270	1	1	240	198	57.1
24	9.84	62.2	4.8	12.0	82.3	600	2	3	468	497	57.1
25	9.20	52.2	4.0	17.5	71.1	298	1	4	244	236	57.1
26	8.28	49.5	3.9	12.0	113.1	546	1	2	413	436	57.1
27	9.31	47.2	4.5	30.2	101.3	170	2	1	124	173	37.1
28	8.19	52.1	3.2	10.8	59.2	176	2	1	156	88	37.1
29	11.65	54.5	4.4	18.6	96.1	248	2	1	217	189	37.1
30	9.89	50.5	4.9	17.7	103.6	167	2	2	113	106	37.1
31	11.03	49.9	5.0	19.7	102.1	318	2	1	270	335	57.1
32	9.84	53.0	5.2	17.7	72.6	210	2	2	200	239	54.3
33	11.77	54.1	5.3	17.3	56.0	196	2	1	164	165	34.3
34	13.59	54.0	6.1	24.2	111.7	312	2	1	258	169	54.3
35	9.74	54.4	6.3	11.4	76.1	221	2	2	170	172	54.3
36	10.33	55.8	5.0	21.2	104.3	266	2	1	181	149	54.3
37	9.97	58.2	2.8	16.5	76.5	90	2	2	69	42	34.3
38	7.84	49.1	4.6	7.1	87.9	60	2	3	50	45	34.3
39	10.47	53.2	4.1	5.7	69.1	196	2	2	168	153	54.3
40	8.16	60.9	1.3	1.9	58.0	73	2	3	49	21	14.3
41	8.48	51.1	3.7	12.1	92.8	166	2	3	145	118	34.3
42	10.72	53.8	4.7	23.2	94.1	113	2	3	90	107	34.3
43	11.20	45.0	3.0	7.0	78.9	130	2	3	95	56	34.3
44	10.12	51.7	5.6	14.9	79.1	362	1	3	313	264	54.3
45	8.37	50.7	5.5	15.1	84.8	115	2	2	96	88	34.3
46	10.16	54.2	4.6	8.4	51.5	831	1	4	581	629	74.3
47	19.56	59.9	6.5	17.2	113.7	306	2	1	273	172	51.4
48	10.90	57.2	5.5	10.6	71.9	593	2	2	446	211	51.4
49	7.67	51.7	1.8	2.5	40.4	106	2	3	93	35	11.4
50	8.88	51.5	4.2	10.1	86.9	305	2	3	238	197	51.4
51	11.48	57.6	5.6	20.3	82.0	252	2	1	207	251	51.4
52	9.23	51.6	4.3	11.6	42.6	620	2	2	413	420	71.4
53	11.41	61.1	7.6	16.6	97.9	535	2	3	330	273	51.4
54	12.07	43.7	7.8	52.4	105.3	157	2	2	115	76	31.4
55	8.63	54.0	3.1	8.4	56.2	76	2	1	39	44	31.4
56	11.15	56.5	3.9	7.7	73.9	281	2	1	217	199	51.4

1	2	3	4	5	6	7	8	9	10	11	12
57	7.14	59.0	3.7	2.6	75.8	70	2	4	37	35	31.4
58	7.65	47.1	4.3	16.4	65.7	318	2	4	265	314	51.4
59	10.73	50.6	3.9	19.3	101.0	445	1	2	374	345	51.4
60	11.46	56.9	4.5	15.6	97.7	191	2	3	153	132	31.4
61	10.42	58.0	3.4	8.0	59.0	119	2	1	67	64	31.4
62	11.18	51.0	5.7	18.8	55.9	595	1	2	546	392	68.6
63	7.93	64.1	5.4	7.5	98.1	68	2	4	42	49	28.6
64	9.66	52.1	4.4	9.9	98.3	83	2	2	66	95	28.6
65	7.78	45.5	5.0	20.9	71.6	489	2	3	391	329	48.6
66	9.42	50.6	4.3	24.8	62.8	508	2	1	421	528	48.6
67	10.02	49.5	4.4	8.3	93.0	265	2	2	191	202	48.6
68	8.58	55.0	3.7	7.4	95.9	304	2	3	248	218	48.6
69	9.61	52.4	4.5	6.9	87.2	487	2	3	404	220	48.6
70	8.03	54.2	3.5	24.3	87.3	97	2	1	65	55	28.6
71	7.39	51.0	4.2	14.6	88.4	72	2	2	38	67	28.6
72	7.08	52.0	2.0	12.3	56.4	87	2	3	52	57	28.6
73	9.53	51.5	5.2	15.0	65.7	298	2	3	241	193	48.6
74	10.05	52.0	4.5	36.7	87.5	184	1	1	144	151	68.6
75	8.45	38.8	3.4	12.9	85.0	235	2	2	143	124	48.6
76	6.70	48.6	4.5	13.0	80.8	76	2	4	51	79	28.6
77	8.90	49.7	2.9	12.7	86.9	52	2	1	37	35	28.6
78	10.23	53.2	4.9	9.9	77.9	752	1	2	595	446	68.6
79	8.88	55.8	4.4	14.1	76.8	237	2	2	165	182	48.6
80	10.30	59.6	5.1	27.8	88.9	175	2	2	113	73	45.7
81	10.79	44.2	2.9	2.6	56.6	461	1	2	320	196	65.7
82	7.94	49.5	3.5	6.2	92.3	195	2	2	139	116	45.7
83	7.63	52.1	5.5	11.6	61.1	197	2	4	109	110	45.7
84	8.77	54.5	4.7	5.2	47.0	143	2	4	85	87	25.7
85	8.09	56.9	1.7	7.6	56.9	92	2	3	61	61	45.7
86	9.05	51.2	4.1	20.5	79.8	195	2	3	127	112	45.7
87	7.91	52.8	2.9	11.9	79.5	477	2	3	349	188	65.7
88	10.39	54.6	4.3	14.0	88.3	353	2	2	223	200	65.7
89	9.36	54.1	4.8	18.3	90.6	165	2	1	127	158	45.7
90	11.41	50.4	5.8	23.8	73.0	424	1	3	359	335	45.7
91	8.86	51.3	2.9	9.5	87.5	100	2	3	65	53	25.7
92	8.93	56.0	2.0	6.2	72.5	95	2	3	59	56	25.7
93	8.92	53.9	1.3	2.2	79.5	56	2	2	40	14	5.7
94	8.15	54.9	5.3	12.3	79.8	99	2	4	55	71	25.7
95	9.77	50.2	5.3	15.7	89.7	154	2	2	123	148	25.7
96	8.54	56.1	2.5	27.0	82.5	98	2	1	57	75	45.7
97	8.66	52.8	3.8	6.8	69.5	246	2	3	178	177	45.7
98	12.01	52.8	4.8	10.8	96.9	298	2	1	237	115	45.7
99	7.95	51.8	2.3	4.6	54.9	163	2	3	128	93	42.9
100	10.15	51.9	6.2	16.4	59.2	568	1	3	452	371	62.9
101	9.76	53.2	2.6	6.9	80.1	64	2	4	47	55	22.9
102	9.89	45.2	4.3	11.8	108.7	190	2	1	141	112	42.9
103	7.14	57.6	2.7	13.1	92.6	92	2	4	40	50	22.9
104	13.95	65.9	6.6	15.6	133.5	356	2	1	308	182	62.9
105	9.44	52.5	4.5	10.9	58.5	297	2	3	230	263	42.9
106	10.80	63.9	2.9	1.6	57.4	130	2	3	69	62	22.9
107	7.14	51.7	1.4	4.1	45.7	115	2	3	90	19	22.9
108	8.02	55.0	2.1	3.8	46.5	91	2	2	44	32	22.9
109	11.80	53.8	5.7	9.1	116.9	571	1	2	441	469	62.9
110	9.50	49.3	5.8	42.0	70.9	98	2	3	68	46	22.9
111	7.70	56.9	4.4	12.2	67.9	129	2	4	85	136	62.9
112	17.94	56.2	5.9	26.4	91.8	835	1	1	791	407	62.9
113	9.41	59.5	3.1	20.6	91.7	29	2	3	20	22	22.9

SMSA data set

This data set provides information for 141 large Standard Metropolitan Statistical Areas (SMSA's) in the United States. A standard metropolitan statistical area includes a city (or cities) of specified population size which constitutes the central city and the county (or counties) in which it is located, as well as contiguous counties when the economic and social relationships between the central and contiguous counties meet specified criteria of metropolitan character and integration. An SMSA may have up to three central cities and may cross state lines.

Each line of the data set has an identification number and provides information on 11 other variables for a single SMSA. The information generally pertains to the years 1976 and 1977, the most recent information available at the time. The 12 variables are:

Variable Number	Variable Name	Description
1	Identification number	1-141
2	Land area	In square miles
3	Total population	Estimated 1977 population (in thousands)
4	Percent of population in central cities	Percent of 1976 SMSA population in central city or cities
5	Percent of population 65 or older	Percent of 1976 SMSA population 65 years old or older

SMSA data set *(continued)*

Variable Number	Variable Name	Description
6	Number of active physicians	Number of professionally active nonfederal physicians as of December 31, 1977
7	Number of hospital beds	Total number of beds, cribs, and bassinettes during 1977
8	Percent high school graduates	Percent of adult population (persons 25 years old or older) who completed 12 or more years of school, according to the 1970 Census of the Population
9	Civilian labor force	Total number of persons in civilian labor force (persons 16 years old or older classified as employed or unemployed) in 1977 (in thousands)
10	Total personal income	Total current income received in 1976 by residents of the SMSA from all sources, before deduction of income and other personal taxes but after deduction of personal contributions to social security and other social insurance programs (in millions of dollars)
11	Total serious crimes	Total number of serious crimes in 1977, including murder, rape, robbery, aggravated assault, burglary, larceny-theft, and motor vehicle theft, as reported by law enforcement agencies
12	Geographic region	Geographic region classification is that used by the U.S. Bureau of the Census, where: 1 = NE, 2 = NC, 3 = S, 4 = W

Data obtained from: U.S. Bureau of the Census, *State and Metropolitan Area Data Book, 1979* (a Statistical Abstract Supplement).

1	2	3	4	5	6	7	8	9	10	11	12
1	1384	9387	78.1	12.3	25627	69678	50.1	4083.9	72100	709234	1
2	4069	7031	44.0	10.0	15389	39699	62.0	3353.6	52737	499813	4
3	3719	7017	43.9	9.4	13326	43292	53.9	3305.9	54542	393162	2
4	3553	4794	37.4	10.7	9724	33731	50.6	2066.3	33216	198102	1
5	3916	4370	29.9	8.8	6402	24167	52.2	1966.7	32906	294466	2
6	2480	3182	31.5	10.5	8502	16751	66.1	1514.5	26573	255162	4
7	2815	3033	23.1	6.7	7340	16941	68.3	1541.9	25663	177355	3
8	1218	2688	0.0	8.8	5255	22137	62.9	1213.3	21524	127567	1
9	8360	2673	46.3	8.2	4047	14347	53.6	1321.2	18350	193125	3
10	6794	2512	60.1	6.3	4562	14333	51.7	1272.7	18221	162976	3
11	4935	2380	21.8	11.0	4071	17752	47.8	1061.2	16120	137479	2
12	3049	2294	19.5	12.1	4005	21149	53.4	967.5	15826	69989	1
13	2259	2147	38.6	9.3	5141	16485	44.6	966.8	14246	138214	3
14	4647	2037	31.5	9.2	3916	12815	65.1	1032.2	14542	112642	2
15	1008	1969	16.6	10.3	4006	16704	55.9	935.5	15953	106646	1
16	1519	1950	31.8	10.5	4094	12545	54.6	906.0	14684	102816	2
17	4326	1832	23.6	7.3	3064	9976	50.4	867.2	12107	106482	3
18	782	1801	28.4	7.8	3119	8656	70.5	915.2	12591	113821	4
19	4261	1683	48.6	9.7	3396	7552	65.3	644.3	10392	112359	4
20	4651	1464	38.8	7.7	3380	8517	67.4	729.2	10375	116861	4
21	2042	1441	24.5	16.5	4071	10039	51.9	681.7	10166	116304	3
22	4226	1427	38.1	9.8	3285	5392	67.8	699.8	10918	91399	4
23	1456	1427	46.7	10.4	2484	8555	56.8	710.4	10104	63695	2
24	2045	1380	37.2	21.4	1949	8863	50.7	543.2	7989	89257	3
25	2149	1375	29.8	10.6	2530	8354	48.4	617.6	9037	68319	2
26	1590	1313	30.1	10.9	2296	9988	50.4	565.7	8411	67965	1
27	27293	1306	25.3	12.3	2018	6323	57.4	510.6	7399	99293	4
28	3341	1293	35.8	10.1	2289	7593	59.9	656.3	9106	81510	2
29	9155	1254	53.8	11.1	2280	6450	60.1	575.2	7766	107370	4
30	1300	1217	47.6	6.8	2794	4989	69.0	610.8	9215	76570	4
31	3072	1144	68.0	9.3	2181	7497	56.0	549.6	7736	61381	2
32	1967	1133	51.1	8.8	2520	8467	45.8	460.5	7038	69285	3
33	3650	1121	34.6	11.1	2358	6224	62.9	539.3	7792	77316	4
34	2460	1087	49.6	8.4	1874	7706	59.9	510.7	6658	62603	2
35	2527	1025	78.7	8.4	1760	7664	46.5	391.1	5582	62694	3
36	2966	970	26.9	10.3	2053	6604	56.3	450.4	6966	54854	1
37	3434	929	28.9	8.3	1844	3215	65.1	422.6	5909	72410	4
38	1392	883	37.2	9.8	1579	6087	46.5	396.8	5705	45642	3
39	2298	886	76.2	9.0	1644	7673	48.2	394.6	5185	52094	3
40	1219	864	31.7	20.6	1396	6158	55.4	352.8	5879	68109	3
41	1708	833	24.0	8.8	1062	5315	56.2	367.5	5489	52606	2
42	8565	822	29.7	7.3	1604	3485	67.6	349.3	4655	49111	4
43	3358	805	35.1	11.3	1649	5512	44.0	359.1	4941	42786	3
44	2624	794	30.4	12.2	1532	4730	55.2	356.5	5094	30771	1
45	2187	777	47.0	10.2	1098	4342	51.9	355.4	5142	46213	2
46	3214	774	47.7	9.4	1285	3459	40.3	401.7	4924	34941	3
47	3491	769	48.5	9.7	1496	5620	59.6	362.3	4798	44513	3
48	4080	773	59.6	9.9	1597	7496	47.3	380.9	4600	33936	3
49	596	723	100.0	6.0	1260	2819	66.0	319.9	5181	46984	4
50	3199	694	80.6	8.7	983	4749	50.8	292.4	4127	43010	3
51	903	661	37.3	9.6	948	4064	55.6	293.3	4102	34725	2
52	2419	647	27.8	9.9	1250	2870	57.8	286.8	3860	30829	1
53	938	644	48.1	7.4	614	3016	50.0	280.9	4177	35106	2
54	1951	629	28.4	14.5	696	4843	47.9	271.5	3667	14868	1
55	1490	624	33.1	11.9	827	3818	47.4	300.2	4144	19090	1
56	5677	610	55.8	10.5	760	3883	56.2	292.0	4035	32146	3
57	1525	597	55.7	8.3	751	3234	44.9	318.5	3777	37070	3
58	2528	593	19.2	10.2	798	3135	55.4	274.1	3489	44442	3
59	312	594	19.5	7.5	769	2463	55.0	298.7	4352	29100	1
60	1537	581	63.8	8.7	1234	5160	62.7	272.6	3725	32271	2
61	1420	576	32.6	9.5	833	2950	54.0	280.8	3553	26645	2
62	47	564	41.9	11.9	745	3352	36.3	258.9	3915	29157	1
63	1023	541	35.1	10.0	639	3144	52.1	234.1	3437	22111	2
64	2115	526	19.9	9.1	676	2296	38.8	253.3	2962	30684	3
65	1182	514	32.4	7.4	518	2515	52.4	216.8	3627	35201	2
66	1165	516	14.5	8.6	746	4277	54.4	237.1	3724	31358	3
67	476	492	8.9	10.9	787	2778	60.1	218.4	3603	24787	1
68	1553	487	50.0	8.0	2207	4931	52.0	257.2	2991	24269	3
69	2023	477	22.1	21.8	752	2317	55.7	194.2	3283	36418	3
70	2766	474	67.9	7.7	679	3873	56.3	224.0	2598	29967	3

1	2	3	4	5	6	7	8	9	10	11	12
71	5966	472	39.5	9.6	737	1907	52.7	246.6	3007	38205	4
72	1863	468	50.4	7.7	674	2989	63.8	194.8	2747	25159	4
73	192	462	60.5	10.8	617	1789	44.1	212.6	3158	27161	1
74	9240	455	67.0	10.3	1123	2347	63.1	183.6	2598	41649	4
75	2277	455	39.5	7.5	512	1788	61.9	221.1	2853	20053	2
76	1630	449	41.9	10.7	724	4395	50.0	198.0	2445	17596	3
77	1617	435	71.0	6.9	518	2031	54.1	197.9	2617	31539	3
78	1057	435	90.7	6.1	479	2551	51.1	163.4	2012	25650	3
79	1624	429	13.4	11.0	832	2938	55.4	207.8	2885	16985	1
80	1676	423	36.6	9.2	505	3297	60.7	156.3	2689	24266	4
81	2818	425	48.5	9.3	540	2694	42.3	172.8	2162	22374	3
82	2866	408	24.9	10.7	427	2864	39.1	169.1	1987	10425	3
83	4883	402	72.4	7.3	873	2236	64.9	185.2	2353	28171	4
84	966	401	24.9	10.6	427	3192	52.2	174.7	2446	15981	2
85	2109	403	41.2	10.3	520	2539	45.2	183.1	2308	16240	3
86	2449	395	68.4	9.6	681	2864	63.2	207.4	2651	25149	2
87	2618	385	31.7	6.1	836	2159	48.0	145.6	1992	25046	3
88	1465	374	30.3	6.8	598	6456	50.6	164.7	2201	26428	3
89	1704	375	52.1	10.5	379	2491	55.6	173.2	2662	18599	2
90	1750	370	49.3	9.7	446	3472	58.2	176.5	2439	16529	2
91	1489	369	58.8	9.5	911	5720	56.5	175.1	2264	26032	3
92	8152	363	22.3	9.1	405	1254	51.7	165.6	2257	28351	4
93	2207	364	57.3	9.7	356	2167	45.5	165.9	2331	19138	3
94	7874	360	44.4	6.9	398	1365	65.2	174.2	2410	33687	4
95	655	364	75.2	6.6	425	3879	51.6	163.0	2088	15623	3
96	1803	362	35.3	10.4	483	2137	53.7	168.9	2666	16405	2
97	2363	356	53.1	10.6	565	2717	49.3	146.4	1996	19212	3
98	1435	352	13.4	11.7	342	1076	44.7	156.8	2165	11273	1
99	946	348	16.4	11.1	366	1455	43.9	163.8	2178	8116	1
100	1136	333	58.6	9.7	448	2630	68.1	171.4	2396	20465	2
101	2658	327	39.0	12.2	365	5430	49.9	136.9	1862	9325	1
102	228	317	31.1	10.2	667	3179	52.8	156.5	2264	19410	1
103	1758	310	56.8	11.5	565	2081	65.3	131.2	1939	17379	4
104	1198	313	55.1	8.0	1171	3877	71.2	172.3	2038	18676	2
105	1412	311	39.2	11.3	436	1837	49.4	154.2	2098	25714	4
106	2071	306	19.9	11.3	470	2531	58.9	133.1	1782	11161	1
107	862	302	26.3	13.4	423	1929	43.3	145.5	2010	7699	1
108	1526	303	71.7	7.7	413	1636	47.1	125.8	1692	20038	3
109	1758	297	33.2	11.6	296	2652	45.3	114.4	1641	12467	3
110	1651	296	64.6	8.9	774	5431	56.1	136.9	1724	14468	3
111	1493	294	64.8	8.9	863	3289	53.7	154.7	1787	15871	3
112	1610	294	59.8	9.5	471	4633	62.9	116.1	1851	18651	4
113	2710	288	63.7	6.2	357	1277	72.8	110.9	1639	18173	4
114	1975	291	46.5	12.6	405	2896	51.5	133.8	1853	12787	2
115	1920	291	49.8	7.8	283	1306	53.2	126.9	1553	12315	3
116	1404	289	38.5	10.0	299	1766	56.2	138.6	1776	11715	2
117	2737	287	45.0	10.5	602	1462	71.3	131.4	1980	18208	4
118	1700	287	18.8	8.0	739	3381	45.9	120.4	1616	14534	3
119	909	277	41.2	11.5	307	1309	54.2	131.9	1762	13722	2
120	1858	277	24.3	13.7	354	1562	46.3	116.9	1507	19133	3
121	3324	275	49.7	8.4	373	929	62.5	120.5	1918	14776	4
122	1697	274	23.8	7.2	338	1610	51.0	105.9	1354	19317	3
123	813	272	46.0	9.8	293	1693	58.4	119.9	1688	10402	1
124	7397	267	47.3	12.5	355	2042	56.2	113.7	1654	12273	2
125	1165	268	43.7	9.4	450	2070	57.5	129.4	1719	16226	2
126	802	268	52.6	9.8	392	1425	52.2	129.6	1816	13230	2
127	1770	268	14.8	12.2	285	2804	44.1	106.7	1537	4205	1
128	495	264	50.7	7.8	220	1177	52.6	119.5	1661	8398	2
129	1255	261	26.0	10.7	458	1646	51.6	113.0	1725	10208	3
130	1148	589	45.3	11.1	891	5790	54.0	277.0	3510	29237	1
131	1509	643	37.6	12.0	1087	4900	51.4	319.6	3982	29058	1
132	2013	254	61.7	9.7	273	1484	50.9	106.7	1412	14446	3
133	711	250	42.4	6.1	1411	3659	67.5	131.0	1790	16228	2
134	471	251	46.3	8.6	219	1128	47.8	105.3	1458	13474	2
135	4552	249	54.4	9.1	329	719	61.9	118.0	1386	15596	4
136	1400	242	50.8	8.0	290	1271	45.7	104.4	1351	10391	3
137	1511	236	38.7	10.7	348	1093	50.4	127.2	1452	16676	4
138	1543	232	39.6	8.1	159	481	30.3	80.6	769	8436	3
139	1011	233	37.8	10.5	264	964	70.7	93.2	1337	14018	3
140	813	232	13.4	10.9	371	4355	58.0	97.0	1589	8428	1
141	654	231	28.8	3.9	140	1296	55.1	66.9	1148	15884	3

SMSA data set *(concluded)*

SMSA Identifications

1 NEW YORK, NY	48 NASHVILLE, TN	95 NEWPORT NEWS, VA
2 LOS ANGELES, CA	49 HONOLULU, HI	96 PEORIA, IL
3 CHICAGO, IL	50 JACKSONVILLE, FL	97 SHREVEPORT, LA
4 PHILADELPHIA, PA	51 AKRON, OH	98 YORK, PA
5 DETROIT, MI	52 SYRACUSE, NY	99 LANCASTER, PA
6 SAN FRANCISCO, CA	53 GARY, IN	100 DES MOINES, IA
7 WASHINGTON, DC	54 NORTHEAST, PA	101 UTICA, NY
8 NASSAU, NY	55 ALLENTOWN, PA	102 TRENTON, NJ
9 DALLAS, TX	56 TULSA, OK	103 SPOKANE, WA
10 HOUSTON, TX	57 CHARLOTTE, NC	104 MADISON, WI
11 ST.LOUIS, MO	58 ORLANDO, FL	105 STOCKTON, CA
12 PITTSBURG, PA	59 NEW BRUNSWICK, NJ	106 BINGHAMTON, NY
13 BALTIMORE, MD	60 OMAHA, NE	107 READING, PA
14 MINNEAPOLIS, MN	61 GRAND RAPIDS, MI	108 CORPUS CHRISTI, TX
15 NEWARK, NJ	62 JERSEY CITY, NJ	109 HUNTINGTON, WV
16 CLEVELAND, OH	63 YOUNGSTOWN, OH	110 JACKSON, MS
17 ATLANTA, GA	64 GREENVILLE, SC	111 LEXINGTON, KY
18 ANAHEIM, CA	65 FLINT, MI	112 VALLEJO, CA
19 SAN DIEGO, CA	66 WILMINGTON, DE	113 COLORADO SPRINGS, CO
20 DENVER, CO	67 LONG BRANCH, NJ	114 EVANSVILLE, IN
21 MIAMI, FL	68 RALEIGH, NC	115 HUNTSVILLE, AL
22 SEATTLE, WA	69 W. PALM BEACH, FL	116 APPLETON, WI
23 MILWAUKEE, WI	70 AUSTIN, TX	117 SANTA BARBARA, CA
24 TAMPA, FL	71 FRESNO, CA	118 AUGUSTA, GA
25 CINCINNATI, OH	72 OXNARD, CA	119 SOUTH BEND, IN
26 BUFFALO, NY	73 PATERSON, NJ	120 LAKELAND, FL
27 RIVERSIDE, CA	74 TUCSON, AZ	121 SALINAS, CA
28 KANSAS CITY, MO	75 LANSING, MI	122 PENSACOLA, FL
29 PHOENIX, AZ	76 KNOXVILLE, TN	123 ERIE, PA
30 SAN JOSE, CA	77 BATON ROUGE, LA	124 DULUTH, MN
31 INDIANAPOLIS, IN	78 EL PASO, TX	125 KALAMAZOO, MI
32 NEW ORLEANS, LA	79 HARRISBURG, PA	126 ROCKFORD, IL
33 PORTLAND, OR	80 TACOMA, WA	127 JOHNSTOWN, PA
34 COLUMBUS, OH	81 MOBILE, AL	128 LORAIN, OH
35 SAN ANTONIO, TX	82 JOHNSON CITY, TN	129 CHARLESTON, WV
36 ROCHESTER, NY	83 ALBUQUERQUE, NM	130 SPRINGFIELD, MA
37 SACRAMENTO, CA	84 CANTON, OH	131 WORCESTER, MA
38 LOUISVILLE, KY	85 CHATANOOGA, TN	132 MONTGOMERY, AL
39 MEMPHIS, TN	86 WICHITA, KS	133 ANN ARBOR, MI
40 FT. LAUDERDALE, FL	87 CHARLESTON, SC	134 HAMILTON, OH
41 DAYTON, OH	88 COLUMBIA, SC	135 EUGENE, OR
42 SALT LAKE CITY, UT	89 DAVENPORT, IA	136 MACON, GA
43 BIRMINGHAM, AL	90 FORT WAYNE, IN	137 MODESTO, CA
44 ALBANY, NY	91 LITTLE ROCK, AR	138 MCALLEN, TX
45 TOLEDO, OH	92 BAKERSFIELD, CA	139 MELBOURNE, FL
46 GREENSBORO, NC	93 BEAUMONT, TX	140 POUGHKEEPSIE, NY
47 OKLAHOMA CITY, OK	94 LAS VEGAS, NV	141 FAYETTEVILLE, NC

Index

A

Addition theorem, 2
Adjusted coefficient of multiple determination, 241–42
All-possible-regressions selection procedure, 421–29
Allocated codes, 351–52
Analysis of variance, 84–86
Analysis of variance models, 343
Analysis of variance table, 89–90
ANOVA table, 89–90
Asymptotic normality, 70
Autocorrelation, 444–48
    remedial measures, 454–60
    test for, 450–54
Autocorrelation parameter, 448
Autoregressive error model; *see* Regression model

B

Backward elimination selection procedure, 435–36
Berkson model, 166–67
"Best" subsets algorithms, 429
Beta coefficient, 262
Biased estimation, 394–95
Binary variable, 330, 354; *see also* Indicator variable
Bivariate normal distribution, 492–96
BMDP, 113, 251, 431
Bonferroni joint estimation procedure
    for inverse predictions, 174
    for mean responses, 158–59, 245
    for prediction of new observations, 159–60, 246–47
    for regression coefficients, 150–54, 243

C

C_p criterion, 426–28
Calibration problem, 174
Central limit theorem, 6
Chi-square distribution, 7–8
    table of percentiles, 520
Cochran's theorem, 92
Coefficient of multiple correlation, 242, 506
    inferences, 506–7
Coefficient of multiple determination, 241, 506
    adjusted, 241–42
    inferences, 506–7
Coefficient of partial correlation, 288–89, 507–8
    first-order, 508
    inferences, 508
    second-order, 508
Coefficient of partial determination, 286–88, 507–8
    inferences, 508
Coefficient of simple correlation, 97–99, 494
    inferences, 502–4
Coefficient of simple determination, 96–97, 501–2
    inferences, 502–4
Column vector, 187
Complementary event, 3
Conditional probability, 3

Conditional probability function, 4
Confidence coefficient, interpretation of, 70–71, 84
Confidence set, 152
Consistent estimator, 9
Contour diagram, 235–37, 494–95
Cook's distance measure, 407–9
Correction for mean sum of squares, 90
Correlation coefficient; *see* Coefficient of multiple correlation; Coefficient of partial correlation; *and* Coefficient of simple correlation
Correlation index, 312
Correlation matrix, 382
  of the independent variables, 381
Correlation model, 491–92
  bivariate normal, 492–96
  multivariate normal, 505
Correlation transformation, 378–79
Covariance
  of two functions of random variables, 6
  of two random variables, 5
Covariance models, 343
Cox, D. R., 176

D

Degrees of freedom, 7–8
Deleted residual, 405–6
Denominator degrees of freedom, 8
Dependent variable, 25, 28
Determinant of matrix, 202
Diagonal matrix, 196–97
Disturbance term, 445
Dummy variable, 34, 330; *see also* Indicator variable
Durbin-Watson test, 450–54
  table of test bounds, 530–31

E

Error mean square, 47
Error sum of squares, 47
Error term, 31
  nonconstancy of error variance, 113–14, 123, 133
  nonindependence of, 116–18, 123, 133
  nonnormality of, 118–20, 123, 133
Error term variance, 31, 46–48, 50
Expected mean square, 90–91
Expected value
  of function of random variables, 5–6
  of random variable, 3
Experimental data, 35
Exponential regression function, 468
Extra sum of squares, 282–86

F

F distribution, 8–9
  table of percentiles, 521–27
Family of estimates, 150
Family confidence coefficient, 150

First differences, 458–59
First-order autoregressive error model, 448–50
  first differences approach, 458–60
  iterative estimation approach, 455–58
  test for autocorrelation, 450–54
First-order regression model, 31, 227, 229–30; *see also* Regression model
Fisher, R. A., 503
Fitted value, 41
  in terms of hat matrix, 401
Forward selection procedure, 435
Full model, 95
Functional relation, 24

G–H

Gauss-Markov theorem, 39–40, 64
Gauss-Newton method, 472–79
General linear regression model, 230–34, 237–38
General linear test, 94–96, 293–96

Hat matrix, 220–21
Heteroscedasticity, 170
Homoscedasticity, 170
Hyperplane, 230

I

Idempotent matrix, 221
Identity matrix, 197
Independence of random variables, 5
Independent variable, 25, 28
Indicator variable, 329–30, 353–54
  in comparing regression functions, 343–45
  as dependent variable, 354–57
  in piecewise linear regression, 346–50
  in time series model, 350–51
Influential observations, 407–9
Instrumental variable, 165–66
Interaction effect, 232–37
  with indicator variables, 335–39
Interaction effect coefficient, 304
Intrinsically linear regression model, 467
Inverse of matrix, 200–204
Inverse prediction, 172–74

J–L

Joint confidence region for regression coefficients, 147–50, 217, 243
Joint probability function, 4

Lack of fit mean square, 129
Lack of fit sum of squares, 128
Lack of fit test, 123–32, 245–46
Least absolute deviations estimation, 410–11
Least squares criterion, 36
Least squares estimation, 10
  control of roundoff errors, 377–82
  multiple regression, 238–39
  nonlinear regression, 470–80
  simple linear regression, 36–40, 44–46, 210–12
  weighted, 167–72, 219–20, 263

Leverage, 402
Likelihood function, 9
Linear dependence, 199–200
Linear effect coefficient, 301
Linear model, 31, 466–67; *see also* General
    linear regression model
  general linear test, 94–96, 293–96
Linear regression model; *see* Regression model
Linearity, test for, 123–32
Logistic regression function, 361–62, 468–69
Logistic transformation, 362
Logit transformation, 362

M

Marginal probability function, 4
Marquardt algorithm, 479–80
Matrix
  addition, 190–91
  with all elements 1, 198
  definition, 185–87
  determinant, 202
  diagonal, 196–97
  dimension, 186
  elements, 186
  equality of two, 189
  hat, 220–21
  idempotent, 221
  identity, 197
  inverse, 200–204
  multiplication by matrix, 192–96
  multiplication by scalar, 192
  nonsingular, 201
  of quadratic form, 215
  random, 205–8
  rank, 200
  scalar, 197–98
  singular, 201
  square, 187
  subtraction, 190–91
  symmetric, 196
  theorems, 204–5
  transpose, 188–89
  vector, 187–88
  zero vector, 198–99
Maximum likelihood estimation, 9–10
  of regression parameters, 50–51
Mean, of population
  estimation of
    difference between two, 14–16
    single, 11
  test concerning
    difference between two, 14–16
    single, 11–12
Mean response, 41
  multiple regression
    estimation, 244
    joint estimation, 245
  simple linear regression
    interval estimation, 75–76, 217

Mean response—*Cont.*
  simple linear regression—*Cont.*
    joint estimation, 157–59
    point estimation, 41–43
Mean square, 46, 88
  expected value of, 90–91
Mean squared error
  of regression coefficient, 395
  total, of n fitted values, 426
Measurement errors in observations, 164–67
Method of steepest descent, 479
Minimum absolute deviations method, 411
Minimum L_1-norm method, 411
Minimum sum of absolute deviations method, 411
Minimum variance estimator, 9
MSE_p criterion, 423–25
Multicollinearity, 271–78, 382–90
  detection of, 390–93
  remedial measures, 393–400
Multiple correlation; *see* Coefficient of multiple
    correlation
Multiple regression; *see* Mean response;
    Prediction of new observation; Regression
    coefficients; Regression function; Regression
    model; *and* Selection of independent
    variables
Multiplication theorem, 3
Multivariate normal distribution, 505

N

Noncentrality parameter, 71
Nonexperimental data, 35
Nonlinear regression model, 468–69
  inferences about parameters, 480–83
  least squares estimation, 470–80
Nonsingular matrix, 201
Normal equations, 38
Normal error regression model; *see* Regression
    model
Normal probability distribution, 6–7
  table of areas and percentiles, 517
Normal probability plot, 118–20
Numerator degrees of freedom, 8

O

Observation, 25
Observational data, 35
Observed value, 41
Orthogonal polynomials, 319
Outlier, 114–16, 123
  identification of, 400–407
Overall F test, 281, 289

P

P-value, 12–13
Paired observations, 15–16
Partial correlation; *see* Coefficient of partial
    correlation
Partial F test, 281, 289

Partial regression coefficient, 229
Piecewise linear regression, 346–50
Point estimator, 38
Polynomial regression model, 300–305
Power of tests for regression coefficients, 71–72
Prediction, of new observation
  inverse, 172–74
  multiple regression, 246–47
  simple linear regression, 76–82, 159–60,
    218–19
Prediction bias, 437
Prediction interval, 77–78
Predictor variable, 25, 28
Probit transformation, 366
Product operator, 2
Pure error mean square, 127–28
Pure error sum of squares, 127

Q

Quadratic effect coefficient, 301
Quadratic form, 215
Quadratic response function, 301
  estimation of maximum or minimum, 317–19
Quantal response, 354

R

R_a^2 criterion, 423–25
R_p^2 criterion, 422–23
Random matrix, 205–8
Random vector, 205–8
Rank of matrix, 200
Reduced model, 95
Regression; *see* Mean response; Prediction of new
  observation; Regression coefficients;
  Regression function; *and* Regression model
Regression, through origin, 160–64
Regression coefficients
  multiple regression, 227–29
    danger in simultaneous tests, 278–82
    interval estimation, 243
    joint estimation, 243
    point estimation, 238–39, 263
    tests concerning, 243, 285–86, 289–93
    variance-covariance matrix of, 242, 263
  partial, 229
  simple linear regression, 33–34
    interval estimation, 65–67, 69–70
    joint estimation, 147–54, 217
    point estimation, 36–40, 50–51, 167–72,
     210–12, 219–20
    tests concerning, 67–68, 71–72, 92–94
    variance-covariance matrix of, 216–17
  standardized, 261–63
Regression curve, 27–28; *see also* Regression
  function
Regression function, 27–28
  comparison of two or more, 343–45
  confidence band, simple linear regression,
    154–57

Regression function—*Cont.*
  confidence region, multiple regression, 244
  estimated regression function, 41–43
  test for fit, 123–32, 245–46
  test for regression relation, 92–93, 240–41
  transformations to linearize, 134–41
Regression mean square, 88
Regression model, 26–29
  effect of measurement errors, 164–67
  first-order autoregressive, 448–50
  general linear, 230–34, 237–38
  multiple, 226–30
    with interaction effects, 232–37, 335–39
    in matrix terms, 237–38
  nonlinear, 468–69
  polynomial, 300–305
  residual analysis for aptness, 111–22
  scope of, 30, 261
  second-order autoregressive, 460
  selection of independent variables
    all possible regressions, 421–29
    backward elimination, 435–36
    forward selection, 435
    ridge regression, 436
    stepwise regression, 430–35
  simple linear
    error term distribution unspecified, 31–35
    in matrix terms, 208–10
    normal error terms, 48–49
    through origin, 160–64
    X is random, 83–84
  uses of, 30–31
Regression sum of squares, 86
  decompositions, 284–86
Regression surface, 227–28
Replication, 124
Residual, 43–44
  deleted, 405–6
  properties of, 110–11
  standardized, 110
  studentized, 405
  studentized deleted, 406
  in terms of hat matrix, 220–21
  variance-covariance matrix of, 221, 402
Residual analysis, 111–22
Residual mean square, 47
Residual plot, 111–13
Residual sum of squares, 47
Response, 41
  binary or quantal, 354
Response function; *see* Regression function
Response surface, 227–28
Response variable, 25, 28
Restricted model, 95
Ridge regression, 394–400
  use for selecting independent variables, 436
Ridge trace, 396–97
Roundoff errors in least squares calculations,
  377–78
Row vector, 188

S

SAS, 259
Scalar, 192
Scalar matrix, 197–98
Scatter diagram, 25
Scatter plot, 25
Scheffé joint estimation procedure
  for inverse predictions, 174
  for prediction of new observations, 159–60,
    246
Scope of model, 30, 261
Second-order autoregressive error model, 460
Second-order regression model, 300–301, 303–5;
  see also Regression model
Selection of independent variables, 29, 417–19
SENIC data set, 533–36
Serial correlation, 444
Simple linear regression model, 31; *see also*
  Regression model
Singular matrix, 201
SMSA data set, 537–41
SPSS, 52
Square matrix, 187
Standard normal variable, 7
Standardized regression coefficient, 261–63
Standardized residual, 110
Statement confidence coefficient, 150
Statistical relation, 25–26
Stepwise regression selection procedure, 430–35
Studentized deleted residual, 406
Studentized residual, 405
Sufficient estimator, 9
Sum of squares, 46
  as quadratic form, 215–16
Summation operator, 1–2
Symmetric matrix, 196

T–U

t distribution, 8
  table of percentiles, 518–19
t test power function charts, 528–29
Third-order regression model, 302; *see also*
  Regression model

Total deviation, 87
Total sum of squares, 85
Total uncorrected sum of squares, 90
Transformations of variables, 134–41
Transpose of matrix, 188–89
Trial, 25

Unbiased estimator, 9
Unrestricted model, 95

V

Variance
  of error term, 31, 46–48, 50
  estimation of
    ratio of two, 17–18
    single, 16
  of function of random variables, 5–6
  of random variable, 4
  test concerning
    ratio of two, 18–19
    single, 16–17
Variance-covariance matrix, 206–7
  of regression coefficients, 216–17, 220, 242,
    263
  of residuals, 221, 402
Variance inflation factor, 391–93
Vector, 187–88
  with all elements 0, 198–99
  with all elements 1, 198
  random, 205–8

W–Z

Weighted least squares, 167–72,
  219–20, 263
Westwood Company, 25–26
Working-Hotelling confidence region
  multiple regression, 244
  simple linear regression, 154–58
  use for joint estimation of mean responses,
    157–58, 245

z' transformation, 503
  table, 532
Zarthan Company, 247

*This book has been set CRT 10 and 9 point Times
Roman, leaded 2 points. Part numbers are 16 point
Helvetica Bold and part titles are 20 point Helvetica
Bold; chapter numbers are 20 point Helvetica Bold
and chapter titles are 18 point Helvetica Bold. The
size of the type page is 30 picas by 47 picas.*